UNDERGRADUATE TEXTS IN CONTEMPORARY PHY

Springer
New York
Berlin
Heidelberg
Hong Kong
London
Milan
Paris
Tokyo

UNDERGRADUATE TEXTS IN CONTEMPORARY PHYSICS

Cassidy, Holton, and Rutherford, Understanding Physics

Enns and McGuire, Computer Algebra Recipes: A Gourmet's Guide to the Mathematical Models of Science

Hassani, Mathematical Methods: For Students of Physics and Related Fields

Holbrow, Lloyd, and Amato, Modern Introductory Physics

Möller, Optics: Learning by Computing, with Examples Using Mathcad

Roe, Probability and Statistics in Experimental Physics, Second Edition

Rossing and Chiaverina, Light Science: Physics and the Visual Arts

OPTICS

Learning by Computing, with Examples Using Mathcad®

K.D. Möller

With 308 Illustrations

K.D. Möller
Department of Physics
New Jersey Institute of Technology
Newark, NJ 07102
USA

Möller, Karl Dieter, 1927–
Optics: learning by computing with examples using MathCAD / Karl Dieter Möller.
p. cm.—(Undergraduate texts in contemporary physics)
Includes bibliographical references and index.
ISBN 0-387-95360-4 (alk. paper)
1. Geometrical optics—Data processing. 2. MathCAD. I. Title. II. Series.
QC381.M66 2002 2003
535′.32′0285—dc21 2002030382

ISBN 0-387-95360-4 Printed on acid-free paper.

Printed in the United States of America.

9 8 7 6 5 4 3 2 1 SPIN 10852182

www.springer-ny.com

Springer-Verlag New York Berlin Heidelberg
A member of BertelsmannSpringer Science+Business Media GmbH

To
colleagues, staff, and students
of the
New Jersey Institute of Technology,
Newark, New Jersey

Preface

This book was written over several years for a one-semester course in optics for juniors and seniors in science and engineering; it uses Mathcad®[1] scripts to provide a simulated laboratory where students can learn by exploration and discovery instead of passive absorption.

The text covers all the standard topics of a traditional optics course, including geometrical optics and aberration, interference and diffraction, coherence, Maxwell's equations, wave guides and propagating modes, blackbody radiation, atomic emission and lasers, optical properties of materials. Fourier transforms and FT spectroscopy, image formation, and holography. It contains step-by-step derivations of all basic formulas in geometrical, wave, and Fourier optics.

The basic text is supplemented by over 170 Mathcad files, each suggesting programs to solve a particular problem, and each linked to a topic in or application of optics. The computer files are dynamic, allowing the reader to see instantly the effects of changing parameters in the equations. Students are thus encouraged to ask "what if . . . " questions to assess the physical implications of the formulas. To integrate the files into the text, applications connecting the formulas and the corresponding computer file are listed and may be assigned for homework. The availability of the numerical Fourier transform makes possible a mathematical introduction to the wave theory of imaging, spatial filtering, holography, and Fourier transform spectroscopy.

The book is written for the study of particular projects but can easily be adapted to a variety of related studies. The threefold arrangements of text, applications, and files make the book suitable of "self-learning" by scientists or engineers who would like to refresh their knowledge of optics. Some files are printed out, and

[1] Mathcad is a registered trademark of MathSoft Engineering & Education, Inc.

all are available on a CD and may well serve as starting points to find solutions to more complex problems as experienced by engineers in their applications.

The book can be used in optical laboratories with faculty–student interaction. The files may be changed and extended to study the assigned projects, and the student may be required to hand in printouts of all assigned applications and summarize in a sentence or two what has been learned. All files are available on the CD as pdf-files. Information on variable range and parameters may be used to print graphs with other computational programs.

I would like to thank Oren Sternberg and Jeffrey Ausiello for their help with the manuscript, Professor Ken Chin for continuous support, Thomas von Foerster of Springer-Verlag for his helpful advice and interest in this book, and my wife for always keeping me in good spirit.

Newark, New Jersey K.D. Möller

Contents

CHAPTER 1

Geometrical Optics

1.1 INTRODUCTION

Geometrical optics uses light rays to describe image formation by spherical surfaces, lenses, mirrors, and optical instruments. Let us consider the real image of a real object, produced by a positive thin lens. Cones of light are assumed to diverge from each object point to the lens. There the cones of light are transformed into converging beams traveling to the corresponding real image points. We develop a very simple method for a geometrical construction of the image, using just two rays among the object, the image, and the lens. We decompose the object into object points and draw a line from each object point through the center of the lens. A formula is developed to give the distance of the image point, when the distance of the object point and the focal length of the lens are known. We assume that the line from object to image point makes only small angles with the axis of the system. This approximation is called the *paraxial theory*. Assuming that the object and image points are in a medium with refractive index 1 and that the lens has the focal length f, the simple mathematical formula

$$\frac{1}{-x_0} + \frac{1}{x_i} = \frac{1}{f} \tag{1.1}$$

gives the image position x_i when the object position x_0 and the focal length are known.

Formulas of this type can be developed for spherical surfaces, thin and thick lenses, and spherical mirrors, and one may call this approach the *thin lens model*.

For the description of the imaging process, we use the following laws.

1. Light propagates in straight lines.
2. The law of refraction,

$$n_1 \sin\theta_1 = n_2 \sin\theta_2. \tag{1.2}$$

The light travels through the medium of refractive index n_1 and makes the angle θ_1 with the normal of the interface. After traversing the interface, the angle changes to θ_2, and the light travels in the medium with refractive index n_2.

3. The law of reflection

$$\theta_1 = \theta_2. \tag{1.3}$$

The law of reflection is the limiting case for the situation where both refraction indices are the same and one has a reflecting surface. The laws of refraction and reflection may be derived from Maxwell's theory of electromagnetic waves, but may also be derived from a "mechanical model" using Fermat's Principle.

The refractive index in a dielectric medium is defined as $n = c/v$, where v is the speed of light in the medium and c is the speed of light in a vacuum. The speed of light is no longer the ratio of the unit length of the length standard over the unit time of the time standard, but is now defined as 2.99792458×10^8m/s for vacuum. For practical purposes one uses $c = 3 \times 10^8$m/s, and assumes that in air the speed v of light is the same as c. In dielectric materials, the speed v is smaller than c and therefore, the refractive index is larger than 1.

Image formation by our eye also uses just one lens, but not a thin one of fixed focal length. The eye lens has a variable focal length and is capable of forming images of objects at various distances without changing the distance between the eye lens and the retina. Optical instruments, such as magnifiers, microscopes, and telescopes, when used with our eye for image formation, can be adjusted in such a way that we can use a fixed focal length of our eye. Image formation by our eye has an additional feature. Our brain inverts the image arriving on the retina, making us think that an inverted image is erect.

1.2 FERMAT'S PRINCIPLE AND THE LAW OF REFRACTION

In the seventeenth century philosophers contemplated the idea that nature always acts in an optimum fashion. Let us consider a medium made of different sections, with each having a different index of refraction. Light will move through each section with a different velocity and along a straight line. But since the sections have different refractive indices, the light does not move along a straight line from the point of incidence to the point of exit.

The mathematician Fermat formulated the calculation of the optimum path as an integral over the optical path

$$\int_{P_1}^{P_2} n ds. \tag{1.4}$$

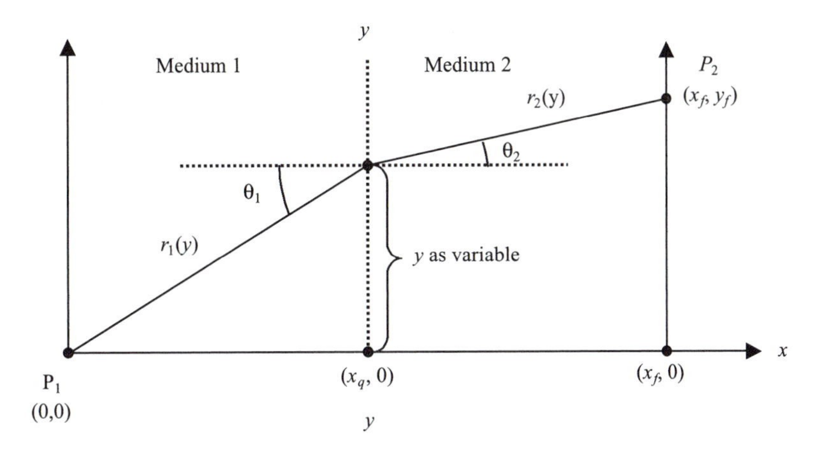

FIGURE 1.1 Coordinates for the travel of light from point P_1 in medium 1 to point P_2 in medium 2. The path in length units and the optical plath are listed.

The optical path is defined as the product of the geometrical path and the refractive index. In Figure 1.1 we show the length of the path from P_1 to P_2,

$$r_1(y) + r_2(y). \tag{1.5}$$

In comparison, the optical path is defined as

$$n_1 r_1(y) + n_2 r_2(y), \tag{1.6}$$

where n_1 is the refractive index in medium 1 and n_2 is the refractive index in medium 2.

The optimum value of the integral of Eq. (1.4) describes the shortest optical path from P_1 to P_2 through a medium in which it moves with two different velocities. It is important to compare only passes in the same neighborhood. In Figure 1.2 we show an example of what should not be compared.

In Figure 1.1, the light ray moves with v_1 in the first medium and is incident on the interface, making the angle θ_1 with the normal. After penetrating into the

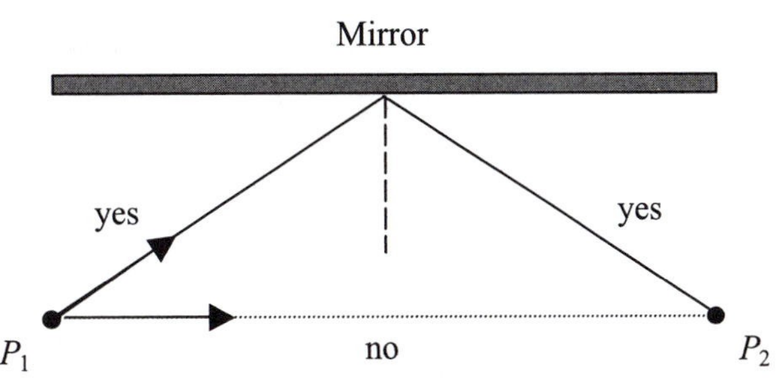

FIGURE 1.2 Application of Fermat's Principle to the reflection on a mirror. Only the path with the reflection on the mirror should be considered.

medium in which its speed is v_2, the angle with respect to the normal changes from θ_1 to θ_2.

Let us look at a popular example. A swimmer cries for help and a lifeguard starts running to help him. He runs on the sand with v_1, faster than he can swim in the water with v_2. To get to the swimmer in minimum time, he will not choose the straight line between his starting point and the swimmer in the water. He will run a much larger portion on the sand and then get into the water. Although the total length (in meter's) of this path is larger than the straight line, the total time is smaller. The problem is reduced to what the angles θ_1 and θ_2 are at the normal of the interface (Figure 1.1). We show that these two angles are determined by the law of refraction, assuming that the velocities are known.

In Figure 1.1 the light from point P_1 travels to point P_2 and passes the point Q at the boundary of the two media with indices n_1 and n_2. The velocity for travel from P_1 to Q is $v_1 = c/n_1$. The velocity for travel from Q to P_2 is $v_2 = c/n_2$. From Eq. (1.4) and Figure 1.1, the optical path is

$$n_1 r_1(y) + n_2 r_2(y), \tag{1.7}$$

where we have

$$\begin{aligned} r_1(y) &= \sqrt{\{x_q^2 + y^2\}} \\ r_2(y) &= \sqrt{\{(x_f - x_q)^2 + (y_f - y)^2\}} \end{aligned} \tag{1.8}$$

and with $r_1(y) = v_1 t_1(y)$ and $r_2(y) = v_2 t_2(y)$ we get for the total time $T(y)$, to travel from P_1 to P_2,

$$T(y) = r_1(y)/v_1 + r_2(y)/v_2. \tag{1.9}$$

Only for the special case that $v_1 = v_2$, where the refractive indices are equal, will the light travel along a straight line. For different velocities, the total travel time through medium 1 and 2 will be a minimum. In FileFig 1.1 we show a graph of $T(y)$ and see the minimum for a specific value of y. In FileFig 1.2 we discuss the case where light is traveling through three media. To determine the optimum conditions we have to require that

$$dT(y)/dy = 0. \tag{1.10}$$

This may be done without a computer. We show it in FileFig 1.3 for two media. Using the expression for $r_1(y)$ and $r_2(y)$ of Figure 1.1, we have to differentiate

$$n_1 r_1(y) + n_2 r_2(y), \tag{1.11}$$

that is,

$$dT(y)/dy = d/dy\{(c/v_1)\sqrt{x_q^2 + y^2} + (c/v_2)\sqrt{(x_f - x_q)^2 + (y_f - y)^2}\} \tag{1.12}$$

and set it to zero. From FileFig 1.3 we get

$$y/(r_1(y)v_1) + (y - y_f)/(r_2(y)v_2) = 0. \tag{1.13}$$

With

$$\sin\theta_1 = y/r_1(y) \quad \text{and} \quad \sin\theta_2 = (y - y_f)/r_2(y) \tag{1.14}$$

we have

$$\sin\theta_1/v_1 = \sin\theta_2/v_2 \tag{1.15}$$

and after multiplication with c, the Law of Refraction,

$$n_1 \sin\theta_1 = n_2 \sin\theta_2. \tag{1.16}$$

FileFig 1.1 (G1FERMAT)

Graph of the total time for travel from P_1 to P_2, through medium 1, with velocities v_1, and medium 2, with v_2. For minimum travel time, the light does not travel along a straight line between P_1 and P_2. Changing the velocities will change the length of travel in each medium.

G1FERMAT

Fermat's Principle

Graph of total travel time: $t1$ is the time to go from the initial position (0, 0) to point (xq, y) in medium with velocity $v1$. $t2$ is the time to go from point (xq, y) to the final position (xf, yf) in medium with velocity $v2$. There is a y value for minimum time. v_1 and v_2 are at the graph.

$$xq := 20 \qquad xf := 40 \qquad yf \equiv 40$$

$$y := 0, .1 \ldots 40.$$

Time in medium 1

Time in medium 2

$$t1(y) := \frac{1}{v1} \cdot \sqrt{(xq)^2 + y^2} \qquad t2(y) := \frac{1}{v2} \cdot \sqrt{(xf - xq)^2 + (yf - y)^2}$$

$$T(y) := t1(y) + t2(y).$$

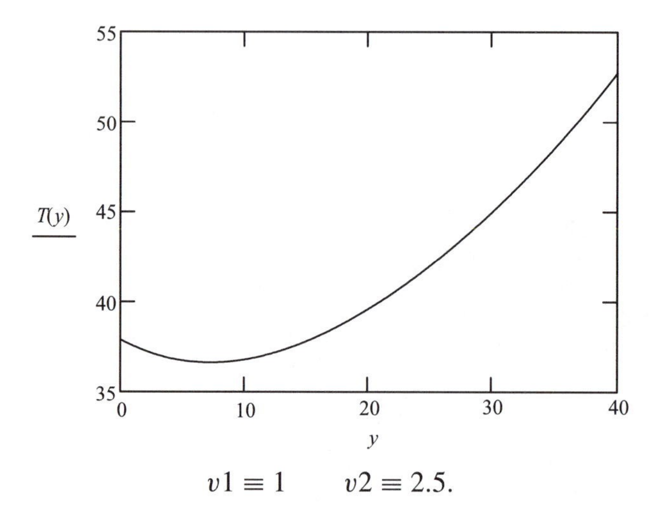

$$v1 \equiv 1 \qquad v2 \equiv 2.5.$$

Changing the parameters $v1$ and $v2$ changes the minimum time for total travel.

Application 1.1.

1. Compare the three choices
   a. $v_1 < v_2$
   b. $v_1 = v_2$
   c. $v_1 > v_2$ and how the minimum is changing.
2. To find the travel time t_1 in medium 1 and t_2 in medium 2 plot it on the graph and read the values at y for $T(y)$ at minimum.

FileFig 1.2 (G2FERMAT)

Surface and contour graphs of total time for traversal through three media. Changing the velocities will change the minimum position.

G2FERMAT is only on the CD.

Application 1.2. Change the velocities and observe the relocation of the minimum.

FileFig 1.3 (G3FERREF)

Demonstration of the derivation of the law of refraction starting from Fermat's Principle. Differentiation of the total time of traversal. For optimum time, the expression is set to zero. Introducing c/n for the velocities.

G3FERREF is only on the CD.

1.3 PRISMS

A prism is known for the *dispersion* of light, that is, the decomposition of white light into its colors. The different colors of the incident light beam are deviated by different angles for different colors. This is called dispersion, and the angles depend on the refractive index of the prism material, which depends on the wavelength. Historically Newton used two prisms to prove his "Theory of Color." The first prism dispersed the light into its colors. The second prism, rotated by 90 degrees, was used to show that each color could not be decomposed any further. Dispersion is discussed in Chapter 8. Here we treat only the angle of deviation for a particular wavelength, depending on the value of the refractive index n.

1.3.1 Angle of Deviation

We now study the light path through a prism. In Figure 1.3 we show a cross-section of a prism with apex angle A and refractive index n. The incident ray makes an angle θ_1 with the normal, and the angle of deviation with respect to the incident light is call δ. We have from Figure 1.3 for the angles

$$\delta = \theta_1 - \theta_2 + \theta_4 - \theta_3 \qquad A = \theta_2 + \theta_3 \tag{1.17}$$

and using the laws of refraction

$$\sin\theta_1 = n\sin\theta_2 \qquad n\sin\theta_3 = \sin\theta_4 \tag{1.18}$$

we get for the angle of deviation, using asin for $\sin^{-1}$

$$\delta = \theta_1 + \text{asin}\,\{(\sqrt{n^2\sin^2(\theta_1)})\sin(A) - \sin(\theta_1)\cos(A)\} - A. \tag{1.19}$$

In FileFig 1.4 a graph is shown of δ (depending on the angle of incidence. A formula may be derived to calculate the minimum deviation δ_m of the prism, depending on n and A. From the Eq. (1.17) and (1.18) we have

$$\delta = \theta_1 - \theta_2 + \theta_4 - \theta_3, \qquad A = \theta_2 + \theta_3, \tag{1.20}$$

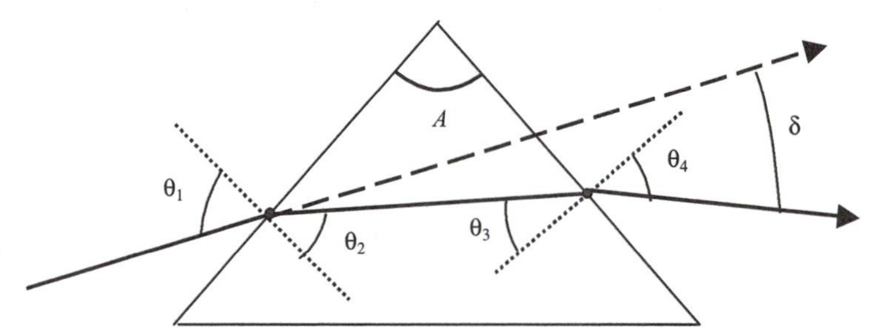

FIGURE 1.3 Angle of deviation δ of light incident at the angle θ_1 with respect to the normal. The apex angle of the prism is A.

and

$$\sin\theta_1 = n\sin\theta_2, \qquad n\sin\theta_3 = \sin\theta_4. \tag{1.21}$$

We can eliminate θ_2 and θ_4 and get two equations in θ_1 and θ_3,

$$\sin\theta_1 = n\sin(A-\theta_3) \tag{1.22}$$
$$n\sin\theta_3 = \sin(\delta + A - \theta_1). \tag{1.23}$$

The differentiations with respect to the angle of Eqs. (1.22) and (1.23) may be done using the "symbolic capabilities" of a computer (see FileFig 1.5). To calculate the optimum condition, the results of the differentiations have to be zero:

$$\cos\theta_1 d\theta_1 + n\cos(A-\theta_3)d\theta_3 = 0 \tag{1.24}$$
$$n\cos\theta_3 d\theta_3 + \cos(\delta + A - \theta_1)d\theta_1 = 0. \tag{1.25}$$

We consider these equations as two linear homogeneous equations of the unknown $d\theta_1$ and $d\theta_3$. In order to have a nontrivial solution of the system of the two linear equations, the determinant has to vanish. This is done in FileFig 1.5, and one gets

$$\cos\theta_1\cos\theta_3 - \cos(A-\theta_3)\cos(\delta + A - \theta_1) = 0.$$

The minimum deviation δ_m, which depends only on n and A, may be calculated from

$$\delta_m = 2\,\mathrm{asin}\,\{n\sin(A/2)\} - A, \tag{1.26}$$

where we use *asin* for $\sin^{-1}$. At the angle of minimum deviation, the light traverses the prism in a symmetric way. Equation (1.26) may be used to find the dependence of prism material on the refractive index n.

FileFig 1.4 (G4PRISM)

Graph of angle of deviation δ_1 as function of θ_1 for fixed values of apex angle A and refractive index n. For fixed A and n the angle of deviation δ has a minimum.

G4PRISM

Graph of the Angle of Deviation for Refraction on a Prism Depending on the Angle in Incidence

$\theta 1$ is the angle of incidence with respect to the normal. $\delta 1$ is the angle of deviation. n is the refractive index and A is the apex angle.

$$\theta 1 := 0, .001 \ldots 1 \qquad n := 2 \qquad A := \left(\frac{2\cdot\pi}{360}\right)\cdot 30$$

$$\delta(\theta 1) := \theta 1 + \operatorname{asin}\left(\sqrt{n^2 - \sin(\theta 1)^2} \cdot \sin(A) - \sin(\theta 1) \cdot \cos(A)\right) - A.$$

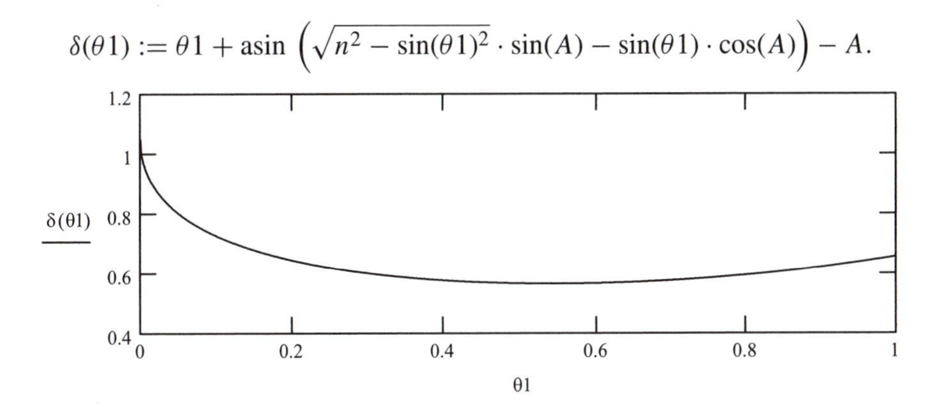

Application 1.4.

1. Observe changes of the minimum depending on changing A and n.
2. Numerical determination of the angle of minimum deviation. Differentiate $\delta(\theta_1)$ and set the result to zero. Break the expression into two parts and plot them on the same graph. Read the value of the intersection point.

FileFig 1.5 (G5PRISMIM)

Derivation of the formula for the refractive index determined by the angle of minimum deviation and apex angle A of prism.

G5PRISMIM is only on the CD.

1.4 CONVEX SPHERICAL SURFACES

Spherical surfaces may be used for image formation. All rays from an object point are refracted at the spherical surface and travel to an image point. The diverging light from the object point may converge or diverge after traversing the spherical surface. If it converges, we call the image point real; if it diverges we call the image point virtual.

1.4.1 Image Formation and Conjugate Points

We want to derive a formula to describe the imaging process on a convex spherical refracting surface between two media with refractive indices n_1 and n_2 (Figure 1.4). The light travels from left to right and a cone of light diverges from the object point P_1 to the convex spherical surface. Each ray of the cone is refracted at the spherical surface, and the diverging light from P_1 is converted to converging light, traveling to the image point P_2. The object point P_1 is assumed to be in a medium with index n_1, the image point P_2 in the medium with index n_2. We assume that $n_2 > n_1$, and that the convex spherical surface has the radius of curvature $r > 0$.

For our derivation we assume that all angles are small; that is, we use the approximation of the paraxial theory. To find out what is small, one may look at a table of $y_1 = \sin\theta$ and compare it with $y_1 = \theta$. The angle should be in radians and then one may find angles for which y_1 and y_2 are equal to a desired accuracy.

We consider a cone of light emerging from point P_1. The outermost ray, making an angle α_1 with the axis of the system, is refracted at the spherical surface, and makes an angle α_2 with the axis at the image point P_2 (Figure 1.4). The refraction on the spherical surface takes place with the normal being an extension of the radius of curvature r, which has its center at C. We call the distance from P_1 to the spherical surface the object distance x_o, and the distance from the spherical surface to the image point P_2, the image distance x_i. In short, we may also use x_o for "object point" and x_i for "image point."

The incident ray with angle α_1 has the angle θ_1 at the normal, and penetrating in medium 2, we have the angle of refraction θ_2. Using the small angle approximation, we have for the law of refraction

$$\theta_2 = n_1\theta_1/n_2. \tag{1.27}$$

From Figure 1.4 we have the relations:

$$\alpha_1 + \beta = \theta_1 \quad \text{and} \quad \alpha_2 + \theta_2 = \beta. \tag{1.28}$$

For the ratio of the angles of refraction we obtain

$$\theta_1/\theta_2 = n_2/n_1 = (\alpha_1 + \beta)/(\beta - \alpha_2). \tag{1.29}$$

We rewrite the second part of the equation as

$$n_1\alpha_1 + n_2\alpha_2 = (n_2 - n_1)\beta. \tag{1.30}$$

The distance l in Figure 1.4 may be represented in three different ways.

$$\tan\alpha_1 = l/x_o, \qquad \tan\alpha_2 = l/x_i, \quad \text{and} \quad \tan\beta = l/r. \tag{1.31}$$

Using small angle approximation, we substitute Eq. (1.31) into Eq. (1.30) and get

$$n_1 l/x_o + n_2 l/x_i = (n_2 - n_1)l/r. \tag{1.32}$$

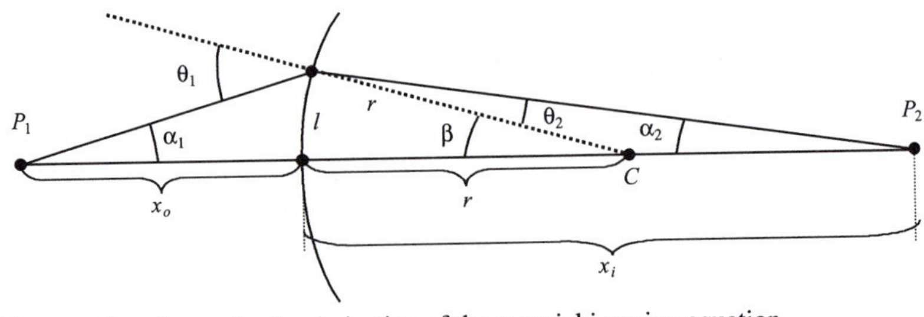

FIGURE 1.4 Coordinates for the derivation of the paraxial imaging equation.

The ls cancel out and we have obtained the image-forming equation for a spherical surface between media with refractive index n_1 and n_2, for all rays in a cone of light from P_1 to P_2:

$$n_1/x_0 + n_2/x_i = (n_2 - n_1)/r. \tag{1.33}$$

So far all quantities have been considered to be positive.

1.4.2 Sign Convention

In the following we distinguish between a convex and a concave spherical surface. The incident light is assumed to travel from left to right, and the object is to the left of the spherical surface. We place the spherical surface at the origin of a Cartesian coordinate system. For a convex spherical surface the radius of curvature r is positive; for a concave spherical surface r is negative. Similarly we have positive values for object distance x_0 and image distance x_i, when placed to the right of the spherical surface, and negative values when placed to the left.

Using this sign convention, we write Eq. (1.33) with a minus sign, and have the equation of "spherical surface imaging" (observe the minus sign),

$$\frac{n_1}{x_0} + \frac{n_2}{x_i} = \frac{n_2 - n_1}{r}. \tag{1.34}$$

The pair of object and image points are called conjugate points.

We may define $\zeta_o = x_o/n_1$, $\zeta_i = x_i/n_2$, and $\rho = r/(n_2 - n_1)$ and have from Eq. (1.34)

$$-1/\zeta_o + 1/\zeta_i = 1/\rho. \tag{1.35}$$

This simplification will be useful for other derivations of imaging equations.

1.4.3 Object and Image Distance, Object and Image Focus, Real and Virtual Objects, and Singularities

When the object point is placed to the left of the spherical surface, we call it a real object point. When it appears to the right of the spherical surface, we call it a virtual object point. A virtual object point is usually the image point produced by another system and serves as the object for the following imaging process. To get an idea, of how the positions of the image point depend on the positions of the object point, we use the equation of spherical surface imaging

$$-n_1/x_o + n_2/x_i = (n_2 - n_1)/r \tag{1.36}$$

or

$$x_i = n_2/[(n_2 - n_1)/r + n_1/x_o],$$

and plot a graph (FileFig 1.6). We choose an object point in air with $n_1 = 1$, a spherical convex surface of radius of curvature $r_1 = 10$, and refractive index $n_2 = 1.5$.

We do not add length units to the numbers. It is assumed that one uses the same length units for all numbers associated with quantities of the equations. When the object point is assumed to be at negative infinity, we have the image point at the image focus

$$x_{\text{if}} = n_2 r/(n_2 - n_1). \tag{1.37}$$

Similarly there is the object focus, when the image point is assumed to be at positive infinity

$$x_{\text{of}} = -n_1 r/(n_2 - n_1). \tag{1.38}$$

We see from the graph of FileFig 1.6 that there is a singularity at the object focus (at $x_o = -20$). To the left of the object focus all values of x_i are positive. To the right of the object focus the values of x_i are first negative, from the object focus to zero, and then positive to the right to infinity.

When $x_o = 0$ we have in Eq. (1.36) another singularity, and as a result we have $x_i = 0$. One may get around problems in plotting graphs around singularities t by using numerical values for x_o that never have values of the singular points.

In FileFig 1.7 we have calculated the image point for four specifically chosen object points, discussed below.

FileFig 1.6 (G6SINGCX)

Graph of image coordinate depending on object coordinate for convex spherical surface, for $r = 10$, $n_1 = 1$ and $n_2 = 1.5$. There are three sections. In the first and third sections, for a positive sign, the image is real. In the middle section, for a negative sign, the image is virtual.

G6SINGCX

Convex Single Refracting Surface

r is positive, light from left propagating from medium with $n1$ to medium with $n2$. xo on left of surface (negative).

Calculation of Graph for xi as Function of xo over the Total Range of xo

Graph for xi as function of xo over the range of xo to the left of xof. Graph for xi as function of xo over the range of xo to the right of xof.

$$r \equiv 10 \qquad n1 := 1 \qquad n2 := 1.5.$$

Image focus

Object focus

$$xif := n2 \cdot \frac{r}{n2 - n1} \qquad xif = 30 \qquad xof := n1 \cdot \frac{r}{n1 - n2} \qquad xof = -20$$

$$xo := -100.001, -99.031 \ldots 100 \qquad xi(xo) := \frac{n2}{\left(\frac{n2-n1}{r}\right) + \frac{n1}{xo}}.$$

$$xxo := -100.001, -99.031 \qquad xxi(xxo) := \frac{n2}{\left(\frac{n2-n1}{r}\right) + \frac{n1}{xxo}}.$$

$$xxxo := -15.001, -14.031 \ldots 50$$

$$xxxi(xxxo) := \frac{n2}{\left(\frac{n2-n1}{r}\right) + \frac{n1}{xxxo}}.$$

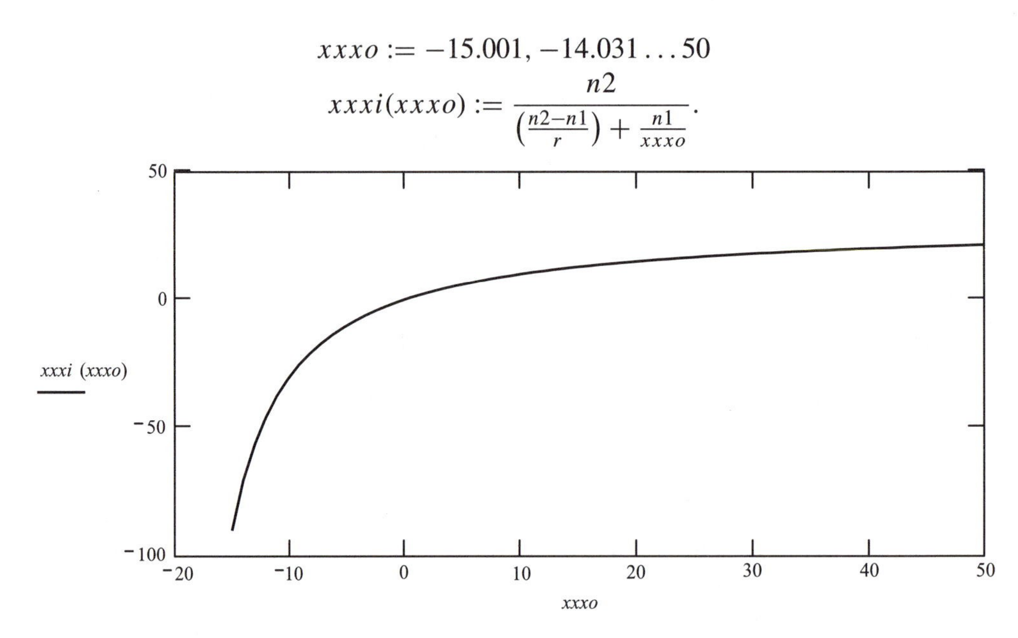

Application 1.6.

1. Change the refractive index and look at the separate graphs for the sections to the left and right of the object focus. To the left of the object focus, x_i is positive. To the right it is first negative until zero, and then positive. What are the changes?
2. Change the radius of curvature, and follow Application 1.

FileFig 1.7 (G7SINGCX)

Convex spherical surface. Calculation of image and object foci. Calculation of image coordinate for four specifically chosen object coordinates.

G7SINGCX

Convex Single Refracting Surface

r is positive, light from left is propagating from medium with $n1$ to medium with $n2$. xo is on left of surface (negative).

Calculation for Four Positions for Real and Virtual Objects, to the Left and Right of the Objects Focus and Image Focus

Calculation of xi from given xo, refractive indices, and radius of curvature. Calculation of magnification

$$r \equiv 10 \qquad n1 := 1 \qquad n2 := 1.5.$$

Image focus | Object focus

$$xif := n2 \cdot \frac{r}{n2 - n1} \qquad xif = 30 \qquad xof := n1 \cdot \frac{r}{n1 - n2} \qquad xof = -20.$$

1. $x1o := -100$

$$x1i := \frac{n2}{\left(\frac{n2-n1}{r}\right) + \frac{n1}{x1o}} \qquad x1i = 37.5 \quad mm1 := x1i \cdot \frac{n1}{x1o \cdot n2} \qquad mm1 = -0.25.$$

2. $x2o := -10$

$$x2i := \frac{n2}{\left(\frac{n2-n1}{r}\right) + \frac{n1}{x2o}} \qquad x2i = -30 \quad mm2 := x2i \cdot \frac{n1}{x2o \cdot n2} \qquad mm2 = 2.$$

3. $x3o := -10$

$$x3i := \frac{n2}{\left(\frac{n2-n1}{r}\right) + \frac{n1}{x3o}} \qquad x3i = 15 \quad mm3 := x3i \cdot \frac{n1}{x3o \cdot n2} \qquad mm3 = 0.5.$$

4. $x4o := 100$

$$x4i := \frac{n2}{\left(\frac{n2-n1}{r}\right) + \frac{n1}{x4o}} \qquad x4i = 25 \quad mm4 := x4i \cdot \frac{n1}{x4o \cdot n2} \qquad mm4 = 0.167.$$

Application 1.7.

1. Calculate Table 1.2 for refractive indices $n_1 = 1$ and $n_2 = 2.4$ (Diamond).
2. Calculate Table 1.2 for refractive indices $n_1 = 2.4$ and $n_2 = 1$.

1.4.4 Real Objects, Geometrical Constructions, and Magnification

1.4.4.1 Geometrical Construction for Real Objects to the Left of the Object Focus

We consider an extended object consisting of many points. A conjugate point at the image corresponds to each point. When using a spherical surface for image formation, a cone of light emerges from each object point and converges to the conjugate image point. Let us present the object by an arrow, parallel to the positive y axis. The corresponding image will also appear at the image parallel to the y axis, but in the opposite direction (Figure 1.5).

The image position and size can then be determined by a simple geometrical construction. In Figure 1.5a we look at the ray connecting the top of the object arrow with the center of curvature of the spherical surface. We call the light ray corresponding to this line the *C-ray* (from center). A second ray, the *PF-ray*, starts at the top of the object arrow and is parallel to the axis along the distance to the spherical surface. It is refracted and travels to the image focal point F_i on the right side of the spherical surface (Figure 1.5c). The paraxial approximation

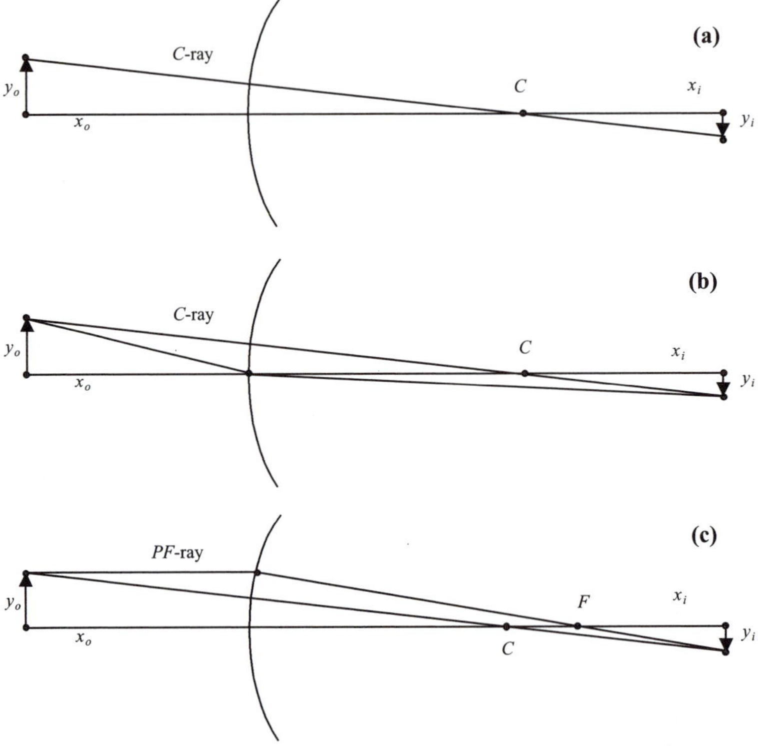

FIGURE 1.5 (a) The C-ray and conjugate points for extended image and object; (b) for the calculation of the lateral magnification we show the C-ray, and the ray from the top of y_0, refracted at the center of the spherical surface, connected to the top of y_i; (c) geometrical construction of image using the C-ray and the FP-ray.

requires that all C-rays and PF-rays have small angles with the axis of the system. The C-ray and the PF-ray meet at the top of the image arrow.

1.4.4.2 Geometrical Construction for Real Object to the Right of the Object Focus

We place the object arrow between the object focus and the spherical surface. From FileFig 1.7, with the input data we have used before, we find that the image position is at -30, when the object position is at -10. The geometrical construction is shown in Figure 1.7b. The C-ray and the PF-ray diverge in the forward direction to the right. However, if we trace both rays back they converge on the left side of the spherical surface. We find the top of an image arrow at the image position, at -30. We call the image, obtained by tracing the diverging rays back to a converging point, a *virtual image*. A virtual image may serve as a real object for a second imaging process.

We have listed in Table 1.1 the image positions for real object positions discussed so far and have indicated for images and objects if they are real or virtual.

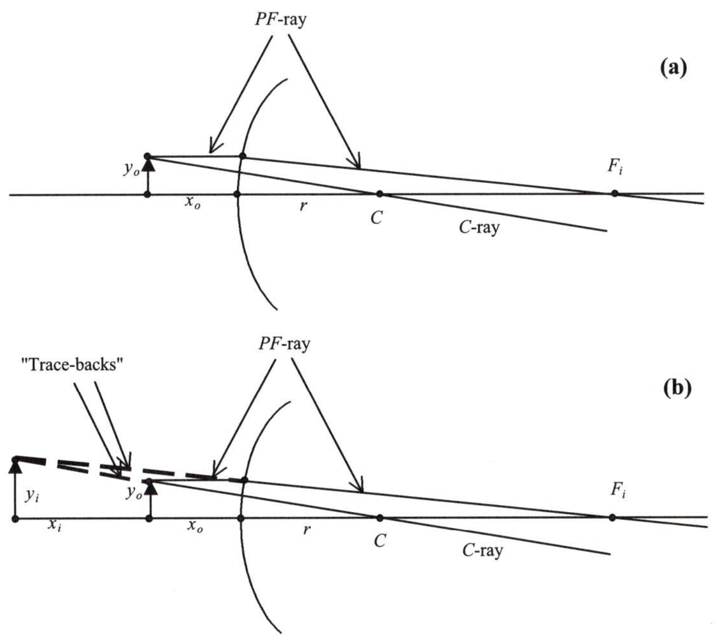

FIGURE 1.6 (a) The C-ray and the PF-ray diverge in the forward direction; (b) they are traced back to the virtual image.

1.4.4.3 Magnification

If we draw a C-ray from the top of the arrow representing the object, we find the top of the arrow presenting the image (Figure 1.6). The *lateral magnification m* is defined as

$$m = y_i/y_o. \tag{1.39}$$

It is obtained by using the proportionality of corresponding sides of right triangles, and taking care of the sign convention

$$-y_i/(x_i - r) = y_o/(-x_o + r). \tag{1.40}$$

For $m = y_i/y_o$ we have

$$m = -(x_i - r)/(-x_o + r). \tag{1.41}$$

Rewritten, eliminating the radius of curvature, one gets with Eq. (1.36),

$$m = y_i/y_o = (x_i/x_o)(n_1/n_2). \tag{1.42}$$

1.4.5 Virtual Objects, Geometrical Constructions, and Magnification

In Figure 1.7 we have made geometrical constructions of virtual objects to the left and right of the image focus. The objects are placed before and after the

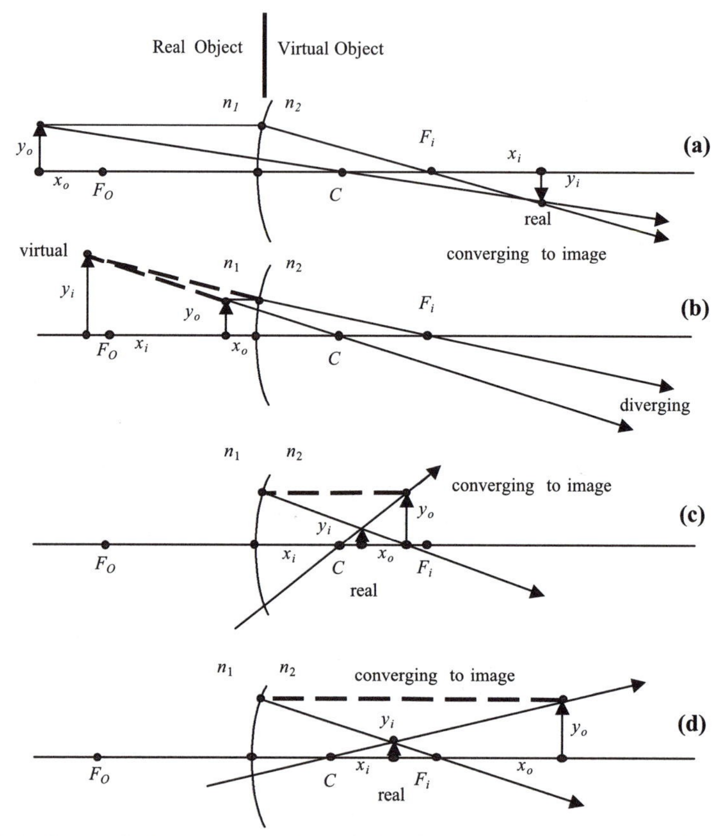

FIGURE 1.7 Geometrical construction of images for the convex spherical surface. The images of real objects are constructed in (a) and (b), for virtual objects in (c) and (d). The light converges to real images in (a), (c), (d). In (b) the light diverges and a virtual image is obtained by "trace back."

image focus. The magnification is obtained from Eq. (1.42) and the calculations are shown in FileFig 1.7.

In FileFig 1.7, we have calculated the four object positions listed in Table 1.1 and shown in Figure 1.7a to d.

1. Real object left of object focus
A real object is positioned to the left of the object focus. The construction uses the C-ray, PF-ray, and image focus. The rays converge to an image point, we have a real image.

2. Real object between object focus and spherical surface
We draw the C-ray and the PF-ray and use the image focus. The rays diverge in a forward direction. We trace both back to a point where they meet. The image is a virtual image.

3. and **4.** Virtual objects.
In Figures 1.7c and 1.7d we consider a virtual object to the right of the spherical surface, one to the left and another to the right of the image focus.

TABLE 1.1 Convex Surface. $r = 10$, $x_{if} = 30$, $x_{of} = -20^a$

x_o	x_i	m	Image	Object
−100	37.5	−.25	r	r
−10	−30	2	vi	r
20	15	05	r	vi
100	25	.0167	r	vi

[a] Calculations with G7SINGCX.

The C-ray is drawn through C in the "forward direction, but the PF-ray is now drawn first "backward" to the surface and then "forward" through the image focus. The C-ray and the PF-ray converge to real images for both positions of the virtual objects.

In Section 1.4 we discussed the case of Eq. (1.36) where $n_1 < n_2$ and r is positive. The case where $n_1 > n_2$ and r is negative will result in a very similar discussion and is considered as an application.

1.5 CONCAVE SPHERICAL SURFACES

The image-forming equation of a convex spherical surface (Eq. (1.34)), is changed for application to a concave spherical surface by changing the radius of curvature to a negative value. We show that this minor change makes image formation quite different.

Again we assume that the refractive index to the left of the surface is smaller than the refractive index on the right ($n_1 < n_2$). The formation of images of extended objects, their magnification, and geometrical construction are similar to the process discussed above for the convex spherical surface.

In FileFig 1.8 we have the graph for the dependence of x_i on x_o. In FileFig 1.9, we determine for four specific positions of x_o, for real and virtual objects, calculations of image positions and magnifications. Observe the difference in the position of object and image focus.

FileFig 1.8 (G8SINGCV)

Graph of image coordinate depending on object coordinate for concave spherical surface, for $r = -10$, $n_1 = 1$, and $n_2 = 1.5$. There are three sections. In the first and third sections, for a negative sign, the image is virtual. In the middle section, for a positive sign, the image is real.

G8SINGCV is only on the CD.

Application 1.8.

1. Observe the singularity at the object focus, which is on the "other side" in comparison to the convex case.
2. Change the refractive index and look at the separate graphs for the sections to the left and right of the object focus. To the left of the object focus, x_i is negative to the left of zero, positive to the right. To the right of the object focus it is negative. What are the changes?
3. Change the radius of curvature, and follow Application 2.

FileFig 1.9 (G9SINGCV)

Concave spherical surface. Calculation of the image and object foci, and image coordinate for four specifically chosen object coordinates.

G9SINGCV is only on the CD.

Application 1.9.

1. Calculate Table 1.2 for refractive indices $n_1 = 1$ and $n_2 = 2.4$ (Diamond).
2. Calculate Table 1.2 for refractive indices $n_1 = 2.4$ and $n_2 = 1$ (Diamond).

The results are listed in Table 1.2, together with the labeling of the real and virtual objects and image.

The geometrical constructions of the four cases calculated in FileFig 1.9 are shown in Figures 1.8a to 1.8d.

1. and **2.** *Real objects.*

A real object is positioned to the left of the spherical surface. The C-ray and PF-ray diverge in a forward direction. The PF-ray is traced back through the image focus (it is on the left). The C-ray and PF-ray meet at an image point. We have virtual images for both positions of the real object.

TABLE 1.2 Concave Surface. $r = -10$, $x_{if} = -30$, $x_{of} = 20$[a]

x_o	x_i	m	Image	Object
−100	−25	.167	vi	r
−20	−15	.5	vi	r
10	30	.2	r	vi
100	−37.5	−.25	vi	vi

[a] Calculations with C9SINGCV.

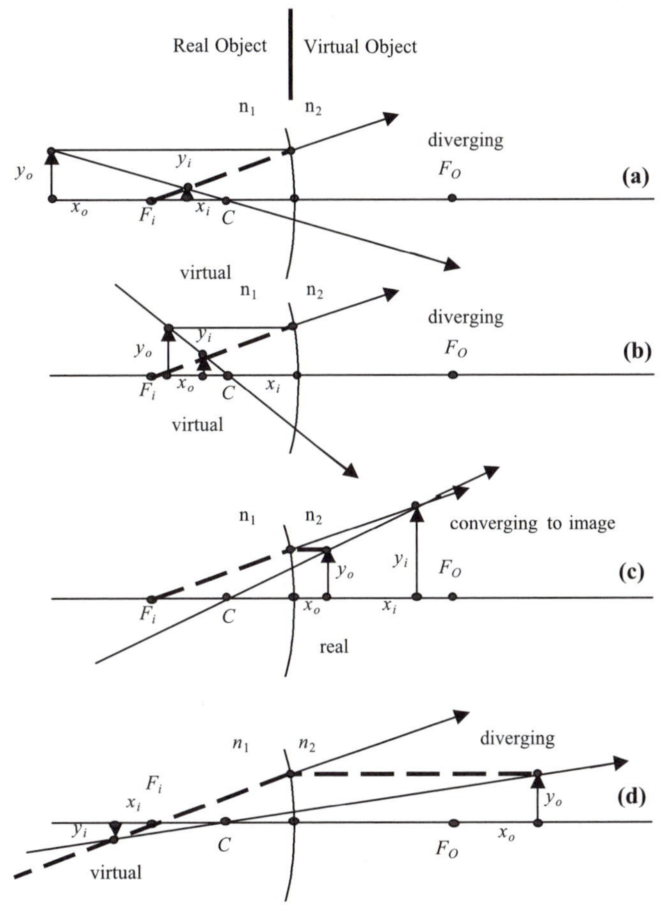

FIGURE 1.8 Geometrical construction of images for the concave spherical surface. The images of real objects are constructed in (a) and (b), for virtual objects in (c) and (d). The light converges to real images in (c). The light diverges in (a), (b), (d), and a virtual image is obtained by "trace back."

3. *Virtual object between spherical surface and object focus.*
   We draw the C-ray and have to trace back the PF-ray to the surface and through the image focus. From there, we extend the ray in a forward direction. The rays converge in a forward direction and we have a real image.
4. *Virtual objects to the right of object focus.*
   The C-ray is drawn through C in a forward direction. The PF-ray is traced back to the surface and then drawn backwards through the image focus. In the backward direction the two rays meet at a virtual image.

Comparing Figures 1.7 and 1.8, one finds that the regions of appearance of real and virtual images are dependent upon the singularities: one when the object distance is equal to the focal length, and the other when the object distance is zero. A virtual image is always found when the C-ray and PF-ray diverge in a forward direction. If we could place a screen into the position of a virtual image, we could not detect it because the rays toward it are diverging.

The case where $n_1 > n_2$ and r is positive is very similar and is discussed as an application in FileFig 1.9.

Applications to Convex and Concave Spherical Surfaces

1. Single convex surface. A rod of material with refractive index $n2 = 1.5$ has on the side facing the incident light a convex spherical surface with radius of curvature $r = 50$ cm.

   a. What is the object distance in order to have the image at $+7$ cm?

   b. What is the object distance in order to have the image at -7 cm?

   c. Assume $r = 25$ cm; make a graph of x_i as a function of x_o for $n_1 = 1$, $n_2 = 1.33$, and do the graphical construction of the image (i) for real objects before and after the object focal point, and (ii) for virtual objects before and after the image focal point.

2. Rod sticks in water, calculation of image distance. A plastic rod of length 70 cm is stuck vertically in water. An object is positioned on the cross-section at the top of the rod, which sticks out of the water and faces the sun. On the other side in the water, the rod has a concave spherical surface, with respect to the incident light from the sun, with $r = -4$ cm. The refractive index of the rod is $n_1 = 1.5$ and of water $n_2 = 1.33$. Calculate the image distance of the object.

3. Single concave surface. A rod of material with refractive index $n_2 = 1.5$ has on one side a concave spherical surface with radius of curvature $r = -50$ cm.

   a. What is the object distance in order to have the image at $+5$ cm?

   b. What is the object distance in order to have the image at -5 cm?

   c. Assume $r = 25$ cm; make a graph of x_i as a function of x_o for $n_1 = 1$, $n_2 = 1.33$, and do the graphical construction of the image (i) for real objects before and after the image focal point, and (ii) for virtual objects before and after the object focal point.

4. Plastic film on water as spherical surface. A plastic film is mounted on a ring and placed on the surface of water. The film forms a spherical surface filled with water. The thickness of the film is neglected and therefore we have a concave surface of water of $n_2 = 1.33$. Sunlight is incident on the surface and the image is observed 100 cm deep in the water. Calculate the radius of curvature of the "spherical water surface."

1.6 THIN LENS EQUATION

1.6.1 Thin Lens Equation

A thin lens has two spherical surfaces with a short distance between them. The thin lens equation is a combination of the imaging equations applied to each of the two surfaces. In the derivation of the final equation, one ignores the distance between the spherical surfaces. The result is an imaging equation, which has the same absolute value for object and image focus. A positive lens has the object focus to the left and the image focus to the right. For the derivation, we assume that the lens has the refractive index n_2, real objects are in a medium with refractive index n_1, and virtual objects are in a medium with refractive index n_3.

To obtain the imaging equation of the thin lens we consider a convex and a concave spherical surface, separated by the distance a. The imaging equation for the first single spherical surface, as given in Eq. (1.35), is

$$-1/\zeta_o + 1/\zeta_i = 1/\rho_1, \tag{1.43}$$

where $\zeta_o = x_o/n_1, \zeta_i = x_i/n_2, \rho_1 = r_1/(n_2-n_1)$, and all distances are measured from the center of the first surface. The imaging equation for the second spherical surface is described by

$$-1/\zeta_o' + 1/\zeta_i' = 1/\rho_2, \tag{1.44}$$

where $\zeta_o' = x_o'/n_2, \zeta_i' = x_i'/n_3, \rho_2 = r_2/(n_3-n_2)$, and all distances are measured from the center of the second surface.

The two surfaces are positioned such that their distance in medium n_2 is "a" (Figure 1.9). To relate this distance to the image distance of the first surface and the object distance of the second surface, we place both at the same point (Figure 1.9). Measured from the first spherical surface the image is at $+\zeta_i$. Measured from the second spherical surface the object is at $-\zeta_o'$. Since ζ_o' and ζ_i are distances divided by the refractive index, we have to do the same with "a".

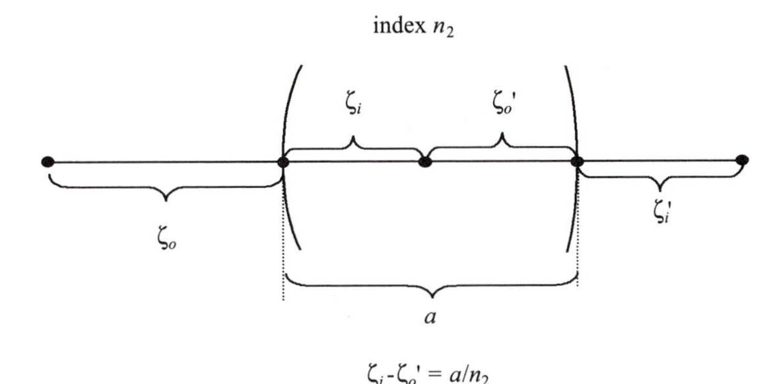

FIGURE 1.9 Coordinates for the derivation of the thin lens equation.

To get the absolute value for a/n_2 we have

$$-\zeta_o' + \zeta_i = a/n_2. \tag{1.45}$$

The relation holds for the coordinates of each lens, and substitution into the equation for Surface 2 results in

$$-1/(-a/n_2 + \zeta_i) + 1/\zeta_i' = 1/\rho_2. \tag{1.46}$$

Adding Eq. (1.46) and the equation for Surface 1, that is, Eq. (1.43), we get

$$-1/\zeta_o + 1/\zeta_i - 1/(-a/n_2 + \zeta_i) + 1/\zeta_i' = 1/\rho_2 + 1/\rho_1. \tag{1.47}$$

The thickness a is now set to zero, two terms cancel each other out, and we obtain

$$-1/\zeta_o + 1/\zeta_i' = 1/\rho_1 + 1/\rho_2. \tag{1.48}$$

Rewriting Eq. (1.48) by using $\zeta_o = x_o/n_1$, $\zeta_i' = x_i'/n_3$, $\rho_1 = r_1/(n_2 - n_1)$, and $\rho_2 = r_2/(n_3 - n_2)$, and setting $x_i' = x_i$, we have

$$-n_1/x_0 + n_3/x_i = (n_2 - n_1)/r_1 + (n_3 - n_2)/r_2. \tag{1.49}$$

The focal length of the thin lens f is defined as

$$1/f = (n_2 - n_1)/r_1 + (n_3 - n_2)/r_2 \tag{1.50}$$

and depends on the refractive indices outside and inside the lens, and on the two radii of curvature. In most cases both sides of the lens have the same refractive index 1; that is, $n_3 = n_1 = 1$. Calling the refractive index of the lens n, we have

$$1/f = (n-1)/r + (1-n)/r'.$$

For a symmetric lens in air we obtain

$$1/f = 2(n-1)/r.$$

Using $n_3 = n_1 = 1$ and the focal length of Eqs. (1.50), we have from Eq. (1.49) the *thin lens equation*,

$$-1/x_o + 1/x_i = 1/f. \tag{1.51}$$

There are positive and negative values for f, associated with positive and negative lenses. For example, a biconvex lens is a positive lens.

1.6.2 Object Focus and Image Focus

When f is positive, that is, for a positive lens, the object focus is on the left and has the coordinate $x_{of} = -f$, and the image focus is at $x_{if} = f$. When f is negative, that is, for a negative lens, the object focus is on the right and has the coordinate $x_{of} = |f|$, and the image focus is $x_{if} = -|f|$.

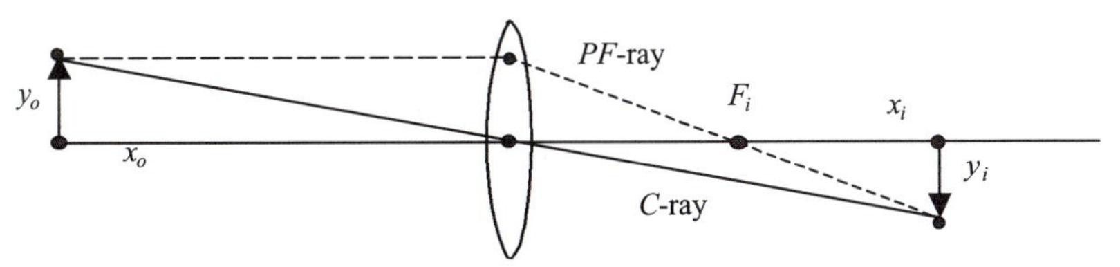

FIGURE 1.10 Graph of C-ray connecting object and image arrows. The length of the object arrow y_0 and image arrow y_i and their distances from the thin lens x_0 and x_i are also indicated.

1.6.3 Magnification

In Figure 1.10 we consider the case of a real object and real image and draw a line from the top of the object arrow through the center of the lens to the top of the arrow on the image arrow. The corresponding light ray is called the chief ray and is again referred to as the C-ray. It passes the lens at the center and therefore is not deviated by refraction. From the two "similar" triangles shown in Figure 1.10 we define the *magnification* m as

$$m = y_i/y_o = x_i/x_o. \tag{1.52}$$

1.6.4 Positive Lens, Graph, Calculations of Image Positions, and Graphical Constructions of Images

In FileFig 1.10 we show a graph of the thin lens equation. The image distance x_i is plotted as a function of x_o for positive f. There is a singularity at the object focus at $-f$. To the left of the object focus, x_i is positive. To the right between the object focus and lens, x_i is negative, and on the right of the lens it is positive. As a result, we have three sections. In the first and third sections, for a positive sign, the image is real. In the middle section, for a negative sign, the image is virtual.

In FileFig 1.11 we have chosen four specific values of object distances and calculate the corresponding image distances and magnifications.

FileFig 1.10 (G10TINPOS)

Graph of image coordinate x_i, depending on the object coordinate x_o for the thin lens equation with $f = 10$.

G10TINPOS

Positive Lens

Focal length f is positive, light from left propagating from medium with index 1 to lens of refractive index n. xo on left of surface (negative).

Calculation of Graph for xi as Function of xo over the Total Range of xo

Graph for xi as function of xo over the range of xo to the left of f. Graph for xi as function of xo over the range of xo to the right of f.

$$f \equiv 10$$

Image focus: f Object focus: $-f$

$$xo := -100.001, -99.031 \ldots 100$$

$$xi(xo) := \frac{1}{\left(\frac{1}{f}\right) + \frac{1}{xo}}.$$

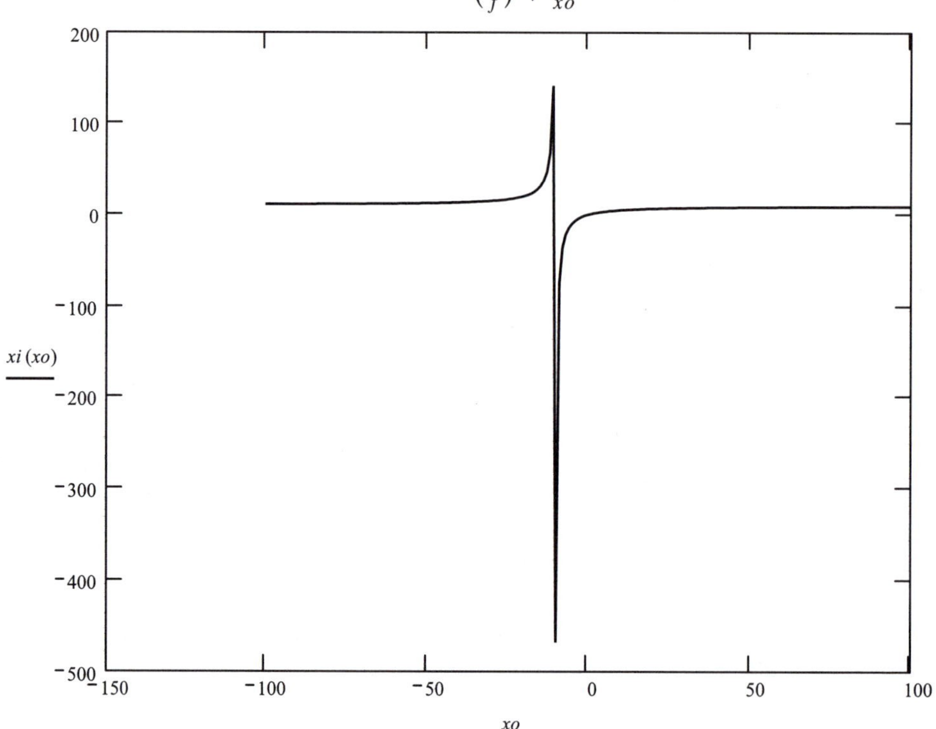

$$xxo := -50.001, -49.031 \ldots -11$$

$$xxi(xxo) := \frac{1}{\left(\frac{1}{f}\right) + \frac{1}{xxo}}.$$

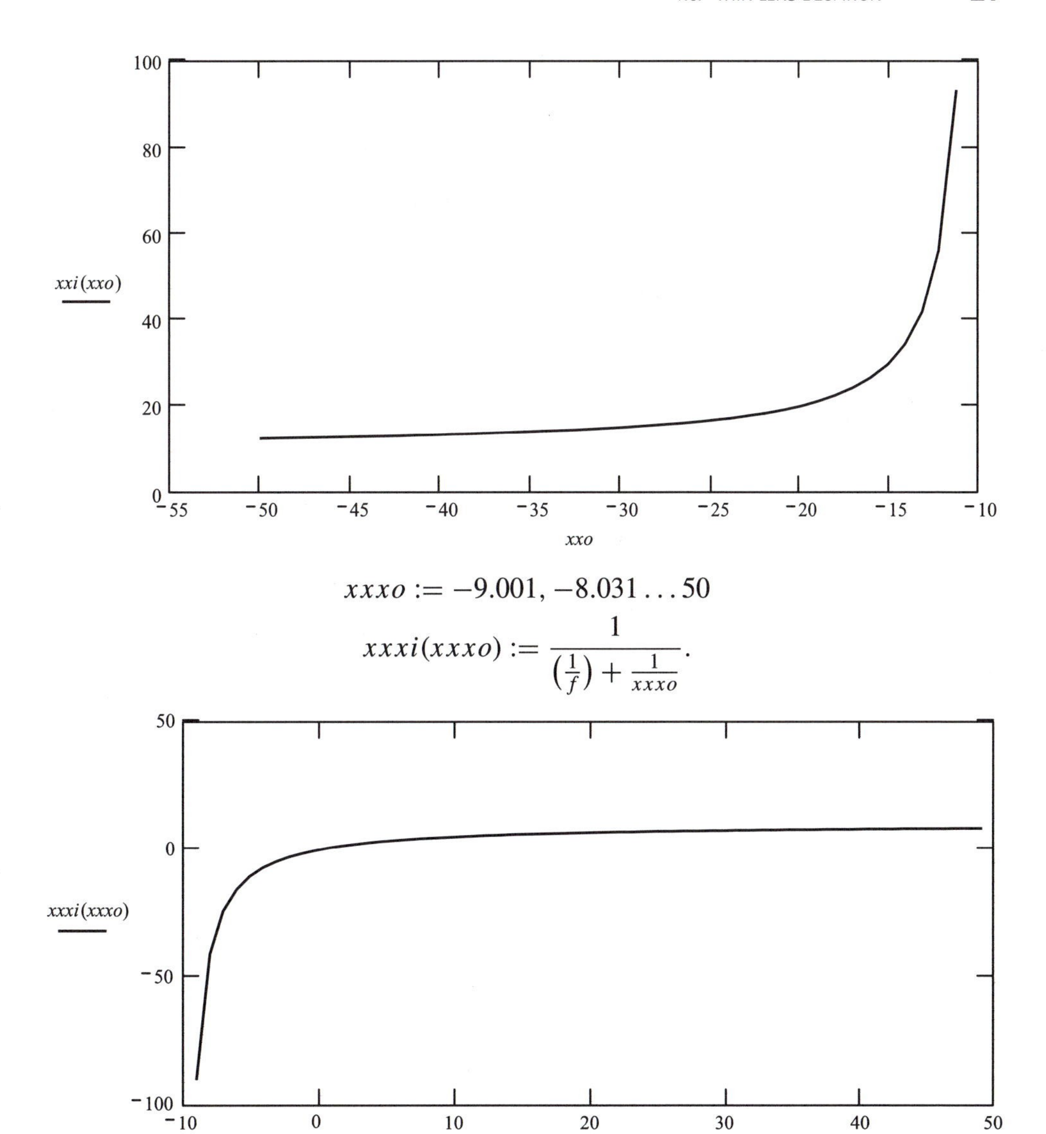

$$xxxo := -9.001, -8.031 \ldots 50$$

$$xxxi(xxxo) := \frac{1}{\left(\frac{1}{f}\right) + \frac{1}{xxxo}}.$$

Application 1.10.

1. Observe the singularity at the object focus, which has the same absolute value as the focal length but with a negative sign. The image focus has a positive sign. Note that they play different roles in the geometrical construction of the image.

2. Change the refractive index and describe what happens.

3. Change the focal length and describe what happens.

FileFig 1.11 (G11TINPOS)

Calculation of image and object foci for $f = 10$. Calculation of image distances x_i and magnification for four specific values of object distance x_o

GG11TINPOS

Positive Lens

Focal length f is positive, light from left propagating from medium with index 1 to lens of refractive index n. xo on left of lens (negative).

Calculation for Four Positions for Real and Virtual Objects, to the Left and Right of the Object Focus and Image Focus

Calculation of xi from given xo and focal length. Calculation of magnification.

$f \equiv 10 \qquad n1 := 1 \qquad n2 : 1.5$

Image focus: f Object focus: $-f$

1. $xo1 := -30$

$$xi1 := \frac{1}{\left(\frac{1}{f}\right) + \frac{1}{xo1}} \quad xi1 = 15 \quad mm1 := \frac{xi1}{xo1} \quad mm1 = -0.5.$$

2. $xo2 := -5$

$$xi2 := \frac{1}{\left(\frac{1}{f}\right) + \frac{1}{xo2}} \quad xi2 = -10 \quad mm2 := \frac{xi2}{xo2} \quad mm2 = 2.$$

3. $xo3 := 5$

$$xi3 := \frac{1}{\left(\frac{1}{f}\right) + \frac{1}{xo3}} \quad xi3 = 3.333 \quad mm3 := \frac{xi3}{xo3} \quad mm3 = 0.667.$$

4. $xo4 := 30$

$$xi4 := \frac{1}{\left(\frac{1}{f}\right) + \frac{1}{xo4}} \quad xi4 = 7.5 \quad mm1 := \frac{xi4}{xo4} \quad mm4 = 0.25.$$

Application 1.11. The distance between the chosen object coordinate and the resulting image coordinate changes with the choice of the object coordinate.

1. Find analytically the condition for the shortest distance between image and object.
2. Make a graph of $y = -x_o + x_i$ depending on x_o and find the minimum.
3. Make a sketch.

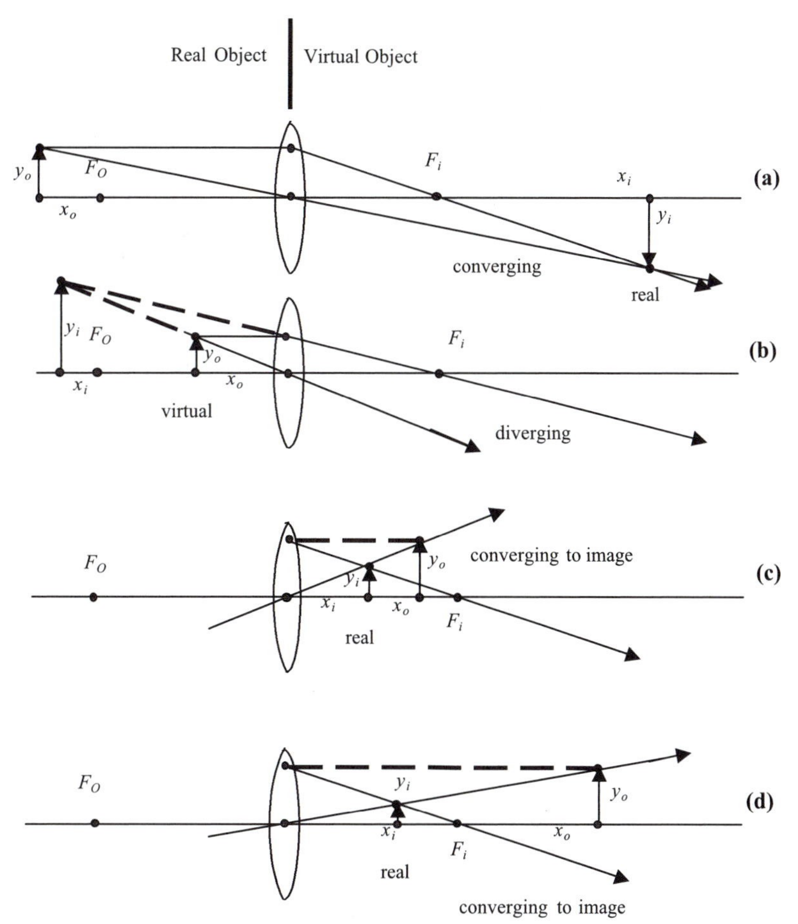

FIGURE 1.11 Geometrical construction of the images for a converging lens with positive f. Real objects for (a) and (b) and virtual objects for (c) and (d). The light converges to real images in (a), (c), (d). The light diverges in (b), and a virtual image is obtained by "trace back."

The geometrical construction of the images for the values calculated in FileFig 1.11 are shown in Figures 1.11a to d.

1. *Realreal objectobject and real image.*
   The object is presented by an arrow of length y_o, placed at the object point x_o. The image point and the length of the arrow presenting the image can be geometrically determined. The C-ray is drawn from the top of the object arrow through the center of the thin lens. The second ray, the PF-ray, is drawn from the object arrow parallel to the axis to the lens, and from there, through the image focus. The two rays meet at the position of the image arrow. In Figure 1.11a, we obtain for a real object a real image.
2. *Real object and virtual image.*
   In Figure 1.11b we place the real object between the object focus and the lens and draw the C-ray and PF-ray. These rays diverge in a forward direction

and both are traced back to the left. They meet at the virtual image. A virtual image is always found when the C-ray and the PF-ray diverge in a forward direction. If we could place a screen into the position of a virtual image, we could not see it, because the rays toward the virtual image are diverging.

3. and **4.** *Virtual object and real images.*
In Figures 1.11c and 1.11d we place the object to the right of the lens. We are considering virtual objects. A virtual object may be produced by the image formed by another optical imaging system. The virtual objects are placed between the lens and the image focus and to the right of the image focus. In both cases we draw the C-ray in a forward direction. The PF-1 ray is drawn first backward to the lens and then forward through the image focus. The C-ray and the PF-ray converge to real images for all positions of the virtual object.

The results of the calculations of the positive thin lens with $f = -10$ are listed in Table 1.3.

1.6.5 Negative Lens, Graph, Calculations of Image Positions, and Graphical Constructions of Images

In FileFig 1.12, we show graphs of the thin lens equation, plotting x_i as a function of x_o for negative f. We see the singularity is at the object focus f, which is now to the right of the lens. To the left of the lens, x_i is negative. Between the lens and the object focus, x_i is positive. To the right of the object focus, x_i is negative. As a result, we have three sections. In the first and third sections, for a negative sign, the image is virtual. In the middle section, for a positive sign, the image is real.

In FileFig 1.13, we have calculated for four specific values of object distance the corresponding image distances and the magnification. In Figure 1.12 we have the geometrical construction of the images for the values calculated in FileFig 1.13.

TABLE 1.3 Positive Lens. $f = 10$, Image Focus 10, Object Focus -10

x_o	x_i	m	Image	Object
−30	15	−.5	r	r
−5	−10	2	vi	r
5	3.3	.67	r	vi
30	7.5	.25	r	vi

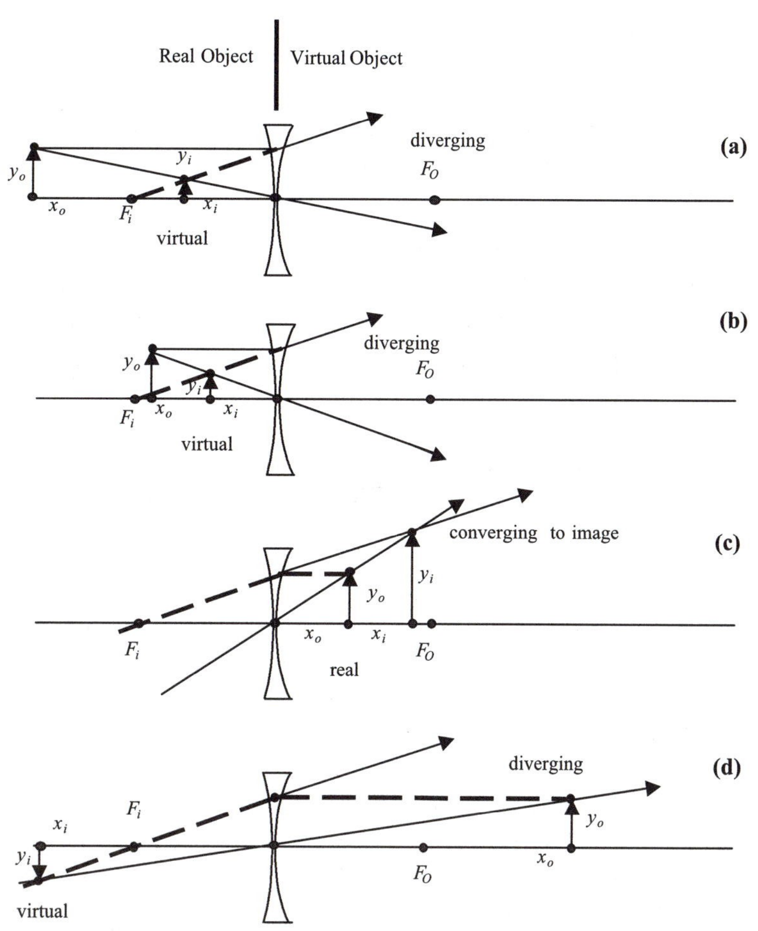

FIGURE 1.12 Geometrical construction of the images for a diverging lens with negative f. Real objects for (a) and (b) and virtual objects for (c) and (d). The light converges to real images in (c). The light diverges in (a), (b), (d), and a virtual image is obtained by "trace back."

FileFig 1.12 (G12TINNEG)

Graph of image coordinate x_i, depending on object coordinate x_o for the thin lens equation with $f = -10$.

G12TINNEG is only on the CD.

Application 1.12.

1. Observe the singularity at object focus, which has the same absolute value as the focal length but with a positive sign. The image focus has a negative sign. Note that they play different roles in the geometrical construction of the image.
2. Change the refractive index and describe what happens.
3. Change the focal length and describe what happens.

FileFig 1.13 (G13TINNEG)

Calculation of image focus and object focus for negative lens. Calculation of image distances x_i and magnification for four specific values of object distance x_o.

G13TINNEG is only on the CD.

Application 1.13. The distance between the chosen object coordinate and resulting image coordinate changes with the choice of the object coordinate.

1. Modify the analytical calculation done in Application FF11 for the condition of the shortest distance between image and object.
2. Make a sketch.

The geometrical construction of the images for the values calculated in FileFig 1.13 are shown in Figures 1.12a to d.

1. and **2.** *Real object to the left of the lens and virtual image.*
The object is presented by an arrow of length y_o, placed at the object point x_o to the left of the negative lens. The image point and the length of the arrow presenting the image can be geometrically determined using the C-ray and the PF-ray. The C-ray is drawn from the top of the object arrow through the center of the thin lens. The PF-ray is drawn from the object arrow parallel to the axis to the lens, and then diverges in a forward direction. It is traced back to the image focus. The two rays meet at the positions of the image arrow. In Figures 1.12a and 1.12b, we obtain for a real object a virtual image. A virtual image is obtained when the C-ray and the PF-ray diverge in a "forward" direction.

3. *Virtual object between lens and object focus.*
In Figure 1.12c, we place the virtual object between the object focus and the lens and draw the C-ray. The PF-ray is first traced back to the lens, then connected to the image focus, and extended in the forward direction. The two rays meet in the forward direction at a real image.

4. *Virtual object on the right side of the object focus.*
In Figure 1.12d, we place the virtual object to the right of the object focus and draw the C-ray. The PF-ray is first traced back to the lens, then connected to the image focus and extended further in the backward direction. The two rays meet in the backward direction for a virtual image.

The results of the calculations of the negative thin lens with $f = -10$ are listed in Table 1.4

For the geometrical construction, we note that the size of the lens does not matter. One uses a plane in the middle of the lens with sufficient extension in the y direction; see Figure 1.13a.

TABLE 1.4 Negative Lens. $f = -10$, Image Focus -10, Object focus 10

x_o	x_i	m	Image	Object
−30	−7.5	.25	vi	r
−5	−3.3	.67	r	r
5	10	2	vi	vi
30	−15	−.5	vi	vi

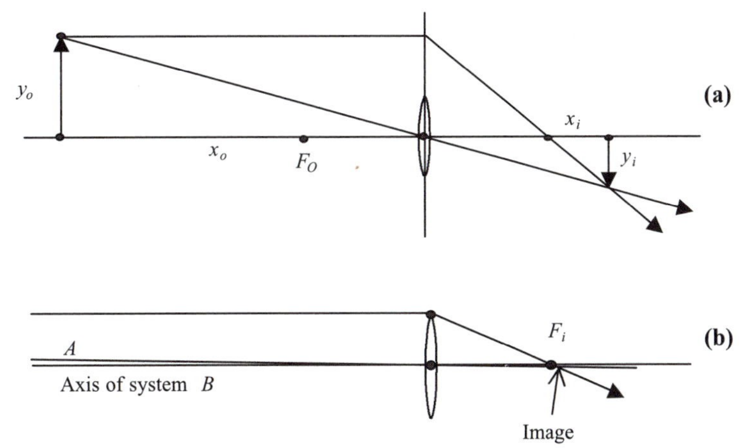

FIGURE 1.13 (a) Image formation of an object larger than the diameter of the lens. The extended plane of the lens is used; (b) image formation for an object at infinity. The axis B of the system is the ray from the center of the object through the center of the lens. the PF-ray is assumed to come from the top of an object at finite distance; the corresponding image is indicated.

If the object is at infinity, one uses for the object distance a finite number so that the image is not exactly at the focal point, where it would have a length equal to zero (Figure 1.13b).

1.6.6 Thin Lens and Two Different Media on the Outside

We go back to the thin lens equation and choose different indices of refraction at the two media on both sides of the lens. We start again from the definitions $\zeta_o = x_o/n_1$, $\zeta_i' = x_i/n_3$, $\rho_1 = r_1/(n_2 - n_1)$ and $\rho_2 = r_2/(n_3 - n_2)$ and have

$$-n_1/x_0 + n_3/x_i = (n_2 - n_1)/r_1 + (n_3 - n_2)/r_2. \tag{1.53}$$

We call the focal length of the thin lens f_n given by

$$1/f_n = (n_2 - n_1)/r_1 + (n_3 - n_2)/r_2 \tag{1.54}$$

and obtain the *thin lens equation*

$$-n_1/x_o + n_3/x_i = 1/f_n. \tag{1.55}$$

This equation is very similar to the spherical surface imaging equation discussed in Section 1.4a just as we found there, we have different values for the object focus and image focus.

For the object focus, when the image point is assumed to be at positive infinity, we have

$$x_{\text{of}} = -n_1 f_n \tag{1.56}$$

and for the image focus, obtained when the object point is assumed to be at negative infinity, we have

$$x_{\text{if}} = n_3 f_n. \tag{1.57}$$

The construction of the images for positive and negative lenses is similar to the procedure for the spherical surfaces and is not discussed further. The value of the focal length for different cases of the refractive indices may be calculated using FileFig 1.14.

FileFig 1.14 (G14TINFOC)

Calculation of the focal length and object and image focus of the thin lens for different combinations of the refractive indices.

G14TINFOC

Focal Length

1. Calculation of focal length of thin lens of refractive index $n2$ in medium with refractive index $n1$.
   First surface: $r1 := -5$. Second surface: $r2 := 5$. r is positive for convex surface, negative for concave surface. Refractive index of lens $n2$: $n2 := 1$. and Refractive index of medium $n1$: $n1 := 1.5$.
2. Graph of focal length of thin lens with index $n2$ depending on refractive index of medium $n1$.
   The range on $n1$ is divided into lower and higher ranges because of singularity. Refractive index of lens $nn2$: $nn2 := 1.5$. Lower range: $nn1 := 1, 1.1 \ldots nn2 - .00001$. Upper range: $nnn1 := nn2 + .1, nn2 + .2 \ldots 4$.

$$ff(nn1) := \frac{1}{\frac{nn2-nn1}{r1} + \frac{nn1-nn2}{r2}} \qquad fff(nnn1) := \frac{1}{\frac{nn2-nnn1}{r1} + \frac{nnn1-nn2}{r2}}.$$

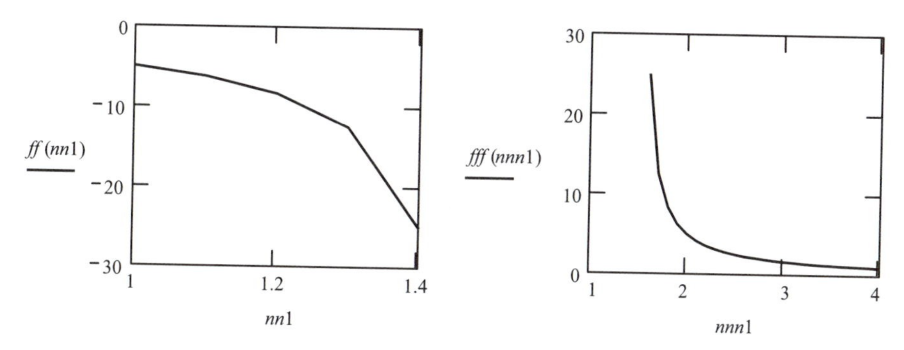

Application 1.14. Consider the case $n_2 > n_1$. What is the result when interchanging n_1 and n_2?

Applications for the Sections on Positive and Negative Lenses

1. Air lens in plastic. A plastic rod is flat on one side and has a spherical surface on the other side. The spherical surface is concave with respect to the incident light, which comes from the flat side. An identical second rod is taken and the two curved ends are put together, forming an air lens by the ends of the two rods. The cross-section of this lens has its thinnest point in the middle. Assume that the radii of curvatures of the spherical surfaces are $r = r' = 10$ cm and the refractive index of the rod is $n = 1.5$. Sunlight is incident on an object on the face of the first rod at 20 cm from the air lens. Find the image distance.
2. Thin lens on water. A lens of refractive index $n = 1.5$ is put on water, one surface in air, the other in water. The lens is a symmetric biconvex lens and has a focal length of $f = 10$ cm in air. The refractive index of water is $n = 1.33$.
   a. Calculate the radii of curvature and the focal length to be used in the above position.
   b. Sunlight is shining on the lens; calculate the image distance in the water.

1.7 OPTICAL INSTRUMENTS

Optical instruments, such as magnifiers, and microscopes, enlarge tiny objects, making it possible to observe objects we can barely see with the naked eye. The magnifier gives us a modest magnification, in most cases less than ten times. The microscope makes it possible to observe objects of about 1 micron diameter, and the telescope enables us to see objects at a far distance in more detail. Our eye is a one lens system and may produce a real image of a real object, like a positive lens (Figure 1.11a). The real image of a real erect object of a positive lens is inverted. However, our brain makes a "correction" (another inversion) and we "see" the object erect, as it is. In discussing optical instruments, we have to take this fact into account when making statements about image formation. For a microscope or astronomical telescope it does not matter much if the final image is erect or inverted. However, for the telescope of a sharpshooter it is important.

From Figures 1.11 and 1.12, we read a simple rule: If the image appears at the same side of the lens as an erect object, it is erect. If it appears on the other side of the lens, it is inverted.

1.7.1 Two Lens System

To obtain the final image distance of a two-lens system, one first applies the thin lens equation to the first lens and determines the image distance. The object distance for the second lens is calculated from the distance between the two lenses and the image distance of the first lens. The thin lens equation is then applied to the second lens and the final image distance for a two-lens system is obtained as the distance from the second lens. The formulas for this procedure are listed in FileFig 1.15.

For graphical constructions one proceeds in the same way. Using C- and PF-rays, one constructs the image of the first lens. The image is taken as an object for the second lens, and C- and PF-rays are used to construct the image formed by the second lens. The existence of the first lens is ignored when going through the second process.

The magnification of the system is the product m of the magnification of each of the two lenses. One has $m = m_1 m_2$ with $m_1 = x_{i1}/x_{o1}$ and $m_2 = x_{i2}/x_{o2}$, where m_1 is calculated with respect to the first lens and m_2 with respect to the second lens.

FileFig 1.15 (G15TINTOW)

Calculation of the final image distance of a two-lens system, for a given object distance of the first lens, focal length, and separation of the two lenses.

G15TINTOW

Two Thin Lenses, Distance Between Lenses: D

1. First lens, $xo1$, $xi1$, $f1$

$$xo1 := -5 \qquad f1 := 6$$

$$xi1 := \frac{1}{\left(\frac{1}{f1}\right) + \frac{1}{xo1}} \qquad xi1 = -30.$$

2. Second Lens, $xo2$, $xi2$, $f2$, and Distance D (Positive Number)

$D := 10 \qquad f2 := 1.85.$

The image distance of the first process is given with respect to the first lens. (Let us assume it is positive.) The object distance must be given with respect to the second lens, taking the distance D between the two lenses into account. (D is negative when counted from the second lens.) Therefore we have

$$xo2 := -D + xi1 \qquad xo2 = -40$$

$$xi2 := \frac{1}{\left(\frac{1}{f2}\right) + \frac{1}{xo2}} \qquad xi2 = 1.94.$$

3. Magnification for each lens and product for the magnification of the system

$$m1 := \frac{xi1}{xo1} \qquad m1 = 6$$

$$m2 := \frac{xi2}{xo2} \qquad m2 = -0.048$$

System

$m1 \cdot m2 = -0.291.$

Application 1.15.

1. Distance between the lenses is larger than $2f$. Calculate the final image distance for two lenses at distance $D = 50$. Assume that the object distance from the first lens is -20. Give the magnification, and make a sketch of object and image, assuming that the object is erect. Consider the following cases.

a. First lens $f_1 = 10$; second lens $f_2 = 10$.

b. First lens $f_1 = 10$; second lens $f_2 = -10$.

c. First lens $f_1 = -10$; second lens $f_2 = 10$.

d. First lens $f_1 = -10$; second lens $f_2 = -10$.

2. Distance between the lenses is smaller than $2f$. Calculate the final image distance for two lenses at distance $D = 6$. Assume that the object distance from the first lens is -20. Give the magnification, and make a sketch of object and image, assuming that the object is erect. Consider the following cases.

a. First lens $f_1 = 10$; second lens $f_2 = 10$.

b. First lens $f_1 = 10$; second lens $f_2 = -10$.

c. First lens $f_1 = -10$; second lens $f_2 = 10$.

d. First lens $f_1 = -10$; second lens $f_2 = -10$.

1.7.2 Magnifier and Object Positions

The size of an image on the retina increases when placed closer and closer to the eye. There is a shortest distance at which the object may be placed, called the near

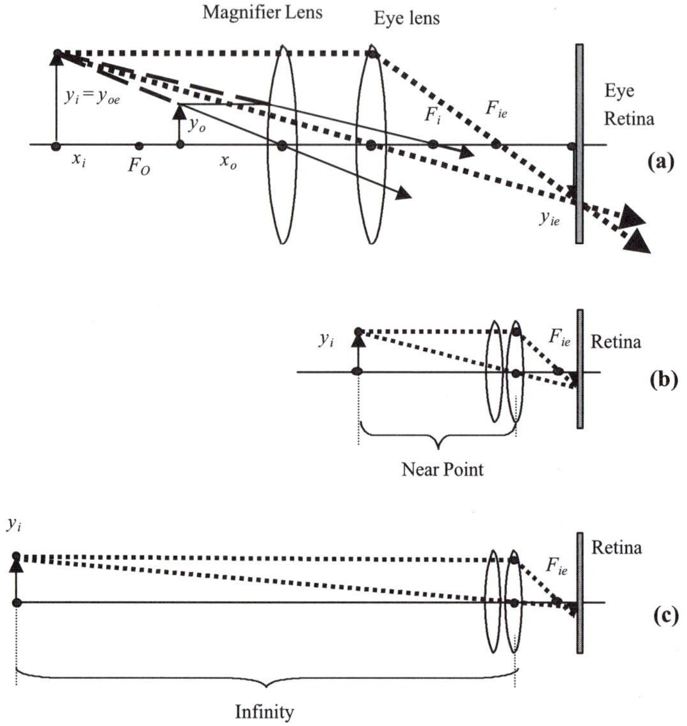

FIGURE 1.14 Two positive lenses in the magnifier configuration: (a) the virtual image y_i of the object y_0 serves as object Y_{oe} for the eye lens. The image y_{ie} (see bold dotted lines) appears on the retina upside down; we see it therefore erect; (b) the object of the eye in the near field configuration; (c) the object of the eye in the infinity configuration.

point at about 25 cm. For shorter distances the eye can no longer accommodate production of an image because the eye–retina distance is fixed. To increase the size of the object one may use a positive lens as a magnifier. In Figure 1.14, we show the magnifier and the eye as a two-thin-lens system. In FileFig 1.16 we show the calculation of the image distance for a two lens system. We assume that the positive lens and the eye are separated by a distance of $D = 1$ cm. Object distance and focal lengths of the lenses are both input data.

From Figure 1.14, we see that the first lens produces a virtual erect image of a real erect object. The second lens (eye) treats the virtual erect image as a real erect object and produces a real inverted image on the retina. The final image on the retina is inverted. However, we "see" it upright because our brain does the conversion. The virtual image of the magnifier lens is the object of the eye lens. The object producing this virtual image may only be positioned with respect to the magnifier in such a way that the virtual image is not closer than the near point, but may have a distance as large as negative infinity. We therefore discuss the two cases: the virtual image is at the near point; and the virtual image is at infinity.

FileFig 1.16 (G16MAG2L)

Calculation of the image distance for a two-lens system consisting of a positive lens and the eye lens. Magnification for each lens and the system.

G16MAG2L is only on the CD.

Application 1.16. Object distance at $x_{o1} = -5$, focal length of first lens $f_1 = 6$, distance D between lens and eye is $D = 0$, focal length of eye $f_2 = 1.85$. Study different resulting magnifications for changes of x_{o1} and f_1.

1.7.2.1 Virtual Image at Near Point

The virtual image produced by the first lens is the real erect object for the second lens (eye), and is assumed to be at the near point (−25 cm). In the first step, we calculate the object distance for the first lens when the image is at −25 cm from the second lens (eye). In the second step we consider the eye. The calculation is shown in FileFig 1.17 where the magnification of the magnifier is given as

$$m_1 = x_{i1}/x_{o1} \tag{1.58}$$

and of the eye as

$$m_2 = x_{i2}/x_{o2}. \tag{1.59}$$

Considering only the magnification m1 of the magnifier, one may use the thin lens equation in order to express m1 in known quantities; that is, f_1 and $x_{i1} = -25$. We have

$$m_1 = x_{i1}/x_{o1} = x_{i1}(1/x_{o1}) = x_{i1}(-1)(1/f - 1/x_{i1}) = (1 - x_{i1}/f_1). \tag{1.60}$$

Neglecting the distance D between magnifier and eye lens, and setting $x_{i1} = -25$, we obtain for the magnification,

$$m_1 = 1 + 25/f_1. \tag{1.61}$$

1.7.2.2 Virtual Image at Infinity

The virtual image produced by the first lens is assumed to be at negative infinity $(-\infty)$. It is the real erect object for the second lens (eye). The calculation is shown in FileFig 1.18 for $f_1 = 12$, and taking for x_{i1} the numerical value of -10^{10}. For the magnification of the magnifier we get, after using the thin-lens equation, similarly done as in Eq. (1.60),

$$m_1 = x_{i1}/x_{o1} = 1 - x_{i1}/f1 = 8.33310^8,$$

This is a meaningless number. In order to discuss the case where the virtual image is at infinity, we have to change our approach and consider angular magnification.

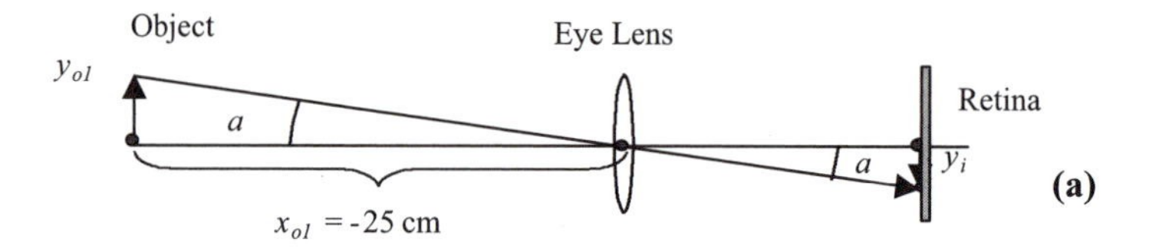

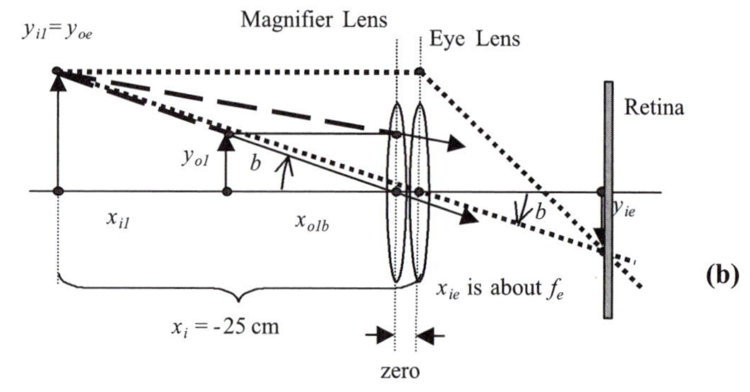

FIGURE 1.15 Angular magnification; (a) object at the near point, seen with the eye lens; object at the near point, seen with magnifier and eye lens.

1.7.2.3 Angular Magnification or Magnifying Power

To avoid the difficulties we encountered in Section 1.7.2.2, where we calculated meaningless numbers for the magnification, we take a different approach and use angular magnification. We compare the angles at the eye by looking at the object with and without a magnifier (Figure 1.15).

The object is positioned at the near point because that gives the largest magnification without a lens. First the eye looks at the object without a magnifier, (Figure 1.15a), where angle α is

$$\alpha = y_{o1}/x_{o1} = y_{o1}/(-25). \tag{1.62}$$

Then we introduce the magnifier and have for the angle β, as shown in Figure 1.15b,

$$\beta = y_{i1}/x_{i1} = y_{o1}/x_{o1} = y_{o1}(1/x_{i1} - 1/f_1), \tag{1.63}$$

where $x_{o1\beta}$ is the object distance when calculating the angle β, and the thin-lens equation was used to eliminate $x_{o1\beta}$.

We define the angular magnification or magnifying power as

$$MP = \beta/\alpha = -25(1/x_{i1} - 1/f_1). \tag{1.64}$$

We now discuss the applications of angular magnification to the cases where the virtual image is at the near point and at infinity.

1. Near point.
The object is at the near point, and assuming $D = 0$ we have $x_{o1} = x_{i1} = -25$ and get

$$MP = 1 + 25/f_1. \tag{1.65}$$

This is the same expression we obtained in Section 1.7.2.1; for the case of the Near point, see Eq. (1.61).

FileFig 1.17 (G17MAGNP)

Calculations of the magnifier in the near point configuration. Assume $D = 0$. First step: Determination of object point for image point at -25 for first lens with $f_1 = 12$, result $x_{o1} = -8.108$. Second step: Determination of x_{i2} for $x_{o2} = -25$ and eye lens $f_2 = 1.85$, result $x_{i2} = 2$. Calculation of magnification.

G17MAGNP is only on the CD.

Application 1.17. Find the resulting magnifications for three choices of f_1.

2. Virtual image at infinity.
We consider the virtual image of lens 1 as the real object of lens 2. We have $x_{i1} = -\infty$, and have for the angular magnification

$$\begin{aligned} MP &= -25(1/x_{i1} - 1/f_1) \\ &= 25/f_1. \end{aligned} \tag{1.66}$$

This value is marked on magnifiers as MP times x. Example: for $f_1 = 5$ we would have $MP = 5x$.
In both cases the object is placed at the near point of the eye without a magnifier, and the resulting angular magnification depends on the focal length of the magnifier.

FileFig 1.18 (G18MAGIN)

Calculations of the magnifier for the "virtual image at infinity" configuration. Assume $D = 0$. First step: Determination of object point for image point $x_{i1} = -10^{10}$, that is, at $(-\infty)$ for the first lens with $f_1 = 12$, result is $x_{o1} = -12$. Second step: Determination of x_{i2} for $x_{o2} = (-\infty)$, for the eye lens $f_2 = 1.85$, result is $x_{i2} = 1.85$. Calculation of magnification.

G18MAGIN is only on the CD.

Application 1.18. Study several resulting magnifications for three choices of f_1.

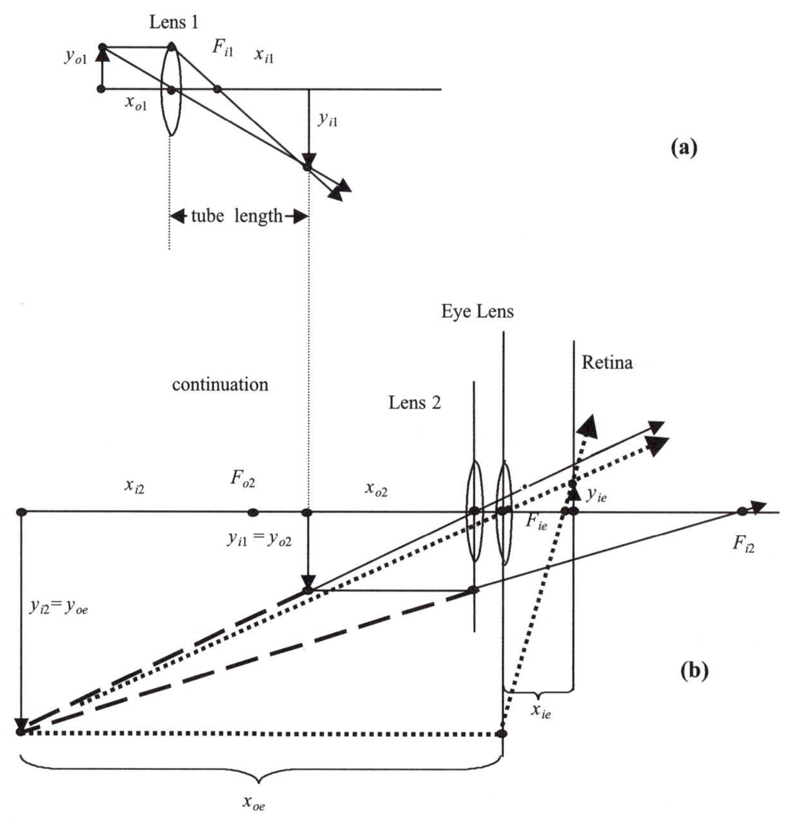

FIGURE 1.16 Microscope as three-lens system of objective, magnifier (ocular), and eye. The object is close to the focal length of the objective lens $L1$ and the image is y_{i1}. The magnifier $L2$ and eye lens act in the magnifier configuration on the image y_{02} produced by $L1$. The image y_{i2} is the object y_{0e} for the magnifier and the image y_{ie} appears on the retina erect; we see it therefore upside down.

1.7.3 Microscope

1.7.3.1 Microscope as Three-Lens System

In a compound microscope, the first lens $L1$ (objective lens) has a short focal length and forms a real inverted image of a real erect object. Then the magnifier configuration is applied, which is the second lens $L2$ (ocular lens) plus the eye lens. See Section 1.7.2 above and Figure 1.16. The final image on the retina is erect, but we see it upside down.

We ignore the eye lens and calculate the final image of a two lens system, using for the image distance x_{i1} the fixed value of tube length 16 cm plus F_{i1} (in cm), see Figure 1.16. The magnification is the product of m_1 of the objective lens, times m_2 of the ocular lens (magnifier). We discuss the following cases where the magnifier is used in (1) the near point configuration, and (2) the virtual image at infinity configuration.

1. Magnification, Near Point Configuration, Magnifying Power

In FileFig 1.19 we calculate the magnification, using $f_1 = 2$, $x_{i1} = 16 + f_1$, $f_2 = 6$, and $x_{i2} = -25$; we have for the magnification

$$m = m_1 m_2 = (x_{i1}/x_{o1})(x_{i2}/x_{o2}) = -41.34. \tag{1.67}$$

The magnifying power MP for the magnifier in the Near point configuration was obtained in Eq. (1.66), and was the same as the magnification $m_1 m_2$. Using the thin-lens equation to x_{o1} and x_{02} we have

$$\begin{aligned} MP &= m_1 m_2 = x_{i1}(-1)(1/f_1 - 1/x_{i1})x_{i2}(-1)(1/f_2 - 1/x_{i2}) \\ &= (1 - [16 + f_1]/f_1)(1 + 25/f_2) \end{aligned} \tag{1.68}$$

and as the result we get $m_1 m_2 = -41.34$. Neglecting f_1 with respect to 16 we have

$$MP \approx (1 - 16/f_1)(1 + 25/f_2) = -36.17. \tag{1.69}$$

The negative magnification indicates that we see the object upside down.

FileFig 1.19 (G19MICNP)

Calculations of the microscope in the near point configuration. The object is close to the focal point of lens 1. Lens 1: $f_1 = 2$ cm; $x_{i1} = +16 + 2$ cm, result $x_{o1} = -2.25$ cm. The magnifier lens L2 is in the near point configuration. Lens 2: $f_2 = 6$ cm, $x_{i2} = -25.008$ cm; $x_{o2} = -4.839$ cm. The angular magnification is also calculated.

G19MICNP is only on the CD.

Application 1.19. Go through all the steps and study the resulting magnification by changing f_1 and f_2.

---

2. Magnification, Virtual Image at Infinity, Magnifying Power

We assume that the virtual image is at infinity; that is, $x_{2i} = -\infty$. The calculations using the direct approach, which is $m = (x_{i1}/x_{o1})(x_{i2}/x_{o2})$, are shown in FileFig 1.20. Using $f_{1=2}$ cm, $x_{i1} = 16 + f_1$, $f_2 = 6$ cm, and $x_{i2} = -10^{10}$ cm, we obtain a meaningless number.

The magnifying power in the near point configuration of the magnifier was obtained in Eq. (1.68) as $MP = (1 - [16 + f_1]/f_1)(1 + 25/f_2)$. The second factor changes for the case where the "virtual image is at infinity," and the result is

$$MP = (1 - [16 + f_1]/f_1)(25/f_2) = -33.333. \tag{1.70}$$

Neglecting f_1 with respect to 16 one has

$$MP = -(16/f_1)(25/f_2) = -29.167. \tag{1.71}$$

One may also disregard the one in the first factor and have $MP = -(16/f_1)(25/f_2)$.

FileFig 1.20 (G20MICIN)

Calculations of the microscope in the "virtual image at infinity" configuration. The virtual image is at infinity; that is, $x_{2i} = -\infty$. Lens 1: $f_1 = 2$ cm; $x_{i1} = +16+2$ cm; result $x_{o1} = -2.25$ cm. Lens 2: $f_2 = 6$ cm; $x_{i2} = -10^{10}$ cm; result $x_{o2} = -6$ cm. The magnification is also calculated neglecting f_1.

G20MICIN is only on the CD.

Application 1.20. Go through all the steps and study the resulting magnification by changing f_1 and f_2.

1.7.3.2 Magnification of Commercial Microscopes

Commercial microscopes give the magnification of the objective and eye lens by a MPx value, similar to the one discussed above for the magnifier. For example, the magnifier power MP of the microscope was approximately $-(16/f_1)(25/f_2)$. Assuming $f_1 = 2$ and $f_2 = 6$, the objective would be marked 8x and the ocular 4x. The magnification of this microscope would be 32 times.

1.7.4 Telescope

1.7.4.1 Kepler Telescope

In a simple telescope, the first lens L1 forms an image of a far away object at a distance close to the focal point f1 of the objective lens (Figure 1.17). The object is considered real and erect, and the image is real inverted. The second lens is the magnifier lens and the eye and magnifier lens are used together in the virtual image at $-\infty$ configuration. In this setup the image of lens 1, which is the object of lens 2, is close to the focal point of f_2, and forms an inverted virtual image at infinity. When we look at this virtual image the final image on the retina is erect, but we see it upside down. The calculations are shown in FileFig 1.21. To find the approximate magnification of the telescope, we do not need to use the concept of magnifying power and can use the calculation of the magnification:

$$m = (x_{i1}/x_{o1})(x_{i2}/x_{o2}), \tag{1.72}$$

where $m_1 = x_{i1}/x_{o1}$ is about f_1/x_{o1}, because the image of lens 1 is close to the focal point. For $m_2 = x_{i2}/x_{o2}$ we have approximately $-x_{i2}/f_2$, because the object for f_2 is close to the focal point. Since x_{o1} and x_{i2} are both large numbers of the same order of magnitude, they cancel each other out and for the

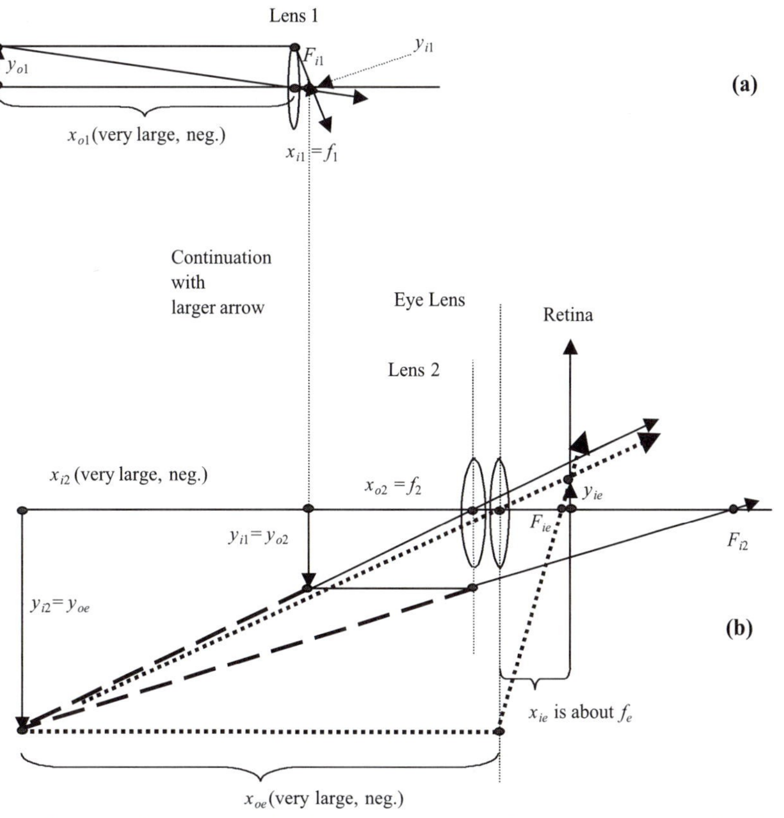

FIGURE 1.17 Optical diagram of a Kepler telescope: (a) the object is far away from the objective lens $L1$ and the image is y_{i1}, located close to $x_{i1} = f_1$; (b) the image y_{i1} is the object for the magnifier $L2$ and eye lens in the magnifier configuration and produces the virtual image $y_{i2} = y_{0e}$. The final image y_{ie} appears on the retina erect, we therefore see it upside down. The distance $f_1 + f_2$ is approximately the length of the telescope.

magnification we have

$$m = m_1 m_2 = -f_1/f_2. \tag{1.73}$$

Note that this is a negative number since f_1 and f_2 are both positive, and the object is "seen" inverted.

To get a large magnification, we need a large value of f_1 and a small one of f_2. The large value of the focal length of the first lens makes powerful telescopes "large."

FileFig 1.21 (G21TELK)

The Kepler telescope is treated as a two-lens system, assuming for x_{o1} and x_{i2} the same large negative numerical values. The magnification is calculated from

$m = (x_{i1}/x_{o1})(x_{i2}/x_{o2})$ and results in $m = m_1 m_2 = -f_1/f_2$. Lens L1: $f_1 = 30$; $x_{o1} = -10^{10}$; $x_{i2} = 30$. Lens L2: $f_2 = 6$; distance $a = f_1 + f_2$; $x_{i2} = -10^{10}$, $x_{o2} = -6$. Calculation of magnification.

G21TELK is only on the CD.

Application 1.21. Study magnifications of 2 and 4 by changing f_1 and f_2 and make a sketch.

1.7.4.2 Galilean Telescope

The Galilean telescope is the combination of a positive lens L1 and a negative lens L2. The positive lens forms a real inverted image of a far-away real erect object (Figure 1.18a). The negative lens replaces the magnifier. The image of lens 1 is the object for lens 2 and is virtual inverted. Lens 2 forms a virtual erect image of it, at negative infinity (Figure 1.18b). The eye looks at the virtual erect image of lens 2 as a real erect object and forms a real inverted image on the retina (we see it erect). The calculation is shown in FileFig 1.22. For the magnification

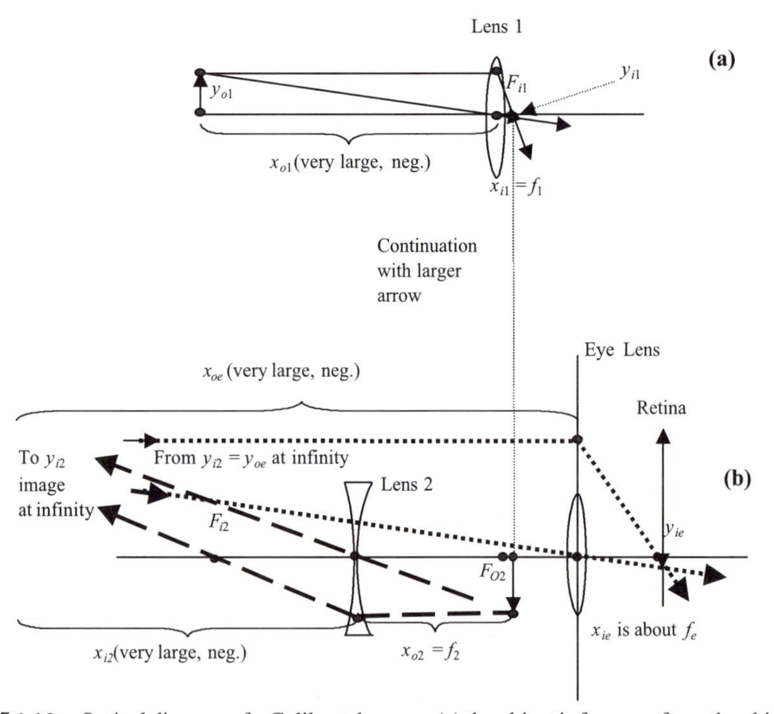

FIGURE 1.18 Optical diagram of a Galileo telescope: (a) the object is far away from the objective lens $L1$ and the image is y_{i1}, located close to $x_{i1} = f_1$; (b) the image of lens 1 is virtual inverted object for lens 2, and lens 2 forms a virtual erect image of it. This virtual erect image is the object of the eye lens and the image y_{ie} appears on the retina upside down, therefore we see it erect.

one gets:

$$m = (x_{i1}/x_{o1})(x_{i2}/x_{o2}), \tag{1.74}$$

where $m_1 = x_{i1}/x_{o1}$ is approximately f_1/x_{o1} because the image of lens 1 is close to the focal point. The magnification of the second lens, $m_2 = x_{i2}/x_{o2}$, is approximately $-x_{i2}/f_2$, because the object of lens 2 is close to the focal point and to the right side of the lens, and f_2 is negative. Since x_{o1} and x_{i2} are both large numbers, of the same order of magnitude, they cancel each other out and we have for the magnification

$$m = m_1 m_2 = -f_1/f_2. \tag{1.75}$$

Note that this is a positive number since f_2 is a negative lens, and the object is seen erect. The Galilean telescope is used for many terrestrial applications in theaters and on ships.

FileFig 1.22 (G22TELG)

The Galilean telescope is treated as a two-lens system with the first lens having a positive focal length and the second lens a negative focal length. For x_{o1} and x_{i2} the same large negative numerical values are assumed. The magnification is calculated as $m = (x_{i1}/x_{o1})(x_{i2}/x_{o2})$ and results in $m = m_1 m_2 = -f_1/f_2$. (Note that the numerical value is positive.) Lens L1: $f_1 = 30$, $x_{o1} = -10^{10}$, $x_{i1} = 30$ Lens L2: $f_2 = -29.99$, $x_{i2} = -9\ 10^4$; $x_{o2} = 30$.

G22TELG is only on the CD.

Application 1.22. Go through all the stages and study magnifications by changing f_1 and f_2.

Applications to Two- and Three-Lens Systems

1. Magnifier. A magnifier lens of $f_1 = 12$cm is placed 8 cm from the eye.
   - **a.** Find the position of x_{o1} for
     - **i.** the near point configuration; and
     - **ii.** the infinite configuration.
   - **b.** Give the magnification and the angular magnification.
2. Microscope. A microscope has a first lens (objective) with focal length .31 cm, a magnifier (ocular) lens of 1.79 cm, and the eye lens is assumed to be $f_e = 2$ cm. The focal length of the objective lens has been chosen so that the image is at about 16 cm. The distance between the lenses is 18 cm and we assume that the eye is in near point configuration. Calculate the magnification of the

first and of the second lenses, and compare the product with the magnifying power, as derived, and its approximation.

3. Microscope (near point). A microscope has a first lens (objective) with focal length .31 cm and a magnifier (ocular) lens of 1.3 cm. We assume that the image of the first lens is at 16 cm and the eye is in the near point configuration.
   a. Find the object distance for the objective lens.
   b. Find the distance from the first image and the magnifier lens.
   c. Find the distance between the lenses (length of microscope).
   d. Find the magnification.
4. Microscope ($-\infty$). A microscope has a first lens (objective) with focal length 2.15 cm and a magnifier (ocular) lens of 6 cm. We assume that the image of the first lens is at 16 cm and the eye is relaxed, looking at $-\infty$.
   a. Find the objective distance for the objective lens.
   b. Find the distance from the first image and the magnifier lens.
   c. Find the distance between the lenses (length of microscope).
   d. Find the magnification.
5. Kepler telescope. Make a suggestion for construction of a Kepler telescope with magnifications of 4 and 10. At what higher number does the construction become unrealistic? Why?
6. Galilean telescope. A Galilean telescope has for the first lens $f_1 = 30$ cm and for the negative lens $f_2 = -9.9$ cm. If x_{o1} is large and the distance a between the two lenses is 20 cm, calculate x_{i2}, the image distance with respect to the negative lens. Calculate the magnification and show that for the object at infinity, one again has $M = -f_1/f_2$. The distance between the two lenses is then $f_1 + f_2$.
7. Laser beam expander. A laser beam of diameter of 2 mm should be expanded to a beam of 20 mm.
   a. A biconvex and a biconcave lens should be used. The beam first passes the biconcave lens of focal length -5 mm. Where should one place the biconvex lens of diameter of 30 mm and focal length of 50 mm?
   b. Two biconvex lenses should be used, one with $f = 5$ mm, the other with $f = 50$ mm. Make a sketch and give approximate values for the diameter of the lenses.

1.8 MATRIX FORMULATION FOR THICK LENSES

1.8.1 Refraction and Translation Matrices

A thick lens has two spherical surfaces separated by a dielectric material of a certain thickness. Previously we ignored the distance between the two surfaces

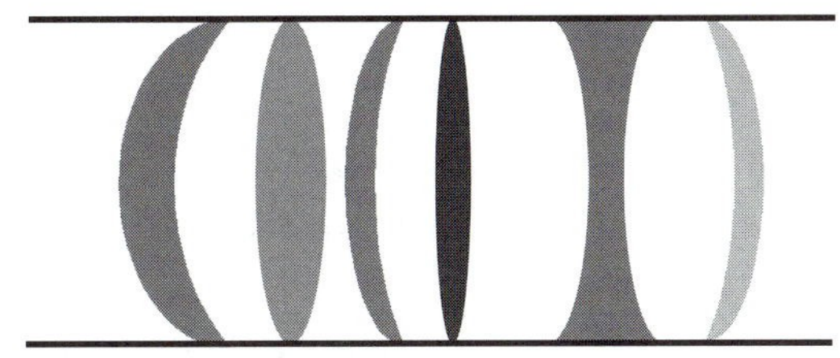

FIGURE 1.19 Multiple lens system. The lenses may have different radii of curvature and different refractive indices.

but now take it into account. One may calculate the image formation of the thick lens by first finding the image produced by the first surface. Then one uses this image as an object in the second imaging process and finds the image produced by the second surface. One could also use this procedure for lens systems with many lenses (Figure 1.19). However, one can develop a mathematical formalism to describe the image formation of a system of lenses by using the thin-lens equation. But one now has to measure the object and image distance from newly determined "principal planes," and not from the center of the thick lens. To do this, we first consider the case of refraction on a spherical surface (Figure 1.20). We want to represent the first surface by an operation which transforms the set of coordinates of the object into the set of coordinates of the first image. We show that this operation can be represented by a transformation matrix, which we call refraction matrix. Then we make a translation to get to the second surface, accomplished by a translation matrix, and the next operation on the second surface is again associated with a refraction matrix. This method is applicable to many different curved surfaces and their separations, having different thickness and refraction indices. The mathematical operation representing the processes of refraction at one and translation between two surfaces is a two-by-two matrix. The matrices are derived by using the paraxial theory, taking as the coordinates the distance from the axis of the point of the ray at the surface and the angle the ray makes with the axes (Figure 1.20a).

We now construct matrices to represent the refraction and translation operations. The matrices act on sets of two coordinates, written in the form of a vector. The initial coordinates (index1) in the plane of the object are acted on, and the result is the set of coordinates (index 2) in the plane of the image. We start from the equation for refraction on a single surface

$$-n_1/x_o + n_2/x_i = (n_2 - n_1)/r \tag{1.76}$$

and rewrite it, using α_1 and l_1 (see Figure 1.20a), as

$$n_1(\alpha_1/l_1) + n_2(-\alpha_2/l_2) = (n_2 - n_1)/r. \tag{1.77}$$

In addition we have for the second coordinate

$$l_1 = l_2. \tag{1.78}$$

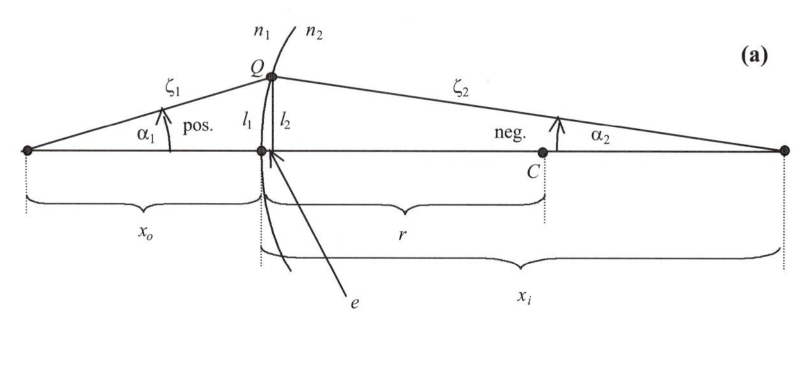

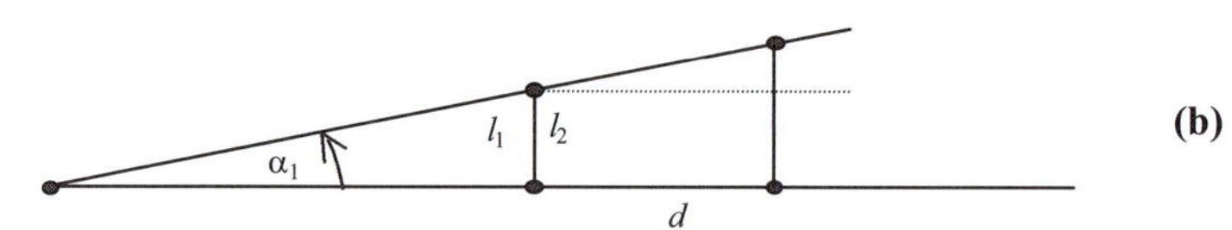

FIGURE 1.20 Coordinates for vector and matrix formulation: (a) the coordinates 1_1 and α_1 are used to form the vectors $I_1 = (l_1, \alpha_1)$, and the coordinates l_2 and α_2 are used to form the vectors $I_2 = (l_2, \alpha_2)$; (b) translation, the dependence of d on α_1, l_1, and l_2.

We define the vectors I_1 of object coordinates and I_2 of image coordinates using for I_1 the coordinates l_1 and α_1, and for I_2 we using l_2 and α_2,

$$I_1 = \begin{pmatrix} l_1 \\ \alpha_1 \end{pmatrix} \qquad I_2 = \begin{pmatrix} l_2 \\ \alpha_2 \end{pmatrix}. \tag{1.79}$$

The two equations (1.77) and (1.78) may be written in matrix notation as

$$\begin{pmatrix} l_2 \\ \alpha_2 \end{pmatrix} = \begin{pmatrix} 1 & 0 \\ -(1/r)(n_2 - n_1)/n_2 & n_1/n_2 \end{pmatrix} \begin{pmatrix} l_1 \\ \alpha_1 \end{pmatrix}. \tag{1.80}$$

For a proof, we may multiply the matrix with the vector and arrive back at Eqs. (1.76) to (1.78). In short notation we may also write

$$I_2 = R_{12} I_1.$$

The matrix R_{12} is called the refraction matrix of a single spherical surface

$$R_{12} = \begin{pmatrix} 1 & 0 \\ -(1/r)(n_2 - n_1)/n_2 & n_1/n_2 \end{pmatrix}. \tag{1.81}$$

For a plane surface, that is, for an infinite large radius of curvature, the matrix of Eq. (1.81) reduces to the refraction matrix of a plane surface

$$R = \begin{pmatrix} 1 & 0 \\ 0 & n_1/n_2 \end{pmatrix}. \tag{1.82}$$

We get the translation matrix T, that is, the translation from one vertical plane to the next over the distance d, by taking into account that $l_2 = l_1 + \alpha_1 d$; see

Figure 1.20b.

$$T = \begin{pmatrix} 1 & d \\ 0 & 1 \end{pmatrix}. \tag{1.83}$$

1.8.2 Two Spherical Surfaces at Distance d and Principal Planes

1.8.2.1 The Matrix

For a thick lens we use the refraction and translation matrices. We apply the refraction matrix corresponding to the first spherical surface, the translation matrix corresponding to the thickness of the lens, and the refraction matrix corresponding to the second spherical surface. We again assume that the light comes from the left, and realize that the sequence of the matrices is the sequence of action on I_1. In other words, the first surface is represented by the matrix on the far right.

First operation:	Refraction on first surface:	Matrix on the right
Second operation:	Translation between the surfaces:	Matrix in the middle
Third operation:	Refraction on the second surface:	Matrix on the left.

For the refraction matrix of a thick lens of thickness d and two different spherical surfaces, we obtain

$$\begin{pmatrix} 1 & 0 \\ -(1/r_2)(n_3 - n_2)n_3 & n_2/n_3 \end{pmatrix} \begin{pmatrix} 1 & d \\ 0 & 1 \end{pmatrix} \begin{pmatrix} 1 & 0 \\ -(1/r_1)(n_2 - n_1)n_2 & n_1/n_2 \end{pmatrix}. \tag{1.84}$$

Multiplication of the three matrices will give us one matrix representing the total action of the thick lens. To do this we define some abbreviations, called refracting powers P_{12}, P_{23}, and P, where P is related to the focal length of the thick lens.

$$P_{12} = -(1/r_1)(n_2 - n_1)/n_2 \tag{1.85}$$

$$P_{23} = -(1/r_2)(n_3 - n_2)/n_3, \quad \text{and} \tag{1.86}$$

$$P = -1/f = P_{23} + dP_{12}P_{23} + (n_2/n_3)P_{12}. \tag{1.87}$$

(From the 2,1 element P we get the focal length of the system.) We obtain the thick-lens matrix as

$$\begin{pmatrix} 1 + dP_{12} & d(n_1/n_2) \\ P & d(n_1/n_2)P_{23} + (n_1/n_3) \end{pmatrix}. \tag{1.88}$$

FileFig 1.23 (G23SYMB3M)

Symbolic calculation of the product of three matrices corresponding to a thick lens of refractive index n_2 and thickness d. The light is incident from a medium with refractive index n_1 and transmitted into a medium with refractive index n_3. The case of the thin lens is derived by setting $d = 0$ and $n_1 = n_3$; one obtains the thin-lens matrix.

G23SYMB3M

Thin-Lens Matrix

Special case of the thin-lens matrix. We start with the symbolic calculation of two surfaces at distance d

$$P12 = (-1/r1)(n2 - n1)/n2 \qquad P23 = (-1/r2)(n3 - n2)/n3$$

$$\begin{bmatrix} 1 & 0 \\ P23 & \frac{n2}{n3} \end{bmatrix} \cdot \begin{bmatrix} 1 & d \\ 0 & 1 \end{bmatrix} \cdot \begin{bmatrix} 1 & 0 \\ P12 & \frac{n1}{n2} \end{bmatrix}$$

$$\begin{bmatrix} 1 + d \cdot P12 & d \cdot \frac{n1}{n2} \\ \frac{(P23 \cdot n3 + P12 \cdot P23 \cdot d \cdot n3 + P12 \cdot n2)}{n3} & \frac{(P23 \cdot d \cdot n3 + n2)}{n3} \cdot \frac{n1}{n2} \end{bmatrix}$$

$$P = P23 + dP12P23 + (n2/n3)P12.$$

We go to the thin lens and set $d = 0$

$$\begin{bmatrix} 1 & 0 \\ P23 & \frac{n2}{n3} \end{bmatrix} \cdot \begin{bmatrix} 1 & 0 \\ 0 & 1 \end{bmatrix} \cdot \begin{bmatrix} 1 & 0 \\ P12 & \frac{n1}{n2} \end{bmatrix}$$

$$\begin{bmatrix} 1 & 0 \\ \frac{(P23 \cdot n3 + P12 \cdot n2)}{n3} & \frac{1}{n3} \cdot n1 \end{bmatrix}.$$

Since $n3$ and $n1$ are set to 1 we have

$$\begin{bmatrix} 1 & 0 \\ (P23 + P12 \cdot n2) & 1 \end{bmatrix}.$$

We set

$$P = (P23 + P12 \cdot n2)$$

and

$$P = \frac{-1}{f}; \quad f \text{ is the focal length of the lens.}$$

With

$$P12 = (-1/r1)(n2 - n1)/n2 \qquad P23 = (-1/r2)(n3 - n2)/n3$$

we obtain for $1/f = -((-1/r2)(1 - n2) + (-1/r1)(n2 - 1))$ and have finally for the thin-lens matrix,

$$\begin{bmatrix} 1 & 0 \\ \dfrac{-1}{f} & 1 \end{bmatrix}.$$

1.8.2.2 Application to the Thin Lens

We demonstrate more about the meaning and significance of the four matrix elements when reducing the matrix to the one corresponding to a thin lens. We use two surfaces close together; that is, we set $d = 0$ (Figure 1.21). The product matrix of Eq. (1.88) reduces to

$$\begin{pmatrix} 1 & 0 \\ P_{23} + (n_2/n_1)P_{12} & 1 \end{pmatrix}. \tag{1.89}$$

Assuming $n_1 = n_3$=1, we have for $P_{23} + (n_1/n_3)P_{12} = -(1 - n_2)/r_2 - (n_2 - 1)/r_1 = -1/f$, where f is the focal length of the thin lens. If we introduce these expressions into Eq. (1.89) and write the matrix with the coordinate vectors as in Eq. (1.80), we get

$$\begin{pmatrix} l_2 \\ \alpha_2 \end{pmatrix} = \begin{pmatrix} 1 & 0 \\ -(1/f) & 1 \end{pmatrix} \begin{pmatrix} l_1 \\ \alpha_1 \end{pmatrix}. \tag{1.90}$$

We label the matrix elements $M_{0,0}$, $M_{0,1}$, $M_{1,0}$, and $M_{1,1}$.

By using the coordinates as done in Eq. (1.77) and (1.78), we want to show that Eq. (1.90) is equivalent to the thin-lens equation. Multiplication yields

$$\begin{aligned} l_2 &= l_1 \\ \alpha_2 &= -l_1/f + \alpha_1. \end{aligned} \tag{1.91}$$

From Figure 1.21 we have $\alpha_2 = -l_1/x_i$, and $\alpha_1 = -l_1/x_o$, and have

$$l_1/(-x_o) + l_1/x_i = l_1/f. \tag{1.92}$$

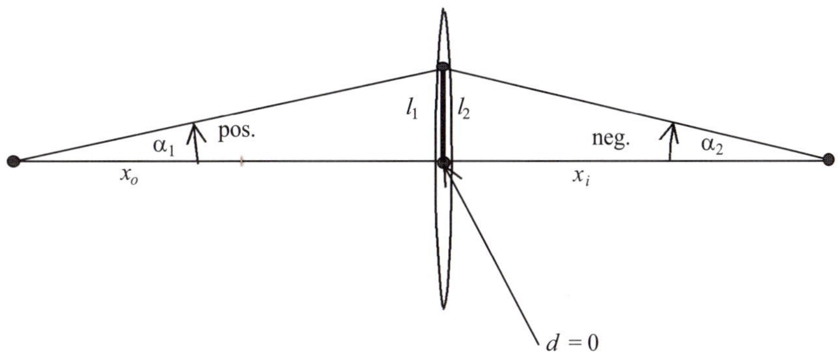

FIGURE 1.21 Coordinates for the thin lens.

We see that if the 0,0 and 1,1 elements are 1 and the 0,1 element is zero, we may obtain the focal length of the thin lens from the 1,0 element; that is, $-1/f = P_{23} + (n_2/n_1)P_{12}$.

We have gone through this example of the thin lens to show how the procedure with the refraction matrix works to get to the object–image relation. We measure x_o and x_i from the surfaces of the thin lens, and apply in the usual way the thin-lens equation, and take the focal length from the 1,0 element.

1.8.2.3 Thick Lens

For a thick lens, the matrix elements 0,0 and 1,1 of Eq. (1.88) are not 1, and the 0,1 element is not zero. To apply a similar procedure to that discussed for the thin lens, we introduce a transformation in order to get the 0,0 and 1,1 element to 1 and the 0,1 element to 0. These three requirements may be obtained by application of a translation. We first translate by $-h$ the plane of the object and at the end we go back by a translation of hh. The introduction of these two new parameters corresponds to the displacements of the points from which we have to count x_o and x_i. We apply these two translations to the thick-lens matrix of Eq. (1.88) and have to calculate

$$\begin{pmatrix} 1 & hh \\ 0 & 1 \end{pmatrix} \begin{pmatrix} 1 + d\,P_{12} & d\,(n_1/n_2) \\ P & d(n_1/n_2)P_{23} + n_1/n_3 \end{pmatrix} \begin{pmatrix} 1 & -h \\ 0 & 1 \end{pmatrix}. \tag{1.93}$$

We rewrite the thick-lens matrix, using the following abbreviations,

$$\begin{pmatrix} 1 & hh \\ 0 & 1 \end{pmatrix} \begin{pmatrix} M_{0,0} & M_{0,1} \\ M_{1,0} & M_{1,1} \end{pmatrix} \begin{pmatrix} 1 & -h \\ 0 & 1 \end{pmatrix}. \tag{1.94}$$

The multiplication is done in FileFig 1.24, and we get as the result

$$\begin{pmatrix} M_{0,0} + hhM_{1,0} & -M_{0,0}h + M_{0,1} + hh(-M_{1,0}h + M_{1,1}) \\ M_{1,0} & -M_{1,0}h + M_{1,1} \end{pmatrix}. \tag{1.95}$$

There are three requirements to be fulfilled, and only two new parameters. We set $M_{0,0} + hhM_{1,0} = 1$ and $-M_{1,0}h + M_{1,1} = 1$, and calculate h and hh. In order to be successful, the introduction of the calculated values of h and hh from these two equations must make the 0,1 element zero. It can be shown analytically that $[-M_{0,0}h + M_{0,1} + hh(-M_{1,0}h + M_{1,1})] = 0$, and numerically as seen in FileFig 1.24.

We have the same form of the matrix as in Eq. (1.89) and find that the (2,1) element has not been changed by the transformation. We have $P = -1/f = M_{1,0}$. As a result of our transformation we have for the parameters h, hh, and the focal length

$$hh = (1 - M_{0,0})/M_{1,0} \tag{1.96}$$

$$-h = (1 - M_{11})/M_{1,0} \tag{1.97}$$

$$P = -1/f = M_{1,0}. \tag{1.98}$$

FileFig 1.24 (G24SYMBH)

Symbolic calculations of the general transformation for a thick lens. Calculation of the two-spherical-surface matrix and displacement matrix with parameters $-h$ and hh. A numerical example is presented for $n_1 = 1$, $n_2 = 1.5$, $n_3 = 1$, $r_1 = 10$, $r_2 = -10$, and $d = 20$.

G24SYMBH

Symbolic Calculations of the Product of Three Matrices Corresponding to a General Thick Lens

1. Symbolic calculation of the matrix for the thick lens

$$\begin{bmatrix} 1 & 0 \\ P23 & \dfrac{n2}{n3} \end{bmatrix} \cdot \begin{bmatrix} 1 & d \\ 0 & 1 \end{bmatrix} \cdot \begin{bmatrix} 1 & 0 \\ P12 & \dfrac{n1}{n2} \end{bmatrix}$$

$$P12 = (-1/r1)((n2 - n1)/n2)$$

$$P23 = (-1/r2)((n3 - n2)/n3)$$

$$\begin{bmatrix} 1 + d \cdot P12 & d \cdot \dfrac{n1}{n2} \\ \dfrac{(P23 \cdot n3 + P12 \cdot P23 \cdot d \cdot n3 + P12 \cdot n2)}{n3} & \dfrac{(P23 \cdot d \cdot n3 + n2)}{n3} \cdot \dfrac{n1}{n2} \end{bmatrix}.$$

2. Determination of h and hh. For simpler calculation we define the matrix

$$\begin{bmatrix} M_{0,0} & M_{0,1} \\ M_{1,0} & M_{1,1} \end{bmatrix}$$

$$M_{0,0} = 1 + d \cdot P12 \qquad M_{0,1} = d \cdot \frac{n1}{n2}$$

$$M_{1,0} = \frac{(P23 \cdot n3 + P12 \cdot P23 \cdot d \cdot n3 + P12 \cdot n2)}{n3} \qquad \frac{(P23 \cdot d \cdot n3 + n2)}{n3} \cdot \frac{n1}{n2}$$

and determine h and hh,

$$\begin{bmatrix} 1 & hh \\ 0 & 1 \end{bmatrix} \cdot \begin{bmatrix} M_{0,0} & M_{0,1} \\ M_{1,0} & M_{1,1} \end{bmatrix} \cdot \begin{bmatrix} 1 & -h \\ 0 & 1 \end{bmatrix}$$

$$\begin{bmatrix} M_{0,0} + hh \cdot M_{1,0} & -h \cdot M_{0,0} - h \cdot hh \cdot M_{1,0} + M_{0,1} + hh \cdot M_{1,1} \\ M_{1,0} & -M_{1,0} \cdot h + M_{1,1} \end{bmatrix}.$$

3. The results for h, hh, and f are

$$hh = \frac{1 - M_{0,0}}{M_{1,0}} \qquad 1 = \frac{-(1 - M_{1,1})}{M_{1,0}} \qquad f = \frac{-1}{M_{1,0}}.$$

4. Numerical calculation

$$P12 := \frac{1}{r1} \cdot \frac{n2 - n1}{n2} \qquad P23 := \frac{1}{r2} \cdot \frac{n3 - n2}{n3}$$
$$P12 = -3.333 \cdot 10^{-11} \qquad P23 = -0.05$$

$$M_{0,0} := 1 + d \cdot P12 \qquad M_{0,1} = d \cdot \frac{n1}{n2}$$
$$M_{0,0} = 1 \qquad M_{0,1} = 6.667$$
$$M_{1,0} := \frac{(P23 \cdot n3 + P12 \cdot P23 \cdot d \cdot n3 + P12 \cdot n2)}{n3}$$
$$M_{1,1} := \frac{(P23 \cdot d \cdot n3 + n2)}{n3} \cdot \frac{n1}{n2}$$
$$M_{1,0} = -0.05 \quad M_{1,1} = 0.667.$$

5. The result for h, hh, and f

$$hh := \frac{1 - M_{0,0}}{M_{1,0}} \qquad h := \frac{-(1 - M_{1,1})}{M_{1,0}} \qquad f := \frac{-1}{M_{1,0}}$$
$$hh = -6.667 \cdot 10^{-9} \qquad h = 6.667 \qquad f = 20.$$

6. The input values are globally defined

$$n1 \equiv 1 \qquad n2 \equiv 1.5 \qquad n3 \equiv 1 \qquad r1 \equiv 10^{10} \qquad r2 \equiv -10 \qquad d \equiv 10.$$

The transformation using the two matrices

$$\begin{pmatrix} 1 & hh \\ 0 & 1 \end{pmatrix} \quad \begin{pmatrix} 1 & -h \\ 0 & 1 \end{pmatrix} \tag{1.99}$$

has the effect that we have to count xo from the point on the axis determined by h, and xi from the point on the axis determined by hh. We do not count from the vertex of the spherical surfaces. If we call the vertex of the first surface V_1 and the vertex of the second surface V_2, we have a similar sign convention as we have used before:

1. if $h > 0$, the point to start calculating x_o is to the right of V_1; otherwise to the left; and
2. if $hh > 0$, the point to start calculating x_i is to the right of V_2; otherwise to the left.

The calculation is shown in FileFig 1.25. The planes perpendicular to the axis at h and hh are called *principal planes*. As a check one finds, that in the approximation of the thin lens, the difference $hh - h = 0$.

We state the general procedure for using the thin-lens equation with the matrix method: One calculates $hh = (1 - M_{0,0})/M_{1,0}$ and $-h = (1 - M_{11})/M_{1,0}$. The focal length f is obtained from $P = -1/f = M_{1,0}$. One measures x_o from h and x_i from hh and applies the thin-lens equation.

FileFig 1.25 (G25SYMBGTH)

Calculation of the general transformation for a thin lens. Calculation of the product of the two-spherical-surface matrix, and the displacement matrix. Determination of the parameters $-h$ and hh. Specialization for the case of the thin lens. Numerical example for $n_1 = 1$, $n_2 = 1.5$, $n_3 = 1.3$, $r_1 = 120$, and $r_2 = -10$.

G25SYMBGTH is only on the CD.

1.8.2.4 Application to the Hemispherical Thick Lens

We consider a thick lens of hemispherical shape (see Figure 1.22). In FileFig 1.26 we present the calculations and for the choice of parameters: $n_2 = 1.5$, $n_1 = n_3 = 1$, $r_1 = 20$, and $r_2 = \infty$.

If we set $n_2 = n = 1.5$, $n_1 = n_3 = 1$, $r_1 = r = d$, and $r_2 = \infty$, we have the result that $P_{12} = -1/3r$, $P_{23} = 0$, $P = -1/2r$; that is, $f = 2r$, $h = 0$, and $hh = -2r/3$. We find for the numerical calculation that the (0,0) and (1,1) elements are equal to zero and the (0,1) element is equal to 1.

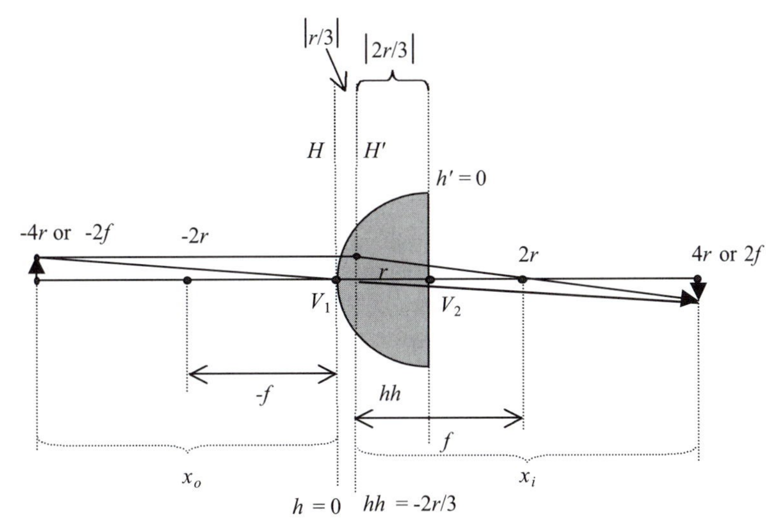

FIGURE 1.22 Coordinates for a hemispherical thick lens of index n. The principal planes are indicated as H and H'.

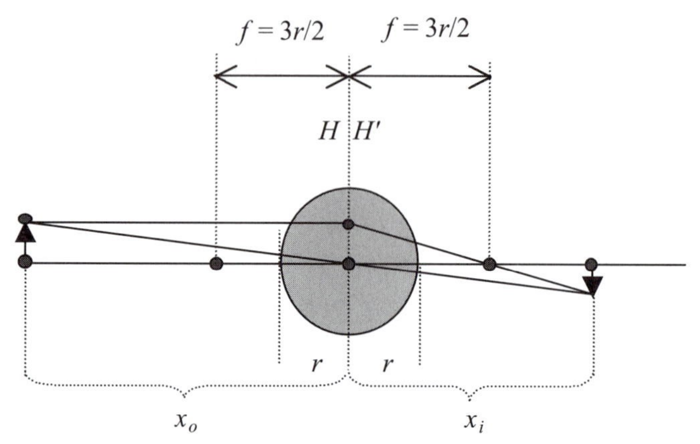

FIGURE 1.23 Coordinates for a spherical thick lens.

FileFig 1.26 (G26HEM)

Calculations of the hemispherical thick lens with curved surface to the left. For the numerical values we take $n_2 = 1.5, n_1 = n_3 = 1, r_1 = 10 = d$, and $r_2 = \infty$.

G26HEM is only on the CD.

Application 1.26. Repeat the calculations for a hemispherical thick lens with curved surface to the right.

1.8.2.5 Application to Glass Sphere

We consider a thick lens of spherical shape (Figure 1.23). In FileFig 1.27 we show the calculations for $n_2 = 1.5 = n$, $n_1 = n_3 = 1$, $r_1 = -r_2 = 10$, and $d = 2r_1 = 20$. The result is $P_{12} = -1/3r$, $P_{23} = -1/2r$, $P = -2/3r$; that is, $f = 3r/2$, $h = r$, and $hh = -r$.

From Figure 1.23 we see that the principal planes are at the center, as expected for a symmetric lens. We have to start at the center to measure x_o and x_i and apply the thin lens equation with focal length $f = 3r/2$. For the numerical calculations we use $r_1 = 10$ and have $h = 10$, and $hh = -10$. The (0,0) and (1,1) elements are zero; the (0,1) element is 1.

FileFig 1.27 (G27SPH)

Calculation of the spherical thick lens. For the numerical values we have chosen $n_2 = 1.5$, $n_1 = n_3 = 1$, $r_1 = 10$, $r_2 = -10$, and $d = 20$.

G27SPH is only on the CD.

Application 1.27. Go over the calculations for two different sets of parameters n_2, n_1, n_3, r_1, d, and r_2.

1.8.3 System of Lenses

1.8.3.1 System of Two Thin Lenses in Air

We now study the application of matrices to the calculation of the final image produced by a system of two lenses. First we consider a system of two thin lenses of focal length f_1 and f_2 and at distance a between them

$$\begin{pmatrix} 1 & 0 \\ -1/f_2 & 1 \end{pmatrix}\begin{pmatrix} 1 & a \\ 0 & 1 \end{pmatrix}\begin{pmatrix} 1 & 0 \\ -1/f_1 & 1 \end{pmatrix}. \tag{1.100}$$

Multiplication yields

$$\begin{pmatrix} (f_1 - a)/f_1 & a \\ -(f_1 - a + f_2)/f_1 f_2 & -(a - f_2)/f_2 \end{pmatrix}. \tag{1.101}$$

Since the (0,0) and (1,1) elements are not zero and the (0,1) element is not 1 we have to apply the transformation to principal planes, as we did for the single thick lens. We have to evaluate (Figure 1.24)

$$\begin{pmatrix} 1 & hh \\ 0 & 1 \end{pmatrix}\begin{pmatrix} (f_1 - a)/f_1 & a \\ -(f_1 - a + f_2)/f_1 f_2 & -(a - f_2)/f_2 \end{pmatrix}\begin{pmatrix} 1 & -h \\ 0 & 1 \end{pmatrix}. \tag{1.102}$$

This is done in FileFig 1.28 and the result is

$$h = -a/Pf_2$$
$$hh = a/Pf_1$$

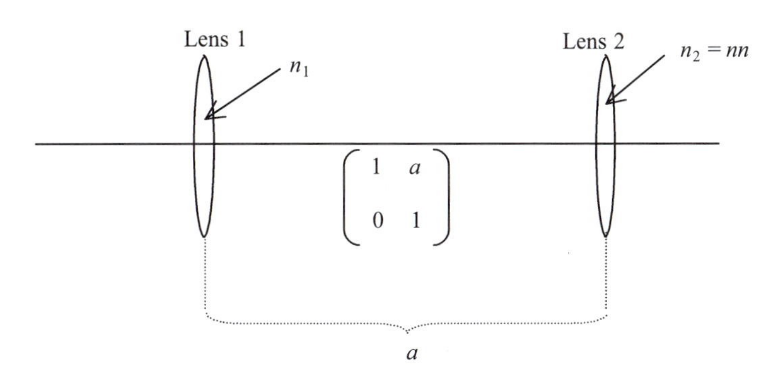

FIGURE 1.24 Coordinates for two lenses in air with corresponding matrices.

$$P = (-1/f_2)(1 - a/f_1) - 1/f_1.$$

For the application to calculate the image from a given object point, focal lengths, and distance between the lenses, we measure x_o from h, x_i from hh, and get the focal length from $-1/f = P$.

FileFig 1.28 (G28SYST2LTI)

Calculation for a system of two thin lenses. For the numerical values we have chosen $f_1 = 10$, $f_2 = 10$, and $a = 100$.

G28SYST2LTI

Symbolic Calculation to Determine the Principal Planes for Two Thin Lenses at Distance a

The matrix (M) as the product of the two lenses and the displacement between them

$$\begin{bmatrix} 1 & 0 \\ -\frac{1}{f2} & 1 \end{bmatrix} \cdot \begin{bmatrix} 1 & a \\ 0 & 1 \end{bmatrix} \cdot \begin{bmatrix} 1 & 0 \\ -\frac{1}{f1} & 1 \end{bmatrix}$$

$$\begin{bmatrix} \frac{(f1-a)}{f1} & a \\ \frac{-(f1-a+f2)}{(f2-f1)} & \frac{-(a-f2)}{f2} \end{bmatrix}.$$

Special case $a = 0$, two thin lenses in contact

$$\begin{bmatrix} 1 & 0 \\ -\frac{1}{f2} & 1 \end{bmatrix} \cdot \begin{bmatrix} 1 & 0 \\ 0 & 1 \end{bmatrix} \cdot \begin{bmatrix} 1 & 0 \\ -\frac{1}{f1} & 1 \end{bmatrix}$$

$$\begin{bmatrix} 1 & 0 \\ \frac{-(f1+f2)}{(f2-f1)} & 1 \end{bmatrix}.$$

Principal planes with h and hh, and $P = (-1/f2)(1 - a/f1) - 1/f1$

$$\begin{bmatrix} 1 & hh \\ 0 & 1 \end{bmatrix} \cdot \begin{bmatrix} \frac{-(-f1+a)}{f1} & a \\ P & \frac{-(a-f2)}{f2)} \end{bmatrix} \cdot \begin{bmatrix} 1 & -h \\ 0 & 1 \end{bmatrix}$$

$$\begin{bmatrix} \frac{(f1-a+hh\cdot P\cdot f1)}{f1} & \frac{(-h\cdot f2\cdot f1+h\cdot f2\cdot a-h\cdot f2\cdot hh\cdot P\cdot f1+f1\cdot a\cdot f2-f1\cdot hh\cdot a+f1\cdot hh\cdot f2)}{(f1\cdot f2)} \\ P & \frac{-P\cdot h\cdot f2+a-f2)}{f2} \end{bmatrix}.$$

If the (1, 1) and 2, 2 elements are one, we have for $hh = a/Pf1$ and $h = a/Pf2$, P is always $-1/f$

$$P = (-1/f2)(1 - a/f1) - 1/f1$$

$$P := \left(\frac{-1}{f2}\right)\cdot\left(1 - \frac{a}{f1}\right) - \frac{1}{f1} \qquad hh := \frac{a}{P\cdot f1} \qquad h := \frac{-a}{P\cdot f2}$$

$$M := \begin{bmatrix} \frac{(f1-a+hh\cdot P\cdot f1)}{f1} & \frac{(-h\cdot f2\cdot f1+h\cdot f2\cdot a-h\cdot f2\cdot hh\cdot P\cdot f1+f1\cdot a\cdot f2-f1\cdot hh\cdot a+f1\cdot hh\cdot f2)}{(f1\cdot f2)} \\ P & \frac{-(P\cdot h\cdot f2+a-f2)}{f2} \end{bmatrix}$$

$$f1 \equiv 10 \qquad f2 \equiv 10 \qquad a \equiv 100$$

$$M = \begin{bmatrix} 1 & 0 \\ 0.8 & 1 \end{bmatrix} \qquad f := \frac{-1}{P}$$

$$hh = 12.5 \qquad h = -12.5 \qquad f = -1.25.$$

Application 1.28. Consider the case where $a = 0$, and compare the resulting focal length f with $1/(1/f_1 + 1/f_2)$.

1.8.3.2 System of Two Thick Lenses

We consider two thick lenses and assume that lens 1 has the refractive index n lens 2 the index nn. We also assume that the radii of curvature of the four spherical surfaces are labeled r_1 to r_4 and that the distance between lens 1 and lens 2 is a. The matrix for the system is obtained from the sequence of three matrices (Figure 1.25).

We start on the right with the thick-lens matrix of the first lens, then the translation matrix, and then to the left the thick-lens matrix of the second lens. The calculation is shown in FileFig 1.29 and one obtains

$$\begin{pmatrix} 1 + d_2 P_{34} & d_2/nn \\ P_2 & d_2(P_{45}/nn) + 1 \end{pmatrix} \begin{pmatrix} 1 & a \\ 0 & 1 \end{pmatrix} \begin{pmatrix} 1 + d_1 P_{12} & d_1/n \\ P_1 & d_1(P_{23}/n) + 1 \end{pmatrix} \tag{1.103}$$

with

$$P_{12} = -(1/r_1)(n-1)/n$$
$$P_{23} = -(1/r_2)(1-n)$$
$$P_{34} = -(1/r_3)(nn-1)/nn$$

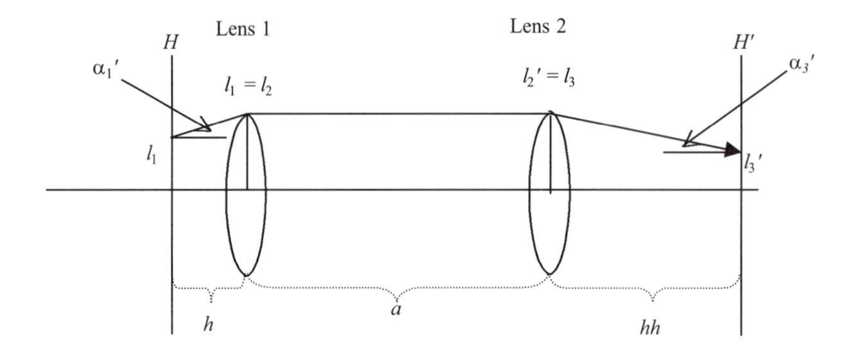

$$\begin{pmatrix} 1 + d_2 P_{34} & d_2/nn \\ P_2 & d_2(P_{45}/nn) + 1 \end{pmatrix} \begin{pmatrix} 1 & a \\ 0 & 1 \end{pmatrix} \begin{pmatrix} 1 + d_1 P_{12} & d_1/n \\ P_1 & d_1(P_{23}/n) + 1 \end{pmatrix}$$

FIGURE 1.25 Coordinates for two thick lenses in air with corresponding matrices.

$$\begin{aligned} P_{45} &= -(1/r_4)(1 - nn) \\ P_1 &= P_{23} + P_{12}P_{23}d_1 + P_{12}n \\ P_2 &= P_{45} + P_{34}P_{45}d_2 + P_{34}nn. \end{aligned}$$

To determine the principal planes of this system, we call M the product of the three matrices in Eq. (1.103), and have to calculate (see FileFig 1.29)

$$\begin{pmatrix} 1 & hh \\ 0 & 1 \end{pmatrix} \quad M \quad \begin{pmatrix} 1 & -h \\ 0 & 1 \end{pmatrix}. \tag{1.104}$$

We have to set in the product matrix the (0,0) and (1,1) elements equal to one, and it follows that the (0,1) element is 0. The result of the transformation is:

$$\begin{aligned} h &= -(1 - M_{1,1})/M_{1,0} \\ hh &= (1 - M_{0,0})/M_{1,0} \\ 1/f &= -M_{1,0}. \end{aligned}$$

To calculate the image distance for a given object distance, we have to measure x_o from h, x_i from hh, and apply the thin-lens equation with focal length f calculated from $-1/f = M_{1,0}$.

For a specific example of a system of two thick lenses we choose a system of two hemispherical lenses. Each lens is one-half of a sphere, and we assume that the distance a is zero. The results are the same as we found in Section 1.8.2.4 for a sphere.

In Figure 1.26, we show the two hemispherical lenses with their refracting powers P_{12} and P_{45}, each of thickness d and a refractive index n with the corresponding matrices. The details of the calculation are shown in FileFig 1.29.

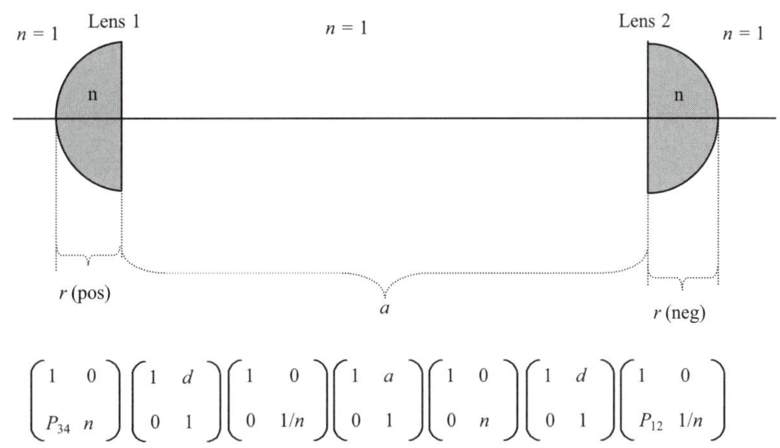

FIGURE 1.26 Two hemispherical lenses at distance a, and the corresponding matrices. The lenses have refractive index n, thickness $d = r$, P_{12} is the refracting power of the first spherical surface, and P_{45} of the last.

FileFig 1.29 (G29SYST2LTC)

Calculation for a system of two thick lenses with refractive indices n and nn at distance a. The choices of the numerical values are $n = 1.5$, $nn = 1.5$, $d_1 = 10$, $d_2 = 10$, $a = 100$, $r_1 = 10$, $r_2 = -10$, $r_3 = 10$, and $r_4 = -10$.

G29SYST2LTC

Symbolic Calculation of the Principal Planes for Two Thick Lenses of Refractive Indices n and nn in Air.

Distance between lenses is a and the thickness of the first is $d1$, of the second $d2$. Radii of curature are $r1$ to $r4$. The matrix of the first lens is on the right.

$$\begin{pmatrix} 1 & 0 \\ P45 & \frac{nn}{1} \end{pmatrix} \cdot \begin{pmatrix} 1 & d2 \\ 0 & 1 \end{pmatrix} \cdot \begin{pmatrix} 1 & 0 \\ P34 & \frac{1}{nn} \end{pmatrix} \cdot \begin{pmatrix} 1 & a \\ 0 & 1 \end{pmatrix}$$

$$\cdot \left[\begin{pmatrix} 1 & 0 \\ P23 & \frac{n}{1} \end{pmatrix} \cdot \begin{pmatrix} 1 & d1 \\ 0 & 1 \end{pmatrix} \cdot \begin{pmatrix} 1 & 0 \\ P12 & \frac{1}{n} \end{pmatrix} \right]$$

$$P12 = -(1/r1)(n-1)/n \qquad P23 = -(1/r2)(n-1)/n$$

$$P34 = -(1/r3)(nn-1)/nn \qquad P45 = -(1/r4)(1-nn)$$

Matrix for the first lens

$$\left[\begin{pmatrix} 1 & 0 \\ P23 & \frac{n}{1} \end{pmatrix} \cdot \begin{pmatrix} 1 & d1 \\ 0 & 1 \end{pmatrix} \cdot \begin{pmatrix} 1 & 0 \\ P12 & \frac{1}{n} \end{pmatrix} \right]$$

$$\begin{bmatrix} 1 + d1 \cdot P12 & \frac{d1}{nn} \\ P23 + P12 \cdot P23 \cdot d1 + P12 \cdot n & \frac{(P23 \cdot d1 + n)}{n} \end{bmatrix}$$

Matrix for the second lens

$$\begin{pmatrix} 1 & 0 \\ P45 & \frac{nn}{1} \end{pmatrix} \cdot \begin{pmatrix} 1 & d2 \\ 0 & 1 \end{pmatrix} \cdot \begin{pmatrix} 1 & 0 \\ P34 & \frac{1}{nn} \end{pmatrix}$$

$$\begin{bmatrix} 1 + d2 \cdot P34 & \frac{d2}{nn} \\ P45 + P34 \cdot P45 \cdot d2 + P34 \cdot nn & \frac{(P45 \cdot d2 + nn)}{nn} \end{bmatrix}$$

For the determination of h and hh

$$\begin{bmatrix} 1 & hh \\ 0 & 1 \end{bmatrix}$$

$$\cdot \begin{bmatrix} 1 + d2 \cdot P34 & \frac{d2}{nn} \\ P45 + P34 \cdot P45 \cdot d2 + P34 \cdot nn & \frac{(P45 \cdot d2 + nn)}{nn} \end{bmatrix} \cdot \begin{bmatrix} 1 & a \\ 0 & 1 \end{bmatrix}$$

$$\cdot \begin{bmatrix} 1 + d1 \cdot P12 & \frac{d1}{n} \\ P23 + P12 \cdot P23 \cdot d1 + P12 \cdot n & \frac{(P23 \cdot d1 + n)}{n} \end{bmatrix} \cdot \begin{bmatrix} 1 & -h \\ 0 & 1 \end{bmatrix}$$

Multiplication results in a very large expression, and we go right away to numerical calculations.

We have for the powers of refraction

$$P12 := -\frac{n-1}{r1 - n} \qquad P23 := -\frac{1-n}{r2}$$

$$P34 := -\frac{nn-1}{r3 \cdot nn} \qquad P45 := -\frac{1-nn}{r4}.$$

The thick lens matrix is then

$$M := \begin{bmatrix} 1 + d2 \cdot P34 & \frac{d2}{nn} \\ P45 + P34 \cdot P45 \cdot d2 + P34 \cdot nn & \frac{(P45 \cdot d2 + nn)}{nn} \end{bmatrix} \cdot \begin{pmatrix} 1 & a \\ 0 & 1 \end{pmatrix}$$

$$\cdot \begin{bmatrix} 1 + d1 \cdot P12 & \frac{d1}{n} \\ P23 + P12 \cdot P23 \cdot d1 + P12 \cdot n & \frac{(P23 \cdot d1 + n)}{n} \end{bmatrix}.$$

The result is

$$\begin{bmatrix} 0.333 & 13.333 \\ -0.667 & 0.333 \end{bmatrix}.$$

We define M as

$$\begin{bmatrix} M_{0,0} & M_{0,1} \\ M_{1,0} & M_{1,1} \end{bmatrix}.$$

For the determination of h and hh we multiply by the two translation matrices

$$\begin{bmatrix} 1 & hh \\ 0 & 1 \end{bmatrix} \cdot \begin{bmatrix} M_{0,0} & M_{0,1} \\ M_{1,0} & M_{1,1} \end{bmatrix} \cdot \begin{bmatrix} 1 & -h \\ 0 & 1 \end{bmatrix}$$

$$\begin{bmatrix} M_{0,0} + hh \cdot M_{1,0} - h \cdot M_{0,0} - h \cdot hh \cdot M_{1,0} + M_{0,1} + hh \cdot M_{1,1} \\ M_{1,0} \qquad -M_{1,0} \cdot h + M_{1,1} \end{bmatrix}$$

$$hh := \frac{1 - (M_{0,0})}{M_{1,0}} \qquad h := \frac{1 - (M_{1,1})}{(-M)_{1,0}} \qquad f := -\frac{1}{M_{1,0}}$$

$$hh = -10 \qquad h = 10 \qquad f = 15.$$

Input Data

$$n \equiv 1.5 \qquad nn \equiv 1.5 \qquad d1 \equiv 10 \qquad d2 \equiv 10 \qquad a \equiv 0$$

$$r1 \equiv 10 \qquad r2 \equiv 10^{10} \qquad r3 \equiv 10^{10} \qquad r4 \equiv -10.$$

Check the form of the final matrix product

$$MM := \begin{bmatrix} 1 & hh \\ 0 & 1 \end{bmatrix} \cdot \begin{bmatrix} M_{0,0} & M_{0,1} \\ M_{1,0} & M_{1,1} \end{bmatrix} \cdot \begin{bmatrix} 1 & -h \\ 0 & 1 \end{bmatrix}$$

$$MM = \begin{bmatrix} 1 & -1.776 \cdot 10^{-15} \\ -0.667 & 1 \end{bmatrix}.$$

Applications to Matrix Method

1. An exercise for matrix multiplication. Draw two cartesian coordinate systems x, y and x', y, the second rotated by the angle θ with respect to the first. Identify the matrix

$$A = \begin{pmatrix} \cos\theta & -\sin\theta \\ \sin\theta & \cos\theta \end{pmatrix}$$

with the rotation of x, y into x', y'.

a. Is this a rotation in the mathematical positive or negative sense?

b. The matrix for rotation in the opposite direction A^{-1} is obtained by substituting for θ the negative value $-\theta$.

c. Show that $A\,A^{-1}$ is the unit matrix.

d. The transposed matrix A^T is obtained from A by interchanging the 2,1 and 1,2 elements. In our case the A^T is equal to A^{-1} and AA^T is the unit matrix.

e. Show that $A\,A$ is the same matrix if we substitute into A the angle 2θ.

2. Noncommutation of matrices. In general two matrices A and B may not be commuted; that is, AB is not equal to BA. We show this in the following example for a different sequence of the same matrices. We consider two hemispherical thick lenses where light is coming from the left. The light hits the first lens $L1$ at a spherical surface of radius of curvature r, then traverses the thickness d, and emerges from a plane surface. The second, $L2$, has the reverse order; first the plane surface, then thickness d, and then the curved surface with the same radius of curvature r. The refractive indices of the lenses are n_2 and outside we assume $n_1 = n_3 = 1$. Make a sketch. See how the two lenses are different. The product matrices for lens 1 and lens 2 are different for the two cases. Compare the position of the principal planes. Compare for the case where $r = \infty$.
3. Calculate, using the matrix method, the position of the two principal planes for a system of two thin lenses, both of focal length f, and a distance f.
4. Consider a concave-convex lens. The first surface has a radius of curvature $r_1 = 20$ cm, the second, a radius of curvature $r_2 = -10$ cm with thickness of $d = 5$ cm.

   a. Calculate the principal planes and focal length and find the image of an object positioned at 5 cm to the left of the first surface.

   b. Find the same result by using twice the imaging equation of a single surface.
5. Thick concentration lens. A thick lens of radius of curvature $-r_1 = r_2 = -5$ mm and thickness of 4 mm is used to concentrate incident parallel light on a detector. Using the matrix method, find the position with respect to the detector plane.
6. Plane-convex and convex-plane lens. The radii of curvature for the convex surface is $r = 10$ cm and for the concave surface $r = -10$ cm and the thickness is 4 cm.

   a. Compare h, hh, and f for both lenses.

   b. An object is placed 100 cm to the left of the first surface. Find the image point for both lenses.

1.9 PLANE AND SPHERICAL MIRRORS

1.9.1 Plane Mirrors and Virtual Images

A two-dimensional object appears in a flat mirror as a virtual and left–right inverted image. First we look at one reflected ray (Figure 1.27a). We observe the law of reflection, which says the angle of incidence has the same absolute value as the angle of reflection.

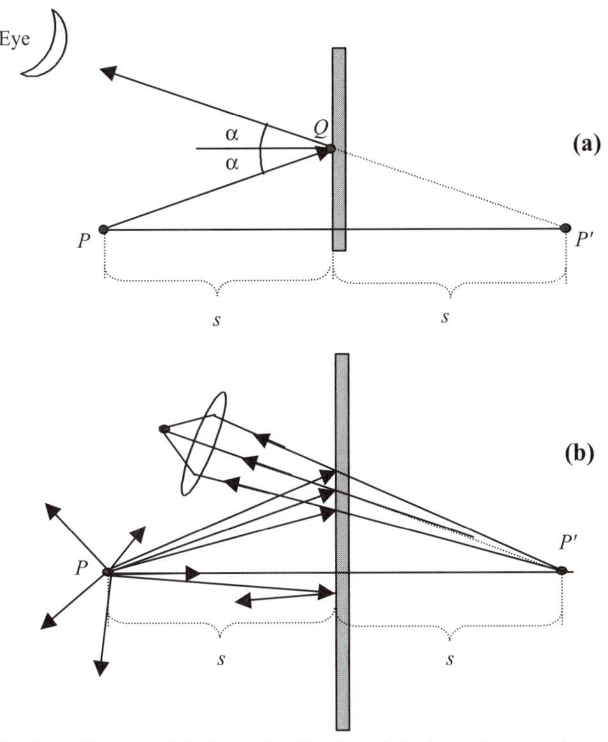

FIGURE 1.27 (a) Coordinates of the law of reflection; (b) virtual image of a real point source produced by a plane mirror. The image is observed by using a lens, which may be the eye lens.

In Figure 1.27b we show the reflection of a cone of light, emerging from a point source. The object point appears to us as it is on the "other side" of the mirror. Now we look at a three-dimensional object, represented by the arrows of a right-handed coordinate system. The virtual image produced by the flat mirror appears as a left-handed coordinate system. This may be seen by comparing the image of our left hand as it appears in a mirror, with our right hand placed before the mirror. Similarly one finds "Ambulance" written on the front of an ambulance truck written in letters from left to right. A driver in a car before the truck can read it "normally" in the rear view mirror.

1.9.2 Spherical Mirrors and Mirror Equation

Spherical concave mirrors of diameters of a few meters are used in astronomical telescopes, replacing the first lens, as discussed in Section 1.7 on optical instruments. A real inverted image is produced by a real erect object. Spherical convex mirrors with much smaller diameters are used for cosmetic applications, where an erect virtual image is formed from an erect object. Our eye uses a positive lens for the image formation on the retina, but "sees" the virtual image erect, as discussed in Section 1.7.

We derive the image-forming equation for spherical mirrors by looking at the image-forming equation of a single spherical surface

$$n_1/(-x_o) + n_2/x_i = (n_2 - n_1)/r. \tag{1.105}$$

By formally setting $n_1 = -n_2$ we get the imaging equation for a spherical mirror

$$n_1/(-x_o) + (-n_1)/x_i = (-n_1 - n_1)/r \tag{1.106}$$

$$n_1/(-x_o) + (-n_1)/xi = (-2n_1)/r, \tag{1.107}$$

where r is the radius of curvature of the spherical surface.

Division by $-n_1$ results in the *spherical mirror equation*

$$\frac{1}{x_o} + \frac{1}{x_i} = \frac{2}{r}. \tag{1.108}$$

1.9.3 Sign Convention

The light is assumed to be incident from the left. The object points x_o are to the left of the mirror, and x_0 is always negative. No positive values are considered. If x_i is negative we have a real image. If x_i is positive we have a virtual image (Figure 1.28).

For a convex spherical mirror, r is positive. For a concave spherical mirror, r is negative.

1.9.4 Magnification

For the magnification (Figure 1.29, we have

$$m = y_i/y_o = -x_i/x_o. \tag{1.109}$$

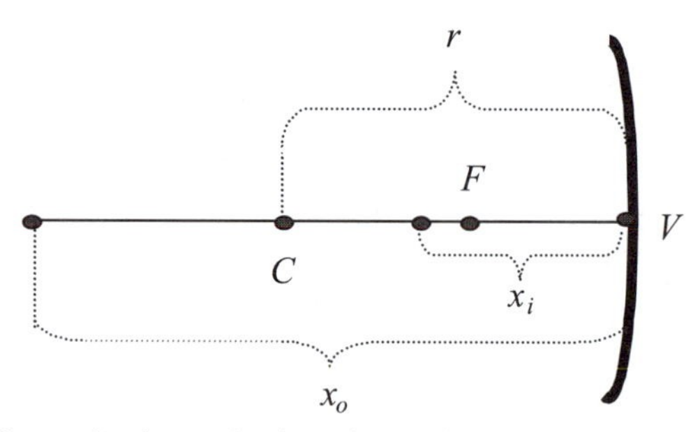

FIGURE 1.28 Coordinates for the production of a real image from a real object.

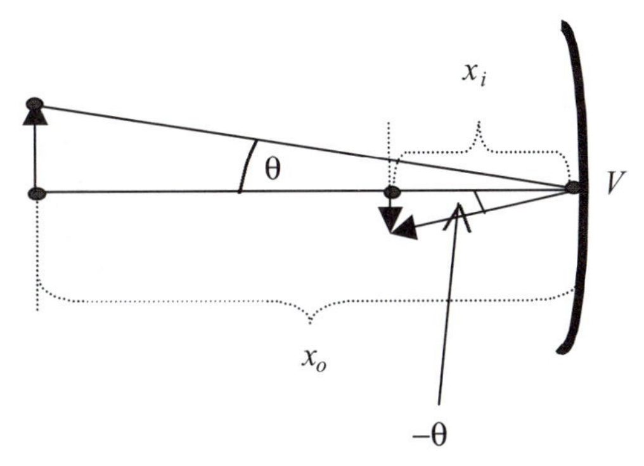

FIGURE 1.29 Magnification with respect to real object and real image.

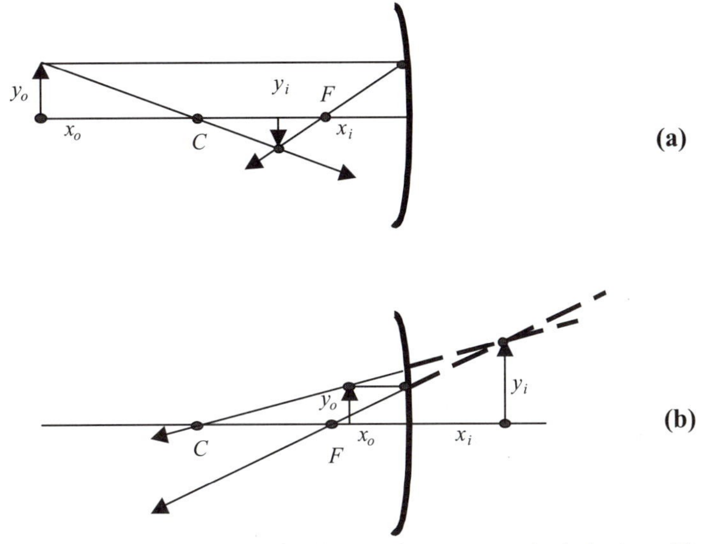

FIGURE 1.30 Geometrical construction of images for concave spherical mirror. The object is: (a) to the left of the focal point; (b) to the right of the focal point.

1.9.5 Graphical Method and Graphs of x_i Depending on x_o

1.9.5.1 Concave Spherical Mirror

Geometrical Construction

1. Choose y_o and draw the PF-ray to the mirror and then reflected back through the focus F, given by $r/2$.
2. The ray incident through the top of the arrow and then going to the center of curvature C is reflected back onto itself (Figure 1.30).
   Both may be extended to the other side of the mirror when x_o is between the focus and mirror.

Graph of x_i as Function of x_o

A concave spherical mirror has a negative radius of curvature. In FileFig 1.30 we calculate the image points for given object points and radius of curvature. The light comes from the left. For $x_{if} = -\infty$, the focus $x_{\text{of}} = r/2$, and since r is negative for a convex mirror, it is to the left of the mirror. This is the only focus we have for a concave mirror. The focus is also a singularity. We obtain real images for x_o to the left and virtual images for x_o to the right.

FileFig 1.30 (G30MIRCV)

Concave spherical mirror. Calculation of image positions from given object positions. Graph for image positions depending on object positions for radius of curvature $r = -50$, that is, $r/2 = -25$, and x_o from -100 to -0.1.

G30MIRCV

Concave Mirror

Raduis of curvature is negative; xo is on left, and is negative. To get around the singularity at $-xo = f$ one chooses the increments such that the value for the singularity does not appear.

$$r := -50$$

$$xo := -60$$

$$xi := \frac{1}{\left(\frac{1}{2}\right) - \frac{1}{xo}} \qquad xi = -42.857$$

$$m := \frac{-xi}{xo} \qquad m = -0.714.$$

Graph

$$xxo := -100, -99.1 \ldots -.1$$

$$xxi(xxo) := \frac{1}{\left(\frac{1}{\frac{r}{2}}\right) - \frac{1}{xxo}}.$$

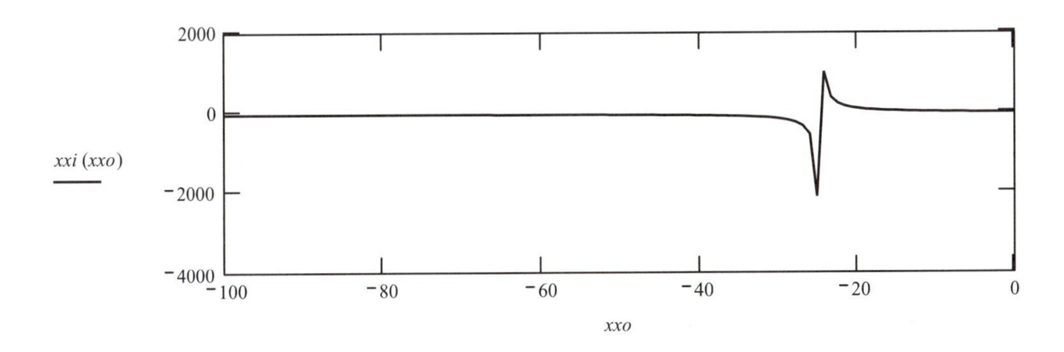

1.9.5.2 Convex Spherical Mirror

Geometrical Construction

1. Choose y_o and draw the PF-ray to the mirror and trace it forward to the focus F'.
2. The ray from the top of the arrow to the center of curvature C is reflected back onto itself (Figure 1.31).

Graph of x_i as Function of x_o

A convex spherical mirror has a positive radius of curvature. We show in FileFig 1.31 a graph of the image points as a function of the object points for x_o from -100 to $-.1$. When the light comes from the left, there is no singularity at $r/2 = x_o$, and we obtain virtual images for all positions of x_o.

FileFig 1.31 (G31MIRCX)

Convex spherical mirror. Calculation of image position from given object position. Graph for image position depending on the object position coordinate for the radius of curvature $r = 50$, that is, $r/2 = 25$, and x_o from -100 to -0.1.

G31MIRCX is only on the CD.

A summary of the image formation and the dependence on the various parameters is given in Table 1.5.

Applications to Spherical Mirrors

1. A corner mirror is made of two flat mirrors, joined together at an angle of 90 degrees. Show that the light incident on one mirror is parallel to the light leaving the other mirror for any angle of incidence.

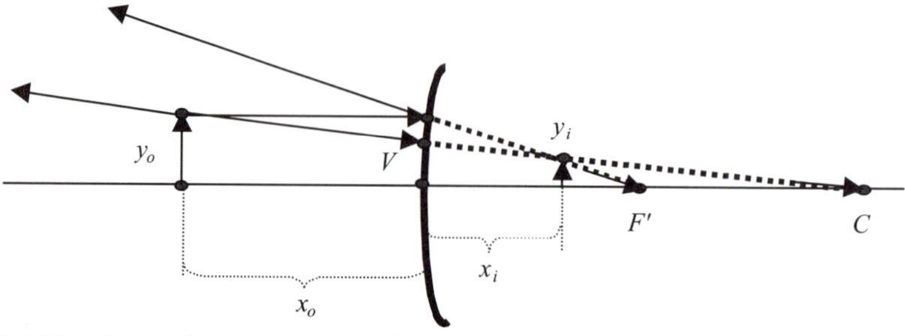

FIGURE 1.31 Geometrical construction of image for convex spherical mirror. The image y_i of object y_o is for any object distance to the right of the mirror (always virtual).

TABLE 1.5

	Concave	Concave	Concave	Convex
Object x_o	Left of f	at f	Right of f	Any
Image x_i	Negative	- Infinity	Positive	Positive
Magnification	Negative		Positive	Positive
	Real		Virtual	Virtual
	Inverted		Upright	Upright

2. Do the geometrical construction of:

a. convex spherical mirrors for (i) object at $-\infty$; (ii) object to the left of focus; and (iii) object to the right of focus.

b. concave spherical mirror with same focal length, for the three positions of x_o about the same values as in **a**.

1.10 MATRICES FOR A REFLECTING CAVITY AND THE EIGENVALUE PROBLEM

The first Ne–He laser used a Fabry–Perot cavity with two flat mirrors at a separation of 1 m. It was very difficult to align this cavity, and the first alignment was done by accident. One of the researchers bumped into the table, causing the flat mirrors to vibrate, and laser action was observed. Later, spherical mirrors were used to construct easy to align cavities.

For our discussion of laser cavities, consisting of two reflecting spherical surfaces, we first look at a periodic lens line, equivalently representing the "round trips" of the light in a reflecting cavity. One section of the lens line is shown in Figure 1.32, where the forward and backward traveling light are shown separately. The next section has the same configuration and the light enters and leaves each section in the same way. The first and third lenses are shared by two sec-

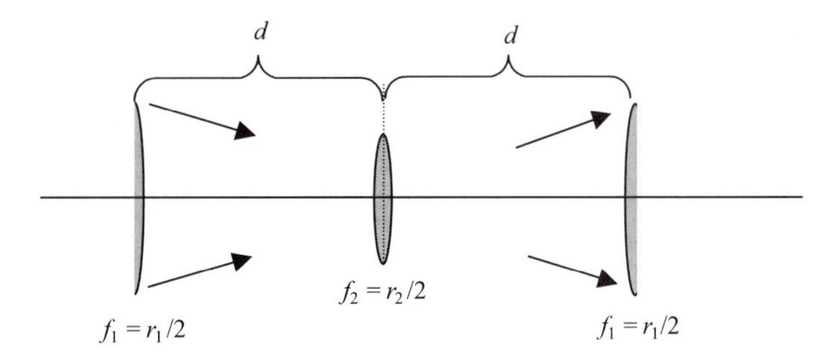

FIGURE 1.32 Unit cell of a lens system of periodic appearance. The light enters each cell in the same way. Such a periodic arrangement may be used to represent the reflection in a mirror cavity. The first and third lenses are only half-lenses and the focal lengths are twice as large. Rays of a possible light path are indicated.

tions. Therefore we have drawn them as half-lenses and assigned to them twice the focal length. The sequence of matrices for the lens line is

$$\begin{pmatrix} 1 & 0 \\ -1/2f_1 & 1 \end{pmatrix}\begin{pmatrix} 1 & d \\ 0 & 1 \end{pmatrix}\begin{pmatrix} 1 & 0 \\ -1/f_2 & 1 \end{pmatrix}\begin{pmatrix} 1 & d \\ 0 & 1 \end{pmatrix}\begin{pmatrix} 1 & 0 \\ -1/2f_1 & 1 \end{pmatrix}. \tag{1.110}$$

We substitute for the focal length of the lenses one-half of the value of the radius of curvature of the corresponding mirrors. We use $f = r/2$ and obtain for the mirror cavity

$$\begin{pmatrix} 1 & 0 \\ -1/r_1 & 1 \end{pmatrix}\begin{pmatrix} 1 & d \\ 0 & 1 \end{pmatrix}\begin{pmatrix} 1 & 0 \\ -2/r_2 & 1 \end{pmatrix}\begin{pmatrix} 1 & d \\ 0 & 1 \end{pmatrix}\begin{pmatrix} 1 & 0 \\ -1/r_1 & 1 \end{pmatrix}. \tag{1.111}$$

The first three matrices describe the travel of the light from the first to the second mirror. The last two matrices describe the travel from the second mirror back to the first.

We introduce the resonator parameters g_1 and g_2,

$$g_1 = 1 - d/r_1 \quad \text{and} \quad g_2 = 1 - d/r_2 \tag{1.112}$$

and calculate the product of the five matrices using FileFig 1.32.

FileFig 1.32 (G32RESGG)

Calculation of the product of the five matrices of the lens line corresponding to a cavity with two reflecting mirrors. Calculation of the eigenvalues of the cavity using g_1, g_2, and d. Graphs of the stability relation.

G32RESGG

Calculation of Resonator Using $g1$, $g2$, and d

$$\begin{pmatrix} 1 & 0 \\ \frac{g1-1}{d} & 1 \end{pmatrix} \cdot \begin{pmatrix} 1 & d \\ 0 & 1 \end{pmatrix} \cdot \begin{bmatrix} 1 & 0 \\ \frac{2\cdot(g2-1)}{d} & 1 \end{bmatrix} \cdot \begin{pmatrix} 1 & d \\ 0 & 1 \end{pmatrix} \cdot \begin{pmatrix} 1 & 0 \\ \frac{g1-1}{d} & 1 \end{pmatrix}$$

$$\begin{bmatrix} -1+2\cdot g1\cdot g2 & 2\cdot d\cdot g2 \\ 2\cdot g1\cdot \frac{(-1+g1\cdot g2)}{d} & -1+2\cdot g1\cdot g2 \end{bmatrix}$$

$$\text{eigenvals}\left[\begin{bmatrix} -1+2\cdot g1\cdot g2 & 2\cdot d\cdot g2 \\ 2\cdot g1\cdot \frac{(-1+g1\cdot g2)}{d} & -1+2\cdot g1\cdot g2 \end{bmatrix}\right]$$

$$\Big[(1,1,1) = -1+2\cdot g1\cdot g2+2\cdot\sqrt{-g1\cdot g2+g1^2\cdot g2^2}$$

$$(1,1,2) = -1+2\cdot g1\cdot g2-2\cdot\sqrt{-g1\cdot g2+g1^2\cdot g2^2}\Big]$$

$$r1 := 1 \qquad r2 := 1 \qquad d := 2$$

$$g1 := 1-\frac{d}{r1} \qquad g2 := 1-\frac{d}{r2}$$

$$\lambda 1 := -1+2\cdot g1\cdot g2+2\sqrt{-g1\cdot g2+g1^2\cdot g2^2}$$

$$\lambda 2 := -1+2\cdot g1\cdot g2-2\sqrt{-g1\cdot g2+g1^2\cdot g2^2}$$

$$\lambda 1 = 1 \qquad \lambda 2 = 1.$$

We set the product $g1g2 = x$ and plot it over the range from -1 to 2.

$$x := -1, -9 \ldots 2$$

$$y(x) := |(2\cdot x-1)+\sqrt{(2\cdot x-1)^2-1}|-1$$

$$yy(x) := |(2\cdot x-1)-\sqrt{(2\cdot x-1)^2-1}|-1$$

5
0
−5
y(x)
−1 0 1 2
x

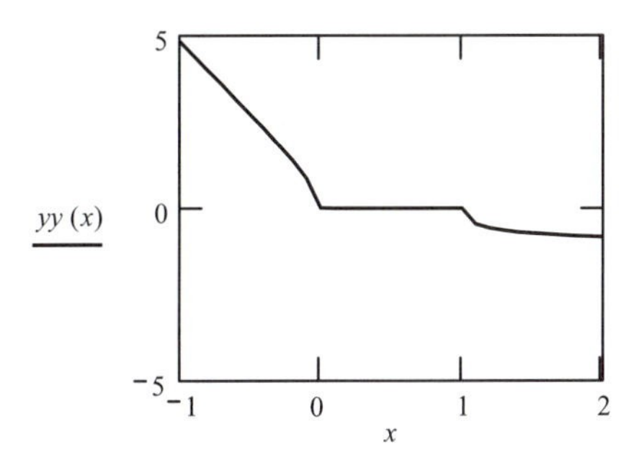

We obtain from FileFig 1.32 the matrix product of the matrices of Eq. (1.111),

$$\begin{pmatrix} -1 + 2g_1g_2 & 2dg_2 \\ 2g_1(-1 + 2g_1g_2)/d & -1 + 2g_1g_2 \end{pmatrix}. \tag{1.113}$$

The round trip in the cavity must have the symmetry of the path of the rays at the beginning and end of a unit cell of the lens line. This corresponds to a mode of oscillation of the cavity. The eigenvalues of this oscillation are obtained from the eigenvalues of the product matrix and the calculation is shown in FileFig 1.32. First the product of the five matrices is calculated using the symbolic method. Then the eigenvalues are obtained

$$\lambda_1 = (2g_1g_2 - 1) + [(2g_1g_2 - 1)^2 - 1]^{1/2} \tag{1.114}$$

$$\lambda_2 = (2g_1g_2 - 1) - [(2g_1g_2 - 1)^2 - 1]^{1/2}. \tag{1.115}$$

The coordinates, used for setting up the matrices, may now be transformed into a new coordinate system. In this coordinate system the matrix describing the round trip in the cavity is a diagonal matrix. When the diagonal matrix is a unit matrix, the light may pass through many round trips and no light will escape. One calls such a resonator stable, and the condition for stability is where the magnitudes of the eigenvalues are equal to 1.

$$|\lambda_1| = |\lambda_2| = 1. \tag{1.116}$$

We may write for Eq. (1.114),

$$\lambda_1 = (2g_1g_2 - 1) + [(2g_1g_2 - 1)^2 - 1]^{1/2} \tag{1.117}$$

or

$$\lambda_1 = (2g_1g_2 - 1) + i[1 - (2g_1g_2 - 1)^2]^{1/2}. \tag{1.118}$$

The real and imaginary parts of Eq. (1.118) must be on a circle of radius 1; that is,

$$|(2g_1g_2 - 1)| \leq 1, \quad \text{or} \quad 0 \leq g_1g_2 \leq 1. \tag{1.119}$$

in agreement with the imaginary part and plotted in FileFig 1.32.

In FileFig 1.33 we show a repetition of the calculations, starting from the five matrices of the cavity in Eq. (1.111), but now in terms of r_1, r_2, and d. In

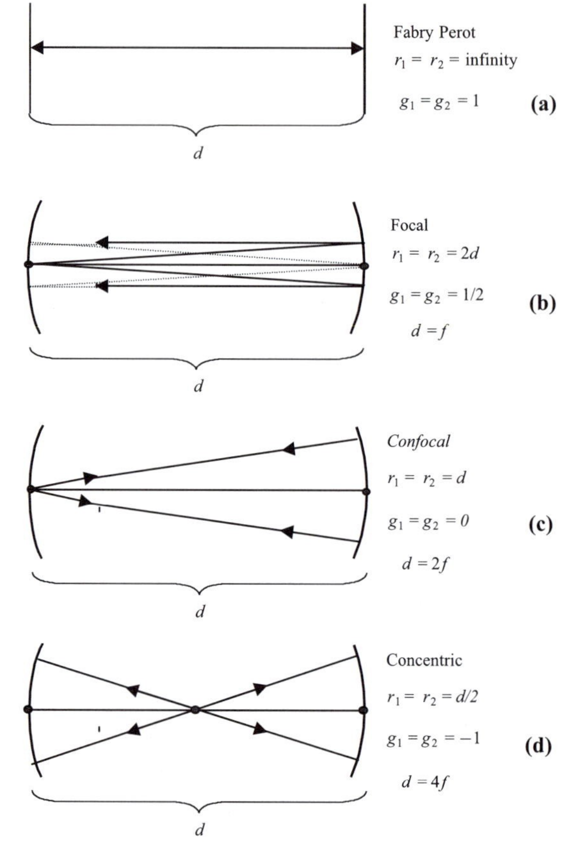

FIGURE 1.33 Schematic of light path for four cavities with different values of radii of curvature and length of cavity. The corresponding values of g_1 and g_2 are indicated: (a) Fabry–Perot; (b) focal; (c) confocal; (d) concentric.

Figure 1.33, we show schematics of the Fabry–Perot, a focal, a confocal and a spherical cavity for values of the parameters r_1, r_2, and d, and also of g_1 and g_2. For both representations one finds that the absolute values of the eigenvalues λ_1 and λ_2 are always 1.

FileFig 1.33 (G33RESCY)

Calculation of the eigenvalues of the cavity with two reflecting mirrors using r_1, r_2, and d. Numerical calculation with $r_1 = 1$, $r_2 = 1$, and $d = 2$.

G33RESCY is only on the CD.

Application 1.33. Use the values of the parameters r_1, r_2, and d for the Fabry–Perot, focal, confocal, and spherical cavities, and find that all are stable cavities.

C H A P T E R

Interference

2.1 INTRODUCTION

In Chapter 1 we described image formation by light, using our model which states that light propagates along straight lines and utilizes the laws of reflection and refraction. We now consider the wave nature of light. In the famous experiment by Thomas Young, one observes on a screen an interference pattern, consisting of bright and not so bright stripes of light. The interpretation of an interference pattern was done by using an analogy to water waves. However, the water wave pattern is observed as an amplitude interference pattern whereas the superposition of light waves, also generated as an amplitude pattern, is observed as an intensity pattern. Historically, Newton associated the light beams of geometrical optics with a stream of particles and some scientists attacked Young in his time, saying that he was diminishing Newton's work. Today we know that light is an electromagnetic wave but, in a complementary way, light is also described by quantum mechanics as an assembly of particles.

In this chapter we use a model for the description of interference phenomena. We assume that there is always one incident wave when two- or more beam interferometry is discussed. After one has taken into account what happens in the experimental setup, the waves leaving the setup appear superimposed. The interference pattern is produced with finite optical path differences. The calculation of the optical path difference and the interpretation of the resulting interference pattern are the main subjects of this chapter. The process of splitting the incident wave into parts involves diffraction, which we neglect in this chapter as a secondary effect and discuss in detail in Chapter 3.

In this chapter we use for the incident wave one harmonic wave, a solution of the scalar wave equation, which is written in Cartesian coordinates as

$$\partial^2 u/\partial x^2 + \partial^2 u/\partial y^2 + \partial^2 u/\partial z^2 = (1/v)^2 \partial^2 u/\partial t^2, \tag{2.1}$$

where v is the phase velocity of light in the medium with refractive index n, related to the speed of light c in vacuum as $v = c/n$. The scalar wave equation follows from Maxwell's theory. It may also be written in spherical coordinates

$$\nabla^2 u + k^2 u = 0, \tag{2.2}$$

where ∇ is the differential operator in spherical coordinates, $k = 2\pi/\lambda$, and λ is the wavelength of the light. A simple solution of this equation is a spherical wave of the type $(e^{ikr})/r$, where r is the distance from the origin to the observation point. The spherical wave propagates from its origin in all directions and its intensity is attenuated by $1/r^2$. We consider such spherical waves only conceptually and approximate them at a large distance by plane waves.

The differential equation of the scalar wave equation is linear and superposition of solutions of the differential equation will again result in a solution. This is part of the superposition principle. In this chapter we only need the superposition of a number of monochromatic waves, each of frequency ν, to result in a monochromatic wave having the same frequency ν.

For our model description we use some results from Maxwell's theory for quantitative expressions of the reflection and transmission coefficients of materials contained in Fresnel's formulas. In particular, we use the results that waves pick up a phase jump of π, when reflected at an optically denser medium, and that they travel in the optically denser medium with wavelength λ/n, where n is the index of refraction. The intensity is calculated either as the time average of the square of the amplitude or the square of the absolute value of the complex representation and may be normalized with an arbitrary constant.

2.2 HARMONIC WAVES

The solution of the scalar wave equation, (Eq. (2.1)), is a function, depending on the space coordinates x, y, z and the time t. In addition, there may be an arbitrary phase factor. We consider harmonic waves in vacuum and in an isotropic and nonconducting medium of index n. However, in most cases, we only need waves depending on one space coordinate and time. We describe the transverse waves by vibrating in the u direction and moving in the x direction, having wavelength λ and time period T.

$$u = A\cos[2\pi(x/\lambda - t/T + \phi)]. \tag{2.3}$$

The amplitude u of the wave varies in the x direction, A is the magnitude of the wave, and ϕ is a phase constant. The first graph of FileFig 2.1 shows the amplitude u, depending on the space coordinate x for three time instances t and three phase constants. The second graph shows the dependence on time for three points in space and three phase constants. The magnitudes A_1 to A_3 and B_1 to

CHAPTER 2

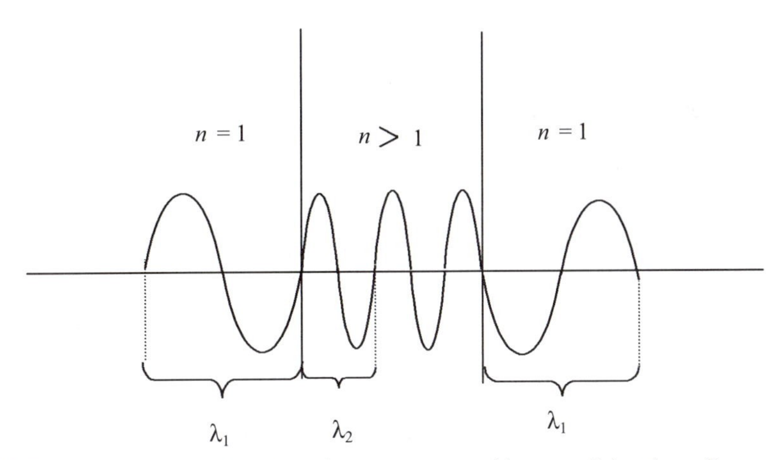

FIGURE 2.1 Change of wavelength as the wave enters and leaves a dielectric medium.

B_3 have been assumed to have the same value, and the three phase constants ϕ_1 to ϕ_3 and Ψ_1 to Ψ_3 are assumed to be different. Comparing the graphs, one observes equivalence of the dependence of the cosine function on x/λ and t/T. Changing the range of variable from x to t, the family of curves depending on x is similar to the one depending on t.

We may modify x/λ and t/T in such a way that they contain phase constants. Then, in the "net" expression $\cos[2\pi(x/\lambda - t/T + \phi)]$ we can not distinguish if ϕ belongs to the space part or the time part. We show below that for our discussions on interference we do not need the time dependence and it is eliminated.

The product of the frequency ν and wavelength λ/n is equal to the phase velocity $v = \omega/k$ of the wave propagating in the medium of refractive index n. The angular frequency $\omega = 2\pi\nu$, and the wave vector $k = 2\pi n/\lambda$, where λ is the wavelength in vacuum. We may then write Eq. (2.3) as

$$u = A\cos(kx - \omega t) \tag{2.4}$$

or

$$u = A\cos k(x - \omega/kt) = A\cos k(x - vt).$$

The phase velocity in vacuum is c, and in an isotropic medium with refractive index n it is c/n. The wavelength of "free space" λ is reduced in the medium to λ/n (see Figure 2.1).

FileFig 2.1 (I1COSWS)

Cosine functions depending on space and time coordinate and one additional phase constant. Graphs are shown for cosine functions depending on the space coordinates for three time instances. This may be interpreted as graphs of the same wave at three consecutive snapshots. Graphs are shown for cosine functions depending on the time coordinates for three points in space.

I1COSWS is only on the CD.

Application 2.1.

1. One may change the phase ϕ and the space coordinate and choose both so there is no resulting change in the graph. Choose $\phi = 2, 4, 6$.
2. One may change the phase ϕ and the time coordinate and choose both so there is no resulting change in the graph. Choose $\phi = 2, 4, 6$.
3. Change ϕ in such a way that there is a shift to smaller values of the position coordinate.
4. Change ϕ in such a way that there is a shift to later values of the time coordinate.

2.3 SUPERPOSITION OF HARMONIC WAVES

2.3.1 Superposition of Two Waves Depending on Space and Time Coordinates

We describe the interference of two waves in a simple way, using the superposition of two harmonic waves u1 and u2. Both waves will propagate in the x direction and vibrate in the y direction.

$$u_1 = A\cos 2\pi[x/\lambda - t/T] \qquad u_2 = A\cos 2\pi[(x-\delta)/\lambda - t/T]. \tag{2.5}$$

We assume that the two waves have an optical path difference δ. At time instance $t = 0$, the wave u_1 has its first maximum at $x = 0$, and u_2 at $x = \delta$ (Figure 2.2). Adding u_1 and u_2 we have

$$u = u_1 + u_2 = A\cos 2\pi[x/\lambda - t/T] + A\cos 2\pi[(x-\delta)/\lambda - t/T]. \tag{2.6}$$

Using

$$\cos(\alpha) + \cos(\beta) = 2\cos\{(\alpha-\beta)/2\}\cos(\alpha+\beta)/2 \tag{2.7}$$

we get

$$u = [2A\cos\{2\pi(\delta/2)/\lambda\}][\cos\{2\pi(x/\lambda - t/T) - 2\pi(\delta/2)/\lambda\}]. \tag{2.8}$$

In FileFig 2.2 we show graphs of the square of Eq. (2.8) for the same time instant t_1 and wavelength λ. We choose a number of optical path differences $\delta_1 = 0$, $\delta_2 = 0.1$, $\delta_3 = 0.2$, $\delta_4 = 0.3$, $\delta_5 = 0.4$, $\delta_6 = 0.5$, corresponding to the ratios of the optical path difference to the wavelength between 0 and $\frac{1}{2}$. One observes that the height of the maxima decreases with increasing δ_1 to δ_6, and shifts to larger values of x.

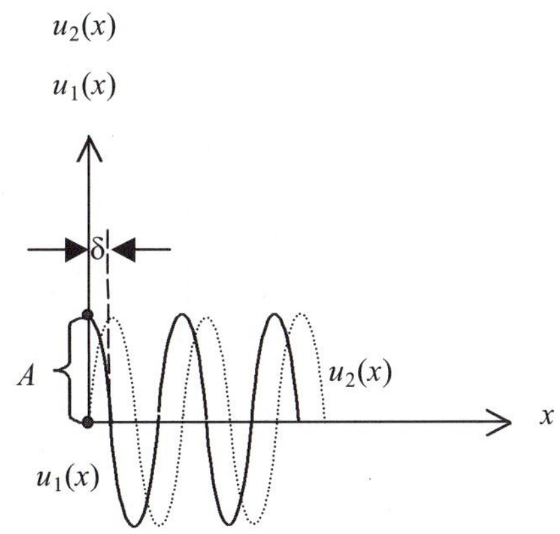

FIGURE 2.2 Two waves with magnitude A and wavelength λ. We have $u_1 = A$ for $x = 0$ and $u_2 = A$ for $x = \delta$.

We now discuss the two factors of Eq. (2.8). The first factor $2A\cos\{2\pi(\delta/2)/\lambda\}$ depends on δ and λ, but not on x and t. One obtains for δ equal to 0 or a multiple integer of the wavelength

$$[2A\cos\{2\pi(\delta/2)/\lambda\}]^2 \quad \text{is } 4A^2 \tag{2.9}$$

and for δ equal to a multiple of half a wavelength

$$[2A\cos\{2\pi(\delta/2)/\lambda\}]^2 \quad \text{is } 0. \tag{2.10}$$

The first factor in Eq. (2.8) may be called the amplitude factor and is used for characterization of the interference maxima and minima.

One has

$$\text{maxima for } \delta = m\lambda, \text{ where } m \text{ is 0 or an integer} \tag{2.11}$$

$$\text{minima for } \delta = m\lambda, \text{ where } m \text{ is } \tfrac{1}{2} \text{ or an integer} \tag{2.12}$$

and m is called the order of interference.

The second factor is a time-dependent cosine wave with a phase constant depending on δ and λ. For the description of the interference pattern this factor is averaged over time and results in a constant, which may be factored out and included in the normalization constant (see below).

In Figure 2.3 we show schematically the interference of two water waves with a fixed phase relation. When the interference factor is zero one has minima, indicated by white strips. They do not depend on time. The maxima oscillate and appear and disappear along the line in the observable direction.

Maxima and minima are shown in FileFig 2.3 as 3-D graphs. The maxima are shown for $\delta = \lambda$, and in the second graph, for $\delta = \lambda/2$, there is just one minimum. The maxima show the time dependence of the second factor for each of the space coordinates. One can estimate that a time average will result in half the maximum value. The minimum is zero. It is zero for all time.

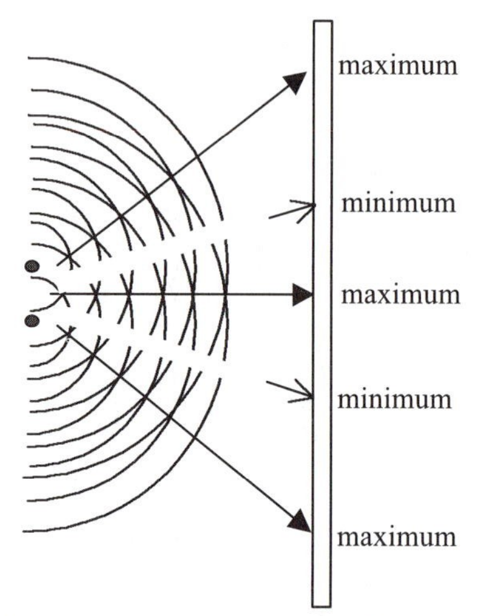

FIGURE 2.3 Schematic of the interference pattern produced by two sources vibrating in phase. At the crossing of the lines, the amplitudes of the waves of both sources are the same and adding. Taking the time dependence into account, the magnitude changes between maximum and minimum. These are the maxima when considering light. Between the maxima we indicate the two lines corresponding to the minima. Along these lines the amplitude of the two waves compensate each other; their sum is zero.

FileFig 2.2 (I2COSSUPS)

Graphs of the superposition of two cosine waves with wavelength $\lambda = 1$, for a number of optical path differences $\delta_1 = 0$, $\delta_2 = 0.1$, $\delta_3 = 0.2$, $\delta_4 = 0.3$, $\delta_5 = 0.4$, $\delta_6 = 0.5$ corresponding to ratios of the optical path difference to the wavelength between 0 and $\frac{1}{2}$.

I2COSSUPS is only on the CD.

Application 2.2.

1. Extend the range of the optical path differences of the six graphs from $\frac{1}{2}$ wavelength to 1 wavelength, and then from 1 wavelength to $\frac{3}{2}$ wavelength and indicate in a list when there is repetition.
2. Make a graph of $y = \cos\{2\pi(\delta/2)/\lambda\}$ for fixed λ as function of δ and make a list of the δ values for minima and maxima. Compare with a list of δ/λ values.

FileFig 2.3 (I3COSGRA)

3-D demonstration of the superposition of two waves for $\delta/\lambda = 1$ corresponding to a maximum, and $\delta/\lambda = 0.5$ corresponding to a minimum. In the graph of the maximum, the amplitude changes in time for a specific spot in space between

0 and $(2A)^2$, and one can estimate that the time average will be half of it. The graph of the minimum is zero for all time and space values.

I3COSGRA

Superposition of Two Cosine Waves

One wave has optical path difference δ with respect to the other. The sum is squared to result in the intensity. We are looking at them time dependence; the graphs are plots in space x and time t. Period T, path difference δ, wavelength λ.

1. Graph for optical path difference corresponding to a maximum

$$\lambda := 1 \qquad A := 1$$

$$N := 40 \qquad i := 0 \ldots N \qquad j := 0 \ldots N$$

$$x_i := -.2 + .05 \cdot i \qquad t1_j := -.2 + .05 \cdot j$$

$$uc(x, t1) := \left[2 \cdot A \cdot \cos\left[2 \cdot \pi \cdot \left(\frac{\delta 1}{2 \cdot \lambda}\right)\right] \cdot \left[\cos\left[2 \cdot \pi \cdot \left(\frac{x}{\lambda} - \frac{t1}{T}\right) - 2 \cdot \pi \cdot \left(\frac{\delta 1}{2 \cdot \lambda}\right)\right]\right]\right]^2$$

$$M_{i,j} := uc(x_i, t1_j) \qquad \delta 1 \equiv 1 \qquad T \equiv 1 \qquad t1 \equiv .1.$$

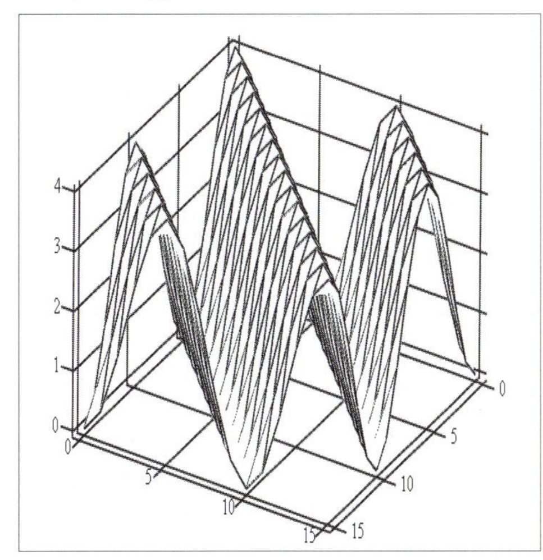

M

2. Graph for optical path difference corresponding to a minimum

$$N := 40$$

$$i := 0 \ldots N \qquad j := 0 \ldots N$$

$$xx_i := -.2 + .04 \cdot i \qquad t1_j := -.2 + .02 \cdot j \qquad \delta 2 \equiv .5$$

$$ud(xx,t1) := \left[2\cdot A\cdot\cos\left[2\cdot\pi\cdot\left(\frac{\delta 2}{2\cdot\lambda}\right)\right]\right.$$
$$\left.\cdot\left[\cos\left[2\cdot\pi\cdot\left(\frac{xx}{\lambda}-\frac{t1}{T}\right)-2\cdot\pi\cdot\left(\frac{\delta 2}{2\cdot\lambda}\right)\right]\right]\right]^2$$

$$M_{i,j} := ud\left(xx_i, t1_j\right) \qquad t1 \equiv .1 \qquad T \equiv 1.$$

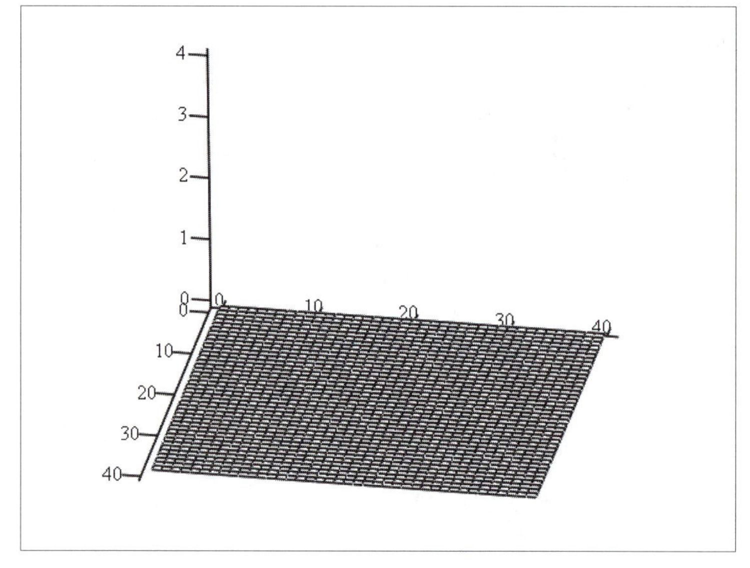

MM

Application 2.3. One may change the wavelength λ such that for $\delta 1/\lambda$ one gets minima, and for $\delta 2/\lambda$ one gets maxima.

2.3.2 Intensities

The interference pattern of water waves is an amplitude pattern. We may observe minima and maxima with respect to the level of the undisturbed water surface. The interference pattern of light shows intensity minima as dark spots in space and maxima as bright spots. An amplitude pattern shows negative amplitudes, but an intensity pattern has only positive or zero values. The amplitude pattern has to be considered first; it produces interference. Then we have to obtain the intensity pattern. We compare the intensity pattern to observations.

For the square of the amplitude of equation (2.8) we have

$$u^2 = [2A\cos\{2\pi(\delta/2)/\lambda\}]^2[\cos\{2\pi(x/\lambda - t/T) - 2\pi(\delta/2)/\lambda\}]^2 \tag{2.13}$$

In Section 2.1 we mentioned that for the intensity we use either the time average of the square of the amplitude or the square of the absolute value, when using complex notation. In this section we compare these two calculations.

2.3.2.1 Use of the Time Average

The time average of the square of a cosine function depending on t, taken over the interval of one period, that is, from 0 to T, is

$$(1/T)\int_0^T \left[\cos t\right]^2 dt = 1/2. \tag{2.14}$$

Using for the cos-function the "second" factor of Eq. (2.8), we get

$$a_v^2 = (1/T)\int_0^T [\cos\{2\pi(x/\lambda - t/T) - 2\pi(\delta/2)/\lambda\}]^2 dt = \tfrac{1}{2} \tag{2.15}$$

As result, we obtain for the intensity of the wave described in Eq. (2.8),

$$I = [2A\cos\{2\pi(\delta/2)/\lambda\}]^2(1/2). \tag{2.16}$$

This is the square of the first factor (Eq. (2.13)) multiplied by 1/2.

2.3.2.2 Complex Notation

Complex notation has its advantages, for example, when adding several harmonic waves. Then, the intensity is calculated by the product $z\,z^*$ and the time dependence is automatically eliminated.

We rewrite the sum of the two real expressions of Eq. (2.5) in complex notation:

$$\begin{aligned} z = z_1 + z_2 = {} & A\exp i(2\pi(x)/\lambda - 2\pi t/T) \\ & + A\exp i(2\pi(x-\delta)/\lambda - 2\pi t/T). \end{aligned} \tag{2.17}$$

Here the real part of z, that is, Re(z), is the amplitude of u of Eq. (2.8). The superposition of the two waves may be written in complex notation as

$$\begin{aligned} z = z_1 + z_2 = {} & A\exp\{-i2\pi t/T + i(2\pi(x)/\lambda)\} \\ & \times \{1 + \exp(-i2\pi\delta/\lambda - 2\pi t/T)\}. \end{aligned} \tag{2.18}$$

In the second factor of Eq. (2.18) we may factor out $\exp -i2\pi\delta/2\lambda$ and then have

$$\begin{aligned} z = {} & A\exp\{-i2\pi t/T + i(2\pi(x)/\lambda)\}\exp(-i2\pi\delta/2\lambda) \\ & \times \{\exp i2\pi\delta/2\lambda + \exp -i2\pi\delta/2\lambda\}. \end{aligned} \tag{2.19}$$

Using

$$\{\exp i2\pi\delta/2\lambda + \exp -i2\pi\delta/2\lambda\} = 2\cos 2\pi\delta/2\lambda \tag{2.20}$$

one gets

$$z = A\exp\{-i2\pi t/T + i(2\pi(x)/\lambda)\}\exp(-i2\pi\delta/2)\{2\cos 2\pi\delta/2\lambda\}. \tag{2.21}$$

For simplicity we collect all phase factors into one term, call

$$\exp\{-i2\pi t/T + i(2\pi(x)/\lambda) - i2\pi\delta/2\lambda\} = \exp i\Psi, \tag{2.22}$$

and have

$$z = [2A\cos(2\pi\delta/2\lambda)]\exp i\Psi. \tag{2.23}$$

For the intensity we use the square of the absolute value of z, that is, the product of z times its complex conjugate z^*,

$$z\,z^* = [2A\cos(2\pi\delta/2\lambda)]\exp i\Psi[2A\cos(2\pi\delta/2\lambda)]\exp -i\Psi \tag{2.24}$$

or

$$z\,z^* = [2A\cos(2\pi\delta/2\lambda)]^2. \tag{2.25}$$

Comparing the intensity in Eq. (2.16) with Eq. (2.25), one sees that they are different by the factor $\frac{1}{2}$. This factor $\frac{1}{2}$ is not significant, since one usually normalizes the results for the calculation of interference and diffraction pattern.

FileFig 2.4 (I4COSINTS)

Graphs of the real part of the superposition of two waves depending on space and time, its square, and the time average of the "second factor." The time-dependent superposition of two waves is given in complex notation. Its real part and the square of the real part are plotted. Then the product zz^* is plotted, depending on x, and appearing at a constant. The "fringe pattern" is plotted (i.e., zz^*), depending on the optical path difference.

I4COSINTS is only on the CD.

2.3.3 Normalization

To compare an interference pattern with observation, one may normalize both. Let us assume we have an interference pattern with a maximum in the center. The scale of the observation data may be changed in such a way that at the center one has the intensity equal to 1. The intensity calculated from the formula will also be changed: one divides by the value calculated at the center, and obtains also for the intensity 1. For example, let us assume that an interference pattern is described by $f(x)$ and that for $x = 0$ we have $f(0) = b$. Then writing

$$I(x) = f(x)/f(0) = I_0 f(x) \tag{2.26}$$

with $I_0 = 1/b$ results in $I(x)$ equals 1 for $x = 0$.

In FileFig 2.4 we have plotted the real part of u and the time average and compared it with zz^*, which is independent of time. We also plot what one may call the "fringe pattern," that is, zz^*, depending on the optical path difference.

2.4 TWO-BEAM WAVEFRONT DIVIDING INTERFEROMETRY

2.4.1 Model Description for Wavefront Division

The classical experiments by Young, Fresnel, and Lloyd were performed to demonstrate the wave theory of light. All three experiments use one incident wave and divide the light from the source into two waves. The two waves are superimposed after an optical path difference is introduced and a pattern showing the maxima and minima is observed. One calls such a pattern, consisting of maxima and minima, a *fringe pattern*. The process of splitting up the incident wave at two small openings into two new waves involves diffraction, considered in detail in Chapter 3. The incident wave we consider is monochromatic, propagates in the x direction, vibrates in the y direction, and has a large lateral extension in the z direction. The experimental setup produces two waves by wavefront division, propagating under an angle with respect to each other (see Figure 2.4a). When we observe the interference pattern at a faraway screen (see Figure 2.4b) the

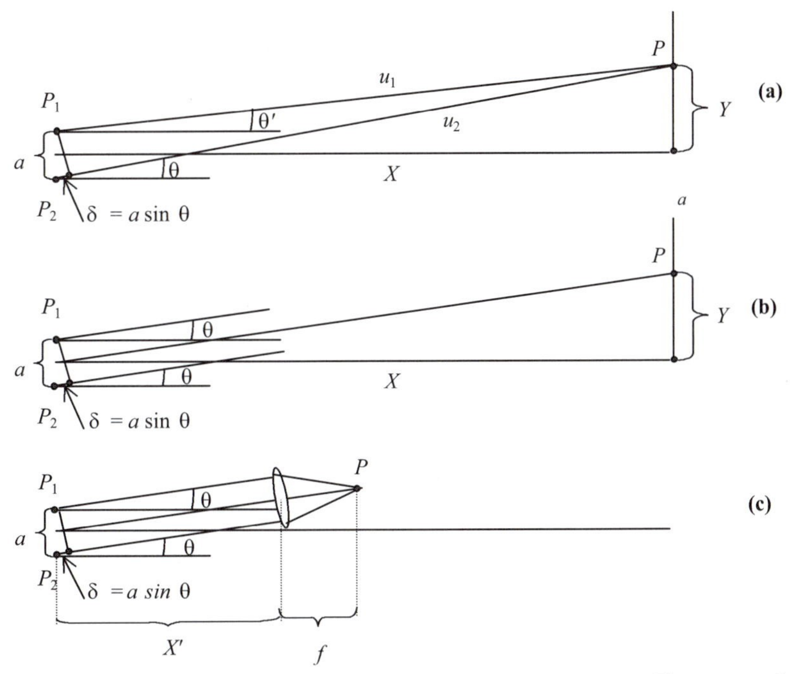

FIGURE 2.4 Observation of interference pattern generated by two sources. The generated waves have a constant phase relation: (a) geometry of the experiment with water waves (see also Figure 2.3); (b) Observation at a screen far away from the source; (c) using a lens to reduce the distance X in (b) to distance X' in (c). In this case only parallel rays meet at one point in the focal plane of the lens.

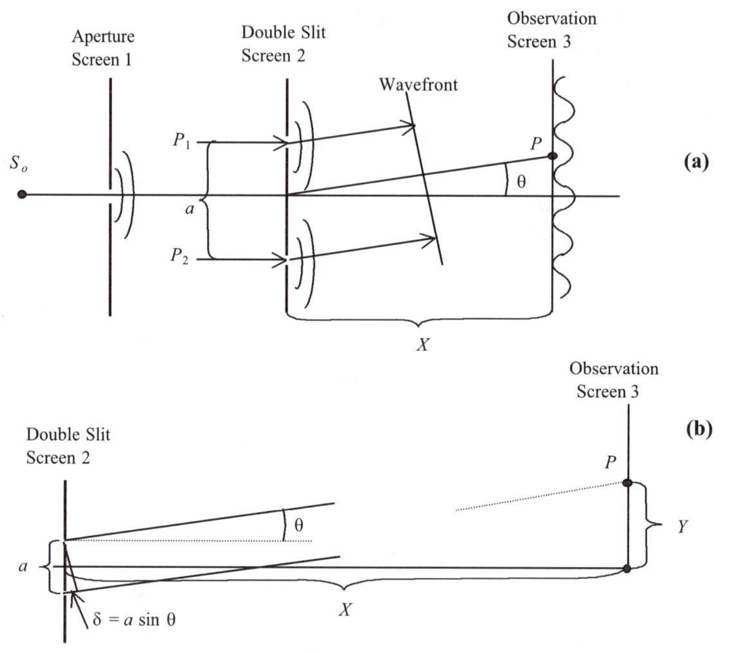

FIGURE 2.5 Schematic of Young's experiment. Light is emitted by the source S_0: (a) according to Huygen's principle, the first aperture A_1 produces a spherical wave which generates two spherical waves with fixed phase relation at S_1 and S_2 and produces a fringe pattern; (b) coordinates for the calculation of the fringe pattern generated by screen 2 and observed on screen 3. The two openings on screen 2 are separated by a. The distance from screen 2 to screen 3 is X and the coordinate on screen 3 is Y.

angles θ and θ' are almost equal. We therefore approximate the propagation by assuming that the two waves move in parallel, and that the optical path difference δ is the same as it was when moving under an angle. We describe both waves in the same coordinate system and use for the intensity of their superposition $I = I_0[A\cos(2\pi\delta/2\lambda)]^2$ (see Eq. (2.25). In Figure 2.4c we show the observation in the focal plane of a lens, giving us the opportunity to have the distance between the experimental setup and the observation screen significantly shorter.

2.4.2 Young's Experiment

The setup of Young's experiment is schematically shown in Figure 2.5a. The light from a source S_0 travels to screen 1, which has a small hole. At the hole, a spherical wave is created and travels a large distance. The wave arrives at screen 2 as "almost" a plane wave. At two small holes on screen 2, P_1 and P_2, the incident plane wave is split into two spherical waves. The two spherical waves travel a large distance and arrive at screen 3 as "almost" plane waves. Their wavefronts are tilted by a small angle with respect to each other, resulting in an optical path difference and superposition results in a fringe pattern. In our model calculation, we assume that the two monochromatic waves, leaving the two openings, travel in parallel in the same direction. They have the angle θ with respect to the symmetry axis and the optical path difference δ (see Figure 2.5),

$$\delta = a \sin \theta. \tag{2.27}$$

In the small angle approximation we have $\delta = aY/X$, and obtain the intensity of the fringes on the observation screen, using Eq. 2.25

$$I(Y) = I_0[\cos\{(\pi a Y)/(X\lambda)\}]^2. \tag{2.28}$$

Constructive interference is observed for

$$\delta = Ya/X = 0,\ \lambda,\ 2\lambda, \ldots. \tag{2.29}$$

and destructive interference for

$$\delta = Ya/X = 0,\ \lambda/2,\ 3\lambda/2,\ 5\lambda/2, \tag{2.30}$$

The graph in FileFig 2.5 shows the intensity for Young's experiment, depending on Y, using the coordinate on the observation screen. One observes that there is a maximum at the center. In Figure 2.6, a photograph shows the interference pattern of Young's experiment and an experimental setup for its observation.

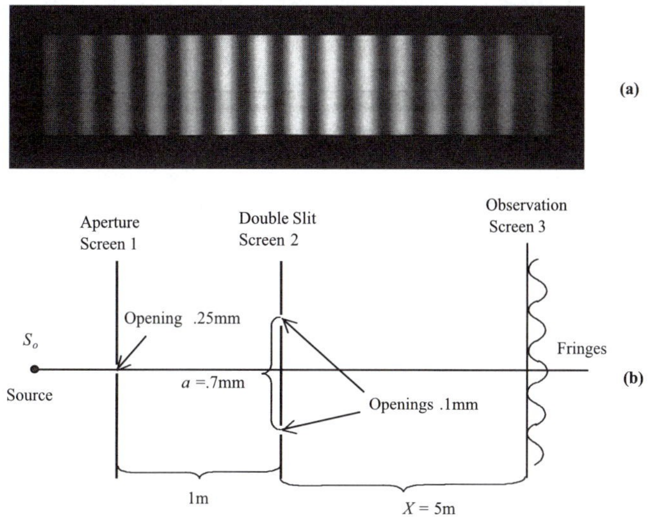

FIGURE 2.6 (a) Photograph of interference pattern observed with Young's experiment (from M. Cagnet, M. Francon, and J.C. Thrierr, *Atlas of Optical Phenomena*, Springer-Verlag, Heidelberg, 1962); (b) physical dimensions for the observation of fringes in Young's experiment, according to R. Pohl.

FileFig 2.5 (I5YOUNGS)

The intensity for Young's experiment (Eq. (2.28)), is plotted for $\lambda = 0.0005$ mm, $a = 0.4$ mm, and $X = 4000$ mm. For this choice of parameters, we see that the separation of the holes on screen 2 (see Figure 2.5), are three orders of magnitude larger than the wavelength and four orders of magnitude smaller than the distance between screens 2 and 3.

I5YOUNGS is only on the CD.

Application 2.5.

1. Change the separation of the source points a to $\frac{1}{2}a$ and $2a$.
2. Change the wavelength λ to $\frac{1}{2}\lambda$ and 2λ.
3. Compare the changes of the ratio a/λ to $\frac{1}{2}$ and 2 times its value.
4. There are no changes if we keep the product $(a/\lambda)(Y/X)$ constant, where Y/X is the angle under which we observe the fringes.

2.4.2.1 Lloyd's Mirror

In the experiment by Lloyd, interference fringes are produced by using one mirror. Two parts of the incident wave are superimposed. One part travels directly from the source to the observation screen, while the other part is incident on the mirror and is reflected under a grazing angle (see Figure 2.7). For the calculation of the optical path difference δ, we use a virtual source S' at distance $d/2$ from the plane of the mirror, but on the "other side" of the mirror. Similarly, as in Young's experiment, we have for the optical path difference $\delta = Ya/X$, and consider the two waves traveling parallel to the observation point. One has to take into account that one of the two superimposed waves has picked up a phase shift

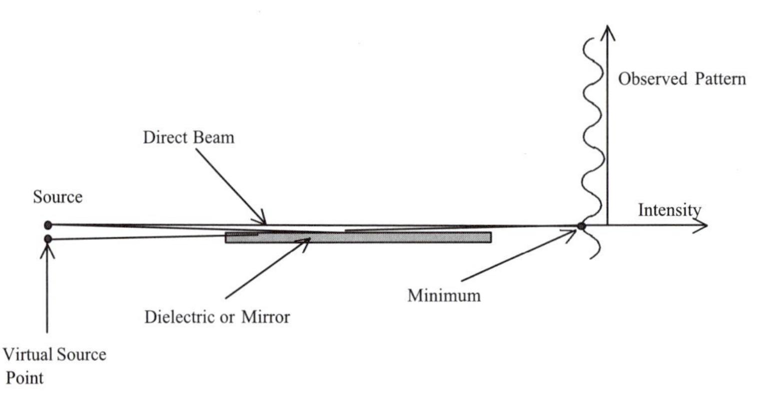

FIGURE 2.7 Optical schematic for Lloyd's mirror experiment drawn with exaggerated size of the angles. One wave from the source travels directly to the observation point; the other is reflected at the mirror. Since the distance of the source from the plane of the mirror is very small the first fringe is about in the plane of the reflecting surface. It is a minimum because of the phase shift.

of π upon reflection on the mirror. This phase shift appears as π in $\cos\{2\pi(x/\lambda - t/T) + \pi\}$, and after superposition, as $\pi/2$ in the amplitude factor used for the intensity. We therefore have

$$I(Y) = I_0[\cos\{\pi(Ya/X\lambda) + \pi/2\}]^2. \tag{2.31}$$

For constructive interference one has

$$\delta = Ya/X + \lambda/2 = 0,\ \lambda,\ 2\lambda, \ldots \tag{2.32}$$

and for destructive interference

$$\delta = Ya/X + \lambda/2 = \lambda/2, 3\lambda/2, 5\lambda/2, \ldots \tag{2.33}$$

The graph in FileFig 2.6 shows the intensity of Lloyd's mirror experiment depending on Y, the coordinate on the observation screen. In comparing Young's experiment to Lloyd's, one has a minimum at the center.

FileFig 2.6 (I6LOYDS)

Intensity of Lloyd's experiment for $\lambda = 0.0005$ mm, $a = .4$ mm, and $X = 4000$ mm. The "first" fringe is dark, that is, a minimum for $Y = 0$, because of the phase shift upon reflection of one of the two waves. The dependence of the fringes on a and λ are the same as in Young's experiment.

I6LOYDS is only on the CD.

2.4.2.2 Fresnel's Double Mirror Experiment

Fresnel's mirror experiment was originally performed to prove the wavelike character of light. Even today it is of some interest since the fringes depend on the tilting angle of the mirrors and may be applied to wavelengths as short as Xrays. The wavefront of the spherical wave, emerging from the source, is divided by two mirrors, which are tilted by a very small angle β (Figure 2.8a). Each part of the incident wave is reflected, and the two wavefronts are tilted by the angle β. The superposition produces the fringe pattern. In our model description we consider the waves at the faraway observation screen as parallel. The optical path difference is the same as calculated when the two waves travel under the small angle β. Making these assumptions, the optical path difference is obtained using the two virtual source points S_1' and S_2' at distance a (see Figure 2.8a). It is similarly done as for Young's experiment (Figure 2.8b). The distance from the virtual sources to the observation screen is $c + f = X$, and the optical path difference $\delta = a(Y/X)$ of the light reflected by the two Fresnel mirrors is

$$\delta = (Y2b\sin\beta)/(c + f). \tag{2.34}$$

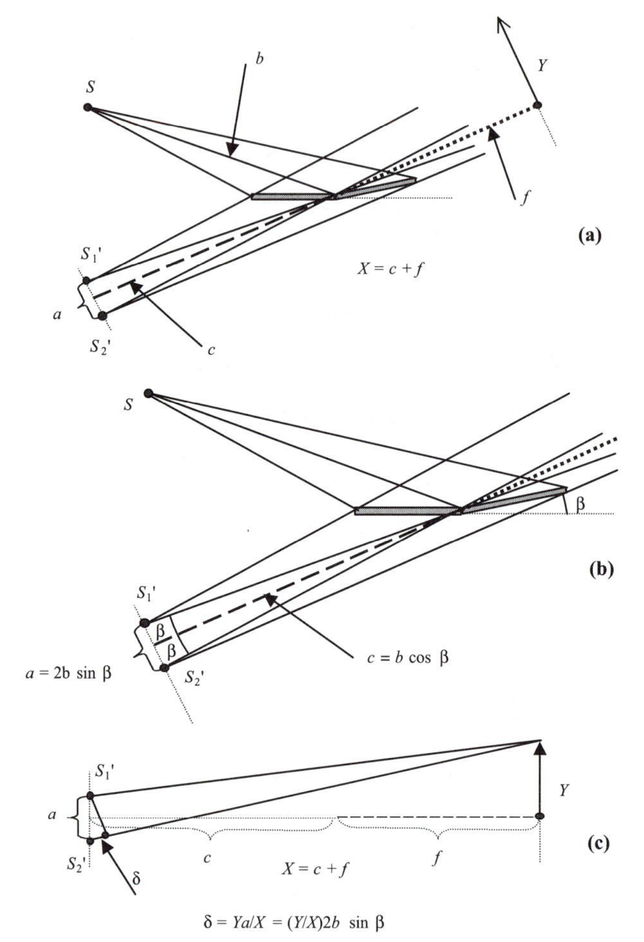

FIGURE 2.8 (a) and (b) are for Fresnel's mirror experiment. S is the source, S_1' and S_2' are virtual source points. They are separated by a and have a tilting angle β. The coordinate on the observation screen is Y. The relation $a = 2b \sin \beta$ is obtained by associating "one b" with one mirror and the "other b" with the other mirror, and remembering that light is reflected by 2β if the mirror is tilted by β; (c) approximate treatment of Fresnel's mirror experiment similar to Young's experiment.

The intensity is

$$I(Y) = I_0[\cos\{(\pi Y 2b \sin \beta)/[(c + f)\lambda]\}]^2, \tag{2.35}$$

and for constructive interference

$$\delta = Y2b \sin \beta/(c + f) = 0,\ \lambda, 2\lambda, \ldots \tag{2.36}$$

and destructive interference

$$\delta = Y2b \sin \beta/(c + f) = \lambda/2,\ 3\lambda/2,\ 5\lambda/2, \ldots \tag{2.37}$$

The graph in FileFig 2.7 shows the intensity of Fresnel's double mirror experiment with a maximum at the center, since both waves are reflected on a mirror and pick up a phase shift π. In FileFig 8 we compare the intensities of Young's, Lloyd's, and Fresnel's double mirror experiments and see the role the phase shift plays.

FileFig 2.7 (I7FREMIRS)

Intensity of Fresnel's double mirror experiment, Eq. (2.35) for $\lambda = 0.0005$ mm, $b = 1000$ mm, $f = 5000$ mm, and for the angle $\beta = 0.0002$.

I7FREMIRS is only on the CD.

Application 2.7.

1. How is the pattern changing when changing λ. How much must λ be changed to double the separation of the maxima or to make them one half.
2. How is the pattern changing when changing β. How much must β be changed to double the separation of the maxima or to make them one half.
3. What angle β has to be chosen to have the maxima separated by 500 microns when using Xrays of 500 Angstrom (1 Å is 10^{-10} m).

FileFig 2.8 (I8FRYOLOS)

Comparison of experiments by Young, Lloyd, and Fresnel. For the choice of parameters, we see that the separation of the "sources" are three orders of magnitude larger than the wavelength, and four orders of magnitude smaller than the distance X between experimental setup and observation screen.

I8FRYOLOS

Fresnel's Mirror, Young's Double Slit, and Lloyd's Mirror

1. Fresnel's mirror
   Y, c, f, b, and λ in mm, β in rad; c is about b for calculation of $X = c + f$. All lengths in mm.

$$\text{Con} := 1 \qquad \lambda \text{ defined above}$$

graph

$$b := 1000 \qquad f := 5000 \qquad c := b\cos(\beta) \qquad \beta \equiv .0002 \qquad Y := -10, -9.99 \ldots 10$$

$$\text{IF}(Y) := \text{Con} \cdot \cos\left(\pi \cdot \frac{Y \cdot 2\frac{b}{c+f} \cdot \sin(\beta)}{\lambda}\right)^2.$$

2. Young's experiment

$$a \equiv .4 \qquad X \equiv 4000$$

$$IY(Y) := \text{Con} \cdot \cos\left(\pi \cdot \frac{Y \cdot a}{\lambda \cdot X}\right)^2.$$

$$\lambda \equiv .0005$$

3. Lloyd's mirror
Same as Young, phase term is added.

$$IL(Y) := \text{Con} \cdot \cos\left(\pi \cdot \frac{Y \cdot a}{\lambda \cdot X} + \frac{\pi}{2}\right)^2.$$

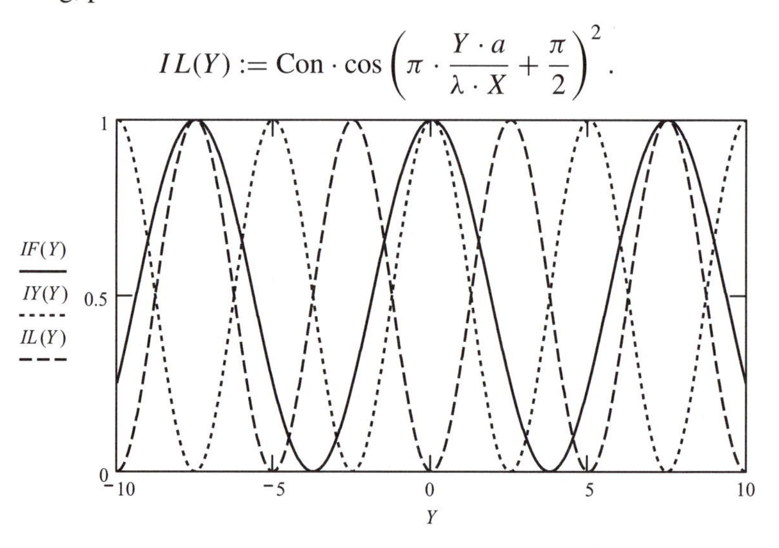

We see that at $Y = 0$ Young has a maximum, Lloyd a minimum. For Young and Lloyd: the position of maxima are changed by changing d and λ; X is considered fixed. For Fresnel, a, b are fixed; position of fringes changes with β and λ.

2.5 TWO-BEAM AMPLITUDE DIVIDING INTERFEROMETRY

2.5.1 Model Description for Amplitude Division

We again assume a monochromatic wave incident on the experimental setup, propagating in the x direction, vibrating in the y direction, and having a large lateral extension in the z direction.

The incident wave is incident on a beam splitter. One part is reflected and the other part is transmitted. One or both parts are manipulated in the experimental setup to pass through a beamsplitter a second time. Fractions of each part are superimposed and travel parallel to the observation screen.

The splitting of amplitude division is different from the splitting discussed for wavefront division. The two waves after wavefront division travel under an angle. After amplitude division, it can be arranged, that both parts travel in parallel in the same direction. The difference between wavefront and amplitude division is related to energy conservation. It is impossible to superimpose two beams in such a way that all the light travels in one direction.

The interference pattern is observed at a faraway screen, and the intensity of the interference pattern is given as $I = I_0[A\cos(2\pi\delta/2\lambda)]^2$ (see Eq. (2.25)), where δ is the optical path difference. The observation may be done in the focal plane of a lens, making the actual distance between the experiment and observation screen much shorter.

2.5.2 Plane Parallel Plate

The interference on a plane parallel plate is the model for the description of interference on thin films, used in various technologies such as coating lenses or mirrors for use with Xrays.

We consider light incident on a plane parallel plate of glass of thickness D and index of refraction $n > 1$, and assume $n = 1$ for the media outside the plate. The incident light, shown in Figure 2.9 at (a), is split at the first interface into a reflected and transmitted part (shown at (b). The transmitted light is reflected and transmitted at the second interface (shown at (c) and the reflected light is again reflected and transmitted at the first interface (shown at (d). We use for further consideration only the light reflected from the first interface (shown as (1)) and the light reflected at the second interface, and then transmitted through the first interface (shown as (2)).

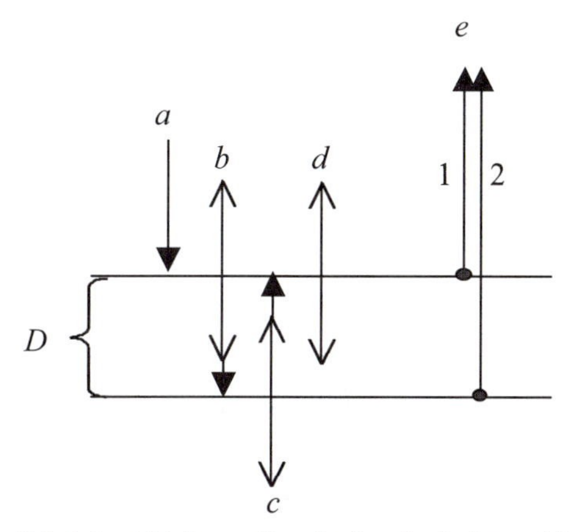

FIGURE 2.9 Plane parallel plate at thickness D and refractive index n; (a) incident light; (b) splitting at first interface; (c) splitting at second interface, (d) splitting at first interface again; (e) the two waves, (1) from b and (2) from c superimposed to generate the interference fringes.

The path difference between the waves (1) and (2) is

$$\delta = 2Dn. \tag{2.38}$$

Both waves (1) and (2) travel in the same direction, and (1) picks up a phase shift of π when reflected at the first interface. For the calculation of the optical path difference we have to multiply 2D by n. In addition, we have to take into account that the reflection on the denser medium, that is, the glass, for wave (1) results in a phase shift of λ, equivalent to $\lambda/2$. The optical path difference is then

$$\delta = 2Dn + \lambda/2, \tag{2.39}$$

where λ is the wavelength outside of the plate. One has for the intensity

$$I = I_0[\cos\{\pi(2Dn + \lambda/2)/\lambda\}]^2 \tag{2.40}$$

or equivalently

$$I = I_0[\cos\{\pi(2Dn)/\lambda + \pi/2\}]^2. \tag{2.41}$$

We have for constructive interference

$$\delta = 2Dn + \lambda/2 = 0,\ \lambda,\ 2\lambda,\ \ldots \tag{2.42}$$

or

$$2Dn = \lambda/2,\ 3/2\lambda,\ 5/2\lambda,\ \ldots \tag{2.43}$$

and for destructive interference

$$\delta = 2Dn + \lambda/2 = \lambda/2,\ 3\lambda/2,\ 5\lambda/2, \ldots \tag{2.44}$$

or

$$2Dn = \lambda,\ 2\lambda,\ 3\lambda, \ldots. \tag{2.45}$$

We see that the appearance of maxima and minima depends on the thickness and the index of refraction of the plane parallel plate. Maxima are obtained for integer numbers of half a wavelength, minima for integer numbers of a wavelength. One may observe the interference pattern on a plane parallel plate by looking at a soap bubble. The film of the bubble is curved, but equally thick over a small area. The colored light we see is produced by individual interference of different wavelengths on the thin film of equal thickness.

The first graph in FileFig 2.9 shows the dependence of the fringes on the thickness of the film for fixed wavelength λ and refractive index $n = 1.5$. The second graph shows the dependence on wavelength for fixed thickness $D = 0.05$ and $n = 1.5$. One observes that there is no interference on the thin film when the wavelength gets too large.

FileFig 2.9 (I9PLANS)

The intensity of two-beam interference on a plane parallel plate of index $n2$ in medium with index $n1 = 1$. The graphs shown are: (i) Dependence on thickness for fixed wavelength $\lambda = 0.0005$ and $n = 1.5$; (ii) Dependence on wavelength for fixed thickness $D = 0.05$ and $n = 1.5$.

I9PLANS is only on the CD.

Application 2.9.

1. Modify the formula to $I(D) = \{\cos(2\pi Dn/\lambda + \lambda/2)\}^2$, for the case that the index of the outside medium is not 1.
2. Consider the following configurations:
   a. an air gap between glass media ($n = 1.5$).
   b. a water film ($n = 1.33$) on glass, and incident light in medium $n = 1$.
   c. a water film on glass and incident light in glass.

   Using a graph for fixed D, start counting maxima at any particular maxima on the graph. Read from the graph the difference in D between, for example, 5 maxima. Recalculate the wavelength using the value of n.
3. When the wavelength exceeds the thickness of the plate, the last fringe is observed for a value of λ, depending on D and n. Find the formula.

2.5.2.1 Wedge-Shaped Air Gap

We consider two glass plates in air with one on top of the other. With a thin object, we produce a wedge shaped air gap of small angle α. As we did for the plane parallel plate, we calculate the optical path difference for the two waves (1) and (2), on the two interfaces of the air gap (Figure 2.10a (side view)). The optical path difference is

$$\delta = 2x \tan\alpha, \tag{2.46}$$

where α is the angle of the wedge and x the distance from the point where the plates are touching. The wave reflected at the lower plate picks up a phase shift of π, equivalent to $\lambda/2$. For the two waves (1) and (2) one has constructive interference

$$2x \tan\alpha + \lambda/2 = 0,\ \lambda,\ 2\lambda,\ 3\lambda, \ldots, m\lambda \tag{2.47}$$

and destructive interference

$$2x \tan\alpha + \lambda/2 = 1/2\lambda,\ 3/2\lambda,\ 5/2\lambda, \ldots, (m + 1/2)\lambda. \tag{2.48}$$

The width $D' = x \tan\alpha$ at the mth maximum is

$$D' = (m - 1/2)\lambda/2, \tag{2.49}$$

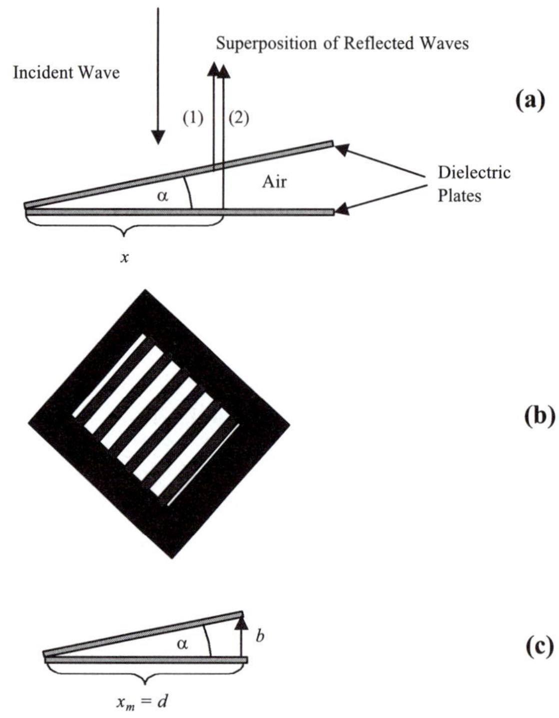

FIGURE 2.10 (a) Optical diagram for a wedge-shaped gap of air between two dielectric glass plates (microscope slides). The gap angle α is assumed to be small. As x increases, the optical path difference between waves (1) and (2) also increases, resulting in an interference pattern; (b) schematic of fringe pattern observed with optical plates flat to a fraction of a wavelength λ; (c) height b at distance $x_m = d$ and $\tan\alpha = b/d$.

where $m = 1, 2, 3, \ldots$, and for destructive interference

$$D' = m\lambda/2, \tag{2.50}$$

where $m = 0, 1, 2, 3, \ldots$. For the intensity we have

$$I = I_0[\cos\{(\pi 2x \tan\alpha)/\lambda + \pi/2\}]^2. \tag{2.51}$$

Interference produced by a wedge-shaped air gap is used to determine the uniformity of a polishing job, schematically shown in Figure 2.10b. A flat plate is used to produce an air gap over a plate, which has been polished. If the maxima and minima are not straight lines, the width of the air gap varies. Deviations of a fraction of a wavelength can be detected.

One can make a simple experiment with two microscope slides and a plastic film. The film is placed at the end of one slide, and a wedge of length approximately 5 cm may be produced with the other microscope slide. Observation of fringes and their corresponding distances makes it possible to determine the thickness of the plastic film.

FileFig 2.10 (I10WEDGES)

The intensity for interference depending on the distance x, of a wedge of $\alpha = 0.002$ rad and $\lambda = 0.0005$ mm. The distance between the maxima is a constant, given as $\lambda/(2\tan\alpha)$. We have also plotted the height depending on the length x for angle α, using a scaling factor a. The first fringe is a minimum, as observed for Lloyd's mirror, because the thickness at origin $D' = 0$.

I10WEDGES is only on the CD.

Application 2.10.

1. From the condition of constructive interference ($2x\tan\alpha + \lambda/2 = 0, \lambda, 2\lambda, \ldots, m\lambda$), show that we have for the difference $\Delta x_m = x_{m+1} - x_m = \lambda/(2\tan\alpha)$, where x_m is the x coordinate at the mth fringe.
2. Since $\Delta y_m/\Delta x_m = \tan\alpha$, where Δy_m is the height difference of the plates between fringes, show that we have $\Delta y_m = \lambda/2$.
3. Assume we observe M fringes over the length $x_m = d$ and want to determine the height b of the gap at that point (see Fig. 2.10a. Since $b/d = \tan\alpha = (\lambda/2)/x_m$, show that we have $b = (\lambda/2)M$.
4. Recalculation of α: produce a graph with $\lambda = .00054$ and $\alpha = .0023$. Find the x coordinate at the 23rd fringe and use for the y coordinate $y = (\lambda/2)23$. Calculate α and compare with input data.
5. Modify the determination of the height at some chosen point x_m for a water film ($n = 1.33$) between glass plates ($n = 1.5$).

2.5.2.2 Newton's Rings

A circular interference pattern may be observed if a spherical surface is placed on a flat surface. The ring pattern is called "Newton's rings" and may be used to determine the radius of curvature of the spherical surface. An experimental setup is shown in Figure 2.11.

A plane convex lens touches a plane parallel plate and an air gap of width D is formed between the lens and the plate. We call the radius of curvature of the spherical surface R and the radius of the rings of the pattern r (Figure 2.12). One has the relation

$$R^2 = r^2 + (R - D)^2. \tag{2.52}$$

After solving a quadratic equation, we have for D(r)

$$D(r) = R - \sqrt{(R^2 - r^2)}. \tag{2.53}$$

The transmitted intensity is

$$I(r) = I_0\{\cos(\pi 2D(r)/\lambda)\}^2. \tag{2.54}$$

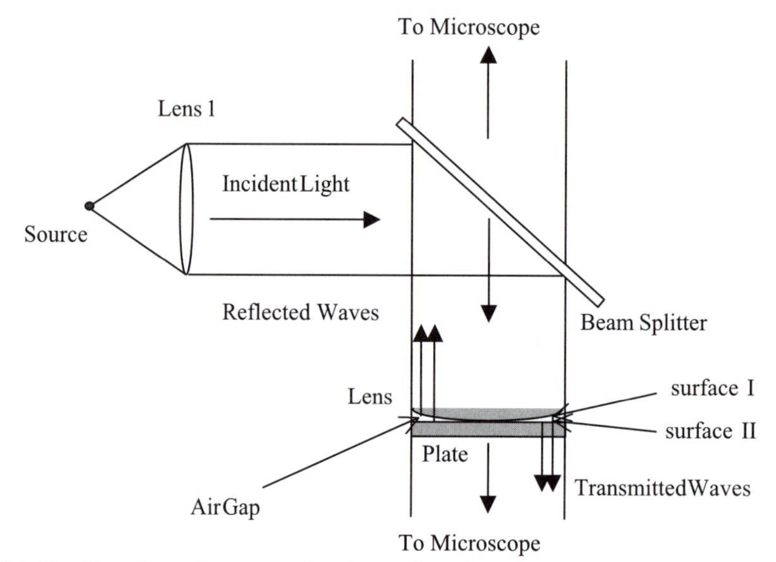

FIGURE 2.11 Experimental setup for th eobservation of Newton's rings. Light is made parallel by lens 1 and is reflected by the beam splitter to the lens-plate assembly. Reflected light from surfaces I and II travels to microscope 1 for observation of fringes in reflection. Transmitted light from surfaces I and II travels to microscope 2 for observation of the tranmission fringes.

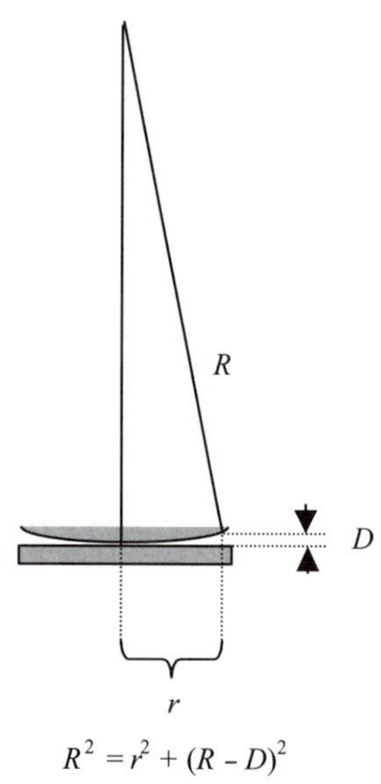

FIGURE 2.12 Coordinates for the calculation of the optical path difference D between rays from surfaces I and II.

In addition the reflected intensity has the term $\pi/2$ because of the reflection on the plane parallel plate of refractive index larger than 1,

$$I(r) = I_0\{\cos(\pi 2D(r)/\lambda + \pi/2)\}^2. \tag{2.55}$$

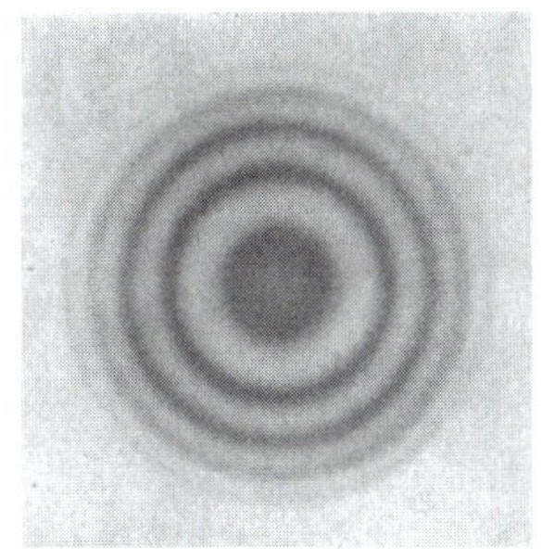
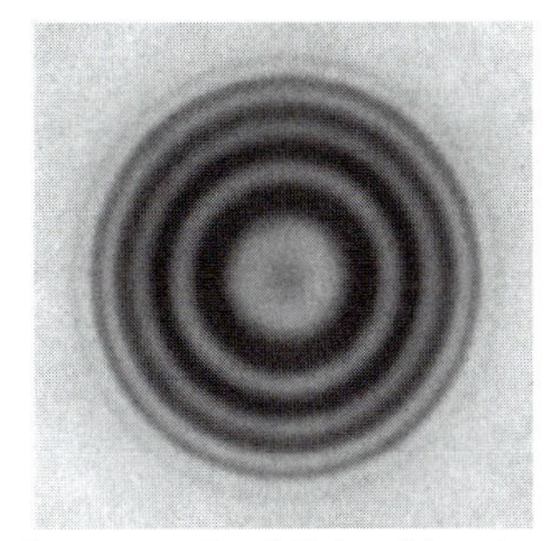

FIGURE 2.13 Newton's rings observed: (a) using transmitted light; (b) using reflected light (from M. Cagnet, M. Francon, and J.C. Thrierr, *Atlas of Optical Phenomena*, Springer-Verlag, Heidelberg, 1962).

For the reflected intensities, one has for constructive interference

$$\delta = D(r) + \lambda/2 = 0,\ \lambda,\ 2\lambda, \ldots, \tag{2.56}$$

and for destructive interference

$$\delta = D(r) + \lambda/2 = \lambda/2,\ 3\lambda/2,\ 5\lambda/2, \ldots. \tag{2.57}$$

For the transmitted intensities one has similar expressions without the $\lambda/2$ term. At the center of the plates, at $D = 0$, one has from Eq. (2.55) for reflected light zero intensity; in other words, we should observe a dark spot, shown in Figure 2.13a. For transmitted light, Eq. (2.54) predicts a bright spot, shown in Figure 2.13b.

FileFig 2.11 (I11NEWTONS)

Intensity for Newton's rings in transmission and reflection depending on the radius r around the center, for $\lambda = 0.0005$ mm and $R = 2000$ mm.

I11NEWTONS is only on the CD.

Application 2.11. Recalculation of R:

1. Show that we have for the mth ring (fringe) a height in the air gap of $m\lambda/2$.
2. Show that we then have to use $(m\lambda/2 - R)^2 = R^2 - r_m^2$ for the calculation of R, assuming that we have read r_m from the graph.

2.5.3 Michelson Interferometer and Heidinger and Fizeau Fringes

2.5.3.1 Michelson Interferometer and Normal Incidence

In 1880, Albert Michelson used the interferometer, named after him, for his famous experiments to show that there is no ether. Today most infrared spec-

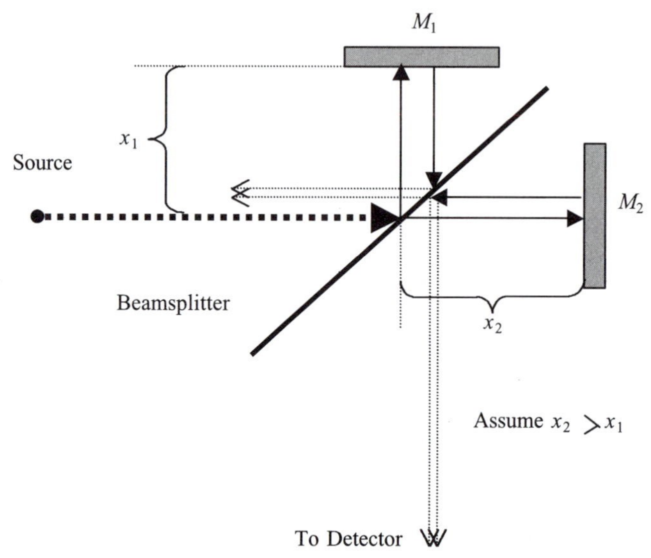

FIGURE 2.14 The Michelson interferometer with arms of unequal lenght x_1 and x_2 in the same medium.

trometers use the Michelson interferometer to obtain an interferogram and the application of the Fourier transform produces the desired spectrum. Chapter 9 discusses Fourier transformation and spectroscopy. In Figure 2.14 we show a schematic of a Michelson interferometer. The amplitude of the incident beam is partly reflected under an angle of 90° toward mirror M_1, and partly transmitted in the direction to mirror M_2. The beam splitter may be a plane parallel plate, and the reflection and transmission properties of such a plate are discussed below, including all multiple reflections. Here we use an idealized beamsplitter and assume that 50% of the incident light will be reflected, that 50% will be transmitted, and no phase shift will be introduced.

The reflected wave travels to mirror M_1, where it is reflected and travels back to the beam splitter for a "second splitting." It has traveled the distance 2x1 between mirror and beam splitter. We call the transmitted part (1), and note that the reflected part travels back to the source. Similarly, the transmitted wave travels to M_2, is there reflected, and travels to the beam splitter for a second splitting. It has traveled the distance 2x2 and we call the reflected part (2); the transmitted part travels back to the source. Parts (1) and (2) are superimposed and travel to the detector. If the distances 2x1 and 2x2 are not the same, we have an optical path difference between (1) and (2) of

$$\delta = 2D = 2(x_2 - x_1), \tag{2.58}$$

where we assume that $x_1 > x_1$. Constructive interference is obtained for

$$\delta = m\lambda \tag{2.59}$$

and destructive interference for

$$\delta = \left(m + \tfrac{1}{2}\right)\lambda, \tag{2.60}$$

where $m = 0, 1, 2, 3$.

The intensity of interference is obtained as

$$I = 4A^2 \cos^2(\pi 2D/\lambda). \tag{2.61}$$

The Michelson interferometer was originally designed to perform exact length measurements. Some time ago, it was used for a now outdated procedure to define the length of the meter by using ^{86}Kr emission.

The first graph in FileFig 2.12 shows the fringes depending on thickness D. The second graph shows the fringes depending on wavelength λ.

FileFig 2.12 (I2MICHDLS)

Intensity of the Michelson interferometer depending on the displacement D of one mirror for wavelength $\lambda = .0005$ mm, and for dependence on λ for $D = .003$.

I12MICHDLS

Michelson Interferometer

Beam splitter is assumed to be a plane parallel plate. Fringe pattern depending on D for wavelength $\lambda = .0005$, and depending on wavelength λ for $D = .003$. The angle $\theta = 0$. All lengths in mm.

1. Dependence on D.

$$\theta := 0 \qquad \lambda; = .0005$$

$$D := 0.027, .02701 \ldots .0325$$

$$I1(D) := \cos\left(\frac{2 \cdot \pi \cdot D \cdot \cos(\theta)}{\lambda}\right)^2.$$

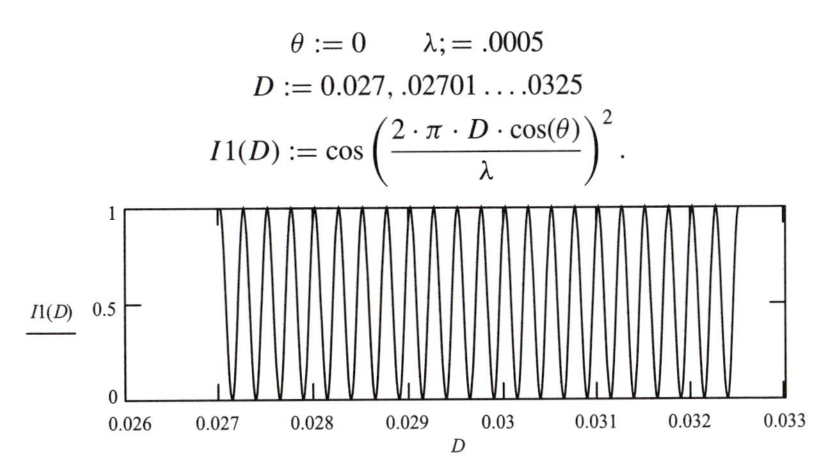

2. Dependence on λ

$$\lambda := .0004, .000401 \ldots .0008$$

$$D := .003$$

$$I2(\lambda) := \cos\left(\frac{2 \cdot \pi \cdot D \cdot (\theta)}{\lambda}\right)^2.$$

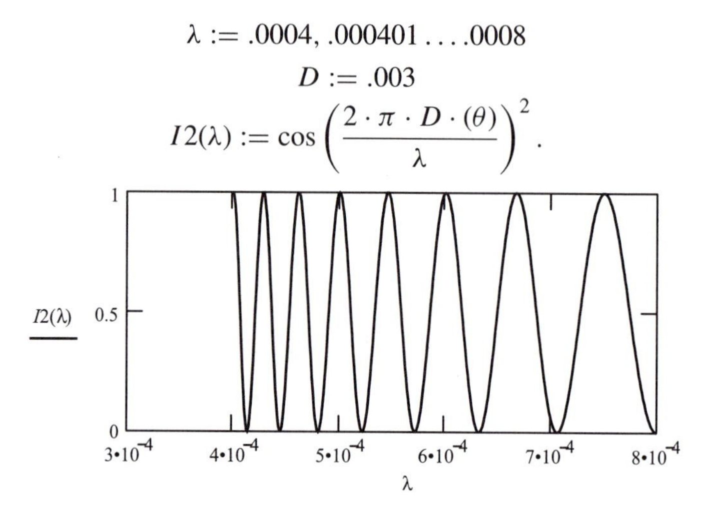

Application 2.12.

1. Resolution depending on displacement D. Add to the graph of the intensity depending on D a graph with a second wavelength $\lambda\lambda = \lambda + \Delta\lambda$; for example, $\lambda\lambda = 0.00052$. Observe that the separation of the fringes gets larger for larger m. For the mth fringe we have the path differences $m\lambda\lambda$ and $m\lambda$. When this difference is $\lambda/2$, we call the two fringes resolved and have $m\lambda\lambda - m\lambda = \lambda/2$, $m\lambda\lambda = \lambda/2$, or $\Delta\lambda/\lambda = 2m$. Compare the formula $\Delta\lambda/\lambda = 2m$ with values read from the graph for choice of $\lambda\lambda$.
2. Add to the graph of the intensity depending on λ a second graph with different D value. The graph shows the change in phase for one wavelength when D is changed. Choose D_1 such that maxima change to minima and D_2 that mimima change to the next maxima. Read from the graph the numerical values and compare with the formula for constructive and destructive interference.

2.5.3.2 Michelson Interferometer, Nonnormal Incidence, Heidinger, and Fizeau Fringes

If the light from the point source fills a cone with opening angle $\Delta\theta$, the distance $x_2 - x_1$ depends on the angle θ and a ring pattern will result in the plane of the observation screen. For the mathematical treatment we fold one beam of the Michelson interferometer over to the other beam as shown in Figure 2.15a. To calculate the path difference of the two beams (1) and (2), we use Figure 2.15b and calculate the path difference using Figure 2.15c. We have for the distances $[ab] = [bc] = D/\cos\theta$, and for the distance $[ac] = [(2D/\cos\theta)(\sin\theta)] = 2D\tan\theta$. The optical path difference δ is then

$$\delta = 2[bc] - [ac]\sin\theta = 2D/(\cos\theta) - 2D\tan\theta\sin\theta = 2D\cos\theta. \quad (2.62)$$

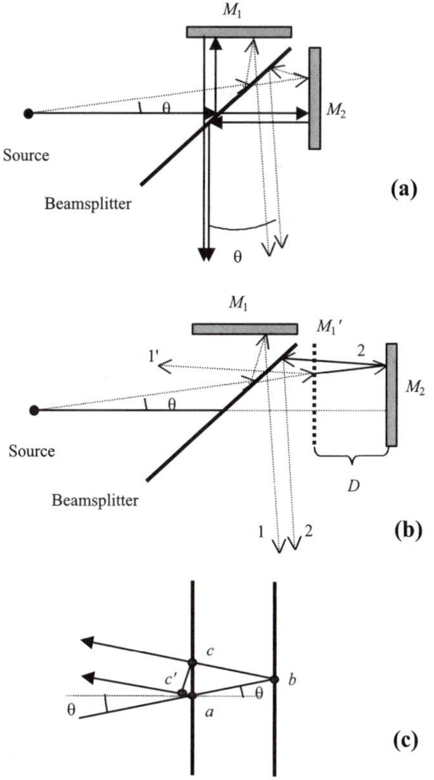

FIGURE 2.15 (a) Light passing through a Michelson interferometer for a light beam incident under the angle θ; (b) light passing through a Michelson interferometer for a light beam incident under the angle θ and one arm folded onto the other; (c) calculation of path difference for a light beam incident under the angle θ.

The intensity of the superimposed two beams is obtained as

$$I_M(\theta, D) = \cos^2(\pi 2D(\cos\theta)/\lambda). \tag{2.63}$$

When a fringe is formed by incident light under the same angle of inclination, one speaks of Heidinger fringes. Heidinger fringes are shown in Figure 2.16. A cone of light was used with a Michelson interferometer for the observation of the ring pattern. In FileFig13. we study graphs of the cross-section of the intensity pattern of Heidinger fringes. They are produced by the Michelson interferometer for the dependence on the angle θ, for fixed wavelength λ and for fixed thickness D. We may also produce with the Michelson interferometer fringes of equal thickness. We fold one beam onto the other beam, as shown in Figure 2.17a, and then consider a similar positioning as discussed for the plane parallel plate. If one of the mirrors of the Michelson interferometer is tilted (Figure 2.17b), we have the same situation as for the wedge-shaped gap (Figure 2.10a). Therefore, the fringe pattern of the Michelson interferometer with one tilted mirror is similar

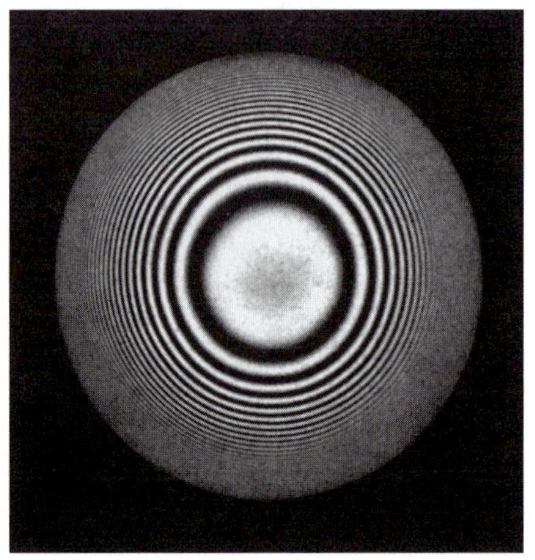

FIGURE 2.16 Heidinger interference fringes observed with a Michelson interferometer. The ring pattern is observed when using a cone of light (from Cagnet, Francon, Thrierr, *Atlas of Optical Phenomena*, Springer-Verlag, Heidelberg, 1962).

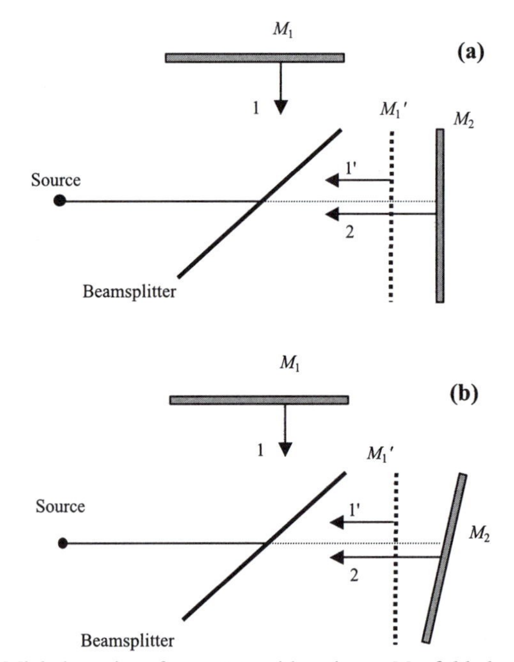

FIGURE 2.17 (a) Michelson interferometer with mirror M_1 folded onto the other arm; (b) Michelson interferometer with mirror M_1 folded onto the other arm and M_2 tilted.

to the wedge-shaped air gap. When a fringe is formed by incident light for the same thickness D, one speaks of Fizeau fringes.

FileFig 2.13 (I3MICHANS)

Ring pattern of intensity of the Michelson interferometer depending on angle θ for two wavelengths λ and $\lambda\lambda$ for fixed D.

I13MICHANS

Michelson Interferometer, Dependence on θ

Fringe pattern depending on angle θ for two fixed wavelengths λ and $\lambda\lambda$ and fixed displacement D. An ideal beamsplitter is assumed. All lengths in mm.

$$\theta := -301, -300 \ldots 3 \qquad \lambda := .0005 \qquad D := .05 \qquad \lambda\lambda := 00052$$

$$IM1(\theta) := \cos\left(\frac{2 \cdot \pi \cdot D \cdot \cos(\theta)}{\lambda}\right)^2 \qquad IM2(\theta) := \cos\left(\frac{2 \cdot \pi \cdot D \cdot \cos(\theta)}{\lambda\lambda}\right)^2.$$

Application 2.13.

1. Observe that the separation of the fringes for wavelengths λ and $\lambda\lambda$ gets smaller for larger angles and that at the center, when one wavelength λ has a maximum, the other has none.

2. Consider one wavelength only and fixed angle θ. To each maximum corresponds an integer m determined by $2D\cos\theta = m\pi$. Use $D/\lambda = x$ for the ratio, and show that the maxima may be numbered by $m(\theta) = 2x\cos(2\pi\theta/360)$. Make graphs for $m(\theta)$ for θ from 0 to 90 and determine the number of rings one has for ratios of $x = 1, 2, 3, 4$. The larger number m belongs to the smallest angle θ. This is different from Young's experiment and similar ones, where the angle is proportional to the order of interference.

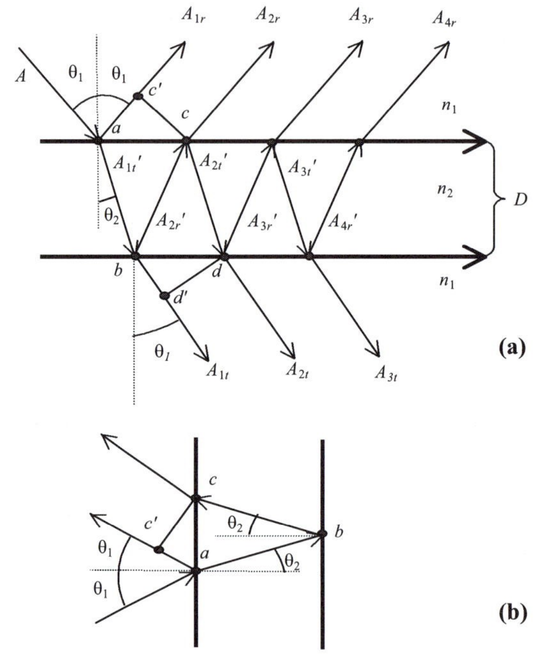

FIGURE 2.18 (a) Geometry for multiple interference at a plane parallel plate of index n_2 and thickness D. The light is incident at an angle θ_1 from a medium with index $n = 1$. Using Snell's law at the first and second surface one may show that the emerging angle is equal to the incident angle θ_1. The reflection angle within the plate is θ_2; (b) geometry for the calculation of the optical path from point a to point c' and c.

2.6 MULTIPLE BEAM INTERFEROMETRY

2.6.1 Plane Parallel Plate

In Section 5.2 we studied the plane parallel plate and considered only two reflected waves (Figure 2.9). We formulated the condition for constructive and destructive interference, but did not investigate the question, "Where does the light travel?" when for destructive interference there is no light in the direction of reflection. Now we want to include in our discussion all internal reflections of the plate and calculate the resulting reflected and transmitted waves as shown in Figure 2.18. We show that the reflected and transmitted intensity is equal to the incident intensity. When we have destructive interference for the reflected light, all light is transmitted and vice versa.

The incident wave is assumed to have magnitude A and makes the angle θ_1 with the normal. The plate has thickness D and a refractive index n_2. The refractive index on both sides of the plate is assumed to be 1. The magnitudes of the reflected and transmitted waves are A_{ir} and A_{it}, where i is 1, 2, 3, ..., respectively. The calculation of the optical path difference is done by using the wave reflected on the first surface and the wave reflected once on the second surface. We assume that both waves propagate in the same direction and calculate the optical path difference, (Figure 2.18b) using the distances $[ac]$ and $[bc]$. One has $[ab] = [bc] = D/\cos\theta_2$ and $[ac] = [(2D/\cos\theta_2)(\sin\theta_2)]] = 2D\tan\theta_2$. The optical path difference between A_{1r} and A_{2r} is then

$$\delta = 2n_2[bc] - [ac]\sin\theta_1 = 2Dn_2/\cos\theta_2 - 2D\tan\theta_2\sin\theta_1. \quad (2.64)$$

Using the law of refraction we get

$$\delta = 2Dn_2/\cos\theta_2 - 2D\tan\theta_2 n_2\sin\theta_2 = 2Dn_2\cos\theta_2 \quad (2.65)$$

The same optical path difference is obtained for transmission. We call r_{12} the amplitude reflection coefficient for a wave incident from medium 1 and reflected at medium 2, with the corresponding intensity R_{12}. Similarly, we call r_{21} for a wave incident from medium 2 and reflected at medium 1, with the corresponding intensity R_{21}. The first index indicates the medium in which the wave travels. The second index indicates the medium at which it is reflected. Similarly, we use τ_{12} for the absolute value of the amplitude of a transmitted wave traveling from medium 1 to 2 and τ_{21} in the opposite direction. We define $\tau_{12} = \sqrt{T_{12}}$ and $\tau_{21} = \sqrt{T_{21}}$, where T_{12} and T_{21} are transmitted intensities. From energy conservation we have that $R_{12} + T_{12} = 1$ and $R_{21} + T_{21} = 1$. The phase difference Δ is

$$\Delta = (2\pi/\lambda)2Dn_2\cos\theta_2 \quad (2.66)$$

and a list of the reflected amplitudes

$$\begin{aligned} A_{1r} &= A\,r_{12} \\ A_{2r} &= A\,\tau_{12}\,r_{21}\,\tau_{21}\,e^{i\Delta} \\ A_{3r} &= A\,\tau_{12}\,r_{21}\,r_{21}\,r_{21}\,\tau_{21}\,e^{i2\Delta} \\ A_{4r} &= A\,\tau_{12}\,r_{21}\,(r_{21}\,r_{21})^2\,\tau_{21}\,e^{i3\Delta} \end{aligned} \quad (2.67)$$

and transmitted amplitudes

$$\begin{aligned} A_{1t} &= A\,\tau_{12}\,\tau_{21} \\ A_{2t} &= A\,\tau_{12}\,r_{21}\,r_{21}\,\tau_{21}\,e^{i\Delta} \\ A_{3t} &= A\,\tau_{12}\,(r_{21}\,r_{21})^2\,\tau_{21}\,e^{i2\Delta} \\ A_{4t} &= A\,\tau_{12}\,(r_{21}\,r_{21})^3\,\tau_{21}\,e^{i3\Delta} \end{aligned} \quad (2.68)$$

For the summation of the reflected amplitudes one gets

$$A_r = Ar_{12} + A\tau_{12}r_{21}\tau_{21}e^{i\Delta}(1 + r_{21}r_{21}e^{i\Delta} + (r_{21}r_{21}e^{i\Delta})^2, \quad (2.69)$$

and for the transmitted amplitudes

$$A_t = A\tau_{12}\tau_{21}(1 + r_{21}r_{21}e^{i\Delta} + (r_{21}r_{21}e^{i\Delta})^2 + \ldots). \quad (2.70)$$

Using the formula for the summation process

$$\sum_{n=0}^{n=N-1} x^N = (1 - x^n)/(1 - x) \tag{2.71}$$

we see that the term of the Nth power can be neglected, since the reflection coefficients are all smaller than 1 and N is a large number. We now have for the reflected amplitude

$$A_r = Ar_{12} + A\tau_{12}r_{21}\tau_{21}e^{i\Delta}/(1 - r_{21}r_{21}e^{i\Delta}) \tag{2.72}$$

and for the transmitted amplitude

$$A_t = A\tau_{12}\tau_{21}/(1 - r_{21}r_{21}e^{i\Delta}). \tag{2.73}$$

We call r the absolute value of the amplitude reflection coefficients r_{12} and r_{21}, and have for $n_2 > n_1$, using Fresnel's formulas (Chapter 5), that $r_{12} = -r$ and $r_{21} = r$; that is, $R_{12} = R_{21}$. As result one has $1 - R_{12} = T_{12} = 1 - r^2 = T_{21} = 1 - R_{21}$ and we may use $\tau_{12}\tau_{21} = 1 - r^2$ and write

$$A_r = -Ar + Ar(1 - r^2)e^{i\Delta}/(1 - r^2e^{i\Delta}) \tag{2.74}$$

$$A_t = A(1 - r^2)/(1 - r^2e^{i\Delta}). \tag{2.75}$$

The transmitted intensity is obtained by multiplication of At with its complex conjugate At*

$$A_tA_t^* = A^2(1 - r^2)^2[1/(1 - r^2e^{i\Delta})(1 - r^2e^{-i\Delta})].$$

One has $(1-r^2e^{i\Delta})(1-r^2e^{-i\Delta}) = 1-r^2e^{i\Delta}-r^2e^{-i\Delta}+r^4 = 1+r^4-r^22\cos\Delta$ and gets

$$A_tA_t^* = A^2[(1 - r^2)^2]/(1 + r^4 - 2r^2\cos\Delta). \tag{2.76}$$

And similarly

$$A_rA_r^* = A^2[2r^2(1 - \cos\Delta)]/(1 + r^4 - 2r^2\cos\Delta). \tag{2.77}$$

Introduction of the normalized intensities

$$I_r = \left|A_rA_r^*/A^2\right| \quad \text{and} \quad I_t = \left|A_tA_t^*/A^2\right| \tag{2.78}$$

results in

$$I_r = [2r^2(1 - \cos\Delta)]/(1 + r^4 - 2r^2\cos\Delta) \tag{2.79}$$

and

$$I_t = [(1 - r^2)^2]/(1 + r^4 - 2r^2\cos\Delta). \tag{2.80}$$

Using the abbreviation

$$g = 2r/(1 - r^2) \tag{2.81}$$

and

and the trigonometric identity

$$\cos\Delta = 1 - 2\sin^2(\Delta/2) \tag{2.82}$$

we obtain

$$I_r = [g^2\sin^2(\Delta/2)]/[(1 + g^2\sin^2(\Delta/2)] \tag{2.83}$$

and

$$I_t = 1/[(1 + g^2\sin^2(\Delta/2)], \tag{2.84}$$

where we recall that $\Delta = (2\pi/\lambda)\delta$, and $\delta = 2Dn_2\cos\theta_2$, the optical path difference of adjacent transmitted and reflected waves. Corresponding to the conservation of energy one has

$$I_r + I_t = 1. \tag{2.85}$$

Depending on the thickness D and the angle of incidence θ_1, the incident intensity is divided between I_r and I_t. If $[\sin\Delta/2]^2 = 0$, we have the condition of constructive interference for transmitted light $I_r = 0$,

$$\delta = 2Dn_2\cos\theta_2 = 0,\ \lambda, 2\lambda, \ldots, m_\lambda. \tag{2.86}$$

If $[\sin\Delta/2]^2 = 1$ we have a minimum of light transmitted. The condition is

$$\delta = 2Dn_2\cos\theta_2 = (1/2)\lambda,\ (3/2)\lambda, \ldots, (m/2)\lambda, \quad m \text{ odd} \tag{2.87}$$

and one has

$$I_t = 1/(1+g) \quad \text{and} \quad I_r = g/(1+g). \tag{2.88}$$

In FileFig 2.14 we show graphs of Eqs. 2.88 for transmitted and reflected intensity, depending on thickness D for fixed wavelength λ and different refractive indices outside of the plate. In Figure 2.19 we show photos of interference fringes for observation in reflection and transmission. The fringes depend on the angle between the incident light and the normal of the surface. They are

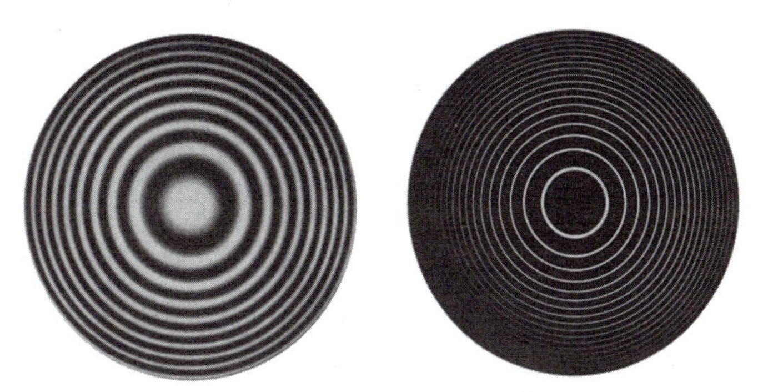

FIGURE 2.19 Interference fringes observed with a plane parallel plate using an extended source: (a) reflection; (b) transmission (from Cagnet, Francon, Thrierr, *Atlas of Optical Phenomena*, Springer-Verlag, Heidelberg, 1962).

Heidinger fringes. The reflection coefficients, used in the graph of FileFig 2.14, are calculated from Fresnel's formulas for a glass plate. For the special case of normal incidence and reflection on the optical denser medium, one has from Fresnel's formulas

$$r = (n_1 - n_2)/(n_1 + n_2). \tag{2.89}$$

In FileFig 2.15 we show graphs, assuming normal incidence, of the transmitted and reflected intensity depending on wavelength for fixed thickness D. In Eq. (2.81) we defined $g = 2r/(1 - r^2)$. We mention here that $\pi g/2$ is called the finesse. It is used for the characterization of the quality of the Fabry–Perot, discussed in the next chapter.

FileFig 2.14 (I14PLANIDS)

Intensity of interference at a plane parallel plate assuming normal incidence. Graph of the reflection and transmission depending on thickness D for fixed wavelength and different values of $n1$, $n2$, and $n3$.

I14PLANIDS

Normal Incidence. Plane Parallel Plate: Reflected and Transmitted Intensity Depending on Thickness for Fixed Wavelength

The reflection coefficients are calculated from Fresnel's formulas for $\theta = 0$. Refractive indices $n1$, $n2$, and $n3$ may all be different and the reflection coefficients for both surfaces are calculated. The calculation of the fringe pattern is done depending on D for fixed λ.

$$\Delta = 2\pi/\lambda\ 2dn2 \qquad \theta 1 := 1$$

$$n1 := 1 \qquad n2 := 1.5 \qquad n3 := 1$$

$$r12 := \frac{n2 - n1}{n2 + n1} \qquad r23 := \frac{n3 - n2}{n3 + n2} \qquad \Delta = (2\pi/\lambda)\ 2dn2\cos\theta 2$$

$$r12 = 0.2 \qquad r23 = -0.2$$

$$\lambda \equiv .0005 \qquad D \equiv .0002, .00021 \ldots .002$$

$$IT(D) := \frac{(1 - r12^2) \cdot (1 - r23^2)}{1 + (r12 \cdot r23)^2 - (2 \cdot r12 \cdot r23) \cdot \cos\left(4 \cdot \pi \cdot \frac{D}{\lambda} \cdot n2\right)}$$

$$IR(D) := 1 - IT(D).$$

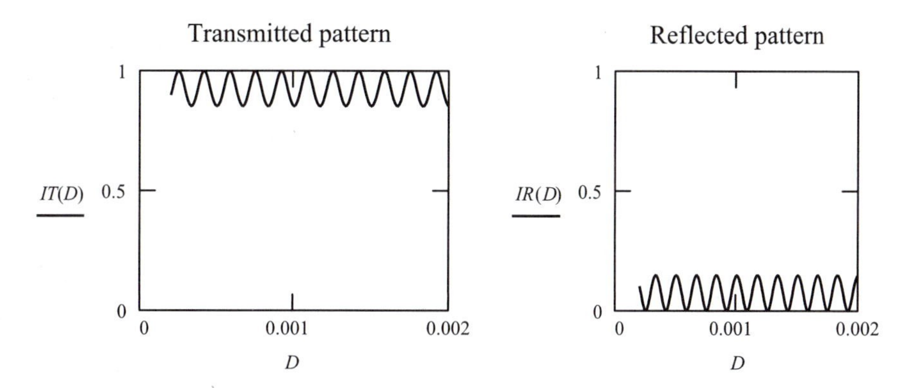

Application 2.14. Consider transmitted and reflected intensity depending on the thickness of the plate.

1. Try out different combinations such as $n1 < n2 < n3$, $n1 < n2 > n3$, and $n1 > n2 < n3$, and see the effect of the phase jump of reflection at the denser medium.
2. Choose arbitrary values for $r < 1$ and observe how the intensity is changing.

FileFig 2.15 (I15PLANIDS)

Intensity of interference at a plane parallel plate assuming normal incidence. Graph of the reflection and transmission depending on wavelength λ for fixed D and different values of $n1$, $n2$, and $n3$.

I15PLANIDS is only on the CD.

Application 2.15. Consider the transmitted and reflected intensity depending on wavelength.

1. Try out different combinations such as $n1 < n2 < n3$, $n1 < n2 > n3$, and $n1 > n2 < n3$, and see the effect of the phase jump of reflection at the denser medium.
2. Find the wavelength for the last fringe, depending on the thickness D of the plate.

2.6.2 Fabry–Perot Etalon

A plane parallel plate with reflecting surfaces on both sides is called an etalon. The reflecting layers may be made of metal or a structure of dielectric films. We show that for a specific wavelength at a specific spacing of two reflecting surfaces, all incident light will be transmitted while a single reflecting surface will transmit only a small amount.

We start from the treatment of the plane parallel plate and assume that one can replace the two interfaces with idealized reflectors, having the reflectivity r.The medium between these two reflectors has refractive index 1. Assuming normal incidence and using the reflectance $R = r^2$ one has from Eq. (2.81),

$$g^2 = 4R/(1 - R)^2. \tag{2.90}$$

For normal incidence, one has for $\Delta/2$,

$$\Delta/2 = 2\pi D/\lambda. \tag{2.91}$$

The reflected and transmitted intensities are obtained from Eqs. (2.83) and (2.84):

$$I_r = g^2 \sin^2(\Delta/2)/(1 + g^2 \sin^2(\Delta/2)) \qquad I_t = 1/(1 + g^2 \sin^2(\Delta/2)). \tag{2.92}$$

There the mathematical form of I_t is called the Airy function.

If $[\sin \Delta/2]^2 = 0$ we have the condition of constructive interference for transmitted light, $I_r = 0$,

$$\delta = 2D = 0, \lambda, 2\lambda, \ldots, m\lambda. \tag{2.93}$$

If $[\sin \Delta/2]2 = 1$ we have a minimum of light transmitted. The condition is

$$\delta = 2D = (1/2)\lambda, (3/2)\lambda, \ldots, (m + 1/2)\lambda, \tag{2.94}$$

where m is an integer. The graph in FileFig 2.16 shows three transmission patterns for three different absolute values of the reflection coefficient. We have chosen $\lambda = .1$ and plotted the transmitted intensity as a function of the spacing D, for $m = 1$ and $r = .7$, .9, and .97, respectively. We see that the width of the transmitted intensity depends on the absolute value of the reflectance r of a single plate and becomes narrower when r gets close to 1. For constructive interference, that is, when $2D = m\lambda$, it follows that $\sin^2 \Delta/2 = 0$. Therefore I_t is 1, independent of the value of r. We may have r so close to 1 that the transmission of a single plate is almost zero, but the transmission of the pair of plates at the right distance will be one. At this distance the Fabry–Perot etalon has a *resonance mode*. In experimental Fabry–Perot etalons, the peak transmission will not be exactly one, due to losses such as absorption in the plates. The Fabry–Perot etalon, using high orders, is applied to investigate with high resolution details of a spectral line in a narrow spectral range. The dependence of the width of the spectral line on the reflection coefficient of the etalon is shown in the graph of FileFig 2.17. The transmittance is plotted depending on the wavelength λ for three different reflection coefficients r and fixed distance D.

FileFig 2.16 (I16FABRYS)

Transmission through a Fabry–Perot depending on separation of plates D for three different reflection coefficients (three different g), $m = 1$, and wavelength $\lambda = 0.1$.

I16FABRYS

Fabry–Perot Transmission Depending on D

Normal incidence. Parameters: reflection coefficient, wavelength λ, refractive index. See for global definition. The finesse $\pi g/2$ is $\lambda/\Delta\lambda$. All lengths in mm.

$$\Delta = (2\pi/\lambda)2D(n2)\cos\theta 2 \qquad D := 0, .001 \ldots 11 \qquad n2 := 1$$

$$g1 := \frac{2 \cdot r1}{1 - r1^2}$$

$$g2 := \frac{2 \cdot r2}{1 - r2^2} \qquad g3 := \frac{2 \cdot r3}{1 - r3^2}$$

$$IT1(D) := \frac{1}{1 + g1^2 \cdot \sin\left(2 \cdot \frac{\pi}{\lambda} \cdot D \cdot n2\right)^2}$$

$$IT2(D) := \frac{1}{1 + g2^2 \cdot \sin\left(2 \cdot \frac{\pi}{\lambda} \cdot D \cdot n2\right)^2}$$

$$IT3(D) := \frac{1}{1 + g3^2 \cdot \sin\left(2 \cdot \frac{\pi}{\lambda} \cdot D \cdot n2\right)^2}$$

$$r1 \equiv .7 \qquad r2 \equiv .9 \qquad r3 \equiv .97 \qquad \lambda \equiv .1.$$

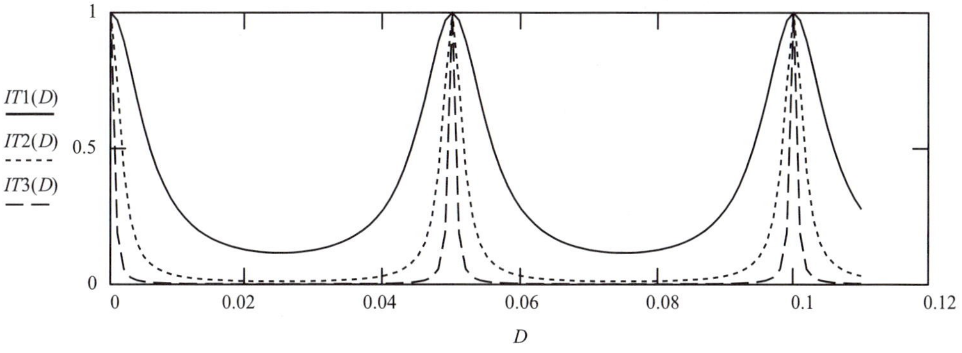

IT1(D)
IT2(D)
IT3(D)
1
0.5
0
0
0.02
0.04
0.06
0.08
0.1
0.12
D

Application 2.16.

1. How do the positions of maxima depend on D and λ?
2. Give a formula for the separation of fringes and verify with the data from the graph.

3. Condsider dependence on θ for θ equal 0.001 to .7 for constant $\lambda = .001$ and $D = .02$. Make a graph and derive for the fringe number $f(\theta) = (2/y)(Dn2\cos\theta)$. Plot $f(\theta)$ and observe that the highest number corresponds to the smallest angle. This is contrary to Young's experiment; see also FileFig 2.13.

FileFig 2.17 (I17FABRYLS)

Transmission through a Fabry–Perot depending on wavelength λ for three different reflection coefficients r (three different g), $m = 1$, and thickness $D = 0.0025$.

I17FABRYLS is only on the CD.

Application 2.17. The bandwidth of a peak is the width at half-height, given by $bw = 2\lambda/\pi g$. Calculate two bandwidths, $bw1$ and $bw2$, with ratio $bw1/bw2 = 5$ and make a graph. Verify the data by reading the bw values from the graph.

2.6.3 Fabry–Perot Spectrometer and Resolution

A Fabry–Perot etalon may be used as a spectroscopic device when varying the spacing between the two reflecting surfaces over a small interval. The result is that the first-order resonance wavelength $\lambda_0 = 2D_0$ is varied around a wavelength interval and therefore we have for different D_i the resonance maximum, corresponding to λ_i. Scanning D gives us a maximum depending on λ and we get the spectral distribution of the incident signal, as one may measure with a grating spectrometer. Two spectral lines of wavelength difference $\Delta\lambda$ may be seen separated or not, depending on the resolution of the Fabry–Perot spectrometer. To calculate the resolution, that is, the wavelength difference $\Delta\lambda = \lambda_2 - \lambda_1$ of the spectral lines of wavelength λ_1 and λ_2, we assume $\lambda_2 > \lambda_1$ (Figure 2.20a). The two resonance lines are considered resolved when the crossing of the "right" side of one line with the "left side" of the other line has the value $\frac{1}{2}$. This condition may be expressed, using Eqs. (2.91) and (2.92), as

$$\begin{aligned} &\{1/(1+g^2\sin^2[(2\pi/\lambda_1)(D-\epsilon)])\} \\ &= \{1/(1+g^2\sin^2[(2\pi/\lambda_2)(D+\epsilon)])\}. \end{aligned} \tag{2.95}$$

From Eq. (2.95) it follows that

$$(D+\epsilon)\lambda_1 = (D-\epsilon)\lambda_2. \tag{2.96}$$

Using $\lambda_1 = \lambda_2 - \Delta\lambda$ and renaming λ_2 as λ, one has $\lambda 2\epsilon = (\epsilon + D)\Delta\lambda$ or

$$\lambda/\Delta\lambda = (D+\epsilon)/2\epsilon. \tag{2.97}$$

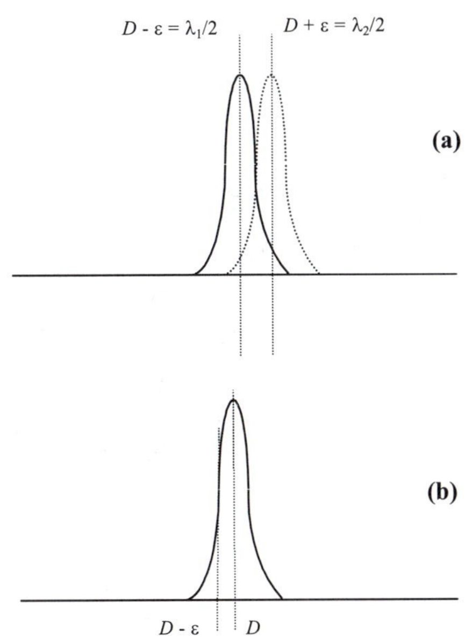

FIGURE 2.20 (a) Spectral lines of wavelength λ_1 at resonance distance $D - \epsilon$ and λ_2 at resonance distance $D + \epsilon$; (b) one spectral line at resonance distance $\lambda = 2D$ with half-height at $D + \epsilon$.

To obtain the value of (we look at a single line at resonance (Figure 2.20b), and get $D = \lambda/2$. The value of ϵ is obtained from one line at half-height

$$\frac{1}{2} = \{1/(1 + g^2 \sin^2[(2\pi/\lambda)(D - \epsilon)])\}. \tag{2.98}$$

We use the sum formula for $\sin(a - b) = \sin(a)\cos(b) - \cos(a)\sin(b)$ and obtain with $\sin(2\pi D/\lambda) = 0$ and $\cos(2\pi D/\lambda) = -1$,

$$1 + g^2[\sin(2\pi\epsilon/\lambda)]^2 = 2. \tag{2.99}$$

Since ϵ is small we may approximate the sine by the angle and have

$$g^2[(2\pi\epsilon/\lambda)]^2 = 1. \tag{2.100}$$

Combining Eqs. (2.97) and (2.99) and assuming $\epsilon \ll D$ one gets

$$\lambda/\Delta\lambda = \pi g D/\lambda \tag{2.101}$$

In first order we have for the resonance distance $D = \lambda/2$, and obtain for $\lambda/\Delta\lambda = \pi g/2$. As mentioned above, $\pi g/2$ is called the finesse F. It characterizes the resolving power of the Fabry–Perot spectrometer. Considering the order m (integer) for the resonance distance of λ, that is, $D = m\lambda/2$, we have more generally for the resolving power $\lambda/\Delta\lambda$

$$\lambda/\Delta\lambda = m\pi g/2. \tag{2.102}$$

It depends only on the reflection coefficient of the single plate. The inverse, $\frac{\Delta\lambda}{\lambda}$ is called the resolution.

FIGURE 2.21 A Fabry–Perot ring pattern obtained using the green line of the mercury spectrum (from Cagnet, Francon, and Thrierr, *Atlas of Optical Phenomena*, Springer-Verlag, Heidelberg, 1962).

In FileFig 2.18 we show a graph of the transmitted intensity of two different wavelengths λ_1 and λ_2 for several orders, using the same value of the finesse. One observes that the resolution may be changed when changing the reflectivity r. In FileFig 2.19 we have calculated for two wavelengths the transmission through a Fabry–Perot, depending on very small angles with respect to the normal. In Figure 2.21 a photo is shown of a ring pattern produced with a Fabry–Perot for the green Hg-line in the visible spectral region.

FileFig 2.18 (I18FABRYRDS)

Graph of Fabry–Perot resonances of two wavelengths λ_1 and λ_2 depending on separation D of the plates.

I18FABRYRDS is only on the CD.

Application 2.18.

1. Make three choices of λ_2 and determine the reflection coefficient r to have them resolved in first-order.
2. Choose λ_1 and λ_2 and determine the reflection coefficient $r2$ to have the lines resolved in second-order, and reflection coefficient $r3$ for the third-order.
3. Introduce $\lambda_2 = \lambda_1 + \Delta\lambda_1 = \lambda_1(1 + 2/m\pi g)$ and make changes to r so that lines are separated for the first-, second-, and third-order.

FileFig 2.19 (I19FABRYAS)

Transmission through a Fabry–Perot depending on angle with the normal. Wavelength $\lambda 1 = 0.0005$, $\lambda 1 = 0.0005025$, thickness $D = 0.01$, reflection coefficient $r = 0.9$, and $m = 1$.

I19FABRYAS is only on the CD.

Application 2.19.

1. Observe that the separation of the fringes changes with m and that the mth fringe is at the center.
2. The resolution is largest for the fringe with the largest m. The two wavelengths $\lambda 1 = 0.00054$ and $\lambda 1 = 0.0005025$ have a difference of 1%. Read from the graph the difference in the angle, give a formula to calculate this difference, and compare.

2.6.4 Array of Source Points

We study the interference pattern produced by a periodic array of N source points, as we would have it when using a grating. We extend Young's experiment to more than two small openings and assume that the distance a between adjacent openings is a constant. In our model description, we assume one incident wave and neglect the diffraction effect in the process of splitting the wave into N waves. We call the openings source points and have N waves traveling with the angle θ to the normal of the array (Figure 2.22). The source points all vibrate coherently. In other words, the amplitudes have maxima and minima at the same time and the optical path difference between adjacent waves is the same. From Young's experiment, we have for superposition of source points 1 and 2,

$$\begin{aligned} u &= u_1 + u_2 \\ &= A\cos(2\pi x/\lambda - 2\pi t/T) + A\cos[2\pi(x-\delta)/\lambda - 2\pi t/T], \end{aligned} \tag{2.103}$$

where $\delta = a\sin\theta$, or in the small angle approximation $\delta = aY/X$. For N apertures of distance a we have

$$\begin{aligned} u = A\cos(2\pi x/\lambda - 2\pi t/T) + A\cos[2\pi(x-\delta)/\lambda - 2\pi t/T] + \cdots \\ \cdots + A\cos\{2\pi[x-(N-1)\delta]\lambda - 2\pi t/T\}, \end{aligned} \tag{2.104}$$

which can be written as

$$u = A\sum_{q=0}^{q=N-1}\cos[2\pi(x-q\delta)/\lambda - 2\pi t/T]. \tag{2.105}$$

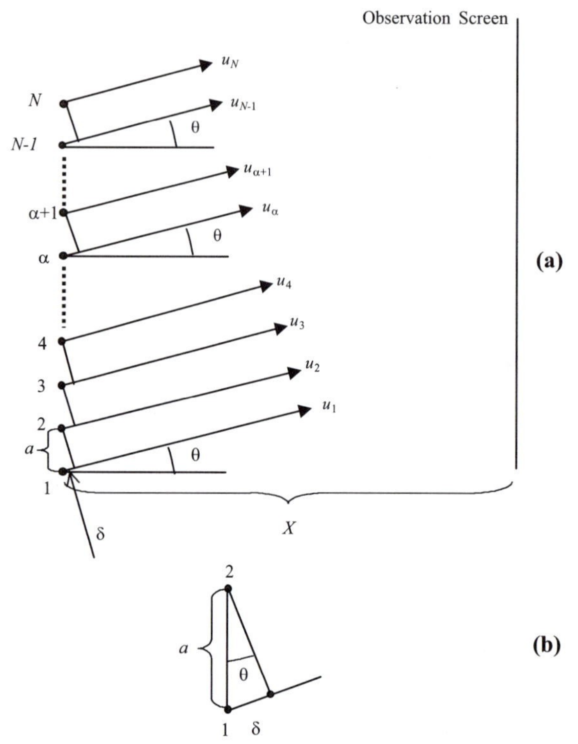

FIGURE 2.22 (a) Waves u_1 to u_N have their origin at source points 1 to N. The source points are spaced by a. The waves travel in the x direction and have the angle θ with respect to the axis of the setup; (b) the optical path difference between adjacent waves is $\delta = a \sin\theta$, or in the small angle approximation, aY/X.

Similar to the discussion in the section on intensities, we introduce complex notation and write

$$u = A \sum_{q=0}^{q=N-1} \exp i[2\pi(x - q\delta)/\lambda - 2\pi t/T] \tag{2.106}$$

or

$$u = A \exp i[2\pi x/\lambda - 2\pi t/T] \sum_{q=0}^{q=N-1} \exp i[2\pi(-q\delta/\lambda)]. \tag{2.107}$$

Using the formula

$$\sum_{n=0}^{n=N-1} x^n = (1 - x^N)/(1 - x) \tag{2.108}$$

we get

$$\begin{aligned} u = {} & A \exp i[2\pi x/\lambda - 2\pi t/T] \\ & \{1 - \exp i[2\pi(-N\delta/\lambda)]\}/\{1 - \exp i[2\pi(-\delta/\lambda)]\}. \end{aligned} \tag{2.109}$$

We note that unlike the case of the plane parallel plate, we can not ignore the Nth power in the summation formula. The expression in brackets of Eq. (2.109) can be rewritten for the numerator as

$$\begin{aligned}&\exp i[2\pi(-N\delta/2\lambda)]\{\exp i[2\pi(N\delta/2\lambda)] - \exp i[2\pi(-N\delta/2\lambda)]\}\\&= \exp i[2\pi(-N\delta/2\lambda)]\{2i \sin 2\pi(N\delta/2\lambda)\}\end{aligned} \tag{2.110}$$

and for the denominator

$$\begin{aligned}&\exp i[2\pi(-\delta/2\lambda)]\{\exp i[2\pi(\delta/2\lambda)] - \exp i[2\pi(-\delta/2\lambda)]\}\\&= \exp i[2\pi(-\delta/2\lambda)]\{2i \sin 2\pi(\delta/2\lambda)\}.\end{aligned} \tag{2.111}$$

We have for the resulting amplitude

$$\begin{aligned}u &= [A \exp i(2\pi(x)/\lambda - 2\pi t/T)] \\&\quad \cdot \{[\exp i[2\pi(-N\delta/2\lambda)] \sin 2\pi(N\delta/2\lambda)\}/[\exp i[2\pi(-\delta/2\lambda)] \sin 2\pi(\delta/2\lambda)]\end{aligned} \tag{2.112}$$

or

$$u = Ae^{i\Psi} \sin[2\pi(N\delta/2\lambda)]/[\sin 2\pi(\delta/2\lambda)] \tag{2.113}$$

where

$$e^{i\Psi} = \exp i[2\pi(x/\lambda) - 2\pi(t/T)] \exp[i2\pi(-N\delta/2\lambda)] \exp[i2\pi(\delta/2\lambda)]. \tag{2.114}$$

We take for the intensity $uu^* = I$, and have

$$I = A^2\{\sin[2\pi(N\delta/2\lambda)]/ \sin 2\pi(\delta/2\lambda)\}^2. \tag{2.115}$$

Substituting $\delta = a \sin\theta$ and taking $A^2 = 1/N^2$ for normalization, we can write I as

$$I = \{\sin(\pi Na \sin\theta/\lambda)/N \sin(\pi a \sin\theta/\lambda)\}^2 \tag{2.116}$$

or for the small angle approximation with $\delta = aY/X$,

$$I = \{\sin(\pi NaY/X\lambda)/N \sin(\pi aY/X\lambda)\}^2. \tag{2.117}$$

Equations (2.116) and (2.117) have their main maxima when both numerator and denominator are zero. This is shown in the first graph of FileFig 2.20, where numerator $y(\theta)$ and denominator $y_1(\theta)$ are plotted separately. One observes between the main maxima $N - 2$ side maxima and $N - 1$ side minima. From the trace of the numerator one sees that two of the side maxima do not appear. They are at the flank of the main maxima, and one side minima is located at the main maxima. The main maxima and side maxima and minima are shown in the second graph of FileFig 2.20. The interference pattern generated by an array of sources has a wide application. It is used in the discussion of Xray diffraction and is used in the discussion of the diffraction grating.

FileFig 2.20 (I20ARRAYS)

Intensity I_A of the array, both numerator and denominator are plotted on one graph depending on angle $\theta = Y/X$, where Y is the coordinate on the observation screen, and X the distance from the experiment to the screen. $N = 5$, wavelength $\lambda = 0.0005$, periodicity constant $a = 0.1$. The main maxima of I_A are obtained for 0/0 of I_A. There are $N - 2$ side maxima, $N - 1$ side minima.

I20ARRAYS

Interference Pattern of N Sources

Parameters: Opening a, wavelength λ, number or lines N. Graph as function of θ, because of small angle $\theta = Y/X$. Normalization to 1. For comparison of maxima, the numerator is plotted separately.

$$\theta := 0, .001 \ldots .5 \qquad \lambda := .0005 \qquad a \equiv .1 \qquad N := 5$$

$$IA1(\theta) := \left(\frac{\sin\left(\pi \cdot N \cdot \frac{a}{\lambda} \cdot \sin\left(2 \cdot \frac{\pi}{360} \cdot \theta\right)\right)}{N \cdot \sin\left(\pi \cdot N \cdot \frac{a}{\lambda} \cdot \sin\left(2 \cdot \frac{\pi}{360} \cdot \theta\right)\right)} \right)^2$$

$$y(\theta) := \sin\left(\pi \cdot N \cdot \frac{a}{\lambda} \cdot \sin\left(2 \cdot \frac{\pi}{360} \cdot \theta\right)\right)^2$$

$$y_1(\theta) := \sin\left(\pi \cdot \frac{a}{\lambda} \cdot \sin\left(2 \cdot \frac{\pi}{360} \cdot \theta\right)\right)^2$$

$$aa \equiv .2 \qquad NN := 5$$

$$IA2(\theta) := \left(\frac{\sin\left(\pi \cdot NN \cdot \frac{aa}{\lambda} \cdot \sin\left(2 \cdot \frac{\pi}{360} \cdot \theta\right)\right)^2}{NN \cdot \sin\left(\pi \cdot \frac{aa}{\lambda} \cdot \sin\left(2 \cdot \frac{\pi}{360} \cdot \theta\right)\right)} \right)^2$$

IA1 (θ)
y(θ)
y1(θ)
0 0.1 0.2 0.3 0.4 0.5
θ

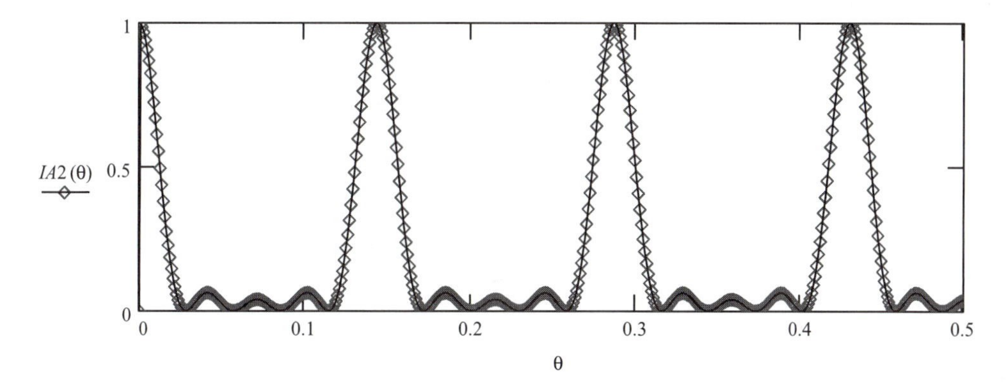

Application 2.20.

1. Observe that the main maxima are at angles when nominator and denominator are both zero.
2. The $N-1$ side minima are at angles when the nominator has minima.
3. There are N side maxima when the nominator has maxima, but only $N-2$ appear.
4. What happens when changing the wavelength?
5. What happens when changing the periodicity constant?
6. What happens when changing N?

2.7 RANDOM ARRANGEMENT OF SOURCE POINTS

In Section 2.6 we assumed for the calculation of the interference of N source points that the optical path differences for all waves are equal. The source points were arranged periodically, having the periodicity distance a. We saw that the incident intensity was redistributed into main maxima and side maxima and minima.

We now study the opposite case where the periodicity constant is no longer a constant but has a random distribution in an interval to be specified. In Figure 2.23 we have the waves u_α propagating in direction θ and the optical path difference between the $(\alpha-1)$th and the αth wave is δ_α (instead of $q\delta$). All δ_α are not the same. From the discussion of the periodic arrangement of N source points we have for the superposition of these waves

$$u = A \exp i(2\pi x/\lambda - 2\pi t/T) \sum_{\alpha=0}^{\alpha=N-1} \exp i(2\pi(-\delta_\alpha/\lambda). \tag{2.118}$$

In this expression δ_α/λ may be larger than 1. One subtracts a wavelength from δ_α until δ'_α/λ is smaller than 1. The value of the trigonometric functions is the same, before and after the reduction process. We call γ_α the reduced value of

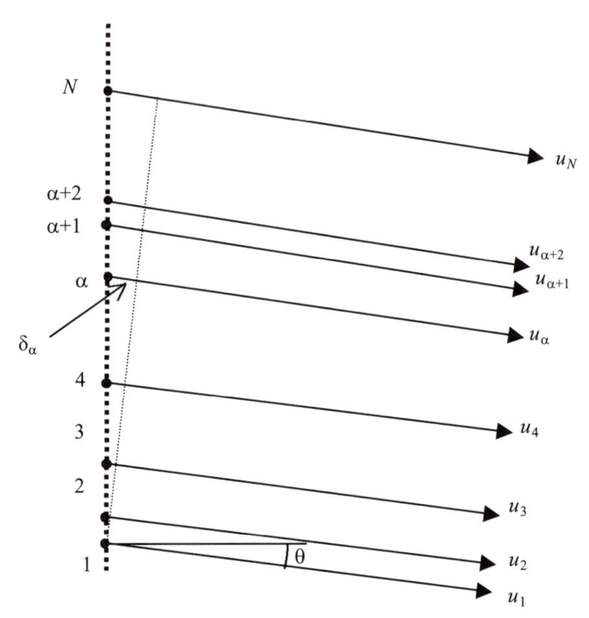

FIGURE 2.23 *N* source points with random spacing between them. THe optical path differences between the waves, or of one wave with respect to a reference wave, are random numbers.

δ_α/λ with values between 0 and 1. Since these γ_α have a random distribution in the interval from 0 to 1, we can not use the summation formula as done in Section 6.4 and furthermore have to consider

$$u = A \exp i(2\pi x/\lambda - 2\pi t/T) \sum_{\alpha=0}^{\alpha=N-1} \exp i(2\pi(-\gamma_\alpha). \tag{2.119}$$

The intensity is calculated from uu^*,

$$I = uu^* = A^2 \left(\sum_{\alpha=0}^{\alpha=N-1} \exp i2\pi(-\gamma_\alpha) \right) \left(\sum_{\beta=0}^{\beta=N-1} \exp i2\pi(\gamma_\alpha) \right). \tag{2.120}$$

The multiplication results for $\alpha = \beta$ in a sum of N values of 1 and then a sum over all terms with $\alpha \neq \beta$ of the random phase angles which are now called γ_γ. The phase angles γ_γ are calculated from $\gamma_\beta - \gamma_\alpha$ and if they are larger that 2π one reduces them until they fall into the interval 0 to 2π. We have with summation over γ,

$$I = [1 + 1 + 1 + 1 + \cdots \left(\sum \exp i2\pi(\gamma_\gamma) \right)]A^2. \tag{2.121}$$

For a large number of the randomly distributed values of γ_γ one can always find another γ'_γ so that the exponents of $\exp i2\pi(\gamma_\gamma)$ and $\exp i2\pi(\gamma'_\gamma)$ cancel. As a result the sum in Eq. (2.121) is zero.

$$\left(\sum \exp i2\pi(\gamma_\gamma) \right) = 0. \tag{2.122}$$

The result for the intensity in the case where the array is not periodic is

$$I_R = A^2 N. \tag{2.123}$$

We compare this result to Eq. (2.117), that is, for the periodic array. For maxima we obtained

$$I_A = A^2 N^2. \tag{2.124}$$

This is an important result for the discussion of phenomena having their origin in periodic and non periodic appearance. In our case of interference, one has for the non periodic case an "incoherent" addition of the waves. The result is an average distribution $I \propto N$, and there is no interference pattern. In the periodic case, the waves add coherently and an interference pattern is observed. The light appears as maxima and minima. In FileFig 2.21 we show the incoherent addition of the waves for the non periodic case. The sum of Eq. (2.122) is plotted depending on the number N_f of randomly positioned openings and approaches zero when choosing large numbers of N_f.

FileFig 2.21 (I21RANDS)

Incoherent addition of N phase factors.

I21RANDS

Addition of Exponential Functions with Random Angles

The real part of the sum of $\exp i\theta$ is plotted.

$$f := 1 \ldots 100 \qquad N_f := f \qquad k := 1, 2, \ldots 1000 \qquad i : \sqrt{-1}$$

$$\theta_k := rnd(2 \cdot \pi) \qquad y_f := \frac{1}{N_f} \cdot \sum_{k=0}^{N_f} e^{i \cdot (\theta_k)}.$$

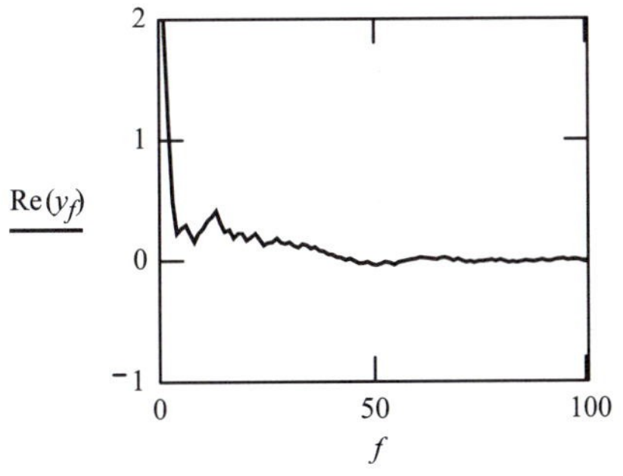

Application 2.21. Change f to a small value and then increase it and observe the changes in the average.

The result for the intensity in the case where the array is not periodic is

$$I_R = A^2 N. \tag{2.123}$$

We compare this result to Eq. (2.117), that is, for the periodic array. For maxima we obtained

$$I_A = A^2 N^2. \tag{2.124}$$

This is an important result for the discussion of phenomena having their origin in periodic and non periodic appearance. In our case of interference, one has for the non periodic case an "incoherent" addition of the waves. The result is an average distribution $I \propto N$, and there is no interference pattern. In the periodic case, the waves add coherently and an interference pattern is observed. The light appears as maxima and minima. In FileFig 2.21 we show the incoherent addition of the waves for the non periodic case. The sum of Eq. (2.122) is plotted depending on the number N_f of randomly positioned openings and approaches zero when choosing large numbers of N_f.

FileFig 2.21 (I21RANDS)

Incoherent addition of N phase factors.

I21RANDS

Addition of Exponential Functions with Random Angles

The real part of the sum of $\exp i\theta$ is plotted.

$$f := 1 \ldots 100 \qquad N_f := f \qquad k := 1, 2, \ldots 1000 \qquad i : \sqrt{-1}$$

$$\theta_k := rnd(2 \cdot \pi) \qquad y_f := \frac{1}{N_f} \cdot \sum_{k=0}^{N_f} e^{i \cdot (\theta_k)}.$$

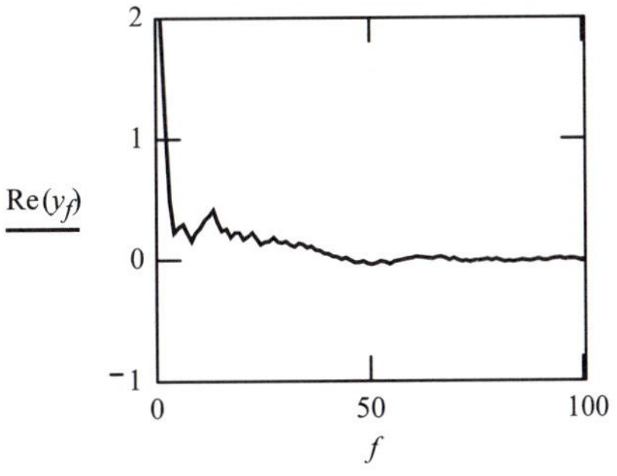

Application 2.21. Change f to a small value and then increase it and observe the changes in the average.

C H A P T E R

Diffraction

3.1 INTRODUCTION

We know *Huygens' Principle* from introductory physics. It tells us that a "new" wavefront of a traveling wave may be constructed at a later time by the envelope of many wavelets generated at the "old" wavefront. One assumes that a primary wave generates fictitious spherical waves at each point of the "old" wavefront. The fictitious spherical wave is called Huygens' wavelet and the superposition of all these wavelets results in the "new" wavefront. This is schematically shown in Figure 3.1. The distance between the generating source points is infinitely small and therefore, integration has to be applied for their superposition.

We discussed in Chapter 2 the superposition of light waves and the resulting interference patterns. In the division process of the incident wave into parts, we neglected the effect of diffraction. In this chapter we take into account interference and diffraction of the wave incident on an aperture. Optical path differences,

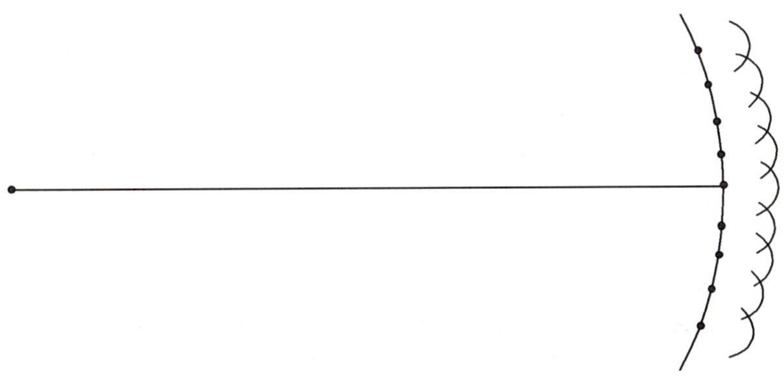

FIGURE 3.1 Schematic of wavefront construction using Huygens' Principle.

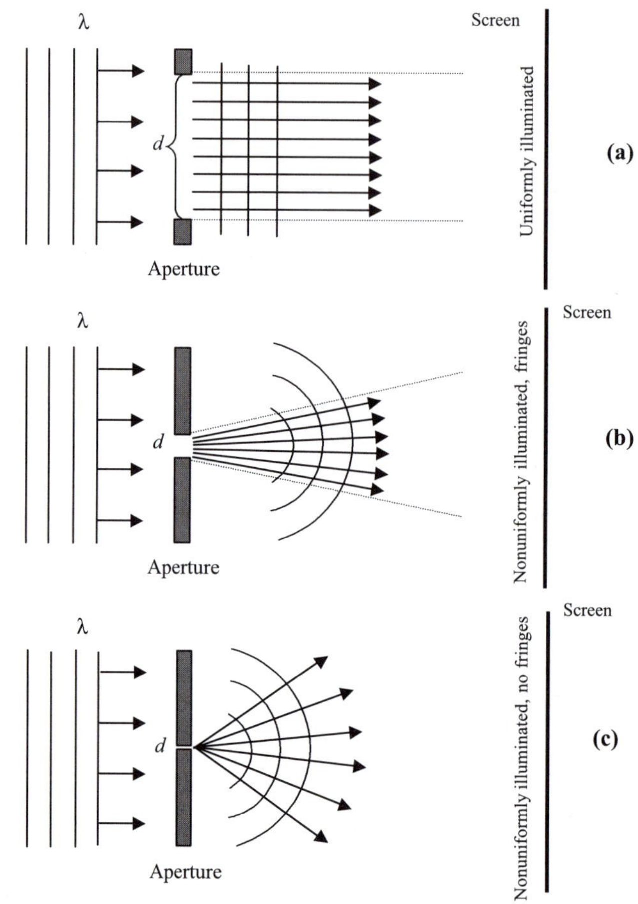

FIGURE 3.2 Conditions for diffraction on a single slit: (a) $d \gg \lambda$, no appreciable diffraction; (b) d of the same order of magnitude of λ, diffraction is observed (fringes); (c) $d \ll \lambda$, nonuniformly illuminated observation screen, but no fringes.

generated between adjacent light waves, are finite for the superposition process of interference and infinitely small for diffraction.

If we apply this division process to an open aperture, the incident wave generates new waves in the plane of the aperture, and these newly generated waves have fixed phase relations with the incident wave and with one another. We assume that all waves generated by the incident wave propagate only in the forward direction, and not backward to the source of light. Let us consider the diffraction on a slit (Figure 3.2). The observed pattern depends on the wavelength and the size of the opening. A slit of a width of several orders of magnitude larger than the wavelength of the incident light will give us almost the geometrical shadow (Figure 3.2a). A slit of width of an order or two larger than the wavelength will bend the light and fringes will occur; see Figure 3.2b. A slit smaller than the wavelength will show an intensity pattern with no fringes and decreasing intensity for larger angles; see Figure 3.2c. All openings will show small deformations of the wavefront close to the edges of the slit (not shown in Figure 3.2).

The model we are using for the description of diffraction is called scalar wave diffraction theory and uses the Kirchhoff–Fresnel integral. All the waves we consider are solutions of the scalar wave equation, as used for the discussion of the interference phenomena in Chapter 2. Here we use spherical waves of the type Ae^{ikr}/r, where A is the magnitude of the wave, r the distance from the origin, and $k = 2\pi/\lambda$. These spherical waves are solutions of the scalar wave equation

$$\nabla^2 u + k^2 u = 0. \tag{3.1}$$

Written in spherical coordinates r, θ, and ϕ one has

$$\nabla^2 = (1/r^2)\{\partial/\partial r(r^2 \partial/\partial r)\} + \text{ (terms in } \theta \text{ and } \phi), \tag{3.2}$$

where we have not explicitly given the terms in θ and ϕ because we only use spherical symmetric solutions and they do not depend on the angular terms.

There is the question of why we should use a summation process based on the idea of Huygen's Principle to describe diffraction theory. Why not solve Maxwell's equations with the appropriate boundary conditions? The mathematical formulation of Huygens' Principle was performed by Gustav Kirchhoff and Augustin Jean Fresnel before Maxwell's theory was developed. It turned out that the use of the Kirchhoff–Fresnel integral for many applications is so much easier than solving Maxwell's equations and applying boundary conditions, that one just continues to use the scalar wave diffraction theory. The wavelength is assumed to be smaller than the aperture opening under consideration.

3.2 KIRCHHOFF–FRESNEL INTEGRAL

3.2.1 The Integral

We assume for the summation process of the Huygens' wavelets, that the primary wave from the source S has amplitude A and travels distance R in the direction of the aperture (Figure 3.3). We disregard the time factor for all waves considered in this chapter. We recall that in Chapter 2 the time factor disappeared when calculating the intensity. At each point of the aperture a Huygens' wavelet is generated, $[(1/\rho)\exp(ik\rho)]$, and travels only in the "forward" direction (Figure 3.3). It has the amplitude of the incident wave, that is, $\{(A/R)\exp(ikR)\}$. We have for a newly generated wavelet

$$[(A/R)\exp(ikR)](1/\rho)\exp(ik\rho)\exp(i\alpha), \tag{3.3}$$

where $\exp(i\alpha)$ is a phase factor related to the generation process. However, it is set to 1.

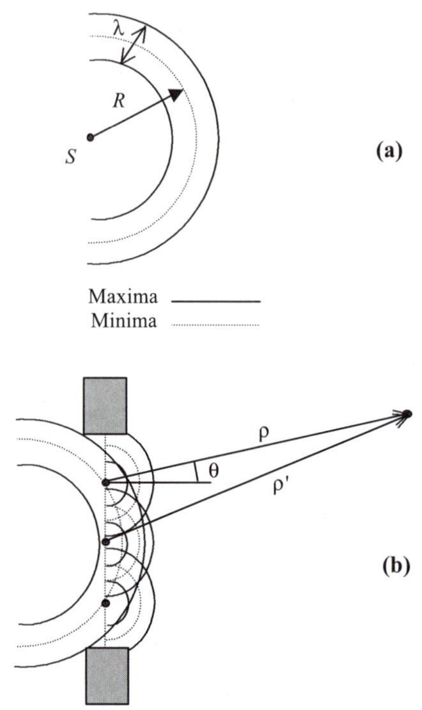

FIGURE 3.3 (a) Maxima and minima of incident wave; (b) three newly generated Huygen's wavelets are shown at the aperture.

From experiments we know that there is an angular dependence of the intensity in the direction of propagation. Therefore we multiply by $\cos\theta$, where θ is the angle to the normal of the aperture, pointing into the forward direction. Integration over all points of the aperture results in the Kirchhoff–Fresnel integral,

$$\int_{\text{opening of aperture}} (A/R)\exp(ikR)(1/\rho)\exp(ik\rho)\cos\theta d\sigma, \tag{3.4}$$

where $d\sigma$ is the surface element for integration over the opening of the aperture. This integral may be derived from the scalar wave equation and Green's Theorem. The derivation yields the factor $\cos\theta$ and shows that the diffracted light is only traveling in the forward direction. However, there are some problems with the boundary conditions. A formulation using Green's function avoids this problem, but is not necessarily better. For more information, see Goodman, 1988, p. 42.

In the following two sections we apply Eq. (3.4) to a special symmetric arrangement of source and observation points, both at large distances from the aperture. We calculate the diffracted intensity only at one observation point on

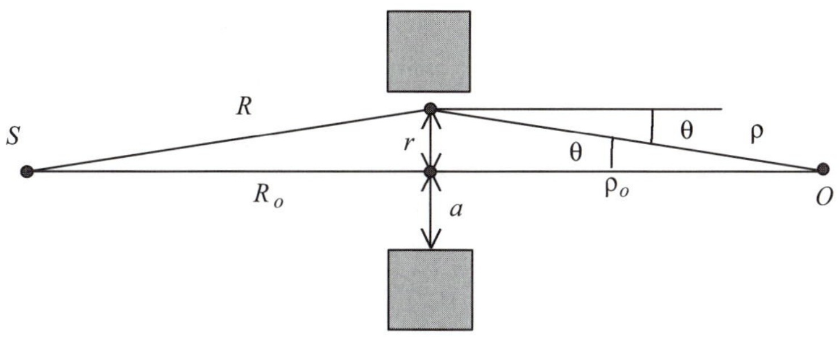

FIGURE 3.4 Coordinates for the circular opening. The source point and the observation point have the same distance from the aperture.

the axis of the system for the diffraction on a round aperture and a round stop. These calculations are taken from Sommerfeld's book on theoretical physics.[1]

3.2.2 On Axis Observation for the Circular Opening

The diffraction on a round opening is important since most lenses, spherical mirrors, and optical instruments have circular symmetry. We consider a round aperture of radius a with equal distance to the aperture of the source point S and the observation point O. In Figure 3.4 we show the coordinates and have $R_0 = \rho_0$, $R = \rho$, $\cos\theta = \rho_0/\rho$, and for the surface element $d\sigma = 2\pi r dr$. The amplitude at the observation screen is then obtained by the integral over the opening

$$u = A \int_{\text{opening}} \{(1/R\rho)\exp(ik(R+\rho))\}\{\rho_0/\rho\}\, 2\pi r dr. \tag{3.5}$$

The integration limits are from ρ_0 to $\sqrt{(a^2+\rho_0^2)}$. We get from $r^2 + \rho_0^2 = \rho^2$ that $rdr = \rho d\rho$ and have

$$u = A 2\pi\rho_0 \int_{\rho_0}^{\sqrt{(a^2+\rho_0^2)}} (1/\rho^2)\exp(ik2\rho)d\rho. \tag{3.6}$$

Integration by parts with $u = 1/\rho^2$, $dv = e^{ik2\rho}$, and $v = (1/2ik)e^{ik2\rho}$ results in

$$(1/\rho^2)(1/2ik)e^{ik2\rho} \Big\|_{\rho_0}^{\sqrt{(a^2+\rho_0^2)}} + (1/ik)\int_{\rho_0}^{\sqrt{(a^2+\rho_0^2)}} (1/\rho^3)e^{ik2\rho}d\rho.$$

Since ρ is large, we retain only the first term with the power of $1/\rho^2$ and have

$$u = (2\pi A\rho_0/2ik)\left(\left(e^{i2k\sqrt{a^2+\rho_0^2}}\right)/(a^2+\rho_0^2) - \exp\{i2k\rho_0\})/\rho_0^2\right). \tag{3.7}$$

[1] *Vorlesungen uber theoretische Physik*, Band IV, by A. Sommerfeld. Dieterich'sche Verlagsbuchhandlung, Wiesbaden, 1950.

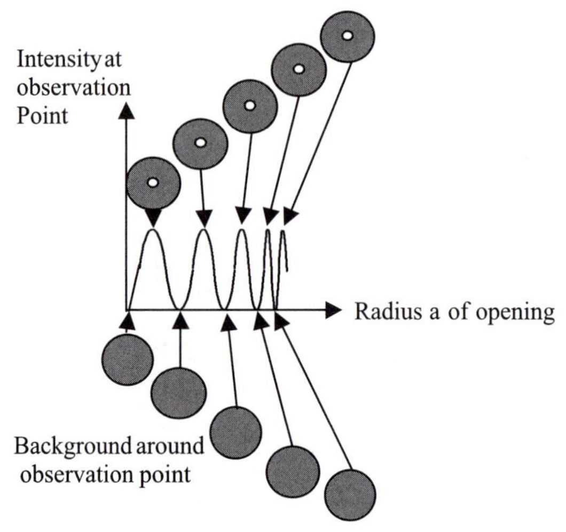

FIGURE 3.5 Diffraction pattern for a circular aperture at the observation point. The intensity has a maximum for certain values of the raduis a, shown as a white spot on the gray background. For other values of a the intensity is zero; only the gray background is shown.

Further simplification is obtained by assuming $\rho_0 \gg a$,

$$\sqrt{a^2 + \rho_0^2} \approx \rho_0(1 + a^2/2\rho_0^2), \tag{3.8}$$

and obtaining for the amplitude

$$u = (2\pi A\rho_0/2ik)(\exp\{i2k\rho_0\})(1/\rho_0^2)[\exp\{ik(a^2/\rho_0)\} - 1]. \tag{3.9}$$

The intensity uu^* is obtained after normalization as

$$I = I_0\lambda^2 \sin^2(ka^2/2\rho_0). \tag{3.10}$$

In Figure 3.5 we show maxima and minima for different radii a of the circular opening, at the center. Our calculation refers to the center spot only and the radius of the corresponding maxima or minima may be read from the graph in FileFig 3.1.

FileFig 3.1 (D1CIRS)

In FileFig 3.1 we show a graph of the intensity at the center. We use $\lambda = 0.0005$ mm, $\rho_0 = 4000$ mm, and radius a of 0.1 to 5 mm. With increasing diameter of the aperture, that is, with increasing a, we have at the center a change from maxima to minima to maxima and so on.

D1CIR is only on the CD.

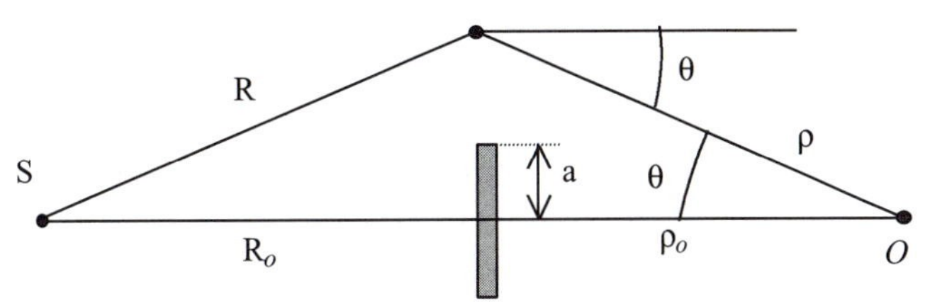

FIGURE 3.6 Coordinates for the circular stop. The source point and the observation point have the same distance from the aperture.

3.2.3 On Axis Observation for Circular Stop

In optics, it is often of interest to study complementary screens or arrays. We apply the Kirchhoff–Fresnel integral, Eq. (3.4), to a circular stop, as shown in Figure 3.4. Similar to the "opening," we must now evaluate

$$u = A2\pi\rho_0 \int_{\sqrt{(a^2+\rho^2)}}^{\infty} (1/\rho^2)\exp(ik2\rho)d\rho. \tag{3.11}$$

Integration by parts yields

$$(1/\rho^2)(1/2ik)e^{ik2\rho} \Big\|_{\sqrt{a^2+\rho_0^2}}^{\infty} + (1/ik)\int_{\sqrt{a^2+\rho_0^2}}^{\infty} (1/\rho^3)e^{ik2\rho}\, d\rho. \tag{3.12}$$

Neglecting the last integral we get

$$u = (2\pi A\rho_0/2ik)[-\{1/(a^2+\rho_0^2)\}]\{\exp(i2k\sqrt{a^2+\rho_0^2})\}. \tag{3.13}$$

Multiplication of u in Eq. (3.13) by u^* yields the intensity, and taking for the normalization $I_0 = A^2\rho_0^2/(a^2+\rho_0^2)$, we have

$$I = I_0\lambda^2/4. \tag{3.14}$$

The intensity of Eq. (3.14) depends only on the wavelength, not on the diameter of the aperture or the distance from it. We have the result that at any point in the shadow of an aperture stop, we will observe a bright spot.

There is the story that Fresnel presented his wave theory of light to the French Academy of Sciences. The famous Poisson questioned the validity and argued that there should be a light spot in the shadow of an illuminated sphere, for example, a steel ball. Another scientist of the Academy, Arago, made the experiment, observed the spot and presented his finding to the Academy in support of Fresnel's theory, but the spot remains the "Poisson spot." A photograph of the Poisson spot and an experimental setup for observation are shown in Figure 3.7.

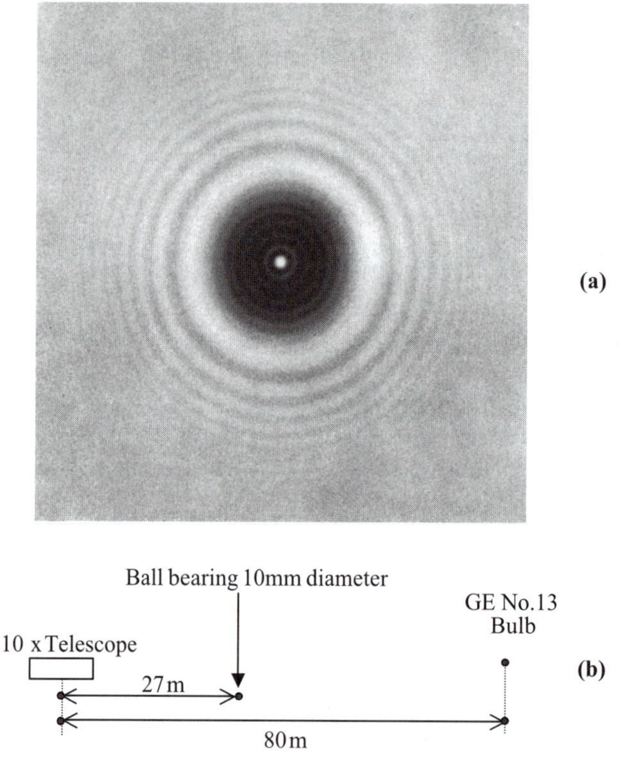

FIGURE 3.7 (a) Photograph of the diffraction pattern produced by a round stop. The Poisson spot appears in the middle (from Cagnet, Francon, Thrierr, *Atlas of Optical Phenomena*, Springer-Verlag, Heidelberg, 1962); (b) parameters for the observation of the Poisson spot (after R. Pohl. Einführung in die Optic, R.W. Pohl, Springer-Verlag, Heidelberg, 1948).

3.3 FRESNEL DIFFRACTION, FAR FIELD APPROXIMATION, AND FRAUNHOFER OBSERVATION

In the first two applications of the Kirchhoff–Fresnel integral, we have assumed that the source of light and the observation point are at large distances from the aperture. How large was not specified. When assuming that the distance is "infinitely large," so large that we essentially have plane waves incident on the aperture, we are at the approximation used in Chapter 2. When observing at a screen similarly far away from the aperture, the waves arriving there are also considered plane waves and are also parallel for their superposition. This is called *far field approximation*. When we use a lens and observe the diffraction pattern in the focal plane, we have Fraunhofer diffraction. The mathematical presentation of far field approximation and Fraunhofer diffraction is the same. In contrast, when the distance from the aperture to the source and observation screen is large but finite, we speak of Fresnel diffraction. We use small angle approximation to show the differences in these approaches to diffraction.

3.3.1 Small Angle Approximation in Cartesian Coordinates

Since the distance from aperture to observation screen is large, we may use small angle approximation for the diffraction angle. We consider the integral in Eq. (3.4),

$$\int_{\text{opening of aperture}} (A/R)\exp(ikR)(1/\rho)\exp(ik\rho)\cos\theta\, d\sigma \tag{3.15}$$

and neglect the $\cos\theta$ factor. The factor $(A/R)\exp(ikR)$ is a constant and can be taken before the integral. Using only a one-dimensional approach for the Y and y directions, we have further to consider,

$$u(Y) = C\int_{\text{opening of aperture}} (e^{ik\rho})/\rho\, dy. \tag{3.16}$$

The coordinates are shown in Figure 3.8, using X for the distance from the aperture to the screen. Since the distance X between the observation screen and the aperture is large, we take ρ in the denominator as a constant, but not in the exponential. We develop ρ using the coordinates of Figure 3.8 and have

$$\begin{aligned} \rho &= \left\{(Y-y)^2 + X^2\right\}^{1/2} \\ &= X + (1/2X)(Y-y)^2. \end{aligned} \tag{3.17}$$

Inserting Eq. (3.17) into (3.16) and including $1/\rho$ and e^{ikX} in a new constant C', we have

$$u(Y) = C'\int_{\text{opening of aperture}} \exp\{ik(1/2X)(Y-y)^2\}\, dy. \tag{3.18}$$

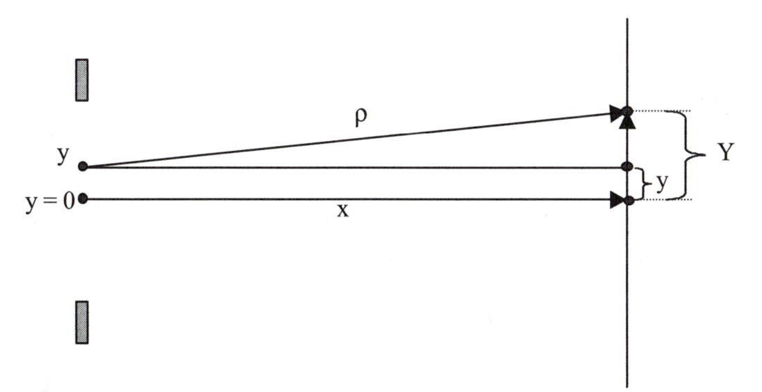

FIGURE 3.8 Coordinates for small angle approximation.

We write the exponent of Eq. (3.18) as

$$ik\{+(1/2X)(y^2) + (1/2X)(Y^2) - (yY)/X\}. \tag{3.19}$$

3.3.2 Fresnel, Far Field, and Fraunhofer Diffraction

3.3.2.1 Fresnel Diffraction

If we do not neglect the quadratic terms in Eq. (3.19), we are back to Eq. (3.18) and have

$$u(Y) = C' \int_{\text{opening of aperture}} \exp\{ik(1/2X)(Y - y)^2\}\, dy. \tag{3.20}$$

This is called Fresnel diffraction. The integral may be expressed using Fresnel's integrals.

3.3.2.2 Far Field Diffraction

In Eq. (3.19) we neglect the quadratic term in y and consider $(1/2X)(Y^2)$ as a constant and include it in C'', we have

$$u(Y) = C'' \int_{\text{opening of aperture}} e^{-ik(yY/X)}\, dy. \tag{3.21}$$

This is the far field approximation.

3.3.2.3 Fraunhofer Diffraction

Fraunhofer diffracton is also far field approximation, but we do not have to go so far, because we observe the pattern in the focal plane of a lens. In this case we have to find the effect on the wavefront when light is focused on the focal plane of a lens with focal length f. We obtain the result that one has the same integral for Fraunhofer diffraction as one has for far field approximation.

Small angle approximation was obtained by considering in the integral the exponent, Eq. (3.19),

$$ik\{+(1/2X)(y^2) + (1/2X)(Y^2) - (yY)/X\}. \tag{3.22}$$

For far field approximation we neglected the quadratic term in y and considered the term in X and Y as a constant.

Now we do not neglect the quadratic term in y and show that this term is compensated by the effect of the lens (for a detailed discussion see Goodman, 1988, p. 78.) We look at the wavefront passing through the lens. The wavefront

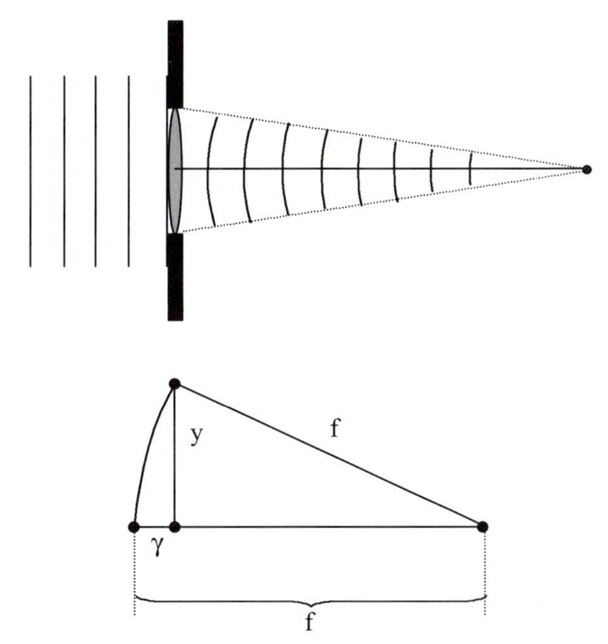

FIGURE 3.9 Coordinates for the calculation of the change of the wavefront by a lens.

is converging to the focal point of the lens (see Figure 3.9). Over the length y we have an increasing phase shift γ, which is calculated from

$$y^2 + (f - \gamma)^2 = f^2. \tag{3.23}$$

Neglecting γ^2 we have for the phase shift

$$\gamma = (y^2)/2f. \tag{3.24}$$

This shift must be subtracted as the phase shift in our integral. We get in the exponent, that is, Eq. (3.22), with $X = f$,

$$ik\{+y^2/2f + (1/2f)(Y^2) - (yY)/f\} - ik(y^2)/2f. \tag{3.25}$$

The new quadratic term in y cancels the old one, which we neglected in the far field approximation. We then have to consider

$$u(Y) = C'' \int e^{-ik(yY/X)}\, dy. \tag{3.26}$$

This integral is the same as obtained in the far field approximation.

3.4 FAR FIELD AND FRAUNHOFER DIFFRACTION

The far field diffraction and Fraunhofer diffraction of the Kirchhoff–Fresnel integral have the same mathematical appearance. The only difference is that in far field approximation the diffraction pattern is observed on a faraway screen, whereas in Fraunhofer diffraction the observation screen is placed at the focal plane of a lens and that may be closer to the aperture.

We now discuss the diffraction pattern of various geometrical shapes of apertures. From Eq. (3.4) we have

$$u(Y) = C \int (e^{ik\rho})\, d\sigma, \tag{3.27}$$

where $d\sigma$ is the surface element of the aperture and $1/\rho$ has been taken before the integral, included in C, which contains all constant terms. Note that in Eq. (3.27) we have not used small angle approximation in the exponent.

3.4.1 Diffraction on a Slit

The diffraction on a slit is important because it is a simple one-dimensional diffraction problem and appears in the diffraction pattern of all types of gratings and in other diffraction-related phenomena. The coordinates for the calculation of the diffraction on a slit are shown in Figure 3.10. We divide the opening into N intervals Δy and sum up all waves traveling in direction θ. In the first step it is assumed that these waves are generated at the limits of all intervals and all adjacent waves have the same optical path difference δ. This is similar to the discussion in Chapter 2 for interference on an array (Eq. (2.107)). The optical path difference between waves of finite steps is $\Delta y \sin\theta$ and replacing $q\delta$ in the sum of Eq. (2.107), we have

$$\sum e^{-ik}(\Delta y \sin\theta). \tag{3.28}$$

Making the step Δ_y infinitesimally small, one gets the integral

$$u(Y) = C \int e^{-ik(y \sin\theta)}\, dy, \tag{3.29}$$

where C includes all constant terms. The integration is from $-d/2$ to $d/2$, and we have to calculate

$$u(Y) = C \int_{y_1=-d/2}^{y_2=d/2} e^{-ik(y \sin\theta)}\, dy, \tag{3.30}$$

or in small angle approximation with $\sin\theta = Y/X$,

$$u(Y) = C \int_{y_1=-d/2}^{y_2=d/2} e^{-ik(yY/X)}\, dy. \tag{3.31}$$

The result of the integration of Eq. (3.31) is

$$u = Cd \sin(\pi d \sin\theta/\lambda)/\{(\pi d \sin\theta)/\lambda\}. \tag{3.32}$$

The normalized intensity is written

$$I = I_O\{\sin(\pi d \sin\theta/\lambda)/\{(\pi d \sin\theta/\lambda)\}^2. \tag{3.33}$$

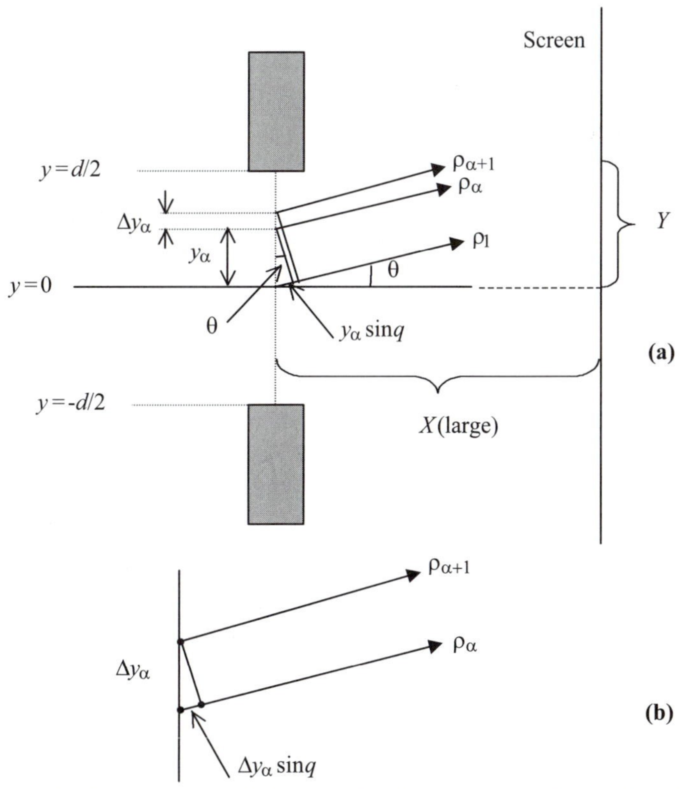

FIGURE 3.10 Coordinates for the calculation of the diffraction on a slit: (a) phase difference of all wavelets with respect to the center one used for the summation process; (b) path difference between ρ_α and $\rho_{\alpha+1}$.

Or in small angle approximation,

$$I = I_O[\{\sin(\pi dY/X\lambda)\}/\{\pi dY/X\lambda\}]^2. \tag{3.34}$$

The diffraction pattern of a slit has a periodic appearance with decreasing intensity of the maxima. The graph in FileFig 3.2 shows three diffraction patterns. The wider ones are for the smaller slit openings. In Figure 3.11 we show a photograph of a diffraction pattern of a slit. In Application FF2 the width of the diffraction pattern with respect to changes in λ and d is studied.

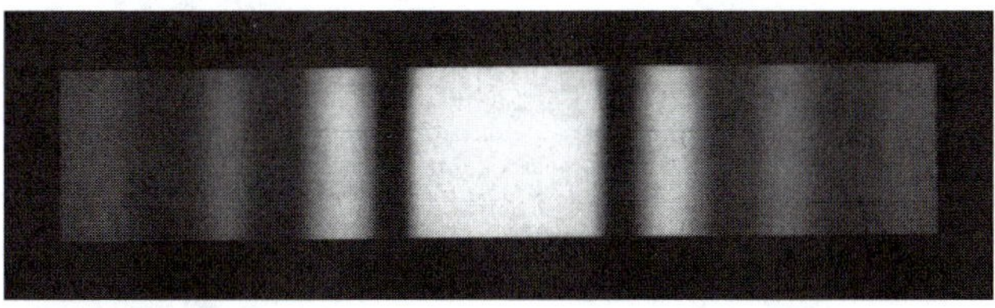

FIGURE 3.11 Diffraction pattern formed by a single slit (from M. Cagnet, M. Francon, J.C. Thrierr, *Atlas of Optical Phenomena*, Springer-Verlag, Heidelberg, 1962).

The main maxima is at $Y = 0$. At that point we have $\sin 0/0$, and a similar discussion to that presented in Chapter 2 results in $I = I_O$. The angle from the center of the slit to the first minimum of the diffraction pattern is called the diffraction angle $\theta = \lambda/d$, and is used when discussing resolution or the Fresnel number for characterizing the losses of a laser cavity. The side maxima are approximately at $Y/X = (m + 1/2)\lambda/d$, which is approximately centered between the minima. An exact determination is done using FileFig 3.3 and Application FF3.

FileFig 3.2 (D2FASLITS)

A graph of the intensity of the diffraction pattern on a slit. By changing the width d, we see that the width of the diffraction pattern is inversely proportional to the width of the slit. By changing λ the width of the diffraction pattern is proportional to the wavelength. This is a general property one observes for diffraction patterns. The minima are at $m\lambda/d$.

D2FASLITS

Diffraction on a Slit of Width d at Wavelength λ

X is distance; slit-screen, Y is coordinate on screen. For small angles, Y/X is proportional to the diffraction angle θ. MCAD notice the singularity at 0. For the graph we get around it using the range $Y = -100.1, -99.1$ to 100.1. All lengths are in mm.

Three slits with different widths $d1$, $d2$, and $d3$:

$$d1 \equiv .08 \qquad d2 \equiv .12 \qquad X := 4000 \quad \lambda \equiv .0005$$

$$I1(\theta) := \left[\frac{\sin\left(\pi \cdot \frac{d1}{\lambda} \cdot \sin\left(\frac{2\cdot\pi}{360} \cdot \theta\right)\right)}{\left[\pi \cdot \frac{d1}{\lambda} \cdot \left(\frac{2\cdot\pi}{360} \cdot \theta\right)\right]}\right]^2$$

$$I2(\theta) := \left[\frac{\sin\left(\pi \cdot \frac{d2}{\lambda} \cdot \sin\left(\frac{2\cdot\pi}{360} \cdot \theta\right)\right)}{\left[\pi \cdot \frac{d2}{\lambda} \cdot \left(\frac{2\cdot\pi}{360} \cdot \theta\right)\right]}\right]^2$$

$$d3 \equiv .16 \qquad I3(\theta) := \left[\frac{\sin\left(\pi \cdot \frac{d3}{\lambda} \cdot \sin\left(\frac{2\cdot\pi}{360} \cdot \theta\right)\right)}{\left[\pi \cdot \frac{d3}{\lambda} \cdot \left(\frac{2\cdot\pi}{360} \cdot \theta\right)\right]}\right]^2$$

$$\theta \equiv -2, -1.99 \ldots 1.$$

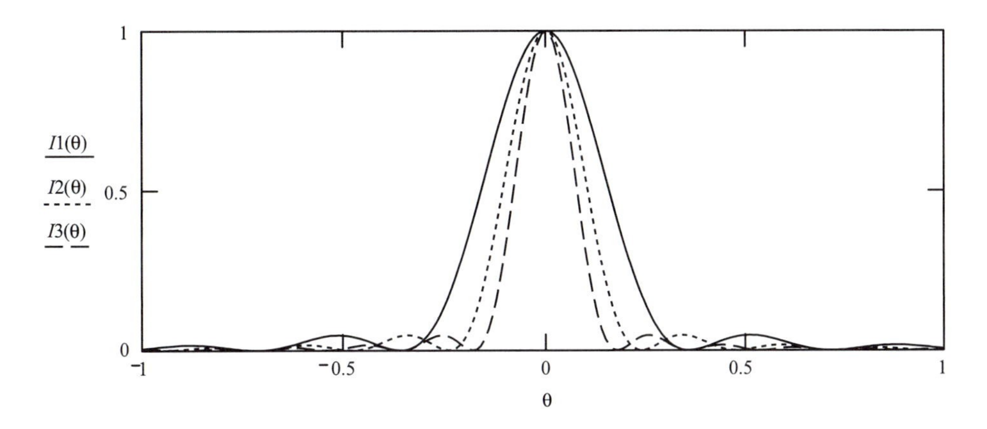

Application 3.2. The dependence of the width of the diffraction pattern on values of λ and d may be studied by changing λ to $\lambda/2$ and 2λ and d to $d/2$ and $4d$ and considering the ratio of d/λ.

FileFig 3.3 (D3FASLITEXS)

Expanded graph of the intensity of the side maxima and minima for the diffraction pattern on a slit, $Y = 18, 19 \ldots 150$, $X = 4000$, and $\lambda = 0.0005$. Numerical determination of the values of the side maxima and minima.

D3FASLITEXS is only on the CD.

Application 3.3.

1. Give the position of the first five minima.
2. Determine the maxima. The values of the secondary maxima are obtained by differentiation of $I = I_O[\sin\{\pi dY/X\lambda\}/\{\pi dY/X\lambda\}]^2$ with respect to Y and setting the resulting expression equal to 0. One may perform the differentiation with a symbolic computer calculation program. One obtains the transcendental equation $\pi yd/\lambda = \tan(\pi yd/\lambda)$, where $y = Y/X$. The solution of the transcendental equation may be obtained by plotting $\pi yd/\lambda$ and $\tan(\pi yd/\lambda)$ on the same graph and using the intersections. Determine the values of the first five intersections, corresponding to the first five side maxima, and compare with the values read from the graph in FF3 and with the approximate formula $Y/X = (m + 1/2)\lambda/d$.
3. Calculate the intensity ratio of the first, second, and third maxima to the zeroth maximum and compare with the theoretical values from the intensity formula for the diffraction on a slit.

3.4.2 Diffraction on a Slit and Fourier Transformation

The integral for the calculation of the diffraction on a slit in small angle approximation (Eq. (3.31)), is

$$u(Y) = C \int_{y_1=-d/2}^{y_2=d/2} \exp -i2\pi(y/\lambda)(Y/X)\, dy, \tag{3.35}$$

where we used $k = 2\pi/\lambda$.We do the following substitutions

$$v = (y/\lambda), \qquad x = Y/X, \qquad a = d/2\lambda \tag{3.36}$$

and have

$$u(x) = C \int_{z_1}^{a_2} \exp[-i2\pi(\nu)(x)]d\nu. \tag{3.37}$$

To write the integral with integration limits from $-\infty$ to ∞ we define the function $Q(\nu)$ as

$$\begin{aligned} Q(v) &= 1 \text{ for } x \text{ between } -a \text{ and } a \\ Q(v) &= 0 \text{ otherwise.} \end{aligned} \tag{3.38}$$

We then have

$$u(x) = C \int_{-\infty}^{\infty} Q(\nu) \exp -i2\pi(\nu)(x)\, d\nu. \tag{3.39}$$

The integral $u(x)$ in Eq. (3.39) is the Fourier transform of $Q(\nu)$. We may integrate and obtain

$$u(x) = C'(\sin 2\pi ax)/(2\pi ax) \tag{3.40}$$

similar to that obtained in Eq. (3.40). We have the result that the Fourier transform of the slit function $Q(v)$ with opening width a is the function $(\sin 2\pi ax)/(2\pi ax)$ which is sometimes called a sinc-function. When $Q(\nu)$ is not the slit function, but given as a numerical function or a complicated analytical function, one can not obtain an analytical expression for $u(x)$ but one can calculate the numerical Fourier transformation. Most computational programs offer Fourier transformation. In FileFig 3.4 we write a step function for x_i with $i = 0$ to 255, assuming that $x_i = 1$ for 0 to d and otherwise 0, and plot x_i as a function of $i/255$. Here we consider only half of the slit and do the Fourier transformation c_j, shown as the graph of c_j depending on $j/255$, plotted from 0 to 0.5. Since we cover with the input data only half of the slit, we get only half of the diffraction pattern. However, because the Fourier transformation is real, and we have used the fast Fourier transformation (FFT), the Fourier transformation c_j shows only $N = 128$ points. The inverse Fourier transformation results again in 256 points. More details on this subject are given in Chapter 9 on Fourier transformations.

FileFig 3.4 (D4FASLITFT)

Fourier transformation of a step function. The step function has been defined for a width of $d = 20$. The number of points to be used is $2^n - 1$. The real and imaginary parts of the Fourier transformation are shown. The Fourier transformation of the Fourier transformation is also calculated and the real part is again a step function, and the imaginary part is 0.

D4FASLITFT is only on the CD.

Application 3.4. For several widths of the step, read off the value of the first zero of the transform. A formula for the value of the first zero may be obtained from Eq. (3.40). Compare with the value of the graph of the Fourier transformation and with Application FF3.

3.4.3 Rectangular Aperture

The diffraction pattern of the rectangular aperture is easy to calculate as an extension from one dimension, as used for the slit, to two dimensions. In the next section we do it in small angle approximation and show that the integral is useful for the calculation of far field diffraction on a round aperture, important for all optical devices and instruments with circular symmetry.

For the calculation of the diffraction pattern of a rectangular aperture with dimensions d in the y direction and a in the z direction, we go back to the integral of Eq. (3.31).

$$u(Y) = \int_{y_1=-d/2}^{y_2=d/2} e^{-ik(yY/X)}\, dy.$$

Integration over the y direction will be extended to include the z direction; see Figure 3.12. As a result, the diffraction pattern is described by a product of two integrals of the type of Eq. (3.31), one over the opening d in the y direction and the other over the opening a in the z direction.

$$u(Y, Z) = \int_{1=-d/2}^{y_2=d/2} \exp -i2\pi(y/\lambda)(Y/X)\, dy \cdot \int_{z_1=-a/2}^{z_2=a/2} \exp -i2\pi(z/\lambda)(Z/X)\, dz. \tag{3.41}$$

We obtain

$$u(Y, Z) = C\{d \sin(\pi dY/X\lambda)/(\pi dY/X\lambda)\{a \sin(\pi aZ/X\lambda)/(\pi aZ/X\lambda)\} \tag{3.42}$$

and we have for the normalized intensity

$$I = I_o[\{\sin(\pi dY/\lambda X)\}/\{\pi dY/\lambda X\}]^2[\{\sin(\pi aZ/\lambda X)\}/\{\pi aZ/\lambda X\}]^2. \tag{3.43}$$

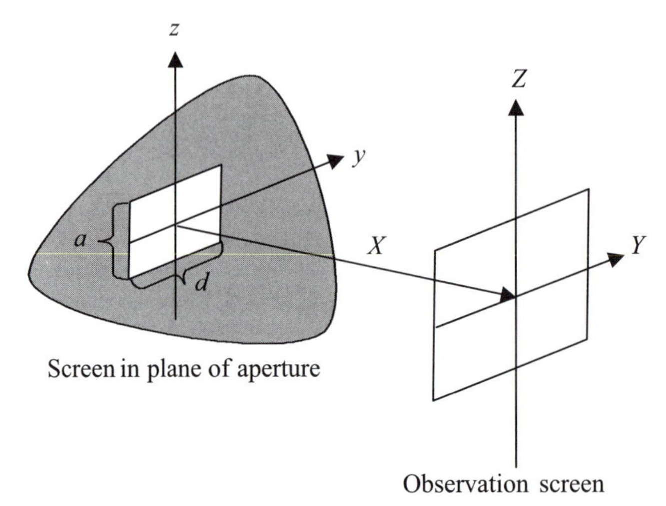

FIGURE 3.12 Coordinates for the calculation of the diffraction pattern of a rectangular aperture.

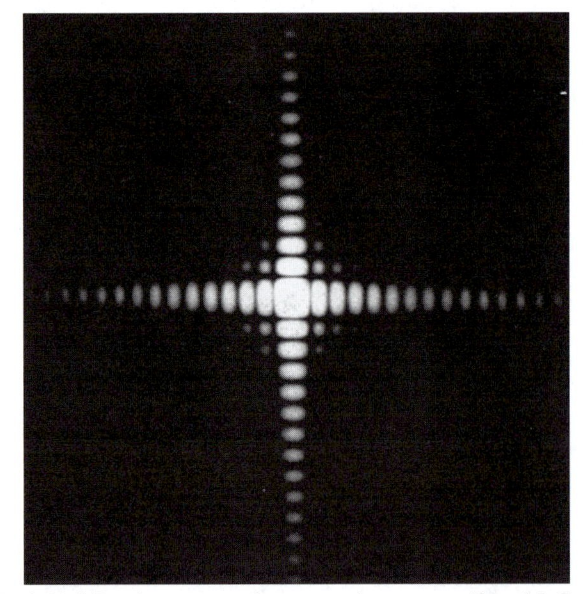

FIGURE 3.13 Far field diffraction pattern of a rectangular aperture (from M. Cagnet, M. Francon, and J.C. Thrierr, *Atlas of Optical Phenomena*, Springer-Verlag, Heidelberg, 1962).

In FileFig 3.5 we have a 3-D graph of the calculated intensity diffraction pattern with maxima and minima in the y and z directions. The diffraction angle, that is, the angle from the center of the aperture to the first minimum, is in the y direction $Y/X = \lambda/d$, and in the z direction $Z/X = \lambda/d$. A photograph of the diffraction pattern is shown in Figure 3.13.

FileFig 3.5 (D5RECTS)

A 3-D chart of the diffraction pattern of a rectangular aperture. By changing the lower limits of x and y and enlarging N, one may get a more densely lined pattern. Changing d and a to larger values will result in a narrower pattern. If d is equal to a, we have a square pattern.

D5FARECTS

Diffraction Pattern of a Rectangular Aperture

The width in the x-direction is d, in the y-direction, a. One may look at the plot from different angles, change colors, and make a contour plot.

$$i := 0 \ldots N \qquad j := 0 \ldots N$$

$$x_i := (-6) + .20001 \cdot i \qquad y_j := 6 + .20001 \cdot j$$

$$\lambda \equiv 4$$

$$M_{i,j} := f(x_i, y_j) \qquad f(x, y) := \left[\frac{\sin\left(2 \cdot \pi \cdot d \cdot \frac{x}{2 \cdot \lambda}\right)}{\left(2 \cdot \pi \cdot d \cdot \frac{x}{2 \cdot \lambda}\right)}\right]^2 \cdot \left[\frac{\sin\left(2 \cdot \pi \cdot a \cdot \frac{y}{2 \cdot \lambda}\right)}{\left(2 \cdot \pi \cdot a \cdot \frac{y}{2 \cdot \lambda}\right)}\right]^2$$

$$N \equiv 60 \qquad d \equiv 3 \qquad a \equiv 2.$$

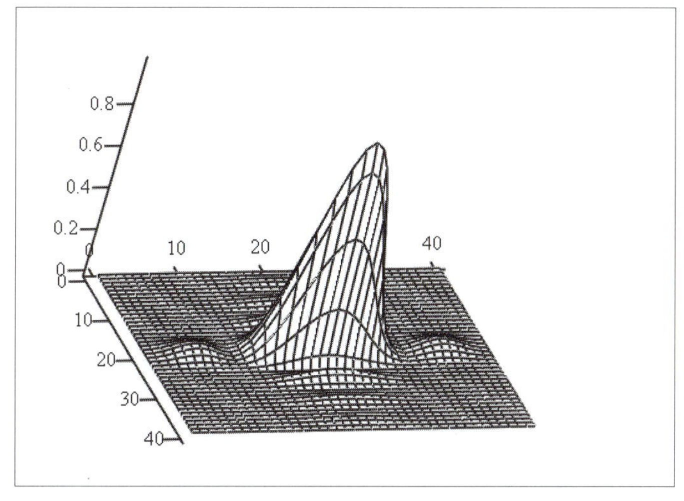

M

Application 3.5.

1. Using a 3-D contour and a 3-D surface plot and making changes in d and a, one may study the intensity of the diffraction of the rectangular aperture for
   - **a.** A square aperture;
   - **b.** A long strip aperture which should have the pattern of a slit on one side.

2. One may change the ratio of d/λ and a/λ and observe a wider or more narrow diffraction pattern.

3.4.4 Circular Aperture

Diffraction on a circular aperture is present on all optical devices and instruments with circular symmetry. Although diffraction seems to be a minor effect, the size of astronomical telescope mirrors is large in order to reduce the limitations of image quality by diffraction. Even the Mount Palomar telescope mirror with a diameter of about 5 m reduces the image quality of a star by diffraction.

For the calculation of the diffraction pattern of a circular aperture we look at the integral for the rectangular aperture (Eq. (3.41)) and integrate over a circular opening

$$u(Y, Z) = \int_{\text{circular opening}} (e^{-ik(zZ+yY)/X})\,dz\,dy. \tag{3.44}$$

Since the problem is circular symmetric, one changes the coordinate system for the mathematical treatment to the coordinate system shown in Figure 3.14.

$$z = r\cos\phi \qquad Z = R\cos\psi \tag{3.45}$$

$$y = r\sin\phi \qquad Y = R\sin\psi, \tag{3.46}$$

where

$$0 \le r \le a, \qquad 0 \le \phi \le 2\pi, \qquad 0 \le \psi \le 2\pi, \tag{3.47}$$

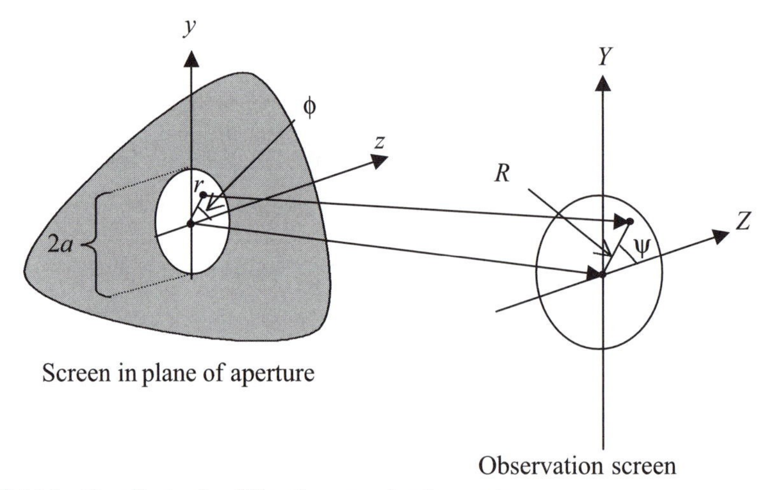

FIGURE 3.14 Coordinates for diffraction on a circular aperture.

and $k = 2\pi/\lambda$. For the integral we obtain

$$u(r, \phi) = u_o \int_{-\pi}^{+\pi} \int_0^a e^{-i2\pi(rR/\lambda X)\cos(\psi-\phi)} r\, dr\, d\phi. \tag{3.48}$$

This integral can be expressed with the Bessel function of zero order

$$J_o(q) = 1/2\pi \int_{-\pi}^{+\pi} e^{iq\cos(\psi-\phi)}\, d\phi, \tag{3.49}$$

where $q = 2\pi(rR/\lambda X)$ and therefore $r = (\lambda X/2\pi R)q$, and we get for Eq. (3.48),

$$u = u_o(\lambda X/2\pi R)^2 2\pi \int_{q=0}^{q=2\pi(aR/\lambda X)} J_o(q) a\, dq. \tag{3.50}$$

Using the relation

$$\int_0^{q'} J_o(q) q\, dq = q' J_1(q') \tag{3.51}$$

one has for the intensity

$$I = I_0\{J_1(2\pi(aR/\lambda X))/(2\pi(aR/\lambda X))\}^2, \tag{3.52}$$

where I_0 is the normalization constant.

The three graphs in FileFig 3.6 show that we have a narrow diffraction pattern for large diameters and vice versa as we found for the diffraction on a slit. The

FIGURE 3.15 Diffraction pattern of a round aperture (from M. Cagnet, M. Francon, and J.C. Thrierr, *Atlas of Optical Phenomena*, Springer-Verlag, Heidelberg, 1962).

angle from the center of the circular aperture to the first minimum of the diffracted intensity (Eq. (3.52)), is the diffraction angle, equal to 1.22 $\lambda/2a$ where $2a$ is the diameter of the opening. Comparing to a slit, we obtain λ/d, where d is the width of the slit. One has for the slit the factor 1, instead of 1.22 for the round aperture. The determination of the factor 1.22 is done in Application FF8. The factor 1.22 appears in catalogues of optical devices to specify limitations by diffraction. A photograph of the diffraction pattern of a round aperture (Airy disc) is shown in Fig. 3.15 and a 3-D graph in FileFig 3.7.

FileFig 3.6 (D6FARONS)

A graph of the intensity of the diffraction pattern of a round aperture. By changing the radius a of the aperture, we see that the width of the diffraction pattern is inversely proportional to the diameter of the round opening. This is a general property one observes for the diffraction pattern, as well as the corresponding Fourier transformation. The size of the pattern is proportional to the wavelength.

D6FARON is only on the CD.

Application 3.6.

1. Do the normalization of Eq. (3.52) by dividing $I(R)$ by $I(R = 0)$.
2. The dependence of the width of the diffraction pattern on values of (and a may be studied by changing λ to $\lambda/2$ and 2λ and $2a$ to $2a/2$ and 2 times $2a$ and considering the ratio of $2a/\lambda$. (The diameter of the round aperture $2a$ is used here for comparison with the slit width d.)

FileFig 3.7 (D7FARON3DS)

A 3-D graph of the round aperture.

D7FARON3DS

3-D Diffraction Pattern of a Round Aperture as a Circular Symmetric Plot Using Two Coordinates

Radius of aperture is a. The coordinate on the observation screen in R. Wavelength λ, distance from aperture to screen, is X. One may look at the plot from different angles, change colors, and make a contour plot.

$$i := 0 \ldots N \qquad j = 0 \ldots N$$

$$x_i := (-40) + 2.0001 \cdot i \qquad y_j := -40 + 2.0001 \cdot j \qquad \lambda \equiv .0005$$

$$R(x, y) := \sqrt{(x)^2 + (y)^2}$$

$$N \equiv 60 \qquad X := 4000 \qquad a \equiv .05$$

$$g(x, y) := \left[\frac{J1\left(2 \cdot \pi \cdot a \cdot \frac{R(x,y)}{X \cdot \lambda}\right)}{\left(2 \cdot \pi \cdot a \cdot \frac{R(x,y)}{X \cdot \lambda}\right)} \right]^2$$

$$M_{i,j} := g(x_i, y_j).$$

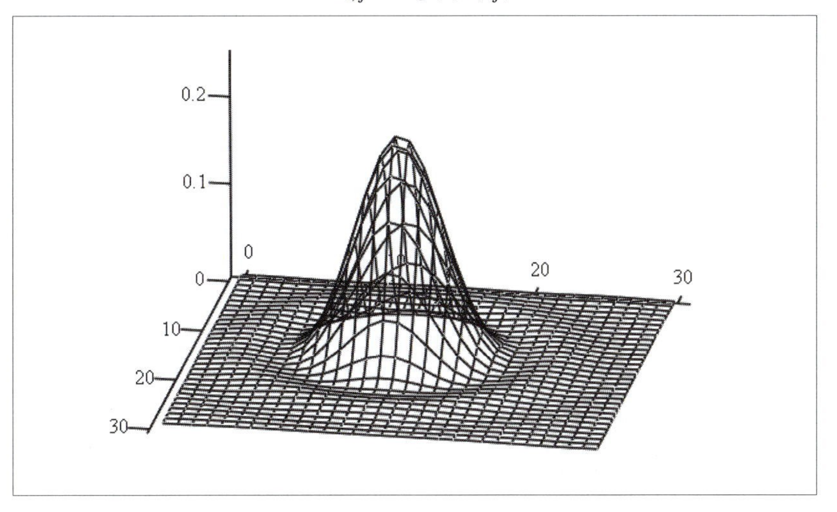

M

Application 3.7. Change the radius of the aperture and wavelength and make both twice as large and reduce both to $\frac{1}{2}$. Consider the ratio of $2a/\lambda$ and change it to twice the value and to $\frac{1}{2}$.

FileFig 3.8 (D8RONEXS)

Graph of diffraction pattern of the round aperture for $R = 3$ to 10, $X = 1000$, and $\lambda = 0.01$ for the determination of the diffraction angle of a round aperture. The first minimum is at $1.22\lambda/2a$.

D8RONEX is only on the CD.

Application 3.8.

1. Modify FileFig 3.8 and plot the Bessel function $J1(q)$ depending on q for 0 to 20. Normalize the Bessel function and determine the first zero at around $q = 3.9$ to five digits.
2. Make a graph of the diffraction pattern as described in FileFig 3.8 and determine the first minimum ($R/X = (\lambda/a)(q/2\pi) = 0.003$).
3. We saw that for a slit the diffraction angle was $Y/X = (\lambda/d)$ or $(Y/X)/(\lambda/d) = 1$. We want to calculate the diffraction angle for the round aperture. Calculate the first zero of $I(R) = J1(2\pi aR/\lambda X)/(2\pi R/\lambda X$ for

$\lambda = 0.01$ mm, $X = 1000$ mm, and $a = 1.5$ mm, and call it RR. Insert it into $R/X = (RR/2\pi)(\lambda/a)$, but leave (λ/a) symbolically. Use the diameter of the opening $d = 2a$ and calculate $(R/X)/(\lambda/d)$ and get the value 1.22.

4. For comparison of the diffraction angle for a slit and a round aperture, plot on the same graph for the same choice of parameters the diffraction pattern of a slit and the diffraction pattern of a round aperture. Take the same value for the width of the slit and the diameter of the aperture; that is, $2a = d$. Compare the values of the first minima and observe that the overall width of the diffraction pattern of the circular aperture is larger than the one for the slit. The height for the first minimum of the round aperture is smaller than that for the slit. Note that the plot for the slit is a linear plot whereas the plot for the round aperture is a radial plot.

3.4.5 Gratings

Gratings are used in spectrometers from the near-infrared to the X-ray region. They are usually reflection gratings with a zig-zag profile and called echelette gratings. Simple transmission gratings may be produced as plastic films as replicas of reflection gratings. We discuss the amplitude transmission grating with larger or smaller transmitting areas at normal incidence or under an angle. We also discuss a transmission echelette grating with the incident light at a specific angle. In the appendix we discuss the step grating because of its potential for use in Fourier transform spectroscopy.

3.4.5.1 Amplitude Grating

Amplitude gratings are periodic structures with alternating transmissive and opaque strips. Similar to what we did for single apertures, we now have to integrate over the open areas of the periodic set of slits (Figure 3.16). The integration over one slit

$$u(\theta) = \int_{y_1=-d/2}^{y_2=d/2} e^{-ik(y\sin\theta)}dy \tag{3.53}$$

is now extended to many slits; that is, we have a summation as

$$\begin{aligned} U(\theta) = &\int_{-d/2}^{d/2} e^{-ik(y_1\sin\theta)}dy_1 + \int_{a-d/2}^{a+d/2} e^{-ik(y_2\sin\theta)}dy_2 \\ &+ \int_{2a-d/2}^{2a+d/2} \ldots dy_3 +, \ldots. \end{aligned} \tag{3.54}$$

We substitute into the second integral $y_1 = y_2 - a$, similarly into the third integral $y_1 = y_3 - 2a$, and so on. Changing the integration variable to y in all integrals

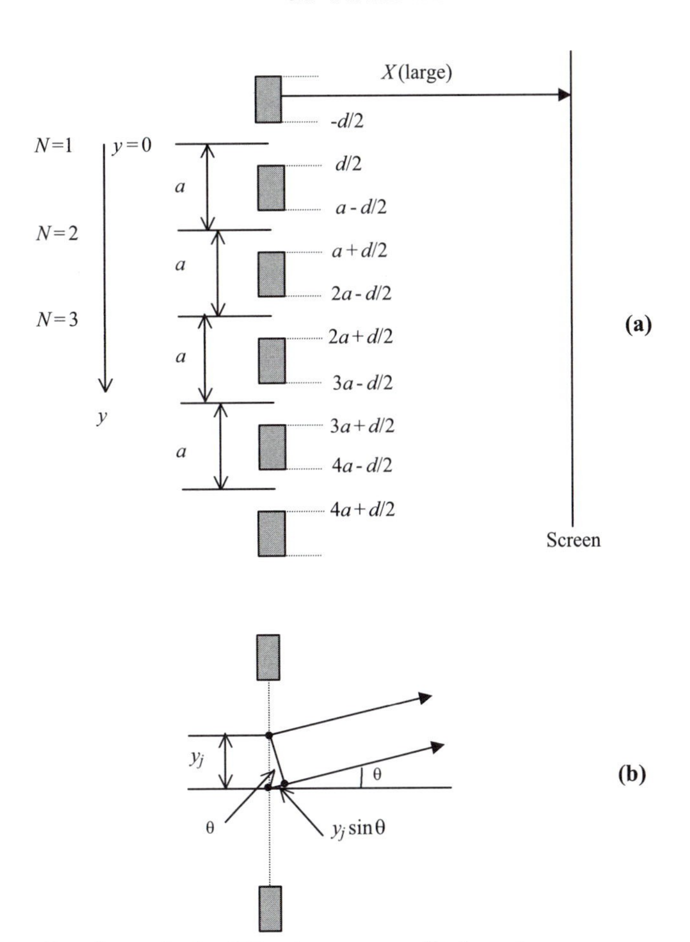

FIGURE 3.16 Coordinates for the diffraction on an amplitude grating.

we get

$$\{1 + e^{-ika\sin\theta} + e^{-ik2a\sin\theta} + e^{-ik3a\sin\theta} + \cdots e^{-ik(N-1)a\sin\theta} \cdot \int_{-d/2}^{d/2} e^{-ik(y\sin\theta)} dy. \tag{3.55}$$

The bracket contains the amplitude array factor or interference factor (Ia) (Chapter 2.4) multiplied by the amplitude diffraction factor (Da), equal to the diffraction on a single slit. Symbolically we have

$$\text{Diffraction amplitude } Pa = \text{Interference factor } Ia^{*}\text{Diffraction factor } Da \tag{3.56}$$

or

$$Pa(\theta) = Ia(\theta)Da(\theta). \tag{3.57}$$

We sum up the interference factor

$$Ia(\theta) = \sum_{m=0}^{N-1} e^{-ikma\sin\theta} = (1 - e^{-ikNa\sin\theta})/(1 - e^{-ika\sin\theta})$$

and have with Eq. (2.116) for the normalized intensity

$$= \{\sin(\pi Na\sin\theta/\lambda)/(N\sin(\pi a\sin\theta/\lambda))\}. \tag{3.58}$$

The intensity is obtained as $P(\theta) = Pa(\theta)Pa(\theta)^*$,

$$P(\theta) = \{[\sin(\pi d\sin\theta\lambda)]/[\pi d\sin\theta/\lambda)]\}^2 \cdot \{\sin(\pi Na\sin\theta/\lambda)/(N\sin(\pi a\sin\theta/\lambda))\}^2, \tag{3.59}$$

and in the small angle approximation

$$P(Y) = \{[\sin(\pi dY/X\lambda)]/[(\pi dY/X\lambda)]\}^2 \cdot \{\sin(\pi NaY/X\lambda)/(N\sin(\pi aY/X\lambda))\}^2. \tag{3.60}$$

We have normalized the resulting pattern by division with $1/N^2$. The intensity $P(\theta)$ of interference and diffraction of the amplitude grating is shown in the graphs of FileFig 3.9 where we used values such as .001 instead of.000 in the numerical calculations when approaching 0/0 at $\theta = 0$.

The first graph shows separately the numerator $y(\theta)$ of the interference factor and on the same graph the intensity $I(\theta)$. The numerator $y(\theta)$ is zero for the main maxima and all side minima of $I(\theta)$. At the main maximum both the numerator and the denominator are zero (i.e., one has 0/0), which results in 1, as discussed for the interference factor in Chapter 2. The second graph of FileFig 3.9 shows the zeroth order (main maximum) of the interference factor, at $\theta = 0$. There are N-1 side minima and N-2 side maxima between main maxima. Two side maxima do not appear. The second graph $P(\theta)$ shows the interference and diffraction factors separately. The interference factor describes the maxima and minima, and the envelope of the diffraction factor limits the intensity of the peaks of the interference factor. The graph in FileFig 3.10 shows the corresponding pattern for a slit opening 10 times smaller than the periodicity constant a.

For integer ratios of a/d, some maxima of the pattern can be seen while others are suppressed. Taking as an example $a/d = 2$, the zeroth and first orders of the pattern are seen, and the second order is suppressed. If a/d has the integral value nth, more orders may be observed and the resulting pattern becomes wider and the nth order is suppressed.

A photograph of the diffraction pattern for the case of $N = 2$, 3, 4, and 5 is shown in Fig. 3.17. One observes one side maximum for $N = 3$, two for $N = 4$, and three for $N = 5$.

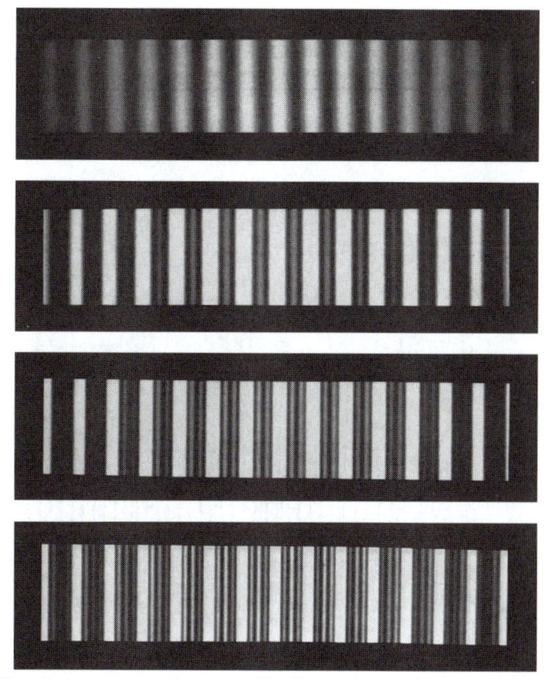

FIGURE 3.17 Diffraction pattern of an amplitude grating: (a) $N = 2$; (b) $N = 3$; (c) $n = 4$; (d) $N = 5$ (from M. Cagnet, M. Francon, and J. C. Thrierr, *Atlas of Optical Phenomena*, Springer-Verlag, Heidelberg, 1962).

FileFig 3.9 (D9FAGRAMPS)

A graph for the amplitude grating is plotted using $\lambda = 0.0005$, $d = 0.001$, $a = 0.002$, and $N = 6$. We have plotted the intensity of the interference and diffraction factor separately as well as the product. The numerator of the interference pattern is shown and, one has 0/0 for all main maxima. The diffraction factor corresponds to a slit pattern. The X scale has its origin at zero and there we have the zeroth order (main maxima) of the interference factor. We have normalized the resulting pattern by division with $1/N^2$ and fixed the $0/0$ problem at $X = 0$ by using for θ values such as 0.001 instead of 0.000. There are N-1 side minima and N-2 side maxima between the two main maxima.

D9FAGRAMPS

Diffraction on an Amplitude Grating at Normal Incidence

Width of openings d, center-to-center distance of strips a, wavelength λ, distance from grating to screen X, and coordinate on screen Y. All distances and wavelengths are in mm; number of lines N. All parameters are globally defined above the graph. $D(A)$ is the diffraction factor. $I(A)$ is the interference factor normalized to 1. The numerator is plotted separately to show where the main maxima are located (0, 0). $P(A)$ is the product of interference and diffraction factor.

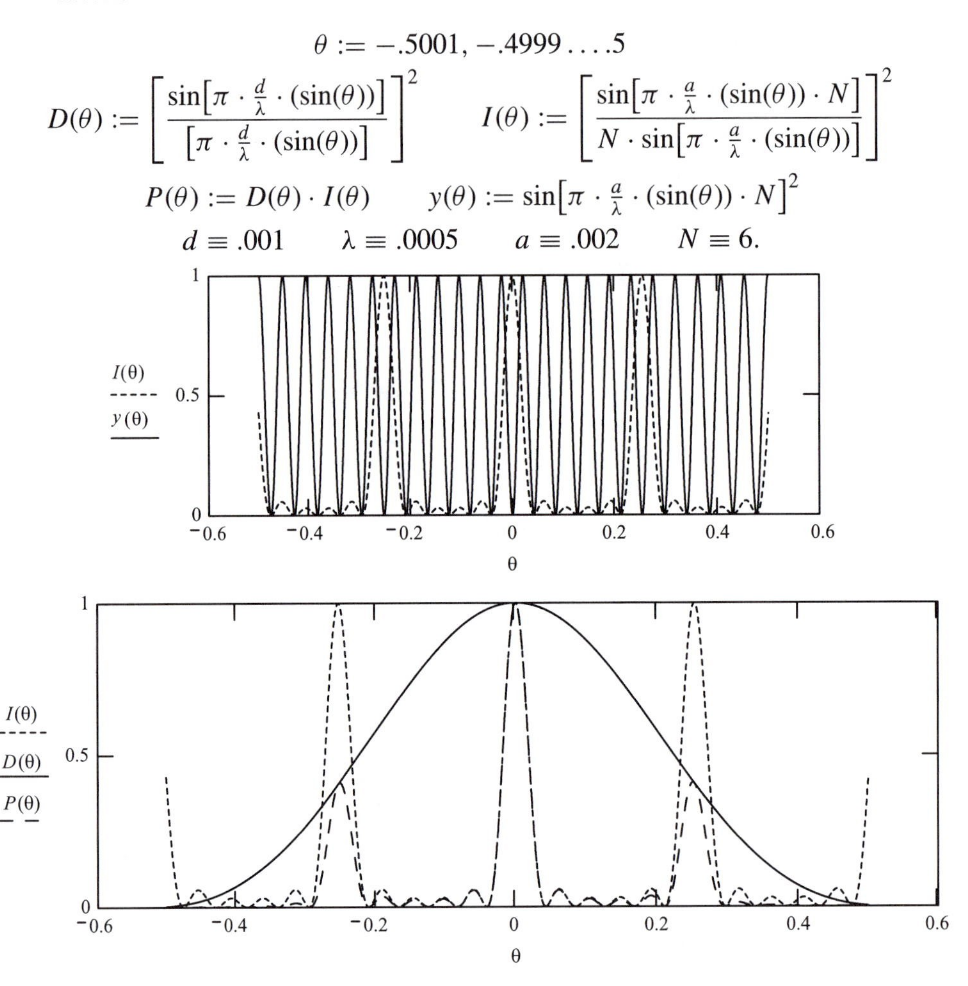

Application 3.9.

1. Change FileFig 3.9 to small angle approximation. Make a graph and initially use (all in *mm*) $d = 0.01$, $\lambda = 0.0005$, $a = 0.02$, $N = 20$, $X = 4000$, and $Y = -200$ to 200. Compare the interference factor $I(Y)$ with the intensity $P(Y)$ and observe;

   a. that $P(Y)$ is 1 for $I(Y)$ at $= 0$;

   b. that the number of the side maxima is N-2 and that the maxima of $I(\theta)$ close to the main maxima do not appear. Show that this is true for other values of N;

   c. that there are $N - 1$ side minima.

2. Use FileFig 3.9 for the amplitude grating with $d = 0.0018$ mm, $\lambda = 0.0005$ mm, $a = 0.0036$ mm, and $\theta = -1.401$ to 1.4.
   a. How is the width of the main maxima dependent on the diffraction angle θ?
   b. What happens to the main maxima if we change d to 0.0009?
   c. For blue light use $\lambda = 0.0004$ mm and for red light $\lambda = 0.0007$ mm. Plot for both the interference and diffraction factor on one graph and compare the two intensity patterns for different values of N.
   d. Use close values of the two wavelengths, (e.g., 0.0005 and 0.00055) and determine when two peaks of different wavelengths and at different orders may be resolved depending on choices of N. (For changes in the angle, we have that 1 rad is about 5 to 6 degrees.)
3. From the interference factor find the numerical values of $\sin\theta$ for which the interference factor is equal to N^2 (or 1 in the normalized case). Express the numerical factor in terms of a, m, and λ; this is the grating equation.

FileFig 3.10 (D10FAGRDSLS)

Graph of the intensity $P(Y)$ of a double slit for $d = 0.02$, $\lambda = 0.0005$, $a = 0.2$, $X = 4000$, $N = 2$, and $Y = -800, -799.9$ to 800. A photograph is shown in Figure 3.18.

D10FAGRDSLS is only on the CD.

Application 3.10.

1. When d is changed from $d = a/2$ to $d \ll a$, what happens to the diffraction factor $D(Y)$ and how is the intensity $P(Y) = D(Y)I(Y)$ affected, where $I(Y)$ is the interference factor?
2. From Figure 3.16 one can obtain $(a \sin\theta) = m\lambda$ with an elementary derivation. This is the grating equation. This equation may also be obtained from the formula of the intensity for the main maxima. Using this formula make

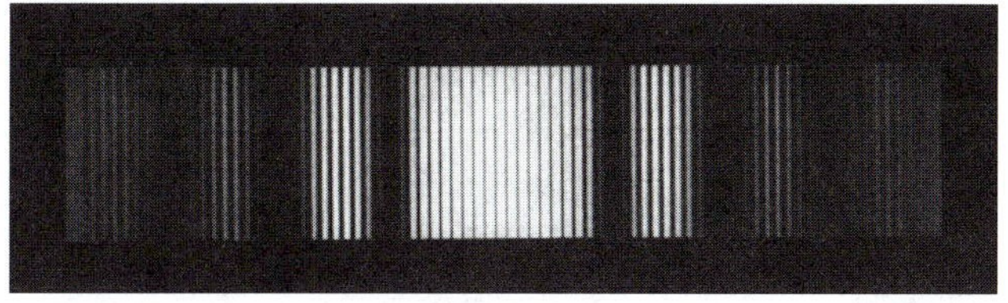

FIGURE 3.18 Diffraction pattern of a double slit where the separation a of the slits is much larger than the width d. The central part of the diffraction pattern shows an interference pattern similar to that observed in Young's experiment (from M. Cagnet, M. Francon, and J. C. Thrieer, *Atlas of Optical Phenomena*, Springer-Verlag, Heidelberg, 1962).

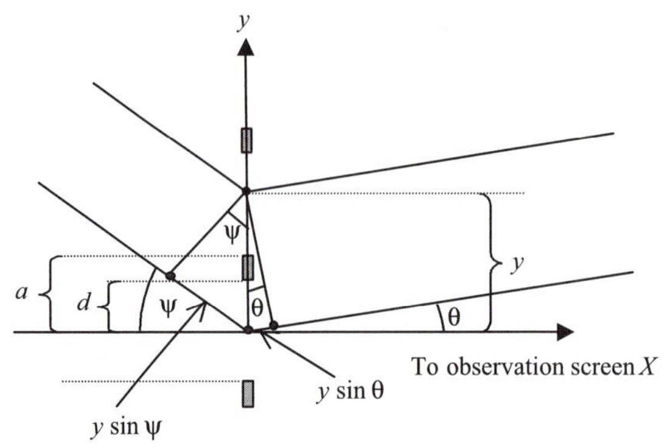

FIGURE 3.19 Coordinates for the amplitude grating. The incident light has angle ψ with the X direction.

a list of the maxima depending on Y, for $d = a/10$, and compare with $P(Y)$ of the graph.

3. Missing order: Use $4d = a$ and start counting the maxima of $P(Y)$ from the center at 0. The fourth maximum of $I(Y)$ is suppressed by the factor $D(Y)$. Use other ratios of a/d.
4. Continue (3) on missing order. Change the wavelength to twice its value and to 1/2 and see what happens to the pattern.

3.4.5.2 Amplitude Grating with Incidence Light at an Angle

An experimental setup may require that the incident light enter the grating not in the direction of the normal but under an angle ψ with respect to the grating surface (Figure 3.19). The only difference now is that the optical path difference (i.e., $a(\sin\theta)$ of the two parallel beams; see Eq. (3.54) is enlarged by the term $a(\sin\theta)$. One obtains for the intensity

$$I(\theta) = \{[\sin(\pi d(\sin\theta + \sin\psi)/\lambda)]/[\pi d(\sin\theta + \sin\psi)/\lambda)]\}^2 \cdot \{\sin[\pi Na(\sin\theta + \sin\psi)/\lambda]/N \sin[\pi a(\sin\theta + \sin\psi)/\lambda]\}^2. \quad (3.61)$$

The interference diffraction pattern is similar to the pattern of the amplitude grating used at normal incidence. The difference is that the main maximum is not at $\theta = 0$. The normalized interference factor is equal to 1 for the main maximum, that is, for $(\sin\theta + \sin\psi) = 0$, or at $\sin\theta = -sin\psi$. The zeroth order is shifted from $\theta = 0$ to $\theta = -\psi$. This is shown in the graph of FileFig 3.11.

FileFig 3.11 (D11FAGRANGS)

A graph is shown of the diffraction pattern of an amplitude grating with the incident light under an angle ψ to the normal.

D11FAGRANG is only on the CD.

Application 3.11. Find the numerical values of ($\sin\theta + sin\psi$) for which the normalized interference factor is equal to N^1. Express the numerical factor in terms of a, m, and λ and compare with the grating equation.

3.4.5.3 Echelette Grating (Phase Grating)

A transmission grating with periodic differences in thickness is called a *phase grating*. When the grating has steps with angle ε, as shown in Figure 3.20, it is called an echelette grating. It may be produced from materials of refractive index n or with a metal surface such as a reflection grating. To calculate the interference diffraction pattern of the echelette grating, we have to determine the optical path difference between equivalent wavelets from neighboring steps and integrate over all the steps. The result is an interference factor, depending on the periodicity constant a of the steps and a diffraction factor, depending on

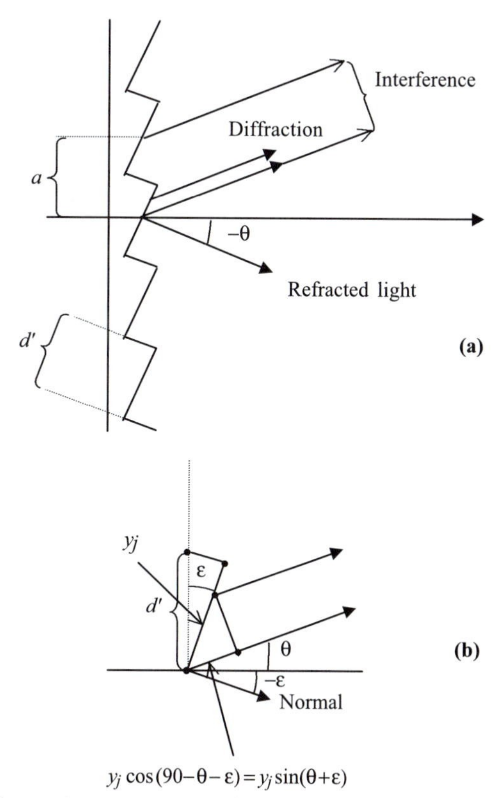

FIGURE 3.20 Coordinates for an echelette grating: (a) refracted and diffracted light and interference; (b) coordinates for the diffraction on one facet. Integration of the diffraction factor over the facet uses the optical path difference $y_i \sin(\theta + \varepsilon)$. The maximum is at $\theta = -\varepsilon$.

the shape of the step. To make the calculation of the diffraction factor easy, we assume that the periodicity constant is equal to the width of the steps d and we choose the angle of incidence such that the transmitted light is refracted in a direction perpendicular to the plane of the step (Figure 3.20a). The angle ε is then the angle between the normal of the facet and the normal of the plane of periodicity. The intensity is again given as the product of the interference and diffraction factor. The interference factor is the same as the one obtained for the amplitude grating, but the diffraction factor is calculated with the optical path difference $y_i \sin(\theta + \varepsilon)$; see Figure 3.20. The intensity is then

$$I(\theta) = \{[\sin(\pi d \sin\{\theta + \varepsilon\}/\lambda)]/[\pi d \sin\{\theta + \varepsilon\}/\lambda]\}^2 \cdot \{[\sin(\pi N a \sin\theta/\lambda)]/N \sin(\pi a \sin\theta)/\lambda]\}^2. \qquad (3.62)$$

The three graphs in FileFig 3.12 demonstrate the dependence on the step angle ε. When $\varepsilon = -.25$ we see that the maximum of the diffraction factor is lined up with the first-order maximum of the interference factor. The zeroth order of the interference factor is at the first minimum of the diffraction factor and is suppressed. For the particular wavelength the grating is said to be blazed. For slightly different wavelengths, the first order would be displaced within the envelope of the diffraction factor. Considering a source emitting many wavelengths, the first order of a range of wavelengths is distributed over the range determined by the diffraction factor.

When changing the step angle ε, one has for $\varepsilon = .25$ that the zeroth order of the diffraction factor is shifted to the -1 order of the interference factor (see the second graph of FileFig 3.12). For $\varepsilon = -.52$, the zeroth order of the diffraction factor is at the $+2$ order of the interference factor (third graph of FileFig 3.12). For other values of (between these values, one finds that the maxima of the interference factor are not lined up with the maxima of the diffraction factor.

FileFig 3.12 (D12FAELGRS)

Graphs of the intensity of diffraction on an echelette grating. Three graphs are shown for three values of ε. There is only one interference factor and three diffraction factors. We see that the diffraction factor is lined up with the first maximum of the interference factor for $\varepsilon = -0.25$. The zero order of the interference factor is at the first minimum of the diffraction factor and is suppressed. If one applies a source emitting many wavelengths, the first-order of a narrow band of different wavelengths is "filling" the area of the diffraction factor. For the wavelength for which both maxima coincide, the grating is said to be "blazed." When choosing $\varepsilon = 0.25$ the zeroth order of the diffraction factor is shifted to the -1 order of the interference factor. For $\varepsilon = -0.52$ we have the zeroth order of the diffraction factor at the +2 order of the interference factor. We may choose other values of ε and find a misalignment.

D12FAELGRS

Diffraction on an Echelette Grating

The graphs for the three different values of ϵ. $D(\theta)$ is the diffraction factor, $I(\theta)$ the interference factor, and $P(\theta)$ the product. The angle in radians of the echelette is ϵ. Diffraction angle θ in radians, wavelength λ, width of openings d, and separation of openings a in mm. N is the number of lines. All parameters are defined globally above the graph.

$$\theta := -.301, -1.299 \ldots 1.3$$

$$D1(\theta) := \left[\frac{\sin\left[\pi \cdot \frac{d}{\lambda} \cdot (\sin(\theta + \epsilon 1))\right]}{\left(\pi \cdot \frac{d}{\lambda} \cdot \sin(\theta + \epsilon 1)\right)}\right]^2 \qquad I(\theta) := \left[\frac{\sin\left[\pi \cdot a \cdot (\sin(\theta)) \cdot \frac{N}{\lambda}\right]}{N \cdot \sin\left[\pi \cdot a \cdot \frac{(\sin(\theta))}{\lambda}\right]}\right]^2$$

$$D2(\theta) := \left[\frac{\sin\left(\pi \cdot \frac{d}{\lambda} \cdot \sin(\theta + \epsilon 2)\right)}{\left[\pi \cdot \frac{d}{\lambda} \cdot \sin(\theta + \epsilon 2)\right)]}\right]^2 \qquad D3(\theta) := \left[\frac{\sin\left[\pi \cdot \frac{d}{\lambda} \cdot (\sin(\theta + \epsilon 3))\right]}{\left(\pi \cdot \frac{d}{\lambda} \cdot \sin(\theta + \epsilon 3)\right)}\right]^2$$

$$P1(\theta) := D1(\theta) \cdot I(\theta) \qquad P2(\theta) := D2(\theta) \cdot I(\theta) \qquad P3(\theta) := D3(\theta) \cdot I(\theta)$$

$$\epsilon 1 \equiv -0.25 \qquad d \equiv 37 \qquad \lambda \equiv 10 \qquad a \equiv 40 \qquad N \equiv 20.$$

$$\epsilon 2 \equiv 0.25.$$

$\epsilon 3 \equiv 0.52.$

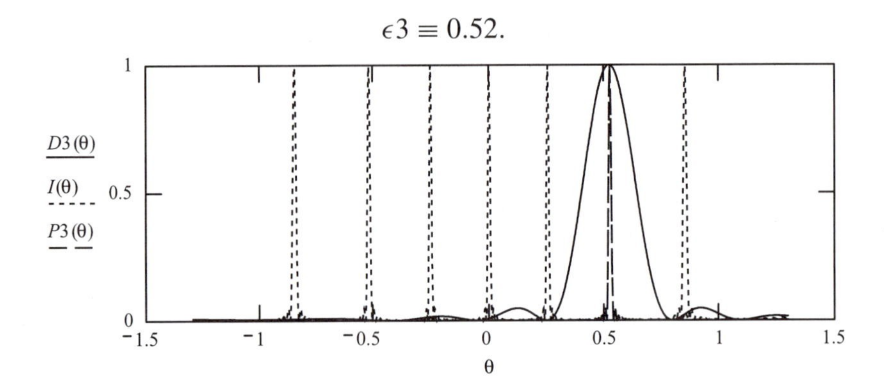

Application 3.12. Use FileFig 3.12 with $d = 0.04$, $a = 0.04$, $\lambda = 0.01$, and $N = 20$.

1. Start from $\varepsilon = 0$ and use for ε positive numbers until the negative first-order of $I(\theta)$ is at the peak of $D(\theta)$. Make another graph using a negative value for ε until the positive first-order of $I(\theta)$ is at the peak of $D(\theta)$.
2. Make three graphs with nine plots, starting with $P1(\theta)$, $D1(\theta)$, and $I1(\theta)$ for $\lambda 1 = 0.015$, then $P2(\theta)$, $D2(\theta)$, and $I2(\theta)$ for $\lambda 2 = 0.01$, and $P3(\theta)$, $D3(\theta)$, and $\lambda 3(\theta)$ for $\lambda 3 = 0.007$. Find a value of ε such that the interference peak of the positive first order of $\lambda 2$ is at the maximum of the diffraction factor. The grating is now "blazed" for the wavelength $\lambda 2 = 0.01$. Make an extra plot only of $P1(\theta)$ to $P3(\theta)$ and see that for the wavelengths $\lambda 1$ and $\lambda 3$ more than one peak appears. These wavelengths do not have maximum intensity and two orders would be observed. The echelette grating would be less efficient.

3.4.6 Resolution

Gratings are used in spectrometers to obtain the spectrum of the incident light. Let us consider incident light containing a spectrum of just two different wavelengths λ_1 and λ_2. As an example let us consider the sodium D lines at 5890 and 5896 A° in the yellow part of the visible spectrum, emitted when the Sodium atom is at a high temperature. One may observe these lines when salt falls into a gas flame and one sees yellow light. The wavelength difference of the two lines is just 6° and we now discuss the question of what has to be required of a grating to observe the two separated lines. We consider two lines of wavelength difference ($\Delta\lambda = \lambda_2 - \lambda_1$. The diffraction angles in the first order will have maxima at $\sin\theta_1 = \lambda_1/a$ and $\sin\theta_2 = \lambda_2/a$ (Figure 3.21a). Assuming $\lambda_1 < \lambda_2$ we have to determine how large $\Delta\lambda = \lambda_2 - \lambda_1$ must be in order to recognize the two spectral lines separately. If $\Delta\lambda$ is large enough, we can recognize these two lines (Figure 3.21b on the left) and if $\Delta\lambda$ is too small, we would have just one peak

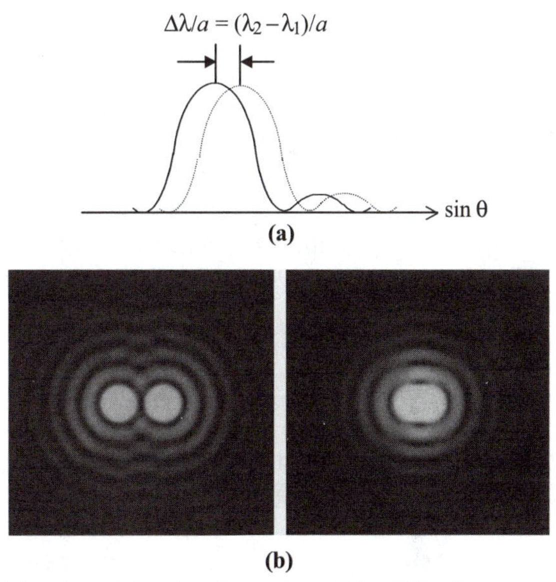

FIGURE 3.21 (a) Closely positioned main maxima of the diffraction patterns produced by the wavelengths λ_1 and λ_2 with $\Delta\lambda = \lambda_2 - \lambda_1$; (b) (left) images of two point sources sufficiently separated; (right) image of two point sources at the limit of resolution. (From M. Cagnet, M. Francon, and J.C. Thrieer, *Atlas of Optical Phenomena*, Springer-Verlag, Heidelberg, 1962.)

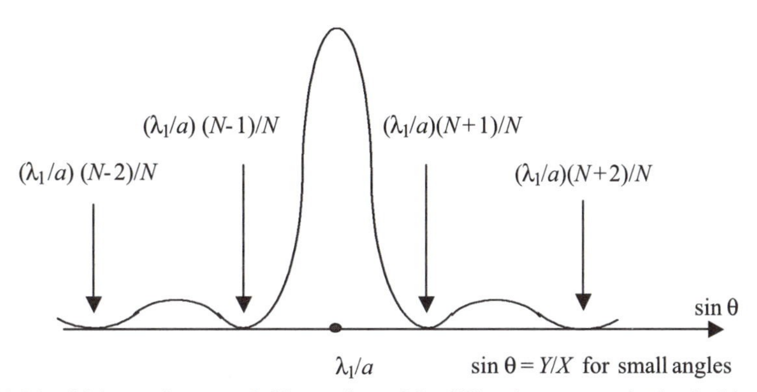

FIGURE 3.22 Main maximum and side maxima of the diffraction pattern obtained with a grating having N lines and using the wavelength λ_1.

(Figure 3.21b on the right). To get a quantitative value for the case where the two lines are just separated, we use the Rayleigh criterion. Which states that the two maxima of a diffraction pattern are considered separated when the maximum of one is at the position of the minimum of the other.

From Figure 3.22 we have for the distance from maximum to minimum of the diffraction pattern of a grating for any order

$$\left(\frac{N-1}{N} - \frac{N}{N}\right)\frac{\lambda_1}{a} = \frac{\Delta\lambda}{a} \tag{3.63}$$

and obtain

$$\Delta\lambda = \lambda_1/N. \tag{3.64}$$

Calling λ_2 just λ and considering the mth diffraction order, we have

$$\lambda/\Delta\lambda = mN. \tag{3.65}$$

Since $\lambda\nu =$ velocity of the light in the medium, where ν is the frequency, one also has

$$\nu/\Delta\nu = mN. \tag{3.66}$$

The ratio $\lambda/\Delta\lambda$ or $\nu/\Delta\nu$ is called the resolving powers of the grating.

The graph in FileFig 3.13 shows two diffracted waves of wavelength λ_1 and λ_2 on a grating. The separation of their maxima depends on N and one may change N to find the separation in agreement with the Rayleigh criterion. Also seen is the dependence of the resolution on the diffraction order. The first graph in FileFig 3.14 shows the diffraction pattern of two round openings of radius a. One is at zero and the other at distance b. One can determine b using the Rayleigh criterion. The second graph shows the two spectral lines in three dimensions.

For our example of the sodium D lines we have from Eq. (3.66) about 1000 for mN. In other words, in first order we need 1000 lines of the grating. If the grating has a periodicity constant of about 1 micron, the width of the grating would be one millimeter. If one could use a larger grating of 10 cm then the number of lines would be 100,000 and in first order, for a wavelength of 5000Å, $\Delta\lambda$ would be 0.02Å.

FileFig 3.13 (D13FAGRRES)

Graph of the diffraction on an amplitude grating for two wavelengths $\lambda_1 = 0.0005$, $\lambda_2 = 0.0006$, openings $d = 0.0001$, periodicity constant $a = 0.002$, and number of lines $N = 6$.

D13FAGRRES is only on the CD.

Application 3.13.

1. Make three choices of λ_2 and determine the N value so that the maximum of λ_2 is on the minimum of λ_1.
2. Choose λ_1 and λ_2 and determine the N value necessary to resolve the two maxima.
3. Consider higher orders m. Choose a pair of lines so close together that they are not resolved in first-order. Determine the order for resolution.

FileFig 3.14 (D14FARES3DS)

Determination of Rayleigh distance and graph of the 3-D-diffraction pattern of two round apertures of radius a and distance b.

D14FARES3DS

Determination of the Wavelength difference for Two Peaks, Resolved According to the Rayleigh Criterion

We call the distance between the maxima b, radius of apertures a, distance between the apertures d, coordinate on the observation screen R, wavelength λ, and distance from aperture to screen X.

1. Determination of Rayleigh distance

$$a \equiv .05 \qquad X \equiv 4000 \qquad R := 0, .1 \ldots 50$$

$$g1(R) := \left[\frac{J1\left(2 \cdot \pi \cdot a \cdot \frac{R}{X \cdot \lambda}\right)}{\left(2 \cdot \pi \cdot a \cdot \frac{R}{X \cdot \lambda}\right)}\right]^2 \qquad gg1(R) := \left[\frac{J1\left(2 \cdot \pi \cdot a \cdot \frac{R-b}{X \cdot \lambda}\right)}{\left(2 \cdot \pi \cdot a \cdot \frac{R-b}{X \cdot \lambda}\right)}\right]^2.$$

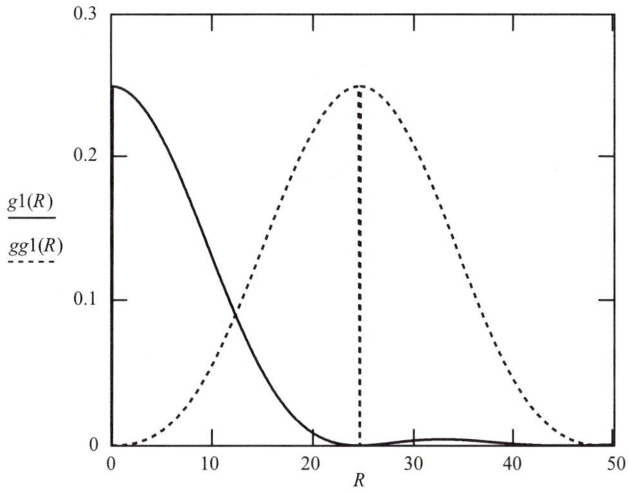

0.3
0.2
0.1
0
$g1(R)$
$gg1(R)$
0 10 20 30 40 50
R

Distance b is assumed to be $b \equiv 24.5$.

2. 3-D Graph of pattern of two round apertures at distance b

$$i := 0 \ldots N \qquad j := 0 \ldots N$$

$$x_i := (-40) + 2.0001 \cdot i \qquad y_j := -40 + 2.0001 \cdot j \qquad \lambda \equiv .0005$$

$$RR(x, y) := \sqrt{(x)^2 + (y)^2} \qquad N \equiv 60 \qquad X := 4000$$

$$g2(x, y) := \left[\frac{J1\left(2 \cdot \pi \cdot a \cdot \frac{RR(x,y)}{X \cdot \lambda}\right)}{\left(2 \cdot \pi \cdot a \cdot \frac{RR(x,y)}{X \cdot \lambda}\right)}\right]^2$$

$$gg2(x,y) := \left[\frac{J1\left(2\cdot\pi\cdot a\cdot\frac{RR(x,y-b)}{X\cdot\lambda}\right)}{\left(2\cdot\pi\cdot a\cdot\frac{RR(x,y-b}{X\cdot\lambda}\right)}\right]^2.$$

$$M_{i,j} := g2(x_i, y_j) + gg2(x_i, y_j).$$

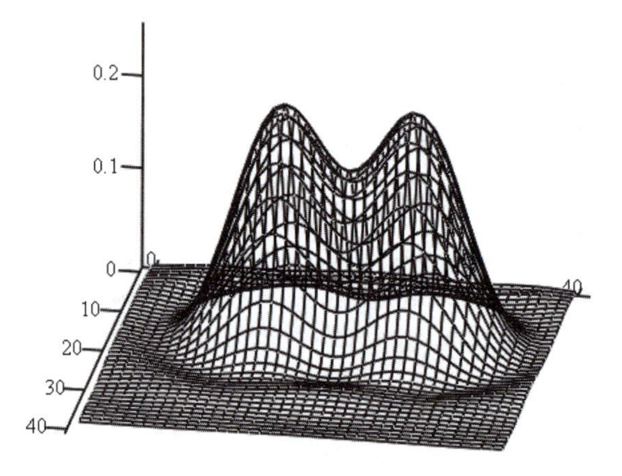

M

3. Calculation of wavelength difference corresponding to b
The diffraction angle is calculated from $b/X = \Delta\theta$. The grating is made of round apertures of diameter a and spaced at distance d. From the grating formula we have for the wavelength difference $\Delta\lambda = d\Delta\theta$ or $\Delta\lambda = (d/X)b$. For $d := .1$, $\Delta\lambda := d\cdot\frac{b}{X}$, $\Delta\lambda = 6.125\times 10^{-4}$.

Application 3.14.

1. Convert to a surface plot and make b such that the maximum of one of the peaks is on the minimum of the other. Then go back to the contour plot.
2. We have used the same wavelength for both peaks. After determining the wavelength resolution, introduce the corresponding wavelengths into the two expressions of the Bessel function and estimate the change.
3. Assume that the angle difference for resolution is given for order m and find the wavelength difference.

3.5 BABINET'S THEOREM

Babinet's Theorem tells us that two complementary screens will give us the same diffraction pattern at the observation screen. Complementary screens are two screens, one of which is transparent at areas where the other is opaque (Figure 3.23). If the two complementary screens are placed together in the same plane, no aperture will exist.

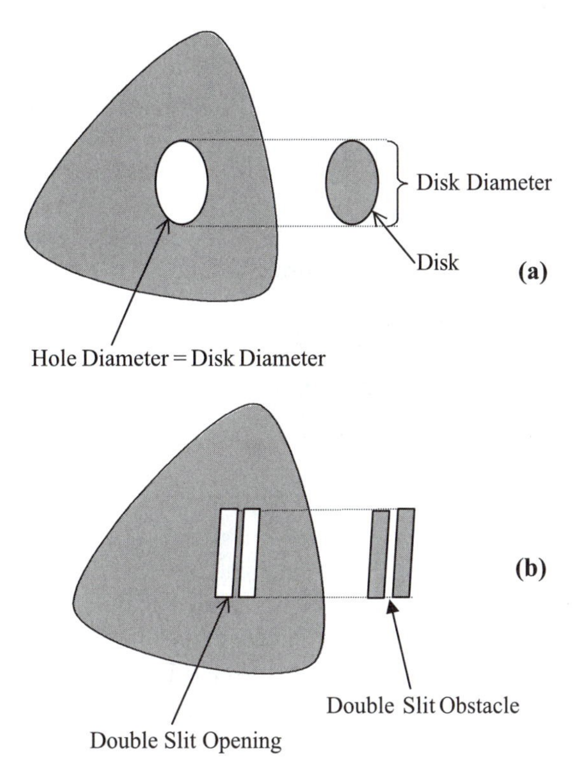

FIGURE 3.23 Examples of complementary screens.

The general appearance of the diffraction pattern on the observation screen is the same, regardless of one screen having larger openings than the other.

Let us consider two gratings, both having periodicity constant a. One has a width of opening d_1 and the other of d_2. These gratings are complementary screens when $d_1 + d_2 = a$. To get the diffraction pattern, we apply the Kirchhoff–Fresnel integral to the openings of each screen. The two integrals must add up to zero since there are no openings to integrate. We assume there is a single wave incident on all openings and use far field approximation,

$$\varphi_1(Y) = C \int_{\text{openings of screen I}} \phi(y)(e^{ik(yY)/X})dy \quad \text{and} \quad \varphi_{II}(Y) = C \int_{\text{openings of screen II}} \phi(y)(e^{ik(yY)/X})dy \tag{3.67}$$

and have for the amplitudes

$$\varphi_I(Y) + \varphi_{II}(Y) = 0 \quad \text{or} \quad \varphi_I(Y) = -\varphi_{II}(Y), \tag{3.68}$$

and for the intensities

$$[\varphi_I(Y)]^2 = [\varphi_{II}(Y)]^2. \tag{3.69}$$

The diffraction pattern of the two screens has the same overall appearance. The first equation tells us that the two diffraction patterns are "out of phase" by 180

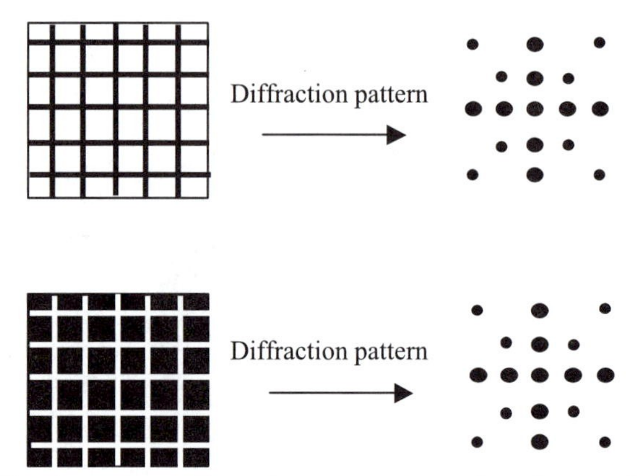

FIGURE 3.24 Diffraction patterns of complementary screens show the same intensity pattern.

degrees and the second one tells us that the diffraction patterns of the two screens are similar.

In Figure 3.24 we show schematically the diffraction pattern of two complementary screens. One screen is made of open squares and the other of black squares. In FileFig 3.15 we consider two amplitude gratings with complementary open areas as complementary screens. The appearance of the diffraction pattern is the same, but the heights of the peaks are different. We assumed that d_1 is different from d_2 and therefore the open areas are different.

FileFig 3.15 (D15FABAGRS)

Two complementary screens are considered. Both are amplitude gratings of periodicity constant a. One has the width d_1 of open strips, the other the width d_2, and since we assume that $a = d_1 + d_2$ the screens are complementary. The diffraction patterns are shown as P_1 for the grid with d_1 and P_2 for the grid with d_2. One observes that the diffraction patterns are similar. Both patterns have peaks at the same location. However, they have different intensities. The different intensities result from the different d_1 and d_2 values and consequently different areas of integration.

D15FABAGRS

Babinet's Theorem

Diffraction on two amplitude gratings, one with width of openings $d1$, the other with width of opening $d2$, and both having center-to-center distance of strips $a = d1 + d2$. Wavelength λ, distance from gratings to screen X, and coordinate on screen Y. All distances and wavelengths area in mm; both have number of lines N. Normal incidence. $D1$ and $D2$ are the diffraction factors, I is the interference factor, normalized to 1. $P(A)$ is the product of interference and diffraction factor. Diffraction pattern of the two complementary screens: one is a grating of width of opening $d1$, the other of $d2$, and the periodicity constant is $a = d1 + d2$.

$$\theta := -.5001, -.4999 \ldots .5$$

$$D1(\theta) := \left[\frac{\sin\left(\pi \cdot \frac{d1}{\lambda} \cdot \sin(\theta)\right)}{\left(\pi \cdot \frac{d1}{\lambda} \cdot \sin(\theta)\right)}\right]^2 \qquad I(\theta) := \left[\frac{\sin\left(\pi \cdot \frac{a}{\lambda} \cdot \sin(\theta) \cdot N\right)}{N \cdot \sin\left(\pi \cdot \frac{a}{\lambda} \cdot \sin(\theta)\right)}\right]^2$$

$$\lambda \equiv .0005 \qquad N \equiv 6 \qquad P1(\theta) := D1(\theta) \cdot I(\theta)$$

$$D2(\theta) := \left[\frac{\sin\left(\pi \cdot \frac{d2}{\lambda} \cdot \sin(\theta)\right)}{\left(\pi \cdot \frac{d2}{\lambda} \cdot \sin(\theta)\right)}\right]^2 \qquad I(\theta) := \left[\frac{\sin\left(\pi \cdot \frac{a}{\lambda} \cdot \sin(\theta) \cdot N\right)}{N \cdot \sin\left(\pi \cdot \frac{a}{\lambda} \cdot \sin(\theta)\right)}\right]^2$$

$$d2 \equiv .001 \qquad d1 \equiv .002 \qquad a \equiv d1 + d2 \qquad P2(\theta) := D2(\theta) \cdot I(\theta).$$

We see that the intensity of the diffraction peaks is different for the two complementary patterns, but the position of the peaks is the same, and that is what Babinet's Principle tells us.

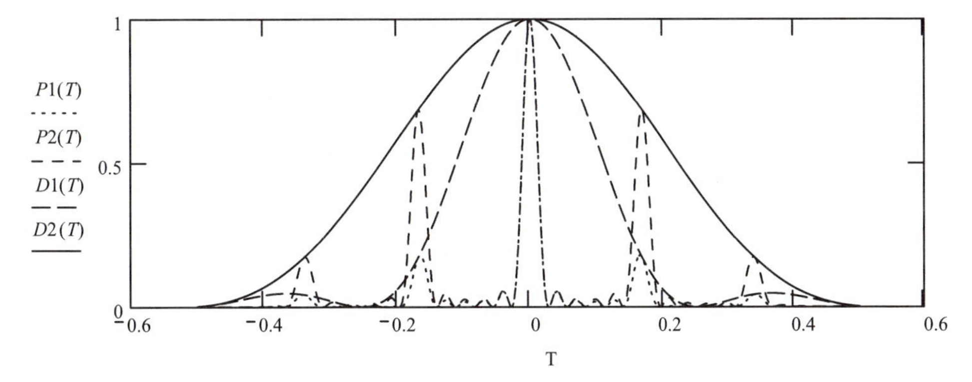

Application 3.15.

1. Keep $a = d1 + d2$ constant and change the width $d1$ to a much smaller value than $d2$. Check how the intensities of the two patterns are affected.
2. For comparison, make $d1$ and $d2$ about equal.
3. Change the constant "a" to see how the pattern is changing.

3.6 APERTURES IN RANDOM ARRANGEMENT

In Chapter 2 we studied the interference pattern of an array and found that the pattern disappears when the array is changed to a random arrangement. We now study the question of what happens to an interference diffraction pattern if we assume a random arrangement.

We consider an amplitude grating and want to describe the changes of the interference diffraction pattern when we change the periodic array into a random arrangement. The diffraction pattern is described by the product of the diffraction and interference factors

$$P_{\text{periodic}} = \{[\sin(\pi d \sin\theta/\lambda)]/[(\pi d \sin\theta/\lambda)]\}^2 \cdot \{[\sin(\pi N a \sin\theta/\lambda)]/(N \sin(\pi a \sin\theta/\lambda)]\}^2. \quad (3.70)$$

As discussed in the chapter on interference, the interference factor will average to a constant when the change to the random array of apertures is done and we are left only with the diffraction factor

$$P_{\text{random}} = \{[\sin(\pi d \sin\theta/\lambda)]/[(\pi d \sin\theta/\lambda)]\}^2. \quad (3.71)$$

The random array of many openings of width d will give us the diffraction pattern of a slit at the observation screen.

In FileFig 3.16 we have a one-dimensional calculation. The first graph shows the interference diffraction pattern of a grating, and the second graph shows what is left if the interference factor disappears. In FileFig 3.17 we show the diffraction pattern of a 2-D grating as a 2-D contour plot. The first graph shows the diffraction pattern of the periodic arrangement and the second graph shows the diffraction pattern of the random arrangement. This is schematically shown in Figure 3.25 for periodic and random arrangements of square apertures. The periodic arrangement shows the interference diffraction pattern, and the random arrangement appears as the superposition of the intensity diffraction pattern of squares.

FileFig 3.16 (D16FAGRRANS)

The product $P1$ of a diffraction and interference factor for a one-dimensional grating is shown. When changing the periodic arrangement of the apertures to a random arrangement, the interference factor is a constant and $P2$ shows the remaining diffraction pattern.
D16FAGRRANS is only on the CD.

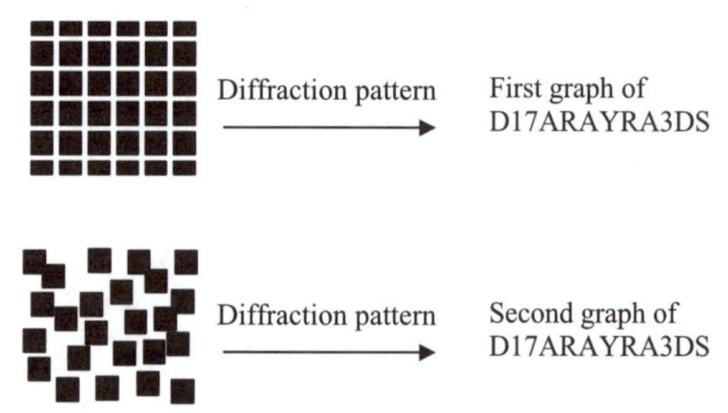

FIGURE 3.25 (a) Diffraction pattern of a periodic array of rectangles; (b) diffration pattern of a random array of rectangles. The resulting pattern is the superposition of the intensity diffraction pattern of rectangles.

FileFig 3.17 (D17ARAYRA3DS)

The product $f(x, y)$ of intensity of diffraction and interference factor for a two-dimensional grating is shown. Three-dimensional plots are shown for regular and random arrays.

D17ARAYRA3DS

3-D Graph of Diffraction Pattern

Periodic array of rectangular apertures compared to the diffraction pattern of rectangular apertures in a random array.

1. Periodic array

$$i := 0 \ldots N \qquad j := 0 \ldots N$$

$$\lambda \equiv 4 \qquad x_i := (-3) + .20001 \cdot i \qquad y_j := -4 + .20001 \cdot j \qquad p := 6$$

$$f(x, y) := \left[\frac{\sin\left(2 \cdot \pi \cdot d \cdot \frac{x}{2 \cdot \lambda}\right)}{\left(2 \cdot \pi \cdot d \cdot \frac{x}{2 \cdot \lambda}\right)}\right]^2 \cdot \left[\frac{\sin\left(2 \cdot \pi \cdot d1 \cdot \frac{y}{2 \cdot \lambda}\right)}{\left(2 \cdot \pi \cdot d1 \cdot \frac{y}{2 \cdot \lambda}\right)}\right]^2 \cdot \left[\left[\frac{\sin\left(2 \cdot \pi \cdot a \cdot \frac{x \cdot p}{2 \cdot \lambda}\right)}{p \cdot \sin\left(2 \cdot \pi \cdot a \cdot \frac{x}{2 \cdot \lambda}\right)}\right]^2 \cdot \left[\frac{\sin\left(2 \cdot \pi \cdot a1 \cdot \frac{y \cdot p}{2 \cdot \lambda}\right)}{p \cdot \sin\left(2 \cdot \pi \cdot a1 \cdot \frac{y}{2 \cdot \lambda}\right)}\right]^2\right]$$

$$M_{i,j} := f\left(x_i, y_j\right) \qquad d \equiv 2 \qquad d1 \equiv 2 \qquad a1 \equiv 4 \qquad a \equiv 4$$

$$N \equiv 40.$$

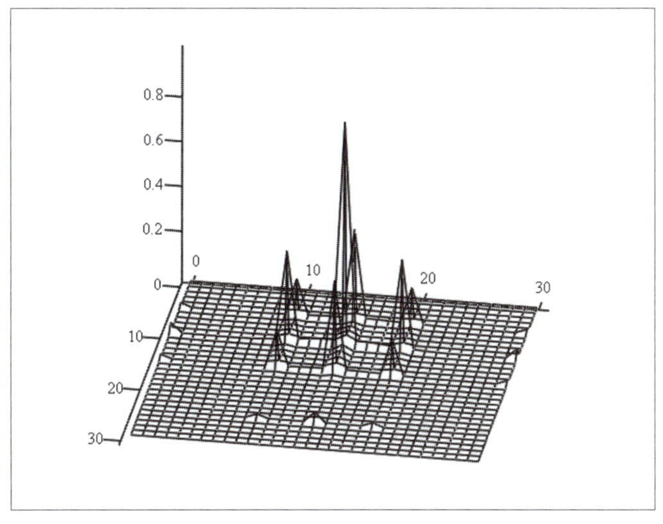

M

2. Random array

$$f1(x, y) := \left[\frac{\sin\left(2 \cdot \pi \cdot d \cdot \frac{x}{2 \cdot \lambda}\right)}{\left(2 \cdot \pi \cdot d \cdot \frac{x}{2 \cdot \lambda}\right)}\right]^2 \cdot \left[\frac{\sin\left(2 \cdot \pi \cdot a \cdot \frac{y}{2 \cdot \lambda}\right)}{\left(2 \cdot \pi \cdot a \cdot \frac{y}{2 \cdot \lambda}\right)}\right]^2$$

$$MM_{i,j} := f1\left(x_i, y_j\right).$$

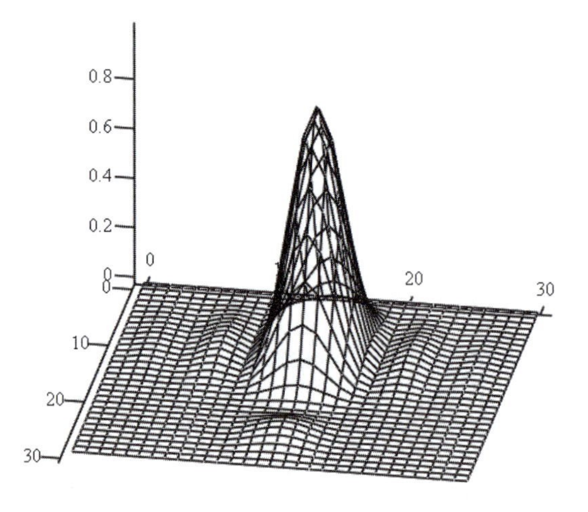

MM

3.7 FRESNEL DIFFRACTION

At the beginning of this chapter we used Fresnel diffraction for the calculation of the Poisson spot. The Kirchhoff–Fresnel integral was applied to a round stop and we found a constant illumination in the shadow of the round aperture on the axis of the system. We are now interested in discussing the diffraction on an edge, which was done by Fresnel using the integrals named after him. The first steps are to study the Fresnel diffraction on a slit and give the definitions of Fresnel's integrals.

3.7.1 Coordinates for Diffraction on a Slit and Fresnels Integrals

We consider the Kirchhoff–Fresnel diffraction integral in small angle approximation; see Eq. (3.20).

$$G(Y) = C \int_{y_1=-d/2}^{y_2=d/2} \exp[ik(1/2X)(Y - y)^2] dy \tag{3.72}$$

and set the constant C equal to 1. To get to the definition of Fresnel's integrals, we change coordinates

$$(y - Y)^2 = \eta^2(\lambda X/2) \tag{3.73}$$

and have

$$\eta = (Y - y)\sqrt{2/\lambda X} \tag{3.74}$$

and for the limits of integration η_1 and η_2,

$$\eta_1 = (Y + d/2)\sqrt{2/\lambda X} \quad \text{and} \quad \eta_2 = (Y - d/2)\sqrt{2/\lambda X}. \tag{3.75}$$

Substituting Eqs. (3.74) and (3.75) into the integral, Eq. (3.72) results in

$$G(\eta) = \int_{\eta_1}^{\eta_2} e^{i(\pi/2)\eta^2} d\eta. \tag{3.76}$$

We may write

$$\int_{\eta_1}^{\eta_2} e^{i(\pi/2)\eta^2} d\eta = \int_{\eta_1}^{\eta_2} \cos[(\pi/2)\eta^2]d\eta + i\int_{\eta_1}^{\eta_2} \sin[(\pi/2)\eta^2]d\eta \tag{3.77}$$

and have for the right side of Eq. 3.77

$$\int_0^{\eta^2} \cos[\pi/2\eta^2]d\eta - \int_0^{\eta_1} \cos[\pi/2\eta^2]d\eta + i\{\int_0^{\eta_2} \sin[\pi/2\eta^2]d\eta - \int_0^{n_1} \sin[\pi/2\eta^2]d\eta\}. \tag{3.78}$$

The Fresnel integrals are defined as:

$$C(\eta') = \int_0^{\eta'} \cos(\pi/2)\eta^2 d\eta, \qquad S(\eta') = \int_0^{\eta'} \sin(\pi/2)\eta^2 d\eta. \tag{3.79}$$

In FileFig 3.18 we show graphs of Fresnel's integrals.

FileFig 3.18 (D18FEFNCS)

The integrals $C(\eta')$ and $S(\eta')$ are plotted as $C(Y)$ and $S(Y)$ for $Y = 0$ to 5.

D18FEFNCS is only on the CD.

3.7.2 Fresnel Diffraction on a Slit

Fresnel diffraction on a slit is calculated using the coordinates of a slit of width d as shown in Fig. 3.26. From Eq. (3.73) and (3.76) we have for the amplitude at the observation point

$$G(Y) = C[\eta_2(Y)] - C[\eta_1(Y)] + i\{S[\eta_2(Y)] - S[\eta_1(Y)]\}, \tag{3.80}$$

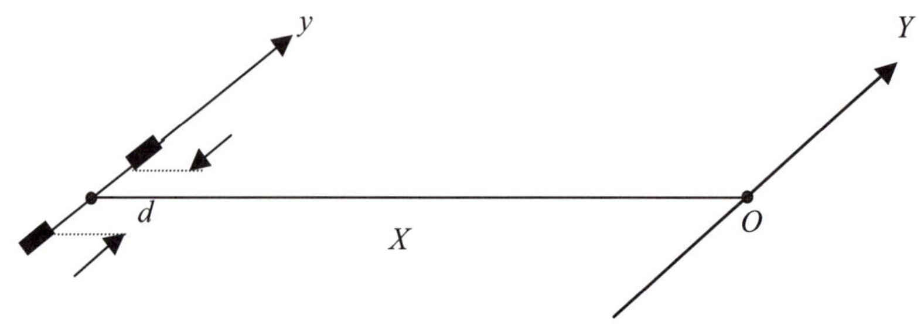

FIGURE 3.26 Coordinates for the calculation of the Fresnel diffraction pattern of a single slit of width d.

where the dependence on Y is

$$\eta_1 = (Y + d/2)\sqrt{(2/\lambda(X)} \quad \text{and } \eta_2 = (Y - d/2)\sqrt{(2/\lambda X)}. \tag{3.81}$$

The intensity is

$$I(Y) = \{C[\eta_2(Y)] - C[\eta_1(Y)]\}^2 + \{S[\eta_1(Y)] - S[\eta_1(Y)]\}^2. \tag{3.82}$$

The graph in FileFig 3.19 shows Fresnel diffraction on a slit. One observes that for the parameters used, the first minimum of the pattern is not zero. By changing d to a smaller value or X to a larger value, one may get to a zero value for the first minimum.

FileFig 3.19 (D19FESLITS)

Fresnel diffraction $I(Y)$ is plotted for a slit of width d at distance $X = 4000$ mm for $\lambda = 0.0005$ mm. These are the same values as used in FileFig 3.3 for far field approximation. For a small slit width, there is no difference. For a larger slit width, the Fresnel diffraction is not zero for the first minimum.

D19FESLITS

Fresnel's Integrals for Calculation of Diffraction on a Slit

All units are in mm, global definition of parameters.
We call $\eta 1$

$$Y := 0, .1 \ldots 10.$$

$$q(Y) := \left(Y + \frac{d}{2}\right) \cdot \sqrt{\frac{2}{\lambda \cdot X}}$$

We call $\eta 2$

$$p(Y) := \left(Y - \frac{d}{2}\right) \cdot \sqrt{\frac{2}{\lambda \cdot X}}$$

$$C_q(Y) := \int_0^{q(Y)} \cos\left(\frac{\pi}{2}\cdot\eta^2\right)d\eta \qquad C_p(Y) := \int_0^{p(Y)} \cos\left(\frac{\pi}{2}\cdot\eta^2\right)d\eta$$

$$S_q(Y) := \int_0^{q(Y)} \sin\left(\frac{\pi}{2}\cdot\eta^2\right)d\eta \qquad S_p(Y) := \int_0^{p(Y)} \sin\left(\frac{\pi}{2}\cdot\eta^2\right)d\eta$$

$$I(Y) := (Cp(Y) - Cq(Y))^2 + (Sp(Y) - Sq(Y))^2.$$

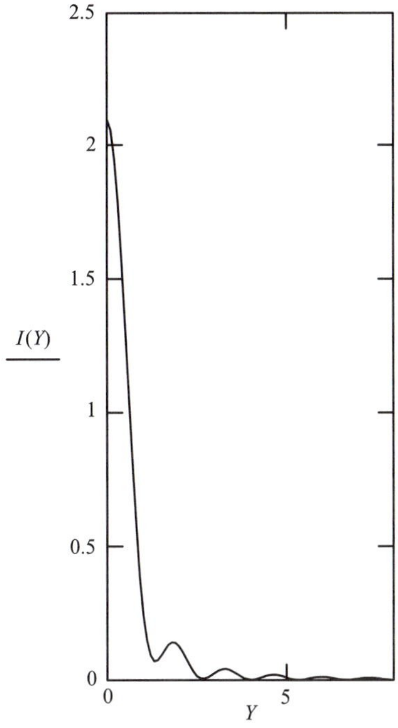

$$\sqrt{\frac{2}{\lambda\cdot X}} = 1$$

$$\mathrm{TOL} \equiv .1$$

$$\lambda \equiv 5\cdot 10^{-4} \qquad X \equiv 4000$$

$$d \equiv 1.5.$$

Application 3.19.

1. Normalize the pattern of the diffraction on a slit using Fresnel diffraction. Make it 1 at the center, and extend the Y range to negative and positive values.
2. Add to the graph the diffraction on a slit using far field approximation. Use the same slit width, wavelength, and distance from aperture to observation screen.
3. For what values of d do we have close agreement?

3.7.3 Fresnel Diffraction on an Edge

Fresnel diffraction on an edge is treated as the diffraction on a large slit with one edge at $y = 0$ and the other at $y = \infty$ (Figure 3.27). For the slit we had the limits for $-d/2$ to $d/2$. Note the negative sign in Eq. (3.74)

$$\eta_1 = (Y + d/2)\sqrt{2/\lambda X}, \qquad \eta_2 = (Y - d/2)\sqrt{2/\lambda X}. \tag{3.83}$$

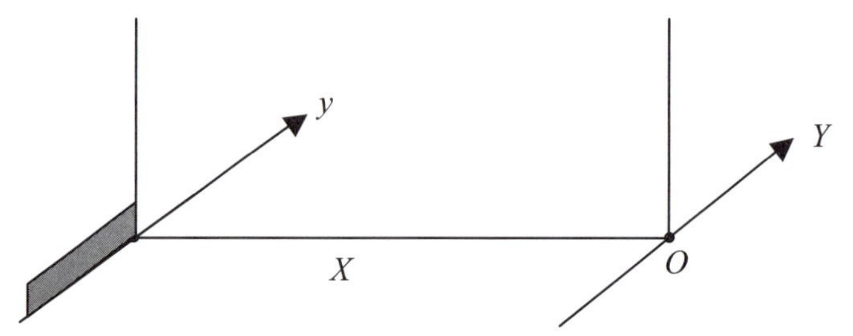

FIGURE 3.27 Coordinates for the calculation of the Fesnel diffraction pattern of an edge treated as a large slit, position of slit from $y = 0$ to $y = (\infty)$. For the value of the integral from 0 to $-\infty$ we use the value $-.5$. The dependence on Y is now only contained in η_1.

The integration limits are now

$$\eta_1 = (Y)\sqrt{2/\lambda X} \qquad \eta_2 = -\infty. \tag{3.84}$$

For the edge presented by a slit with one side at $y = 0$ and the other side at $y = \infty$, we have

$$u(Y) = \int_{\eta_1}^{-\infty} e^{-i(\pi/2)\eta^2} d\eta = C(\eta) + iS(\eta)|_{\eta_1}^{-\infty}. \tag{3.85}$$

The integrals $C(\eta)$ and $S(\eta)$ have to be taken from 0 to $-\infty$. At $-\infty$, both are $-.5$, and we get for the intensity (not normalized)

$$I(Y) = \{-.5 - C(\eta_1)\}^2 + \{-.5 - S(\eta_1)\}^2. \tag{3.86}$$

The first graph in FileFig 3.20 shows the intensity diffraction pattern (Eq. (3.86)) and Figure 3.28 shows a photograph of the diffraction on an edge. The second graph in FileFig 3.20 shows how the diffraction on an edge is derived from the diffraction on a large slit.

FIGURE 3.28 Coordinates for the calculation of the Fresnel diffraction pattern of an edge treated as a large slit, position of slit from $y = 0$ to $y = (\infty)$. For the value of the integral from 0 to $-\infty$ we use the value $-.5$. The dependence on Y is now only contained in η_1. (From M. Cagnet, M. Francon, and J.C. Thrieer, *Atlas of Optical Phenomena*, Springer-Verlag, Heidelberg, 1962.)

FileFig 3.20 (D20FEEDGES)

The intensity $I(Y)$ for the diffraction on an edge is shown for the range from $Y = -5$ mm to $Y = +15$ mm. To show that this is one side of a large slit, the diffraction pattern of a large slit is shown as $II(Y)$.

D20FEEDGES

Fresnel's Integrals for Calculation of Diffraction on an Edge

All units are in mm, global definition of the parameters.

$$Y := -4, -3.95 \ldots 15 \qquad \text{TOL} \equiv .001 \qquad \sqrt{\frac{2}{\lambda \cdot X}} = 1 \cdot 10^{-4}.$$

We treat the diffraction at an edge as diffraction on a large slit. One side is set at $d = 0$, the other at $d = -\infty$. This translates into

$$\text{For } p(Y) = -\text{infinte; we have } Cp(Y) = Sp(Y) = -.5$$

$$q(Y) := (Y) \cdot \sqrt{\frac{2}{\lambda \cdot X}}.$$

We take $q(Y)$ equal Y, square root is for scaling $q(Y) := Y$.

$$Cq(Y) := \int_0^{q(Y)} \cos\left(\frac{\pi}{2} \cdot \eta^2\right) d\eta \qquad Sq(Y) := \int_0^{q(Y)} \sin\left(\frac{\pi}{2} \cdot \eta^2\right) d\eta$$

$$I(Y) := (Cq(Y) - (-.5))^2 + (Sq(Y) - (-.5))^2$$

$$X \equiv 4000$$

$$\lambda \equiv 5 \cdot 10^4.$$

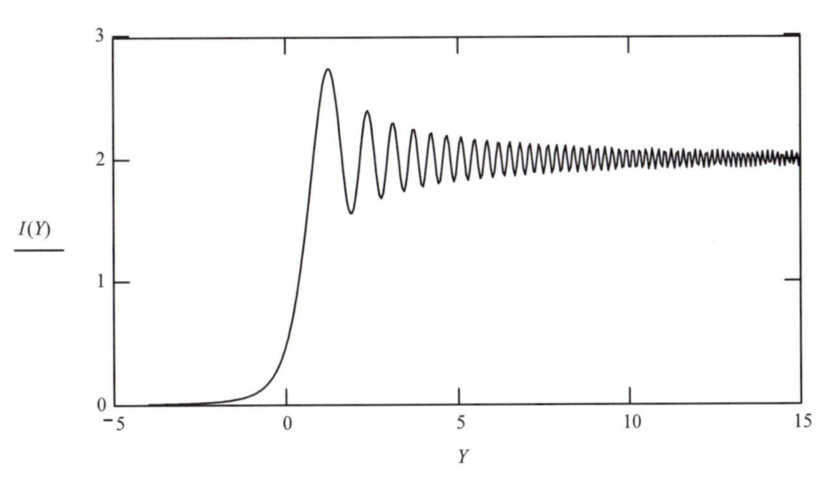

To see that we actually derived this from a large slit, we treat a large slit with positions at 0 and 10.

$$p(Y) : (Y - (10))$$

$$Cp(Y) := \int_0^{p(Y)} \cos\left(\frac{\pi}{2} \cdot \eta^2\right) d\eta \quad Sp(Y) := \int_0^{p(Y)} \sin\left(\frac{\pi}{2} \cdot \eta^2\right) d\eta$$

$$I1(Y) := (Cq(Y) - Cp(Y))^2 + (Sq(Y) - Sp(Y))^2.$$

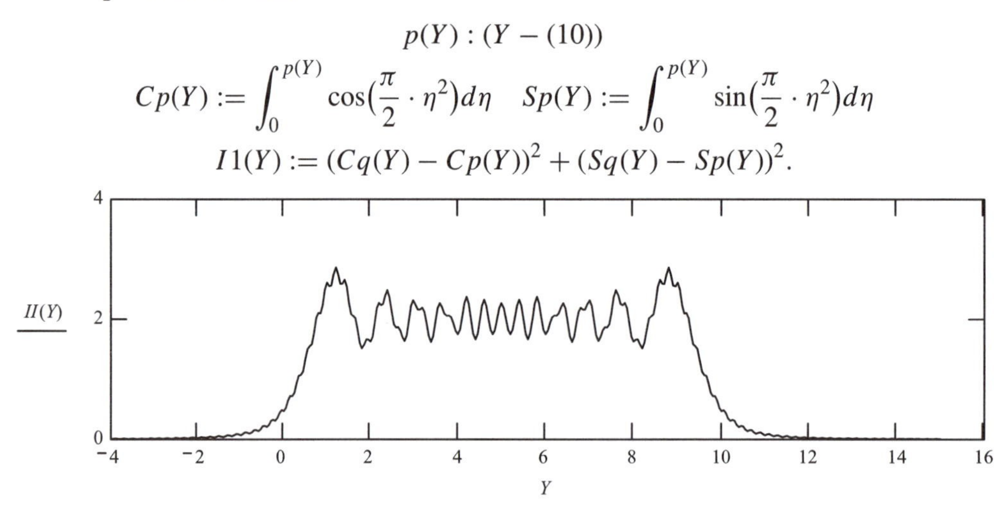

Application 3.20. Vary the width of the slit by choosing values other than 10 in $P(Y) = Y - 10$ and compare the diffraction pattern of the slit with the pattern of the edge.

APPENDIX 3.1

A3.1.1 Step Grating

A grating with a rectangular reflecting surface is called a step grating. The grating has the periodicity constant a, and the reflecting surfaces of length $a/2$ are positioned in two planes. Such a grating may be produced by using two sets of interpenetrating gratings, shown schematically in Figure A3.29. The distance H between planes I and II may be varied, and the corresponding interference diffraction pattern, depending on the height of H, is called an interferogram. Assuming that the incident light contains many wavelengths, the application of a Fourier transformation to the interferogram results in a spectrum.

We discuss here the step grating as shown in Figure A3.29b for a fixed step height H. The incident light is reflected at planes I and II and travels in direction θ. At a faraway screen, one observes an interference diffraction pattern. We calculate the diffraction pattern on the array of N steps, having width d, step height H, and periodicity constant a.

Each of the two interpenetrating gratings produces the pattern of an amplitude grating and, in addition, we have the interference of the light from planes I and II in the direction of the observation screen. The optical path difference is $\delta = d \sin\theta + H$ and in order to get the intensity of the interference (see

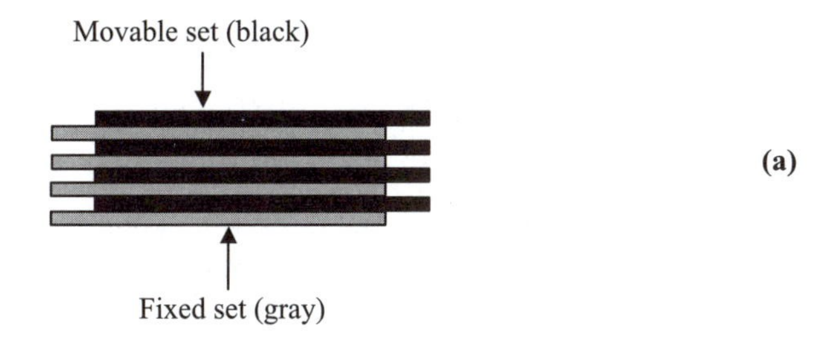

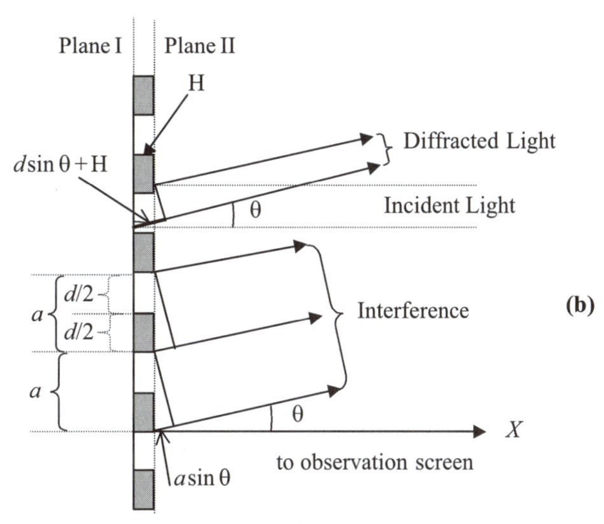

FIGURE A3.29 (a) Schematic of a step grating; (b) coordinates for the calculation of the diffraction on a step grating. Only the emerging diffracted light is shown for the calculation of the path difference. The two gratings in planes I and II interfere with each other in the X direction.

Chapter 2, Eq. (2.8)), we have to insert this path difference δ into $[\cos(\pi\delta/\lambda)]^2$. The intensity $P = uu^*$ of the diffraction and interference of each grating and the interference of the two gratings with each other is then the product of the diffraction factor

$$D(\theta) = \{[\sin(\pi d \sin\theta/\lambda)]/[(\pi d \sin\theta/\lambda)]\}^2, \tag{A3.1}$$

the interference factor

$$I(\theta) = \{[\sin(\pi N a \sin\theta/\lambda)]/[N \sin(\pi a \sin\theta/\lambda)]\}^2, \tag{A3.2}$$

and the step factor

$$ST(\theta) = [\cos\{(\pi/\lambda)(d \sin\theta + H)\}]^2. \tag{A3.3}$$

The diffracted intensity is

$$P(\theta) = \{[\sin(\pi d \sin\theta/\lambda)]/[(d \sin\theta/\lambda)]\}^2 \{\sin \pi N a \sin\theta/\lambda)/N \sin(\pi a \sin\theta/\lambda)\}^2 \cdot [\cos\{\pi/\lambda)(d \sin\theta + h)\}]^2. \quad \text{(A3.4)}$$

The diffraction pattern depends not only on d and a, but also on H. The graph of FileFig A3.21 shows $P(\theta)$ for several values of H and also the diffraction factor $D(\theta)$. The diffraction factor supplies the envelope and the interference factor supplies the lines. The zeroth order is at the center and the first order is within the envelope. Since we necessarily have 1:2 for the ratio d/a, the second order is suppressed by the zeros of the diffraction factor. Variation of H will redistribute the intensity between the zeroth and first order. This may be understood from energy conservation. We assume that the third and higher orders may be neglected. Then for $H = 0$, the incident light is reflected into the zeroth order, that is, in the direction of reflection on the mirror surface. When $H = \lambda/4$, no light can travel in the direction of the zeroth order. In other words, no light may be reflected on the surface of the grating facets and all light travels into the first order. For $H = \lambda/2$ we have again reflection on the grating facets.

For H being a multiple of the wavelength, all light is diffracted into the zeroth order and for H being a multiple of half a wavelength, all light is diffracted into the first order. The graph in FileFig A3.21 shows the intensity pattern $P(\theta)$ for two different wavelengths. One observes only one peak for the zeroth order, but two peaks for the first order. The zeroth order changes its intensity depending on H. Recording the intensity depending on H will reveal the interferogram.

FileFig A3.21 (DA1FAGSTEP1S)

There are four intensity patterns $P1$ to $P4$, each for a different value of H. The values of H are presented as $H = n\lambda$. When n is an integer, we have all light in the zeroth order. For noninteger values of n, we subtract from λ all full wavelengths and look at the remaining fraction n'. For example, if $n = 10.5$, we look at $n' = 0.5$. The optical path difference is now half a wavelength and all the light is diffracted in the first orders. For values such as 0.125, 0.25, and 0.375 there is light partly diffracted into the zeroth and the first order.

DA1FAGSTEP1S is only on the CD.

Application A3.21.

1. Find the values of n for which the patterns of constructive and destructive interference are repeating for values of n from 0 to 2. Observe that the path difference between constructive and destructive interference is $\lambda/2$, and the successive constructive or destructive interference pattern is λ. These length differences correspond to the length differences produced by H.

2. A lamellar grating is an interferometer with adjustable length H. It produces the optical path difference by increasing H in small steps. It is usually operated in reflection and then the length is only 1/2 the length needed to produce the same optical path difference compared to the case of transmission, discussed in (1).

FileFig A3.22 (DA2FAGRSTEP2S)

There are four intensity patterns $P1$ to $P4$, each for a different value of $H1$ to $H4$. Each two are written for the same wavelength. The values of $H1$ and $H2$ have been chosen such that $P1$ and $P2$ show the constructive interference pattern for $\lambda 1$ and $\lambda 2$, and $H3$ and $H4$ that P3 and P4 show the destructive interference pattern for $\lambda 1$ and $\lambda 2$.

DA2FAGSTEP2S is only on the CD.

Application A3.22.

1. Get FileFig 3.A2 on the screen and save it. Then modify $P1$ to $P4$ such that all H have the same height. Now we can simulate the lamellar grating interferogram for two wavelengths $\lambda 1 = 0.0005$ and $\lambda 2 = 0.0007$ by changing from $H = 0.00005$ in steps of 0.00005 and see how the constructive and destructive interference patterns change on the observation screen for the two wavelengths. An interferogram can be obtained by observing the center and recording the sum of the intensities for the two wavelengths.
2. We go back to FileFig 3.A2. The four patterns show constructed and destructed interference for different settings of the lamellar grating. In other words, all four patterns using $H1$ to $H4$ are different. For the two wavelengths, we have the constructed interference at the center. Destructive interference appears at different length Y from the center on the observation screen. Since the detector should only observe the zero order, one has to choose the size such that the first orders are not detected.
3. What is the diameter of the detector area when the smallest wavelength is $\lambda 1 = 0.001$ and the largest wavelength $\lambda 2 = 0.004$?
4. What are the changes of the detector area, when changing a to $2a$ or $\frac{1}{2}a$?
5. What happens when changing N.

APPENDIX 3.2

A3.2.1 Cornu's Spiral

A graph of the Fresnel integrals S versus C is called a Cornu spiral (FileFig 3.A3). One can graphically obtain a diffraction pattern from Cornu's spiral. As an example we discuss the diffraction on a slit. The intensity is given by

$$I(Y) = \{C[\eta_2(Y)] - C[\eta_1(Y)]\}^2 + \{S[\eta_2(Y)] - S[\eta_1(Y)]\}^2 \tag{A3.5}$$

and depends on the points $\eta_1(Y)$ and $\eta_2(Y)$, assuming that d, λ, and X are given. $I(Y)$ is the square of the geometrical distance on the Cornu spiral between the points η_1 and η_2. The diffraction pattern is obtained by plotting (distance)^2 as a function of Y for η_1 and η_2. By the division of all values by (distance)^2 for $Y = 0$ one can obtain a normalized pattern.

FileFig A3.23 (DA3FECOR)

Graph of $S(Y)$ as function of $C(Y)$. This graph is called Cornu's spiral.

DA3FECOR is only on the CD.

A3.2.2 Babinet's Principle and Cornu's Spiral

We consider two complementary screens, I and II. Screen I may just have a hole and screen II has a stop of the same size as the hole. If added together, no light may pass. Babinet's principle tells us that complementary screens generate similar diffraction patterns. In FileFig 3.A3 we have plotted one-half of the Cornu spiral for η from 0 to ∞. Let us consider a slit and the complementary screen, a stop of the same width. The diffraction pattern of screen I for one point Y is obtained by measuring the length between the corresponding points η_1 and η_2 (as discussed

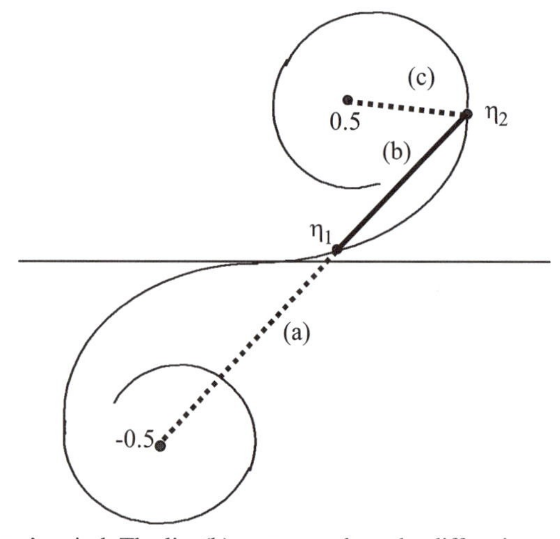

FIGURE A3.30 Cornu's spiral. The line(b) corresponds to the diffraction pattern of a slit and (a) and (c) of a stop.

above) which we call length (b) (see Figure A3.30). The diffraction pattern of screen II may be obtained by using the distances (a) from $-\infty$ to η_1 and (c) from η_2 to ∞. If we want to get the diffraction pattern on a different point Y, the points η_1 and η_2 are displaced in correspondence with the choice of the Y-value and will affect the diffraction pattern for both screens in the same way.

The distances (a) and (c) and the distance (b) from η_1 to η_2 represent the diffraction pattern for the addition of the two complementary screens for which no diffraction pattern can be observed.

C H A P T E R

Coherence

4.1 SPATIAL COHERENCE

4.1.1 Introduction

In the chapter on interference, we always considered only one incident wave and assumed that it was emitted by a distant point source. Recalling Young's experiment, the incident wave generated two new waves at the two openings at distance a of a double slit screen. The two new waves had a fixed phase relation and their superposition generated the interference pattern. The two waves were called coherent waves. In our model description we used two monochromatic waves with a fixed phase relation. The superposition generated the amplitude interference pattern, and the corresponding intensity pattern could be related to observations.

In this chapter we study waves emitted independently from several distant sources and assume that there is no fixed phase relation among them. Using for the analysis the double slit aperture of Young's experiment, one may observe an intensity interference pattern for specific distances of the source points between them. However, one may also choose other specific distances, and one will not find an intensity interference pattern. It is common to call the light of the same sources "spatially coherent" when an intensity interference pattern is observed and "spatially incoherent" when the intensity interference pattern disappears.

4.1.2 Two Source Points

Let us look at a double star, where each star emits its light independently of the other. The waves of both are incident on a screen of two slits of width d and separation a of Young's experiment. In Figure 4.1, this is shown (without diffraction) for one source point S_1 positioned on the axis and for a second source point S_2 at distance s from the first.

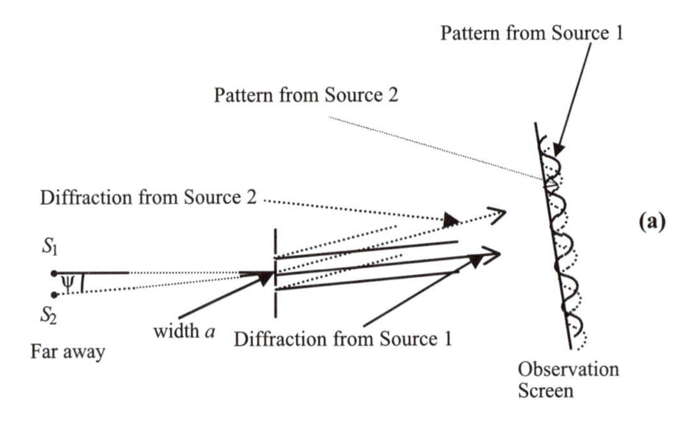

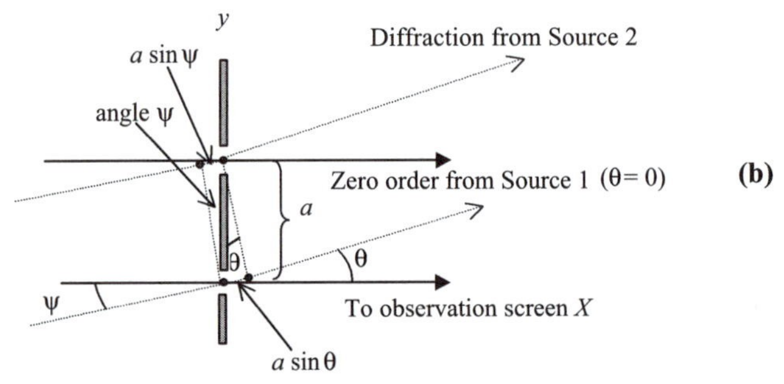

FIGURE 4.1 (a) Superposition of two of Young's experiments using separate sources S_1 and S_2. The distance between the sources is s and both "use" the same separation a of the double slit. The distance from the sources to the aperture is Z, and to the observation screen is X; (b) light from S_2 has the angle ψ with the axis and is diffracted into the angle θ.

For each source point we have an intensity diffraction pattern and for the two source points we look at the superposition of two intensity patterns. We assume in our model calculation that each source point generates a monochromatic wave of wavelength λ and both are incident on the double slit aperture. The light from source point S_1 produces on the observation screen the intensity

$$I(\theta, 0) = I_0\{\sin[(\pi d/\lambda)\sin\theta]/[(\pi d/\lambda \sin\theta]\}^2\{\cos[(\pi a/\lambda)\sin\theta]\}^2, \quad (4.1)$$

where θ is the diffraction angle and Io is the normalized intensity. Equation (4.1) is obtained from the discussion of the double slit (Chapter 3, FileFig 3.10). For source point S_2, the axis of the double slit experiment is rotated by the angle ψ around the point at the center between the two openings. As a result, the diffraction angle is counted from the new axis and we have to use $d(\sin\theta - \sin\psi)$ and $a(\sin\theta - \sin\psi)$ in the diffraction and interference factors of Equation (4.1), respectively. For the intensity of the light from point S_2 we have

$$I(\theta, \psi) = I_0\{\sin[(\pi d/\lambda)(\sin\theta - \sin\psi)]/[(\pi d/\lambda)(\sin\theta - \sin\psi)]\}^2 \cdot \{\cos[(\pi a/\lambda)(\sin\theta - \sin\psi)]\}^2. \quad (4.2)$$

The calculation of the optical path difference is similar to the calculation of the optical path difference for a grating, illuminated under an angle to the normal (see Chapter 3, Eq. (3.61)). In this model of spatial coherence, monochromatic light from each point source uses the same double slit aperture and generates an intensity fringe pattern. The waves from each source point have no fixed phase relation between each other and each produces an intensity fringe pattern of its own. The intensities of these two fringe patterns are superimposed. Whether fringes can be observed depends not only on the separation s (and consequently on ψ) and the wavelength, but also on the separation a of the two slits in the double slit arrangement. However, this separation "a" is assumed to be constant.

In FileFig 4.1 we calculate the superposition of the intensity pattern depending on the separation of the two source points. The separation s is taken in "common length units" as discussed in Chapter 1. The first graph shows the intensity interference pattern for both source points at the same spot, that is, for $s = 0$. The second graph shows the reduced interference pattern for the distance between the two source points of $s = 1.5$. The third graph shows the disappearance of the intensity pattern at the distance of $s = 2.25$ and the fourth graph shows the reappearance at $s = 2.6$. We see that the superposition of the intensity pattern, produced by the two sources with incoherent light, cancel for the specific distance between the two source points of $s = 2.25$.

When fringes are observed of the superposition of the two intensity fringe patterns, one calls the light producing the fringe pattern spatially coherent. When no fringes are observed, the light is called spatially incoherent.

FileFig 4.1 (C1COH2S)

Graphs are shown for the superposition of the intensities $I(\theta, 0)$ and $I(\theta, \psi)$ for two point sources at variable distances s as a function of the angle θ. Parameters used are the separation of the two openings $a = 1$ mm, opening of the slits $d = 0.05$mm, wavelength $\lambda = 0.0005$ mm, distance from source to double slit $Z = 9000$ mm, and distance from aperture to observation screen $X = 4000$ mm. Four distances are used, $s = 0$, $s = 1.5$mm, $s = 2.25$mm, and $s = 2.6$mm, of separation $s = Z\psi$ of the two source points, corresponding to four values of ψ. Fringe patterns are observed for separations s smaller than 2.25 mm. For $s \cdot \psi = \lambda/2$, that is, $s = 2.25$ mm, we have for the first time disappearance of fringes; that is, the maxima of $I(\theta, \psi)$ are at the minima of $I(\theta, 0)$. For s larger than 2.25 mm the fringe pattern reappears.

C1COH2S

Intensity of Two Sources Separated by s

Superposition of two double slit patterns. The slits have width d and separation a; one pattern is untilted with $\psi = 0$, the other tilted by $\psi = x/Z$, and distance from sources to slit is Z. Distance from slit to screen is X, coordinate on screen is Y, $Y/X = \theta$. By enlarging ψ, starting from 0, one finds the first fringe disappearance. If ψ is further enlarged, the fringes reappear, but now the minima are not zero. Another point of view: fringes may disappear for constant s and changing a.

$\theta \equiv -.006, -.00599, .006 \qquad d \equiv .05 \qquad a := 1 \qquad Z \equiv 9000 \qquad \lambda := .0005$

$$I1(\theta) := \frac{\sin\left[(\pi)\cdot\frac{d}{\lambda}\cdot\sin(\theta)\right]^2}{\left[\pi\cdot\frac{d}{\lambda}\cdot(\sin(\theta))\right]^2}\cdot\cos\left[\pi\cdot\frac{a}{\lambda}\cdot(\sin(\theta))\right]^2$$

$$II1(\theta) := \frac{\sin\left[\pi\cdot\frac{d}{\lambda}\cdot(\sin(\theta))\right]^2}{\left[\pi\cdot\frac{d}{\lambda}\cdot(\sin(\theta))\right]^2}\cdot\cos\left[\pi\cdot\frac{a}{\lambda}\cdot(\sin(\theta)+\sin(\psi 1))\right]^2 \qquad s1 \equiv 0 \qquad \psi 1 \equiv \frac{s1}{Z}.$$

$$II2(\theta) := \frac{\sin\left[\pi\cdot\frac{d}{\lambda}\cdot(\sin(\theta))\right]^2}{\left[\pi\cdot\frac{d}{\lambda}\cdot(\sin(\theta))\right]^2}\cdot\cos\left[\pi\cdot\frac{a}{\lambda}\cdot(\sin(\theta)+\sin(\psi 2))\right]^2 \qquad s2 \equiv 1.5 \qquad \psi 2 := \frac{s2}{Z}.$$

$$II3(\theta) := \frac{\sin\left[\pi \cdot \frac{d}{\lambda} \cdot (\sin(\theta))\right]^2}{\left[\pi \cdot \frac{d}{\lambda} \cdot (\sin(\theta))\right]^2} \cdot \cos\left[\pi \cdot \frac{a}{\lambda} \cdot (\sin(\theta) + \sin(\psi 3))\right]^2 \qquad s3 \equiv 2.25 \qquad \psi 3 := \frac{s3}{Z}.$$

I1(θ) II3(θ)

1, 0.5, 0

−0.006, −0.004, −0.002, 0, 0.002, 0.004, 0.006

θ

$$II4(\theta) := \frac{\sin\left[\pi \cdot \frac{d}{\lambda} \cdot (\sin(\theta))\right]^2}{\left[\pi \cdot \frac{d}{\lambda} \cdot (\sin(\theta))\right]^2} \cdot \cos\left[\pi \cdot \frac{a}{\lambda} \cdot (\sin(\theta) + \sin(\psi 4))\right]^2 \qquad s4 \equiv 2.6 \qquad \psi 4 := \frac{s4}{Z}.$$

I1(θ) II4(θ)

1.5, 1, 0.5, 0

−0.006, −0.004, −0.002, 0, 0.002, 0.004, 0.006

θ

Application 4.1.

1. Change d to 0.1 and determine the s value for disappearance of fringes. Save the changed values for future comparisons.
2. Change λ to 0.0006 and determine the s value for disappearance of fringes. Save the changed values for future comparisons.
3. Change a to 1.2 and determine the s values for disappearance of fringes. Save the changed values for future comparisons.

4.1.3 Coherence Condition

We have seen that the two source points produce superimposed intensity patterns for certain distances s between them. In our case we found fringes for distance $s = 0$, and for a larger distance $s = 2.25$mm, the fringes disappeared. For a small separation s of the two source points, one would observe fringes similar to the pattern produced by one source point. The coherence condition tells us how large one can make the separation s and still have a fringe pattern similar to the one produced by one source point only. For the discussion, we assume that the openings of the two slits are very small and therefore we may omit diffraction. We then have for the pattern generated by sources S_1 and S_2,

$$I_1(\theta, 0) = I_0\{\cos[(\pi a/\lambda)\sin\theta]\}^2 \tag{4.3}$$

$$I_2(\theta, \psi) = I_0\{\cos[\pi a/\lambda)(\sin\theta - \sin\psi)]\}^2. \tag{4.4}$$

The center of the pattern of I_1 is at $(a \sin\theta) = 0$ and the first minimum is at $(a \sin\theta) = \lambda/2$. We want to find the magnitude of the angle ψ for which we will see only one fringe pattern. If the center of the pattern of I_2 is at the minimum of I_1 (i.e., for $(a \sin\psi) = \lambda/2$), the angle ψ is too large and we are at the position where fringes disappear for the first time. Therefore, in order to observe just one fringe pattern of the superposition of I_1 and I_2 we have to require that

$$(a \sin\psi) \ll \lambda/2. \tag{4.5}$$

Equation (4.5) may also be written in good approximation as $s \cdot a/Z$, where a/Z is the opening angle ω from the source points (Figure 4.2) and one has

$$s \cdot a/Z = s \cdot \omega \ll \lambda/2. \tag{4.6}$$

The opening angle ω seen from the middle of the two source points is given as a/Z and the product of this angle times the separation s of the source points must be small compared to half of the wavelength. We state this as

$$\text{(source size) times (opening angle } \omega\text{) must be} \ll \lambda/2. \tag{4.7}$$

This is called the *coherence condition.*

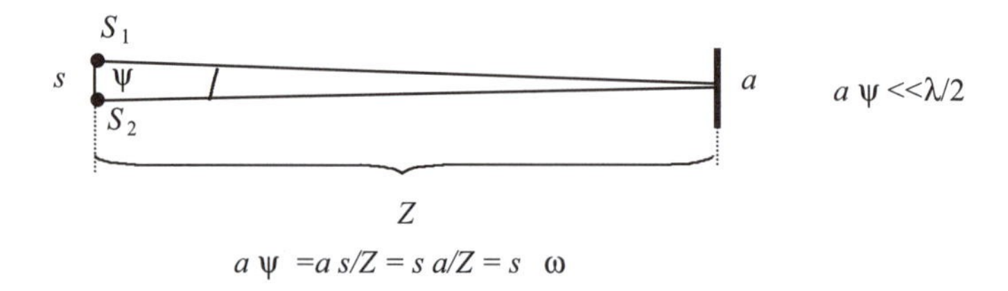

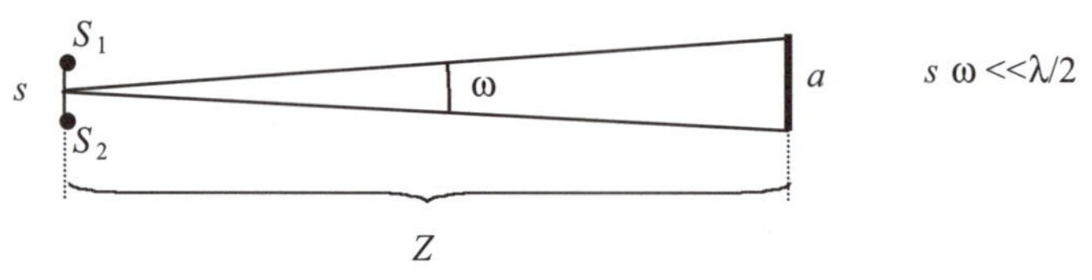

FIGURE 4.2 The condition $a \sin\psi = a(s/Z) \ll \lambda/2$ is equivalent to $sa/Z = s \sin\omega \ll \lambda/2$ and tells us that the product of the distance between the source times the opening angle must be similar to that of a pointlike source.

4.1.4 Extended Source

The discussion of the double star was the first step in investigating the question of how far away a source has to be to qualify as a point source, even if it has a finite diameter. To discuss an extended source we look at a line source made of a sequence of point sources. The distance between the source points is considered infinitesimally small, but there is no fixed phase relation between the light of one source point with respect to any other. We apply to the line source what we have done for two source points, and now have to integrate $I(\theta, \psi)$ over the angle ψ from 0 to $\psi_s = s/Z$

$$I_0 = \int_0^{s/Z} \{\sin[(\pi d/\lambda)(\sin\theta - \sin\psi)]/[(\pi d/\lambda)(\sin\theta - \sin\psi)]\}^2 \cdot \{\cos[\pi a/\lambda)(\sin\theta - \sin\psi)]\}^2 d\psi. \quad (4.8)$$

The integration may be interpreted as the superposition of the intensity fringe pattern, produced by all source points between S_1 and S_2. The upper integration limit, that is, the angle s/Z, is taken as such a numerical value that we can compare the angles with calculations of the fringes for the two point sources. The first graph in FileFig 4.2 shows the intensity interference pattern for a line source of length $s = 1$ mm. The second graph for the length of $s = 1.5$ mm again shows a fringe pattern. The third graph shows the disappearance of the intensity pattern at a length of $s = 4.5$, and the fourth graph shows the reappearance at $s = 5$. We see that the integrated intensity pattern is produced by an extended source of length $s = 4.5$ mm. All fringes cancel for $s = 4.5$ mm. This value is twice as large as $s = 2.25$ mm, the value we found for the two source points.

We may associate the length of a line source with the diameter of a circular source. The disappearance of the interference pattern occurs when the diameter of the circular source is twice the distance of the two source points. In Figure 4.3 we show an experimental setup to observe a fringe pattern of an extended source.

The coherence condition (Eq. (4.7)) may be applied to the extended source as well. One has that the area s times the solid angle a/Z of the source must be small compared to $(1/2)\lambda$. This condition must be obeyed when setting up Young's experiment. The source point is the illuminated hole of the first screen and must be so small and far away that the wave at the next two holes has a fixed phase relation. The size of the hole must be determined by the opening angle a/Z of the experimental setup. On the other hand, a star may be very large, but if the star is so far away that the product s times the opening angle ω (see Figure 4.2) is small compared to $\lambda/2$, the coherence condition is met. As the result, the light from the star is like parallel light when entering the double slit. Fringes may be observed and the light of the "large star" is called spatially coherent.

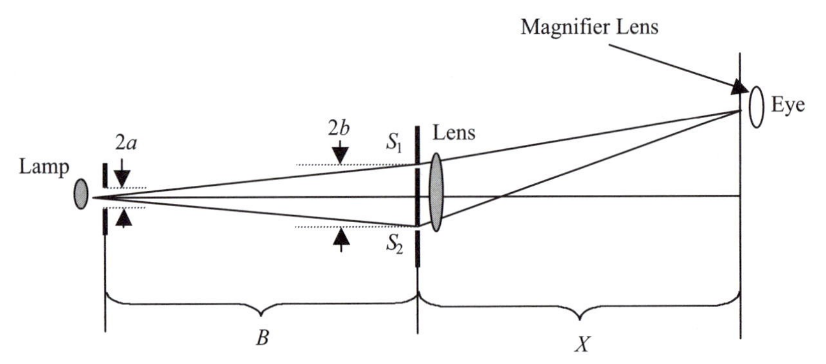

FIGURE 4.3 Laboratory setup for the observation of fringes from an extended source. Experimental values: $B = 20$ m, $X = 1$ m, width of S_1 and $S_2 = .4$ mm, $b = 6$ mm, $\lambda = .00057$ mm. At $2a = 2$ mm, the fringes disappear for the first time, and the upper limit of the experiment is $2a = 4$ mm. (Adapted from Einführung in die Optic, R.W. Pohl, Springer-Verlag, Heidelberg, 1948.)

FileFig 4.2 (C2COHEX)

Graphs are shown for the superposition of the integrated intensities $I(\theta, \psi_0)$ depending on the upper limit of the integration values ψ_0 of the angle $\psi_0 = s/Z$. Four values of ψ_0 are used, corresponding to $s = 1$ mm, $s = 1.5$ mm, $s = 4.5$ mm, and $s = 5$ mm. Parameters used are the separation of the two openings $a = 1$ mm, opening of the slits $d = 0.05$ mm, wavelength $\lambda = 0.0005$ mm, distance from source to the double slit $Z = 9000$ mm, and distance from aperture to observation screen $X = 4000$ mm. By choosing the same values for a and λ as used in FileFig 4.1, we find fringes disappear for the first time for the size of the source of $s = 4.5$ mm, that is, twice as large as found for the distance between the two point sources. Fringe patterns are observed for separations s smaller than 4.5 mm. For $s \cdot \psi = \lambda/2$, that is, $s = 4.5$ mm we have disappearance of fringes for the first time. For s larger than 4.5 mm the fringe pattern reappears.

C2COHEX

Intensity of an Extended Source

Width is s and interference diffraction is on a double slit. Slit openings are d and separation a, distance from source to slit Z, from slit to screen X, coodinare on screen is Y, and small angle approximation $Y/X = \theta$.

$$a := 1 \quad d \equiv .05$$

$$0 \le \psi \le 2 \quad k \equiv 0..200 \quad \theta_k \cdot \equiv .01 - k \cdot .0001 \quad \lambda := .0005 \quad Z = 9000$$

$$s1 \equiv 1 \qquad \psi_1 \equiv \frac{s1}{Z}$$

$$I1_k := \int_0^{\psi_1} \frac{\sin\left[\pi \cdot \frac{d}{\lambda} \cdot (\sin(\theta_k) + \sin(\psi))\right]^2}{\left[\pi \cdot \frac{d}{\lambda} \cdot (\sin(\theta_k) + \sin(\psi))\right]^2} \cdot \cos\left[\pi \cdot \frac{a}{\lambda} \cdot (\sin(\theta_k) + \sin(\psi))\right]^2 d\psi.$$

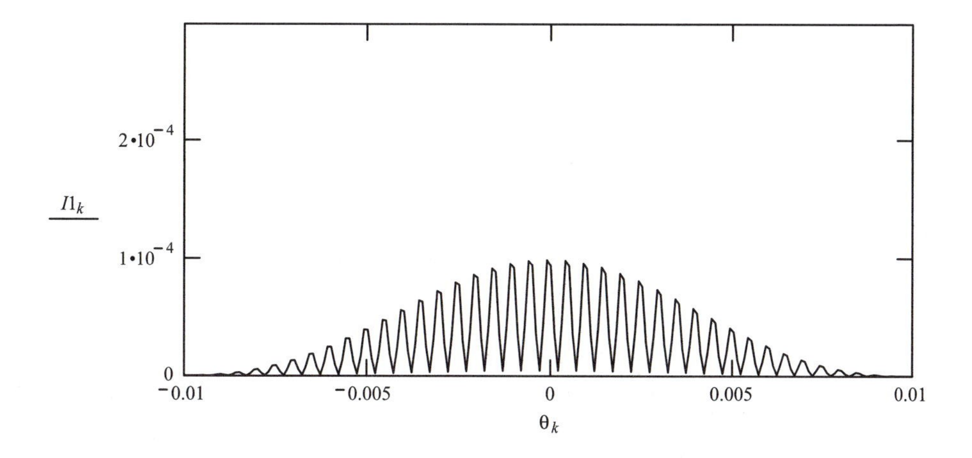

$$s2 \equiv 1.5 \qquad \psi_2 \equiv \frac{s2}{Z}$$

$$I2_k := \int_0^{\psi_2} \frac{\sin\left[\pi \cdot \frac{d}{\lambda} \cdot (\sin(\theta_k) + \sin(\psi))\right]^2}{\left[\pi \cdot \frac{d}{\lambda} \cdot (\sin(\theta_k) + \sin(\psi))\right]^2} \cdot \cos\left[\pi \cdot \frac{a}{\lambda} \cdot (\sin(\theta_k) + \sin(\psi))\right]^2 d\psi.$$

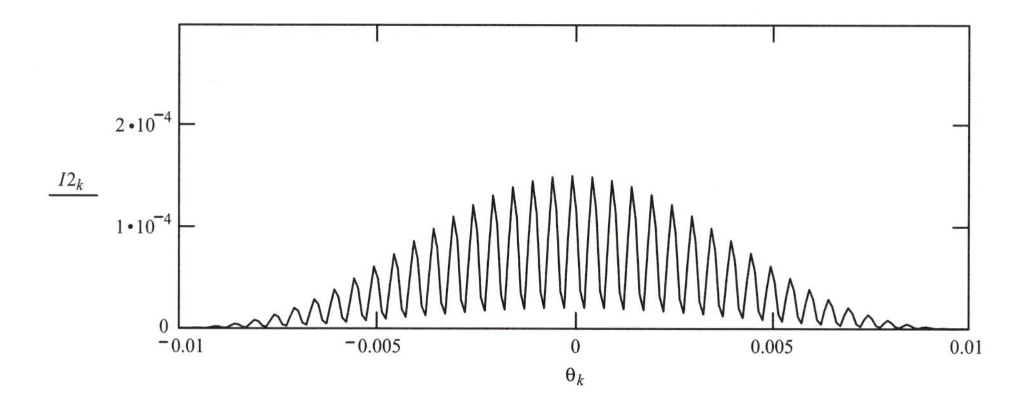

$$s3 \equiv 4.5 \qquad \psi_3 \equiv \frac{s3}{Z}$$

$$I3_k := \int_0^{\psi_3} \frac{\sin\left[\pi \cdot \frac{d}{\lambda} \cdot (\sin(\theta_k) + \sin(\psi))\right]^2}{\left[\pi \cdot \frac{d}{\lambda} \cdot (\sin(\theta_k) + \sin(\psi))\right]^2} \cdot \cos\left[\pi \cdot \frac{a}{\lambda} \cdot (\sin(\theta_k) + \sin(\psi))\right]^2 d\psi.$$

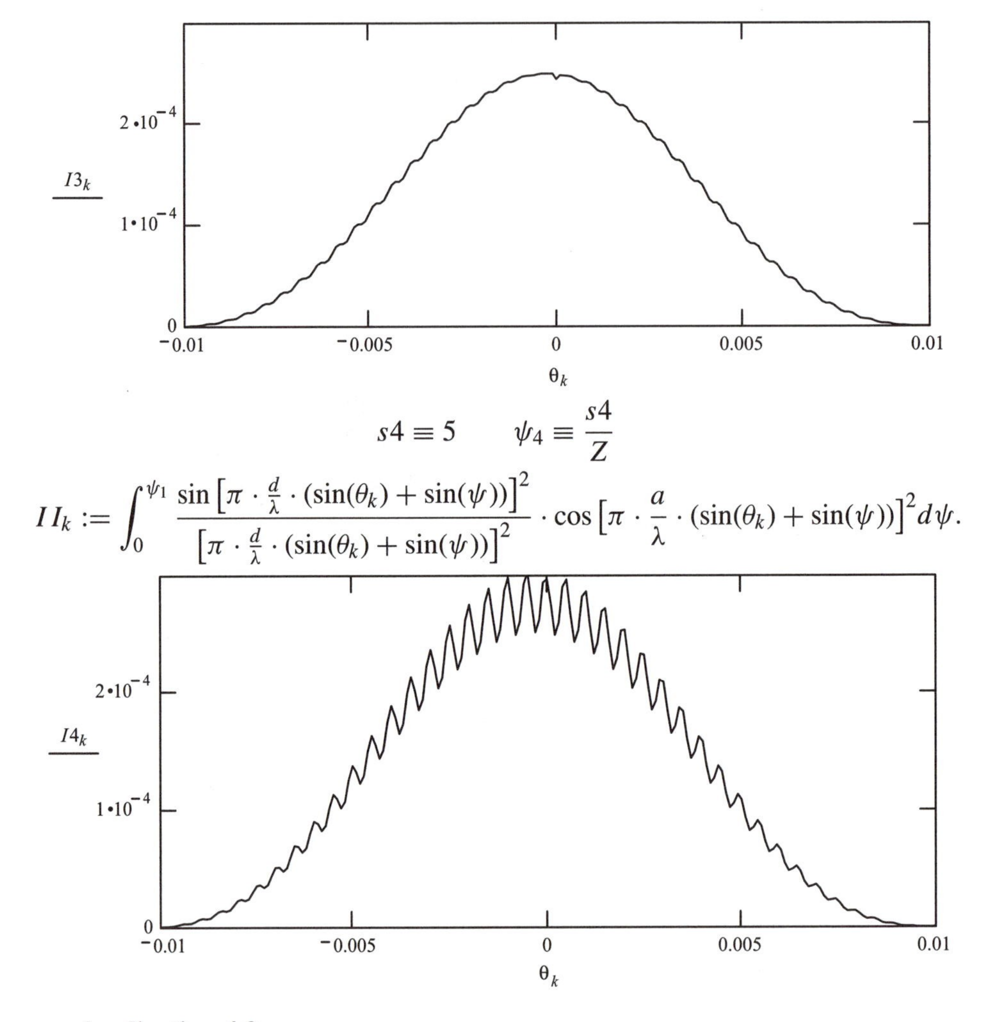

$$s4 \equiv 5 \qquad \psi_4 \equiv \frac{s4}{Z}$$

$$I I_k := \int_0^{\psi_1} \frac{\sin\left[\pi \cdot \frac{d}{\lambda} \cdot (\sin(\theta_k) + \sin(\psi))\right]^2}{\left[\pi \cdot \frac{d}{\lambda} \cdot (\sin(\theta_k) + \sin(\psi))\right]^2} \cdot \cos\left[\pi \cdot \frac{a}{\lambda} \cdot (\sin(\theta_k) + \sin(\psi))\right]^2 d\psi.$$

Application 4.2.

1. Change d to 0.1 and determine the s value for disappearance of fringes. Compare to FileFig 4.1.
2. Change λ to 0.0006 and determine the s value for disappearance of fringes. Compare to FileFig 4.1.
3. Change a to 1.2 and determine the s values for disappearance of fringes. Compare to FileFig 4.1.

4.1.5 Visibility

4.1.5.1 Visibility for Two Point Sources

We have discussed appearance and disappearance of fringe patterns for two point sources at variable distances, and for extended sources of variable diameters. To measure how well the fringes may be seen, Michelson has defined the visibility of fringes as the absolute value of

$$V = \left| \frac{I_{\text{tot,max}} - I_{\text{tot,min}}}{I_{\text{tot,max}} + I_{\text{tot,min}}} \right|. \tag{4.9}$$

The intensities are taken in the nominator and denominator of one fringe. For the pattern shown in the first graph of FileFig 4.2, one has $I_{\text{max}} = 1$ and $I_{\text{min}} = 0$ and obtains for the visibility, $V = 1$. When considering the superposition of two intensity patterns, one has to determine I_{max} and I_{min} depending on the angle ψ. We saw that the angle ψ is small because Z is large and applied the small angle approximation (i.e., $\sin\theta = Y/X$ and $\sin\psi = Y'/X$). From Eqs. (4.3) and (4.4), disregarding the diffraction factors, we have for the total intensity I_{tot} (i.e., the sum of Eqs. (4.3) and (4.4)),

$$I_{tot}(Y) = \{\cos[(\pi a/\lambda X)(Y)]\}^2 + \{\cos[(\pi a/\lambda X)(Y - Y')]\}^2. \tag{4.10}$$

Using $2\cos^2\alpha/2 = 1 + \cos\alpha$, we get

$$I_{tot}(Y) = 1 + 1/2\{\cos[(2\pi a/\lambda X)(Y)]\} + 1/2\{\cos[(2\pi a/\lambda X)(Y - Y')]\}, \tag{4.11}$$

and with $\cos\alpha + \cos\beta = 2\cos\{\alpha + \beta/2\}\cos\{\alpha - \beta)/2\}$ one has

$$I_{tot}(Y) = 1 + \{\cos[(2\pi a/\lambda X)(Y'/2)]\}\{\cos[(2\pi a/\lambda X)(Y - Y'/2)]\}. \tag{4.12}$$

The minimum and maximum values of I_{tot} depend only on Y, since Y' is fixed by the angle ψ for which we want to determine the visibility. The maximum value of $\{\cos(2\pi a/\lambda X)(Y - Y'/2)\}$ is 1 because that is the maximum of the cos-function and similarly the minimum value is -1. We have therefore

$$\begin{aligned} I_{\text{tot,max}} &= 1 + \{\cos[(2\pi a/\lambda X)(Y'/2)]\} \\ I_{\text{tot,min}} &= 1 - \{\cos[(2\pi a/\lambda X)(Y'/2)]\} \end{aligned} \tag{4.13}$$

and using the definition of Eq. (4.9) one obtains for the visibility

$$V = |\cos(\pi a/\lambda X)(Y')|. \tag{4.14}$$

For small values of $\psi = Y'/X = s/Z$ we have a maximum. The first zero of the visibility is for $\pi a Y'/\lambda X = \pi/2$; that is, $Y'/X = s/Z = \lambda/2a$.

In FileFig 4.3 we have plotted Eq. (4.14), using similar values to those in FileFig 4.1. We observe that the visibility is first zero for $s = 2.25$ mm, the value we determined in FileFig 4.1 for the first disappearance of the fringes. For larger values of s we found reappearance of fringes (see the fourth graph of FileFig 4.1).

The interval from $s = 0$ to the value of s for which the visibility is first zero is called the *coherence interval*.

FileFig 4.3 (C3VIS2S)

Graph of the visibility for two point sources.

C3VIS2S is only on the CD.

Application 4.3.

1. Change λ to 0.0006 and determine the s value for $V = 0$. Compare to FileFig 4.1.
2. Change a to 1.2 and determine the s value for $V = 0$. Compare to FileFig 4.1.

4.1.5.2 Visibility for an Extended Source (Line Source)

The visibility for an extended source is obtained by determination of $I_{\text{tot,max}}$ and $I_{\text{tot,min}}$ of the total integrated intensity. Again, disregarding the diffraction factor, one has

$$I_{tot}(Y) = I_0 \int_0^{s/Z} \{\cos[(\pi a/\lambda)(\sin\theta - \sin\psi)]\}^2 d\psi. \tag{4.15}$$

We calculate in small angle approximation the integral

$$I_{tot}(Y) = I_0 \int_0^{Y'/X} \{\cos[(\pi a/\lambda)(Y/X - Y''/X)]\}^2 d(Y''/X). \tag{4.16}$$

First one has to go through the steps of Eqs. (4.10) to (4.12) and determine the maxima and minima. The second step is the integration of Eq. (4.14) and the result is

$$V = |\{\sin(\pi a/\lambda X)(Y')\}/(\pi a/\lambda X)(Y')|, \tag{4.17}$$

where $Y'/X = s/Z$. The first minimum of the visibility is obtained at $(\pi a/\lambda X)Y' = \pi$, (that is for $Y'/X = \lambda/a$). This is twice the value we found for the two point sources at distance s. In FileFig 4.4 we have plotted, for similar values to those used in FileFig 4.2, Eq. (4.17) as a function of the length s of the extended source. One finds the first disappearance of the fringes at $s = 4.5$, the same value determined in FileFig 4.2.

FileFig 4.4 (C4VISEX)

Graph of the visibility for an extended source.

C4VISEX is only on the CD.

Application 4.4.

1. Change λ to 0.0006 and determine the s value for $V = 0$. Compare to FileFig 4.2.
2. Change a to 1.2 and determine the s value for $V = 0$. Compare to FileFig 4.2.

4.1.6 Michelson Stellar Interferometer

In the discussion of the intensity pattern produced by Young's experiment for the two source points and the extended source, we assumed a fixed separation a of the two openings. We studied the change in the intensity pattern depending on the separation s of the source points and the diameter s of the extended source. Michelson was interested in the measurement of the diameter of a star. In the Young's experiment we discussed above, the separation a of the two slits was fixed and the distance between the two source points s was varied. Application of this experiment to the measurement of the diameter of a star is very difficult because both parameters s and a are fixed. Michelson wanted to modify Young's experiment by making the separation a variable. However, since the diameter he wanted to measure was so small, the setup of Young's experiment had to be modified.

We start with Young's experiment for two source points. The fringe pattern of the superposition of the two patterns on the observation screen depends on the distance of the two source points. This is shown in Fig. 4.4a. The intensities of the two source points are for small angles

$$u_I = A^2\{\cos(\pi a\theta/\lambda)\}^2 \tag{4.18}$$

$$u_{II} = A^2\{\cos(\pi a(\theta - \phi)/\lambda)\}^2. \tag{4.19}$$

The angle ϕ is what we wish to determine. In the setup of Figure 4.4a the displacement of the intensity pattern of source I with respect to source II is limited by the size of the distance a between the two slits. The modification is shown in Figure 4.4b. A new parameter h of the large distance between the first two mirrors is introduced. Changing the length h produces a change of the fringe pattern. From the variation of the fringes for two values of h one can determine the angle ϕ. In Figure 4.4b we show the modified setup and the two intensity patterns are now given as

$$u_{\mathrm{I}} = A^2\{\cos(\pi a\theta/\lambda)\}^2 \tag{4.20}$$

$$u_{\mathrm{II}} = A^2\{\cos[\pi(a\theta - h\phi)/\lambda]\}^2. \tag{4.21}$$

The superposition $u_I + u_{II}$ will show maxima and minima depending on h for fixed ϕ and a. We may calculate the angle ϕ from the observed h values for a resulting maximum and the following minimum. For a resulting maximum we

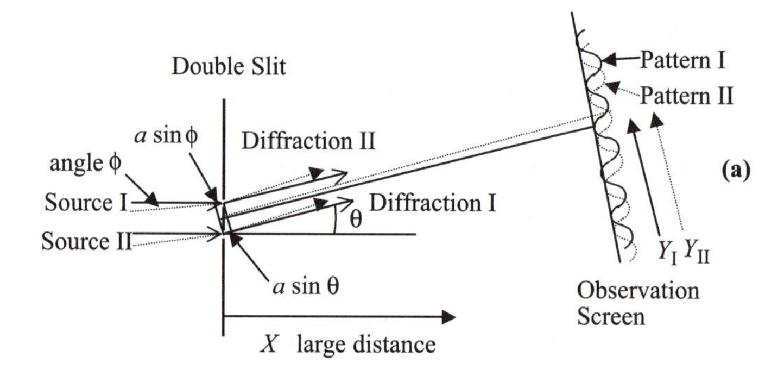

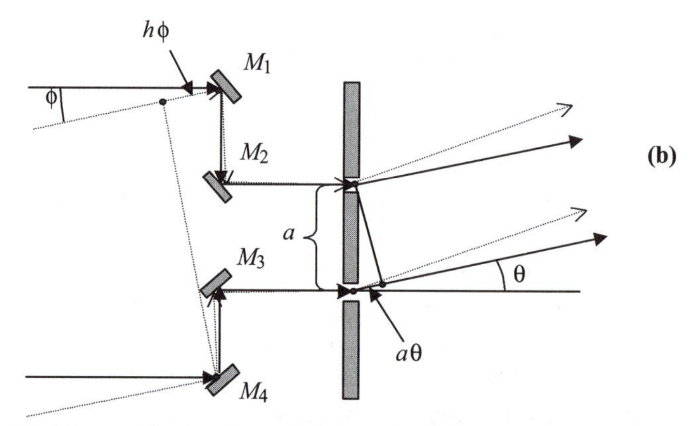

FIGURE 4.4 Michelson stellar interferometer: (a) light waves from two stars, I and II, forming an angle ϕ when they arrive at a double slit of width a. The waves from each star produce a fringe pattern on a screen at distance X, described by the coordinate Y_I and Y_{II}; (b) four mirrors are added, M_1 to M_4, to produce a new optical path difference $h\phi$ for the incident light, where h is adjustable. The angle θ is the diffraction angle.

have

$$[\pi(a\theta - h_1\phi)/\lambda] = (\pi/2)(2m) \tag{4.22}$$

and for the following minimum

$$[\pi(a\theta - h_2\phi)/\lambda] = (\pi/2)(2m + 1). \tag{4.23}$$

The difference for $h_2 - h_1 = \lambda/2\phi$ and ϕ is obtained since $h_2 - h_1$ can be measured. This type of modified interferometer was applied to measure the angular diameter of Betelgeuse in the Orion constellation. At the Mt. Wilson observatory an interferometer was used with a distance h of the two mirrors of 302 cm (121 in.). The angle was determined to be 22.6×10^{-8} rad. The distance to the star was known from parallax measurement and the diameter was determined to be about 300 times that of the sun. A simulation with a numerical example is given in FileFig 4.5. The graph shows a plot of the interference pattern for assumed values of h and ϕ. We determine the two values of h for observance and disappearance of fringes. Application of $h_2 - h_1 = \lambda/2\phi$ results in the angle and that must be the angle we had to assume for the simulation.

FileFig 4.5 (C5MICHSTS)

Graphs are shown of the resulting intensity interference pattern of u_I and u_{II}, depending on values of h over the range $Y = -20$ to 20. The distance from source to interferometer is $X = 4000$, wavelength $\lambda = 0.0005$, separation in the interferometer $a = 0.5$. For the simulation we have to use a value for the angle we actually want to determine and choose $\phi = 0.00005$. We then can determine the h values of the minima and maxima and they must satisfy the relation $h_2 - h_1 = \lambda/2\phi$.

C5MICHSTS

Michelson's Stellar Interferometer

Diffraction angle is Y/X, wavelength λ, and angle to be determined is Φ. Interferometer distance of mirrors $M1$ and $M4$ is h. In the real setup we change h to go from fringe pattern to no fringe pattern. From the difference of these two values we calculate the angle Φ. In this simulation we choose an angle Φ and show that the fringe pattern changes for the two values of h we determine. Example h equals 100 and 95.

$Y := -30, -29.9, 30 \qquad \Phi \equiv .00005 \qquad X := 4000 \qquad \lambda := .0005 \qquad d \equiv .5$

$$uI(Y) := \cos\left[\pi \cdot d \cdot \left(\frac{Y}{X \cdot \lambda}\right)\right]^2 \qquad uII(Y) := \cos\left(\pi \cdot \frac{\frac{Y}{X} \cdot d - h \cdot \Phi}{\lambda}\right)^2.$$

$$h \equiv 95$$

This is an indication of the presence or absence of fringes.

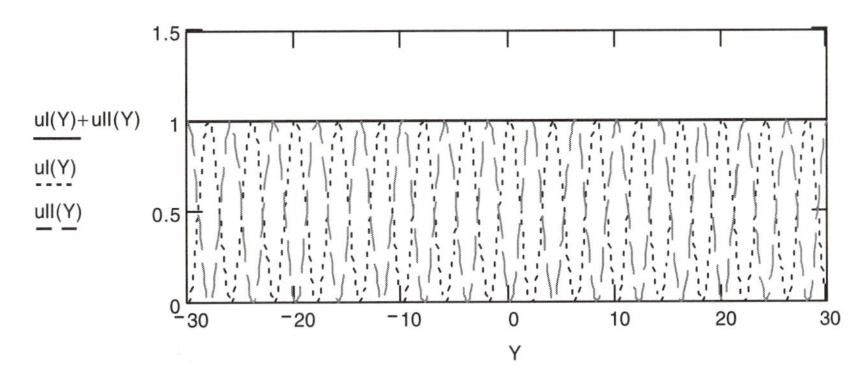

Application 4.5. Change the wavelength to 0.00055 and find ϕ from observed values of $h2 - h1$.

4.2 TEMPORAL COHERENCE

4.2.1 Wavetrains and Quasimonochromatic Light

We have studied in Chapter 2 the superposition of monochromatic waves and their amplitude and intensity pattern. In this section we examine finite wavetrains and their superposition. Monochromatic waves are infinitely long. The sum of a number of monochromatic waves with wavelengths in a certain wavelength interval results in a periodic wave. However, when integrating over the wavelength interval, one obtains a finite wavetrain, (see Figure 4.5). The wavetrain appears with decreasing amplitude for large distances. The length of the wavetrain $\Delta x = l_c$ is proportional $1/\Delta\nu$, where $\Delta\nu$ is the frequency interval corresponding to the wavelength interval $\Delta\lambda$ of the wavetrain. The average wavelength of the wavetrain is called λ_m. The reciprocity of Δx and $\Delta\nu$ comes from Fourier transformation theory. The "window" in the space domain is l_c and $\Delta\nu$ is the "window" in the frequency domain. The product Δx and $\Delta\nu$ is a constant and appears in modified form in quantum mechanics as the "uncertainty relation." Wavetrains which satisfy the condition

$$\Delta\lambda/\lambda_m \ll 1 \tag{4.24}$$

are called quasimonochromatic light.

To get an idea of how the waveform appears for quasimonochromatic light, we show in the first graph of FileFig 4.6 the superposition of four waves having wavelength $\lambda = 1.85, 1.95, 2.05$, and 2.15 for medium wavelength $\lambda_m = 2$. In the second graph we show the waveform for the integration over the same wavelength interval.

$$u(Y) = \int_{\lambda_1}^{\lambda_2} \{\cos(2\pi x/\lambda)\} d\lambda. \tag{4.25}$$

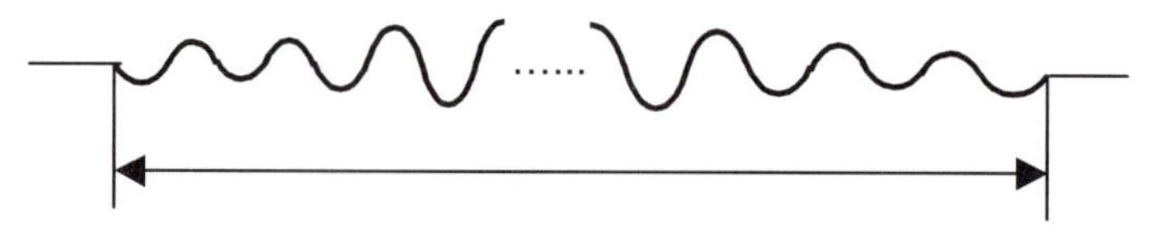

FIGURE 4.5 Schematic of a wavetrain of finite length l_c. One refers to l_c as the coherence length.

FileFig 4.6 (C6SUPERS)

First graph: Sum of four waves with wavelength $\lambda = 1.85, 1.95, 2.05$, and 2.15 for medium wavelength $\lambda_m = 2$. Second graph: Integration over the wavelength range from $\lambda = 1.85$ to 2.15.

C6SUPERS is only on the CD.

Application 4.6.

1. Extend the x coordinate to larger ranges to see more of the periodicity for the "summation" case and the decrease for the integration case.
2. Study the waveform for different wavelength intervals for both cases.
3. Extend the sum of four to a larger sum of different wavelengths, but keep the wavelength interval constant. Compare with the integration case.

4.2.2 Superposition of Wavetrains

In Chapter 2 we studied interference fringes produced by monochromatic light. For the magnitude of the superposition of two monochromatic waves with optical path difference δ, we have used $\cos(\pi\delta/\lambda)$.

Interference fringes may be observed for quasimonochromatic light of narrow width of wavelength $\Delta\lambda$. In FileFig 4.7 we show the amplitude pattern of the superposition of two wavetrains. The interval of integration is $\lambda = 1.85$ to 2.15 and three optical path differences are considered, $\delta = 0, \frac{1}{2}\lambda_m$, and λ_m.

$$I(Y) = \int_{\lambda_1}^{\lambda_2} [\{\cos(2\pi(x - \lambda_m a)/\lambda)\} + \{\cos(2\pi(x)/\lambda)\}]d\lambda. \tag{4.26}$$

The optical path difference of $\delta = 0$ corresponds to constructive interference for $\delta = (1/2)\lambda_m$ to destructive interference and for $\delta = \lambda_m$ again to constructive interference. The resulting amplitude of the case of destructive interference is not zero, but much smaller than for constructive interference. We see that the interference pattern decreases for larger and larger values of x. In FileFig 4.8 we have calculated the intensity pattern, corresponding to the cases of constructive interference, $\delta = 0$, and $\delta = \lambda_m$, and destructive interference, $\delta = \frac{1}{2}\lambda_m$.

FileFig 4.7 (C7COHTEMS)

Amplitude pattern of superposition of two wavetrains. Graphs 1 to 3: Integration of waves over wavelength range from $\lambda = 1.85$ to 2.15, having optical path differences of $\lambda = 0, \frac{1}{2}\lambda_m$, and λ_m, respectively.

C7COHTEMS is only on the CD.

Application 4.7. Change the wavelength interval to smaller values and approach

in 1: the corresponding case of the monochromatic wave;
in 2: the corresponding monochromatic case for destructive interference;
in 3: the corresponding monochromatic case for constructive interference.

FileFig 4.8 (C8COHINTS)

Intensity pattern of superposition of two wavetrains. Graphs 1 to 3: Integration of waves over wavelength range from $\lambda = 1.85$ to 2.15, having optical path differences of $\delta = 0, \frac{1}{2}\lambda_m$, and λ_m, respectively.

C8COHINTS is only on the CD.

Application 4.8. Change the wavelength interval to smaller values and approach

in 1: the corresponding case of the monochromatic wave;
in 2: the corresponding monochromatic case for destructive interference;
in 3: the corresponding monochromatic case for constructive interference.

4.2.3 Length of Wavetrains

The length of finite wavetrains (see Fig. 4.5) may be determined with a Michelson interferometer. The incident light is divided at the beam splitter into two parts traveling to M_1 and M_2, respectively (see Fig. 4.6a). Each part is reflected by a mirror and travels back to the beamsplitter (Fig. 4.6b). At the beam splitter, the reflected part of the light from M_1 and the transmitted part of the light from M_2 are superimposed and travel to the detector (Fig. 4.6c). When the mirror in the Michelson interferometer is displaced, the light traveling to the detector shows a superposition pattern of two wavetrains of finite length. First, the well-known pattern of the superposition of two monochromatic waves appears. At a certain large distance this pattern disappears. The wavetrain from one arm of the Michelson interferometer will "miss" the other wavetrain because of the finite length of the wavetrains (see Fig. 4.6d). As an example we may look at the emission of light from ^{86}Kr at 6056.16 Å. Since the wavetrain has a length of about 1 m and displaces the one mirror by $1/2$ meter because of reflection, the interference is not observable. For this reason one calls the length of the wavetrain the coherence length. For most atomic emission processes the coherence length is much smaller whereas resonance in laser cavities may produce a much longer coherence length of the order of 10^5 m.

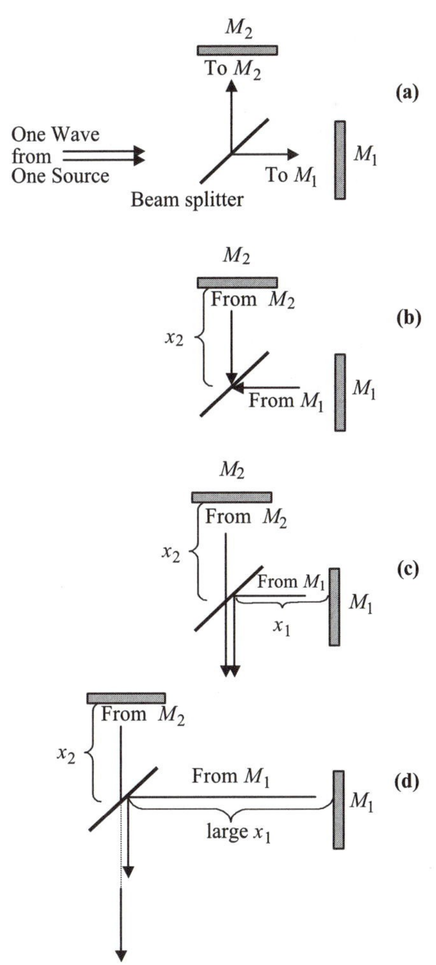

FIGURE 4.6 Splitting of the incident light in a Michelson interferometer: (a) incident light; (b) reflected light traveling to beam splitter; (c) two waves traveling to the detector; (d) for large displacements of M_1, a finite wavetrain "misses" recombination with its "counterpart."

APPENDIX 4.1

A4.1.1 Fourier Transform Spectrometer and Blackbody Radiation

Blackbody radiation contains a large band of wavelengths and has a coherence length with respect to the medium wavelength λ_m, of only a few wavelength's. If we consider a series of filters with smaller and smaller band width, the coherence length of the light passing these filters has increasingly larger and larger values. Michelson's interferometer may be used for Fourier transform spectroscopy, as discussed in Chapter 9. When using blackbody radiation, the total bandwidth of the incident light has to be limited to the interval from 0 to a highest frequency, determined by the sampling theorem. The Fourier transform process analyzes the light and determines the intensity of the resolution width $\Delta\nu$, which is equal to 1/2 L where L is the length of the interferogram in meters. This length L may be increased to larger values and consequently $\Delta\nu$ is decreased. Therefore the length of the corresponding wavetrain l_c is increasing. This process comes to a halt when the signatures of the interferogram are obscured by noise.

In comparison, the size of the coherence length of the atomic emission of ^{86}Kr has its limitation in a time-limited emission process whereas when using blackbody radiation in Fourier transform spectroscopy, the coherence length is limited by the available signal-to-noise level.

C H A P T E R

Maxwell's Theory

5.1 INTRODUCTION

In Chapter 2, we discussed the wave theory of light, developed a model to superimpose waves, and described the resulting interference pattern in terms of intensity depending on wavelength. The model was based on the scalar wave equation, but in addition we made reference to electromagnetic theory. We needed to take into account that a light wave changes its wavelength when traveling through a medium of refractive index n and also used Fresnel's formulas. Electromagnetic theory is described by Maxwell's equations. The first hint that light is electromagnetic radiation came from electromagnetic experiments not involving visible light. In the analysis of the experiment a constant appeared which had the value of the speed of light. From relativity theory we know that the speed c of light is a fundamental constant and the ultimate limit of speed. We derive from Maxwell's theory and the laws of reflection and refraction, as we assumed in the chapter on geometrical optics, that light travels in straight lines. We also may derive from Maxell's equations what we used in the chapters on interference and diffraction and obtained from the scalar wave equation.

In this chapter we describe light by electrical and magnetic field vectors and discuss the polarization of light. At each point in space there are two field vectors vibrating in the perpendicular direction: taking boundary conditions into account we derive Fresnel's formulas. Electromagnetic theory is the basis for the description of all optical phenomena as long as quantum effects are not involved.

Maxwell's equations are a mathematical formulation of the electromagnetic laws of Faraday, Ampere, and Gauss. Maxwell analyzed the mathematical structure of these experiments, added some terms suggested by similarities in appearance of the electric and magnetic fields, and formulated the four equations bearing his name. Today we write Maxwell's equations in vector notation and call **B** the magnetic field vector. This point is well explained by Feynman in his *Lecture Notes*, Volume II, pp. 32–34.

The four Maxwell's equations may be written as

$$\begin{aligned} \nabla \times \mathbf{E} &= -\partial \mathbf{B}/\partial t \\ c^2 \nabla \times \mathbf{B} &= \mathbf{j}/\varepsilon_0 + \partial \mathbf{E}/\partial t \\ \nabla \cdot \mathbf{E} &= \rho/\varepsilon \\ \nabla \cdot \mathbf{B} &= 0, \end{aligned} \tag{5.1}$$

where **E** is the electrical field vector, **B** the magnetic field vector, **j** the current density vector, ρ the charge density, and $\varepsilon_0 = 8.854 \times 10^{-12}$ F/m the permittivity of vacuum. The mathematical form of the differential vector operator ∇ and its scalar "square" ∇^2 is given in Appendix 5.1.

For light propagating in a vacuum, we have $\mathbf{j} = 0$ and $\rho = 0$ and Maxwell's equations are reduced to

$$\begin{aligned} \nabla \times \mathbf{E} &= -\partial \mathbf{B}/\partial t \\ c^2 \nabla \times \mathbf{B} &= +\partial \mathbf{E}/\partial t \\ \nabla \cdot \mathbf{E} &= 0 \\ \nabla \cdot \mathbf{B} &= 0. \end{aligned} \tag{5.2}$$

From this set of equations we arrive at the wave equations for the vectors **E** and **B** as shown in Appendix 5.1.

$$\partial^2 \mathbf{E}/\partial x^2 + \partial^2 \mathbf{E}/\partial y^2 + \partial^2/\mathbf{E}/\partial z^2 = (1/c^2)\partial^2 \mathbf{E}/\partial t^2 \tag{5.3}$$

$$\partial^2 \mathbf{B}/\partial x^2 + \partial^2 \mathbf{B}/\partial y^2 + \partial^2/\mathbf{B}/\partial z^2 = (1/c^2)\partial^2 \mathbf{B}/\partial t^2. \tag{5.4}$$

5.2 HARMONIC PLANE WAVES AND THE SUPERPOSITION PRINCIPLE

5.2.1 Plane Waves

We consider a nondispersive medium. A plane wave solution of the wave equation for the electrical field components vibrating in the y direction and propagating in x direction may be written as

$$E_y = E_{yo} \cos\{2\pi(x/\lambda - t/T)\}, \tag{5.5}$$

where E_{yo} is the magnitude of the electrical field, λ the wavelength, t the time, and T the period of vibration. Equation (5.5) may be rewritten, introducing the wave vector $k = 2\pi/\lambda$, and the angular velocity $\omega = 2\pi/T$. Using exponential notation we have

$$E_y = E_{yo} \exp i(kx - \omega t). \tag{5.6}$$

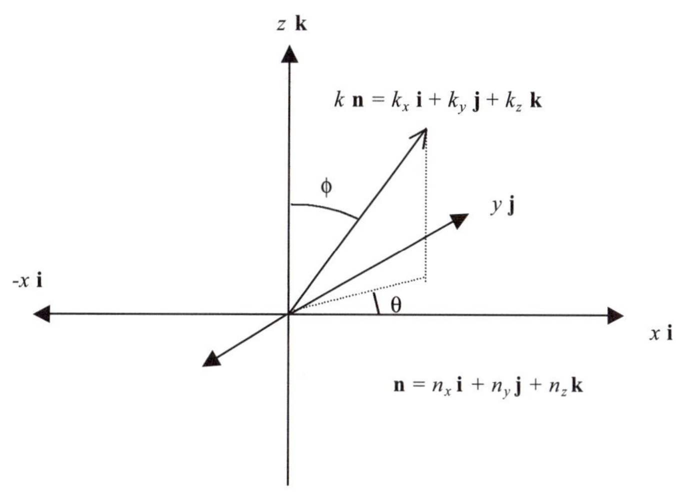

FIGURE 5.1 Coordinate system for a wave with wave vector $k\mathbf{n} = 2\pi/\lambda\mathbf{u}$, traveling in x, y, z space. Using k here must be distinguished from $\mathbf{k}$, the unit vector in the z direction. The vector $\mathbf{r}$ points from 0 to a point in space with the coordinates (x, y, z).

Introduction of Equation (5.6) into the wave equation results in

$$\omega/k = v \qquad (\text{here } v = c), \tag{5.7}$$

where ω/k is the phase velocity.

To extend the propagation to any direction in x, y, z space we use vector notation. For the description of the plane waves we choose in the x, y, z space the coordinate axes $x\mathbf{i}$, $y\mathbf{j}$, and $z\mathbf{k}$, where $\mathbf{i}$, $\mathbf{j}$, $\mathbf{k}$ are unit vectors and a point in x, y, z space is given as $\mathbf{r} = x\mathbf{i} + y\mathbf{j} + z\mathbf{k}$. (Note that k is used for the wave vector and for the unit vector in z direction.) When solving Eq. (5.4) by "separation of variables" one obtains for the wave vector in the direction of propagation $k\mathbf{n} = k_x\mathbf{i} + k_y\mathbf{j} + k_z\mathbf{k}$, where $\mathbf{n}$ is a unit vector and $k = 2\pi/\lambda$ (Figure 5.1). We need to find the components k_x, k_y, and k_z for a wave moving in x, y, z space in direction $\mathbf{n}$. To do this we evaluate the dot product $k(\mathbf{n} \cdot \mathbf{i})$, $k(\mathbf{n} \cdot \mathbf{j})$, and $k(\mathbf{n} \cdot \mathbf{k})$ and obtain for k_x, k_y, and k_z (Figure 5.1),

$$\begin{aligned} k_x &= (2\pi/\lambda) \cdot \sin\phi\cos\theta \\ k_y &= (2\pi/\lambda) \cdot \sin\phi\sin\theta \\ k_z &= (2\pi/\lambda) \cdot \cos\phi. \end{aligned} \tag{5.8}$$

For the special case of the E_y component moving in the x direction we have $\theta = 0°$ and $\phi = 90°$, and get (Chapter 2),

$$E_y = E_{yo} \exp i\{2\pi x/\lambda - 2\pi t/T\}. \tag{5.9}$$

For the case of E_y moving in the x-z plane, one has $\theta = 0°$,

$$k\mathbf{n} = k_x\mathbf{i} + k_z\mathbf{k} = (2\pi/\lambda)(\sin\phi\mathbf{i} + \cos\phi\mathbf{k}) \tag{5.10}$$

and obtain

$$E_y = E_{yo} \exp i\{kx \sin\phi + kz\cos\phi - 2\pi t/T\} \tag{5.11}$$

For the general case of **E** and **B** propagating in three dimensions one gets

$$\mathbf{E} = \mathbf{E}_0 \exp i\{k(n_x x + n_y y + n_z z) - \omega t\} \tag{5.12}$$

and

$$\mathbf{B} = \mathbf{B}_0 \exp i\{k(n_x x + n_y y + n_z z) - \omega t\}.$$

5.2.2 The Superposition Principle

The superposition principle is an important principle in physics. Adding n solutions of the wave equations we may write

$$E(x, y, z, t) = \sum_i^n E_i(x, y, z, t). \tag{5.13}$$

Since the wave equation is a linear second-order differential equation and the $E_i(x, y, z, t)$ are solutions to it, it follows that $E(x, y, z, t)$ is also a solution.

5.3 DIFFERENTIATION OPERATION

Inserting **E** and **B** into Maxwell's equations, one finds, the following as a result of differentiation with respect to time and space coordinates.

5.3.1 Differentiation "Time" $\partial/\partial t$

One obtains for operating $(\partial/\partial t)$ on the function $\exp i\{k(n_x x + n_y y + n_z z) - \omega t\}$:

$$-i\omega \exp i\{k(n_x x + n_y y + n_z z) - \omega t\}, \tag{5.14}$$

that is, multiplication by $-i\omega$ of the exponential function.

5.3.2 Differentiation "Space" $\nabla = \mathbf{i}\partial/\partial x + \mathbf{j}\partial/\partial y + \mathbf{k}\partial/\partial z$

Differentiation with respect to x, that is, $(\partial/\partial x)$ on the function $\exp i\{k(n_x x + n_y y + n_z z) - \omega t\}$, results in

$$ikn_x \exp i\{k(n_x x + n_y y + n_z z) - \omega t\} \tag{5.15}$$

and operating with ∇ on the function $\exp i\{k(n_x x + n_y y + n_z z) - \omega t\}$ one gets

$$ik\mathbf{n} \exp i\{k(n_x x + n_y y + n_z z) - \omega t\}, \tag{5.16}$$

that is, multiplication by $ik\mathbf{n}$ of the exponential function.

The results of differentiation under Sections 5.3.1 and 5.3.2 may be stated as: the differentiations of plane wave solutions of **E** and **B** with respect to time or space coordinates are equivalent to multiplications by

$$\text{Time coordinate: } (\partial/\partial t) \to -i\omega$$
$$\text{Space coordinate: } \nabla \to ik\mathbf{n}.$$

Substitution of these operations into the set of Maxwell's equations for a vacuum results in

$$k\mathbf{n} \times \mathbf{E} = i\omega\mathbf{B} \tag{5.17a}$$
$$c^2 k\mathbf{n} \times \mathbf{B} = -i\omega\mathbf{E} \tag{5.17b}$$
$$k\mathbf{n} \cdot \mathbf{E} = 0 \tag{5.17c}$$
$$k\mathbf{n} \cdot \mathbf{B} = 0. \tag{5.17d}$$

From Eq. (5.17a) and (5.17b) one sees that the vectors **n**, **E**, and **B** form a mutual orthogonal triad. Since **n** points into the direction of propagation, **E** and **B** are perpendicular to **n** and vibrate perpendicular to each other and to the direction of propagation. The field vectors **E** and **B** are both transverse waves (see Eqs. (5.17c) and (5.17d)) and have no component in the direction of propagation. For the phase velocity ω/k, we obtain c for the wave traveling in a vacuum. The components of **E** and **B** are related by

$$|\mathbf{B}| = (1/c)|\mathbf{E}|. \tag{5.18}$$

5.4 POYNTING VECTOR IN VACUUM

The Poynting vector **S** is defined as the time rate of flow of electromagnetic energy per unit area

$$\mathbf{S} = \varepsilon_0 c^2 \mathbf{E} \times \mathbf{B} = (1/\mu_0)\mathbf{E} \times \mathbf{B}. \tag{5.19}$$

We have in MKS units: watts per square meter. The direction of the Poynting vector is parallel to the wave vector $k\mathbf{n}$. If one takes **E** in the x direction and **B** in the y direction, the vector product of Eq. (5.19) is in the z direction

$$\mathbf{S} = \mathbf{k}(1/\mu_0)E_0 B_0 \cos^2\{k(n_x x + n_y y + n_z z) - \omega t\}, \tag{5.20}$$

where we used

$$\mathbf{E} = \mathbf{i}E_0 \cos\{k(n_x x + n_y y + n_z z) - \omega t\}$$

and

$$\mathbf{B} = \mathbf{j}B_0 \cos\{k(n_x x + n_y y + n_z z) - \omega t\}.$$

In the case of vacuum one has $B_0 = (1/c)E_0$ and gets

$$\mathbf{S} = \mathbf{k}(1/c\mu_0)E_0 E_0 \cos^2\{k(n_x x + n_y y + n_z z) - \omega t\}.$$

Since detectors cannot follow a time variation of optical frequencies, the time average of the $\cos^2$ function must be used. The integral over one period T, divided by T, results in 1/2 (see Chapter 2). Therefore, one has for the time average of the Poynting vector

$$\langle S \rangle = (1/2)(1/c\mu_0)E_0^2. \tag{5.21}$$

One observes that the flow of energy is proportional to the square of the amplitude of the electrical field vector. Using complex notation, one may write

$$\mathbf{S} = \mathbf{k}(1/c\mu_0)E_0 E_0 \exp i\{k(n_x x + n_y y + n_z z) - \omega t\}.$$

The absolute value of $\mathbf{S}$ is calculated from the square root of $\mathbf{SS}^*$. As a result the exponential factor is eliminated and one has

$$|S| = (1/c\mu_0)E_0^2. \tag{5.22}$$

Equation (5.21) for the time average carries the factor 1/2, although that factor is not present when taking the absolute value. However, the flow of energy is still proportional to the square of the amplitude of the electrical field vector. Compare the factor 1/2 with what has been discussed in Chapter 2 on complex notation and intensities.

5.5 ELECTROMAGNETIC WAVES IN AN ISOTROPIC NONCONDUCTING MEDIUM

In the following we consider isotropic materials which do not have an optic axis. In contrast, in Section 5.7 we discuss birefringent materials which have an optic axis.

A harmonic wave vibrating in the y direction and propagating in the x direction may be represented as

$$E_y = E_{yo}\cos(2\pi\{x/\lambda - t/T\}). \tag{5.23}$$

Using $k = 2\pi/\lambda$ and $\omega = 2\pi\nu$, where ν is the frequency $\nu = 1/T$, we have for the product $\lambda\nu = c$, and for the phase velocity $\omega/k = c$. One may write Eq. (5.23) as

$$E_y = E_{yo}\cos(kx - \omega t). \tag{5.24}$$

In an isotropic, nonconductive medium, the phase velocity ω/k is the same in all directions and is called v. The index of refraction n is defined as $n = c/v$. We call the wavelength in the medium λ_n and have $\lambda_n\nu = v$. One may write for the wave in the medium

$$E_y = E_{yo}\cos(2\pi x/\lambda_n - \omega t). \tag{5.25}$$

Since the frequency is the same in vacuum or in the medium, one has $\nu = c/\lambda_n = v/\lambda_n$, or the wavelength in the medium $\lambda_n = \lambda(v/c) = \lambda/n$. Therefore we may also write

$$E_y = E_{yo}\cos(2\pi xn/\lambda - \omega t). \tag{5.26}$$

As the refractive index n is always larger than one, the wavelength λ in vacuum is reduced to $\lambda_m = \lambda/n$ in the medium. (Also see "optical path difference" in Chapters 2 and 3.) For harmonic waves, the proportionality of B and E changes from Eq. (5.18) for a vacuum to

$$|\mathbf{B}| = (1/v)|\mathbf{E}| = (n/c)|\mathbf{E}|. \tag{5.27}$$

Consequently the Poynting vector S has for this case the average value

$$\langle S\rangle = (1/2)(n/c\mu_0)E_0^2 \tag{5.28}$$

and the absolute value is

$$|S| = (n/c\mu_0)E_0^2. \tag{5.29}$$

5.6 FRESNEL'S FORMULAS

5.6.1 Electrical Field Vectors in the Plane of Incidence (Parallel Case)

We consider an interface in the X–Z plane of two nonconducting isotropic media with refractive indices n_1 and n_2. A wave is incident on the interface and we assume that its electrical field vector $\mathbf{E}$ vibrates in the plane of incidence,

$$\mathbf{E}_{\|} = \mathbf{E}_{\|o}\exp i(k(\mathbf{n}\cdot\mathbf{r}) - \omega t), \tag{5.30}$$

and its magnetic field vector is perpendicular to the plane of incidence,

$$\mathbf{B}_{\perp} = \mathbf{B}_{\perp o}\exp i(k(\mathbf{n}\cdot\mathbf{r}) - \omega t). \tag{5.31}$$

The propagation vector $k\mathbf{n}$ is in the plane of incidence and in our coordinate system, $\mathbf{B}_{\perp}$ is in the Z direction. The sign is determined with the right-hand rule for the triad of $\mathbf{n}$, $\mathbf{E}$, and $\mathbf{B}$. To determine the analytical expression of the exponent in Eqs. (5.30) and (5.31), the unit vectors $\mathbf{n_i}$, $\mathbf{n_r}$, and $\mathbf{n_t}$, pointing in the direction of propagation of the incident, reflected, and transmitted light (Figure 5.2), are calculated as

$$\mathbf{n_i} = (\mathbf{n_i}\cdot\mathbf{i})\,\mathbf{i} + (\mathbf{n_i}\cdot\mathbf{j})\,\mathbf{j} \tag{5.32}$$

$$\mathbf{n_r} = (\mathbf{n_r}\cdot\mathbf{i})\,\mathbf{i} + (\mathbf{n_r}\cdot\mathbf{j})\,\mathbf{j} \tag{5.33}$$

$$\mathbf{n_t} = (\mathbf{n_t}\cdot\mathbf{i})\,\mathbf{i} + (\mathbf{n_t}\cdot\mathbf{j})\,\mathbf{j} \tag{5.34}$$

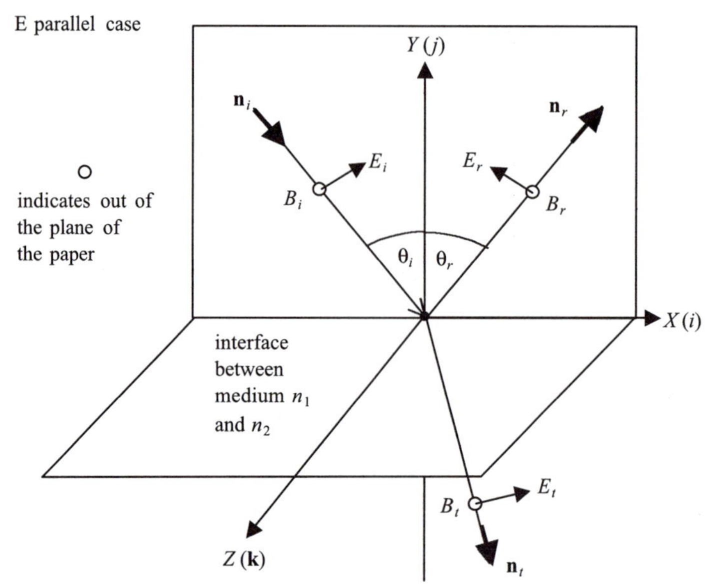

FIGURE 5.2 Coordinates for the derivation of Fresnel's formulas for the case where the electrical field vectors are parallel to the plane of incidence. All magnetic field vectors point in the Z direction. The choice of the direction of the E vectors is such that for normal incidence E_i and E_r point in the opposite direction (Born and Wolf convention).

and for the dot products one gets

$$(\mathbf{n_i} \cdot \mathbf{i}) = \sin\theta_i \qquad (\mathbf{n_r} \cdot \mathbf{i}) = \sin\theta_r \qquad (\mathbf{n_t} \cdot \mathbf{i}) = \sin\theta_t \tag{5.35}$$

$$(\mathbf{n_i} \cdot \mathbf{j}) = -\cos\theta_i \quad (\mathbf{n_r} \cdot \mathbf{j}) = \cos\theta_r \quad (\mathbf{n_t} \cdot \mathbf{j}) = -\cos\theta_t. \tag{5.36}$$

The incident wave, using $\mathbf{n_i}(X\mathbf{i} + Y\mathbf{j})$ may be written as

$$E_i = E_{io} \exp i[k_1(\sin\theta_i \mathbf{i} - \cos\theta_i \mathbf{j}) \cdot (X\mathbf{i} + Y\mathbf{j}) - \omega t]. \tag{5.37}$$

After calculating the dot products and proceeding similarly for the reflected and transmitted components of the electrical field E_r and E_t, we list all components as

$$E_i = E_{io} \exp i\{k_1(\sin\theta_i X - \cos\theta_i Y) - \omega t\} \quad \text{incident} \tag{5.38}$$

$$E_r = E_{ro} \exp i\{k_1(\sin\theta_r X + \cos\theta_r Y) - \omega t\} \quad \text{reflected} \tag{5.39}$$

$$E_t = E_{to} \exp i\{k_2(\sin\theta_t X - \cos\theta_t Y) - \omega t\} \quad \text{transmitted,} \tag{5.40}$$

where $k_1 = 2\pi n_1/\lambda$ and $k_2 = 2\pi n_2/\lambda$. The refractive index n_1 is the refractive index of the medium of the incident and reflected waves, and n_2 the refractive index of the medium of the transmitted wave. Note that both are refraction indices and not components of the vector $\mathbf{n}$.

The boundary condition of electromagnetic theory for a dielectric interface requires that the tangential component of the field vectors **E** and **B** are continuous. We have to take the amplitudes of the incident and reflected wave and set it equal to the amplitude of the transmitted wave. (This should not be confused with energy conservation, which is when the intensity of the incident wave is equal to the intensity of the reflected plus transmitted wave.) The field components E_i, E_r, and E_t are not parallel to the X-axis and their directions have been chosen similar to those by Born and Wolf (II Ed. 1964, p.38–41). To apply the boundary conditions, one has to use the projections on the X-axis (see Figure 5.2), and has for $Y = 0$, using Eqs. (5.38) to (5.40),

$$\cos\theta_i E_{io} \exp i[k_1(\sin\theta_i X) - \omega t] - \cos\theta_r E_{ro} \exp i[k_1(\sin\theta_r X) - \omega t] = \cos\theta_t E_{to} \exp i[k_2(\sin\theta_t X) - \omega t]. \tag{5.41}$$

This has to be true for all times; that is,

$$k_1 \sin\theta_i = k_1 \sin\theta_r = k_2 \sin\theta_t. \tag{5.42}$$

As a result one has the *law of reflection*,

$$\theta_i = \theta_r, \tag{5.43}$$

and with $k_1 = 2\pi n_1/\lambda$ and $k_2 = 2\pi n_2/\lambda$, the law of refraction,

$$n_1 \sin\theta_i = n_2 \sin\theta_t. \tag{5.44}$$

We have also to consider

$$\cos\theta_i E_{io} - \cos\theta_r E_{ro} = \cos\theta_t E_{to}. \tag{5.45}$$

Since E_{or} and E_{ot} are unknowns, we need a second equation to determine their values. This can be obtained from the applications of the boundary conditions to the magnetic field vectors $\mathbf{B}_\perp$. The three vectors perpendicular to the plane of incidence $\mathbf{B}_i$, $\mathbf{B}_r$, $\mathbf{B}_t$, which correspond to $\mathbf{E}_i$, $\mathbf{E}_r$, and $\mathbf{E}_t$, all point in the Z direction (Figure 5.2). The phase of the three $\mathbf{B}$ vectors is the same as it was for the three $\mathbf{E}$ vectors. The application of the boundary condition results in

$$B_i + B_r = B_t \tag{5.46}$$

From Eq. (5.27), one has $\mathbf{B} = (1/v)\mathbf{E}$. For a medium with refractive index $v = c/n$ one has $\mathbf{B} = (n/c)\mathbf{E}$. Insertion into Eq. (5.46) results in

$$n_1 E_{io} + n_1 E_{ro} = n_2 E_{to}. \tag{5.47}$$

Dividing Eqs. (5.45) and (5.47) by E_{io}, which normalizes by the magnitude of the incident wave, and calling the reflected amplitude $r_\parallel = E_{ro}/E_{io}$, and the transmitted amplitude $t_\parallel = E_{to}/E_{io}$, one has two equations for $r_\parallel$ and $t_\parallel$:

$$\cos\theta_r r_\parallel + \cos\theta_t t_\parallel = +\cos\theta_i \tag{5.48}$$

and

$$n_1 r_\parallel - n_2 t_\parallel = -n_1. \tag{5.49}$$

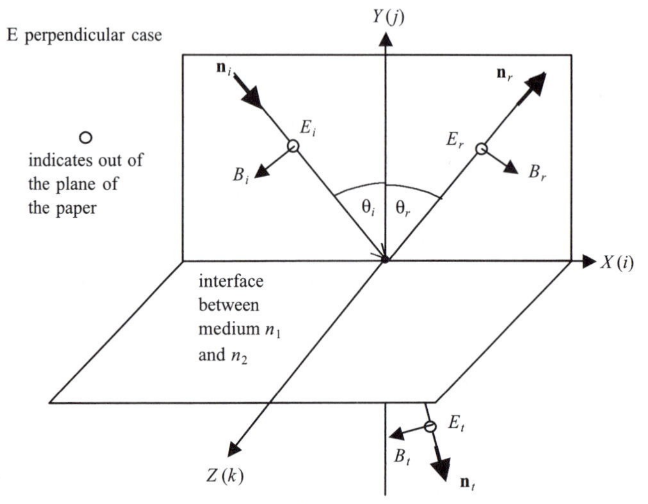

FIGURE 5.3 Coordinates for the derivation of Fresnel's formulas for the case where the electrical field vectors are perpendicular to the plane of incidence. All electrical field vectors point in the Z direction. The choice of direction of the B vectors is such that for normal incidence B_i and B_r point in the opposite direction (but E_i and E_r are parallel; Born and Wolf convention).

Renaming θ, both the angle of incidence θ_i and the angle of reflection θ_r, and calling the angle of refraction $\theta_t = \theta''$, we solve Eqs. (5.48) and (5.49) for $r_{\|}$ and $t_{\|}$ (FileFig 5.1):

$$r_{\|} = (n_2 \cos\theta - n_1 \cos\theta'')/(n_2 \cos\theta + n_1 \cos\theta'') \tag{5.50}$$

$$t_{\|} = (2n_1 \cos\theta)/(n_2 \cos\theta + n_1 \cos\theta''). \tag{5.51}$$

These are Fresnel's formulas for the case where the **E** vector of the incident light is in the plane of incident, called the parallel case, the TE-case, or p-polarization.

5.6.2 Electrical Field Vector Perpendicular to the Plane of Incidence (Perpendicular Case)

In Figure 5.3, we show $\mathbf{E}_i$, $\mathbf{E}_r$, and $\mathbf{E}_t$ pointing out of the plane of incidence, parallel to the Z-axis. The corresponding magnetic field vectors B_i, B_r, and B_t are also indicated. They are in the plane of incidence and need to be projected onto the X-axes. Similar to the discussion above, after application of the boundary conditions, we have for the magnetic field vectors,

$$-\cos\theta_i B_{io} + \cos\theta_r B_{ro} = -\cos\theta_t B_{to}, \tag{5.52}$$

or, rewritten using the electrical field vector,

$$-n_1 \cos\theta_i E_{io} + n_1 \cos\theta_r E_{ro} = -n_2 \cos\theta_t E_{to}. \tag{5.53}$$

A second relation is obtained for the electrical field vectors

$$E_{io} + E_{ro} = E_{to}. \tag{5.54}$$

We divide Eqs. (5.53) and (5.54) by E_{io} and call E_{ro}/E_{io} the reflected amplitude $r_\perp$, and E_{to}/E_{io} the transmitted amplitude $t_\perp$. After rearranging the two equations for $r_\perp$ and $t_\perp$ we have

$$n_1 \cos\theta_r r_\perp + n_2 \cos\theta_t t_\perp = +n_1 \cos\theta_i \tag{5.55}$$

$$r_\perp - t_\perp = -1. \tag{5.56}$$

After solving for $r_\perp$ and $r_\perp$ one obtains

$$r_\perp = (n_1 \cos\theta - n_2 \cos\theta'')/(n_1 \cos\theta + n_2 \cos\theta'') \tag{5.57}$$

$$t_\perp = (2n_1 \cos\theta)/(n_1 \cos\theta + n_2 \cos\theta''). \tag{5.58}$$

These are Fresnel's formulas for the case where the **E** vector of the incident light is pointing out of the plane of incidence, called the perpendicular case, the *TM-case*, or *s-polarization*.

FileFig 5.1 (M1FRFOR)

We may use the symbolic operations of a computer program to solve for Fresnel's formulas for the p-case (parallel) and s-case (perpendicular).

M1FRFOR is only on the CD.

5.6.3 Fresnel's Formulas Depending on the Angle of Incidence

Using the law of refraction, we can eliminate θ'' and express Fresnel's formulas depending on the angle of incidence. Application of

$$cos\theta'' = \sqrt{1 - (\sin\theta'')^2} = \sqrt{(1 - [(n_1/n_2)\sin\theta]^2}$$

results in r_p and t_p for the parallel (p) case

$$r_p = \frac{(n_2 \cos - n_1\sqrt{1 - [(n_1/n_2)\sin\theta]^2})}{(n_2 \cos\theta + n_1\sqrt{1 - [(n_1/n_2)\sin\theta]^2})} \tag{5.59}$$

$$t_p = \frac{(2n_1 \cos\theta)}{(n_2 \cos\theta + n_1\sqrt{1 - [(n_1/n_2)\sin\theta]^2})}, \tag{5.60}$$

and r_s and t_s for the perpendicular (s) case

$$r_s = \frac{(n_1 \cos\theta - n_2\sqrt{1 - [(n_1/n_2)\sin\theta]^2})}{(n_1 \cos\theta + n_2\sqrt{1 - [(n_1/n_2)\sin\theta]^2})} \tag{5.61}$$

$$t_s = \frac{(2n_1 \cos\theta)}{(n_1 \cos\theta + n_2\sqrt{(1 - [(n_1/n_2)\sin\theta_2)}}. \tag{5.62}$$

5.6.4 Light Incident on a Denser Medium, $n_1 < n_2$, and the Brewster Angle

5.6.4.1 Phase Shifts

We mentioned in Section 6.1 that we have chosen the sign of the reflected electrical vectors at the boundary in the same way as done by Born and Wolf (1964). At normal incidence, this sign convention results in a negative sign for the reflection coefficient of the p-case. For the s-case there is a positive sign. Since there is no distinction between the p-case and s-cases for normal incidence, how can this be explained?

A light wave reflected at a denser medium picks up a phase shift of π, and we may write for the reflection coefficients: absolute value times the complex phase factor; that is, for the p- and s-cases, $r_p = |r_p|e^{i\phi_p}$ and $r_s = |r_s|e^{i\phi_s}$, respectively. In the book by Born and Wolf (1964) the phase factor for the p-case is included in r_p; that is, for normal incidence $\phi_p = \pi$, $e^{i\phi_p} = -1$, and r_p is $-|r_p|$. In the s-case the phase factor is not included; that is $\phi s = 0$, $e^{i\phi_s} = 1$, and r_s is $|r_s|$. Graphs of $|r_p|$, $|r_s|$, ϕ_p, and ϕ_s are shown in FileFig 5.2 for light incident from a less dense medium to a more dense medium. Note that ϕ_p is the argument of r_p, and ϕ_s of r_s.

5.6.4.2 Brewster's Angle

In FileFig 5.2 one observes from graphs of r_p and r_s, depending on the angle of incidence θ, that for a specific angle, called the *Brewster angle*, r_p is zero and r_s is not. From Eq. (5.59) we have for the Brewster angle

$$(n_2 \cos\theta_B - n_1\sqrt{1 - [(n_1/n_2)\sin\theta_B]^2}) = 0. \tag{5.63}$$

We abbreviate $a = n_2/n_1$ and square Eq. (5.63):

$$a^2\cos^2\theta_B - n_1 = 1 - (\sin^2\theta_B/a^2) = \cos^2\theta_B + \sin^2\theta_B - (\sin^2\theta_B/a^2) \tag{5.64}$$

or

$$(a^2 - 1)\cos^2\theta_B = (1 - a^{-2})(\sin^2\theta_B) = [(a^2 - 1)/a^2]\sin^2\theta_B.$$

Since $a^2 > 1$ we can cancel $(a^2 - 1)$ and obtain

$$\tan\theta_B = n_2/n_1. \tag{5.65}$$

This is the tangent of the Brewster angle.

FileFig 5.2 (M2FRN2L)

Fresnel's formulas for $n_1 < n_2$. Graphs are shown of the absolute value and the argument of r_p, r_s, and t_p, t_s, depending on the angle of incidence θ. The choice of

$n_1 = 1$ and $n_2 = 1.5$ presents the case for light incident on a material like glass. At the Brewster angle, one sees that the phase angle changes for the parallel case from 0 to 180 degrees. For the perpendicular case there is no Brewster angle and no change. The transmission coefficients are not undergoing any phase shifts.

M2FRN2L

Amplitudes

Fresnel's formulas as function of angle of incidence for first medium 1, second medium 2, and $n1 < n2$.

1. Reflection coefficients
Absolute value and imaginary parts for p-case (parallel) and s-case (perpendicular).

$$rp(\theta) := \frac{\left(\frac{n2}{n1}\right)\cdot\cos\left(2\cdot\frac{\pi}{360}\cdot\theta\right) - \sqrt{1-\left[\left(\frac{n1}{n2}\right)\cdot\sin\left(2\cdot\frac{\pi}{360}\cdot\theta\right)\right]^2}}{\left(\frac{n2}{n1}\right)\cdot\cos\left(2\cdot\frac{\pi}{360}\cdot\theta\right) + \sqrt{1-\left[\left(\frac{n1}{n2}\right)\cdot\sin\left(2\cdot\frac{\pi}{360}\cdot\theta\right)\right]^2}}$$

$$rs(\theta) := \frac{\cos\left(2\cdot\frac{\pi}{360}\cdot\theta\right) - \left(\frac{n2}{n1}\right)\cdot\sqrt{1-\left[\left(\frac{n1}{n2}\right)\cdot\sin\left(2\cdot\frac{\pi}{360}\cdot\theta\right)\right]^2}}{\cos\left(2\cdot\frac{\pi}{360}\cdot\theta\right) + \left(\frac{n2}{n1}\right)\cdot\sqrt{1-\left[\left(\frac{n1}{n2}\right)\cdot\sin\left(2\cdot\frac{\pi}{360}\cdot\theta\right)\right]^2}}$$

$$\theta \equiv \ldots 1, \ldots 2 \ldots 90 \qquad n1 \equiv 1 \qquad n2 \equiv 1.5.$$

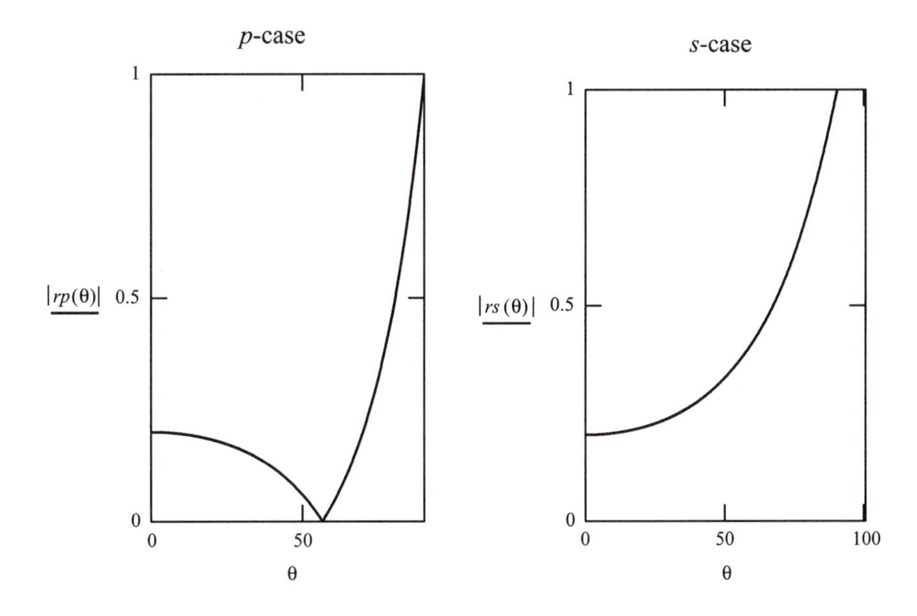

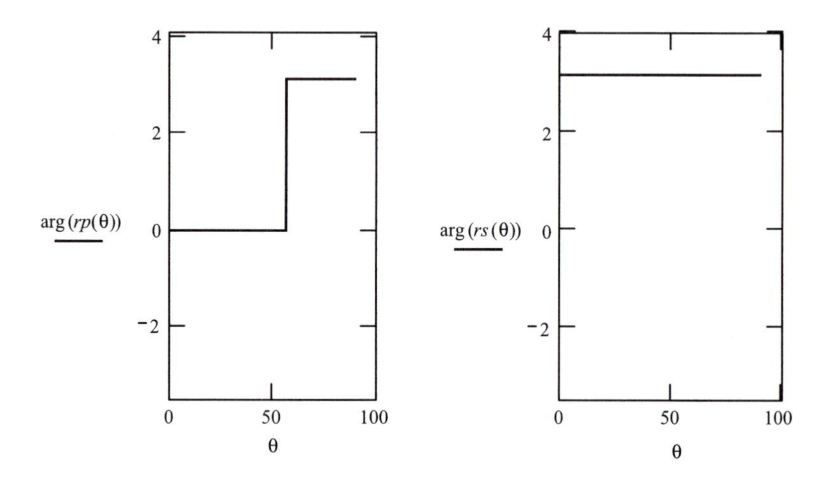

2. Transmission coefficient
Absolute value and imaginary part for p-case and s-case.

$$tp(\theta) := \frac{2 \cdot \cos\left(2 \cdot \frac{\pi}{360} \cdot \theta\right)}{\left(\frac{n2}{n1}\right) \cdot \cos\left(2 \cdot \frac{\pi}{360} \cdot \theta\right) + \sqrt{1 - \left[\left(\frac{n1}{n2}\right) \cdot \sin\left(2 \cdot \frac{\pi}{360} \cdot \theta\right)\right]^2}}$$

$$ts(\theta) := \frac{2 \cdot \cos\left(2 \cdot \frac{\pi}{360} \cdot \theta\right)}{\cos\left(2 \cdot \frac{\pi}{360} \cdot \theta\right) + \left(\frac{n2}{n1}\right) \cdot \sqrt{1 - \left[\left(\frac{n1}{n2}\right) \cdot \sin\left(2 \cdot \frac{\pi}{360} \cdot \theta\right)\right]^2}}$$

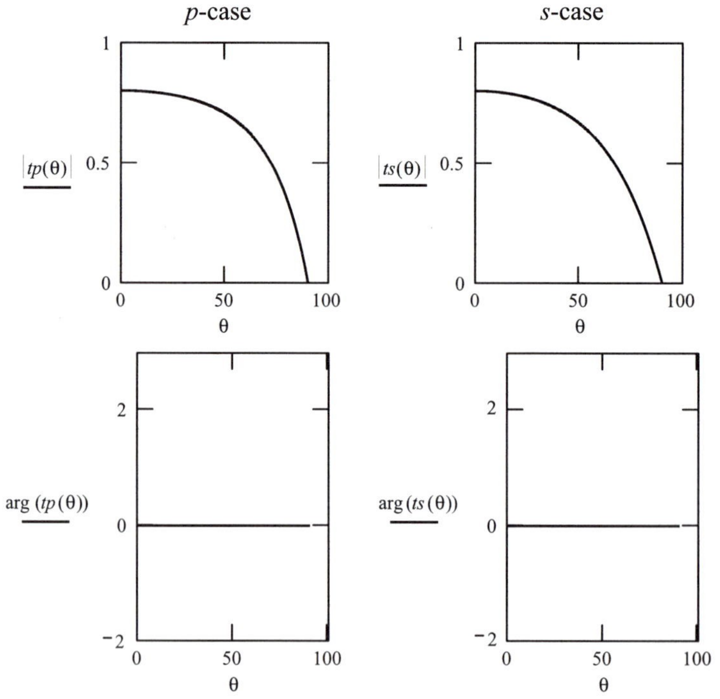

p-case　　s-case

|tp(θ)|　|ts(θ)|

arg (tp(θ))　arg (ts(θ))

Application 5.2.

1. Study, for a large angle, the changes of the Brewster angle and the slope of the reflection coefficient. Choose n_1 from a value close to n_2 to values much smaller than n_2. Save the case n_1 smaller than n_2 for FileFig 5.4.
2. Make graphs for $n_1 = 1$ and n_2 for two different refractive indices. Study the changes of the argument, the Brewster angle, and the slope of the reflection coefficient at large angles, that is, at grazing angle incidence. Make the difference of n_2 and n_2' larger and smaller.
3. Look at the transmission coefficients for the p- and s-cases and plot them on one graph. Study the difference of the two for different refractive indices n_2.

5.6.5 Light Incident on a Less Dense Medium, $n_1 > n_2$, Brewster and Critical Angle

5.6.5.1 Brewster Angle

In FileFig 5.3 we have plotted the reflection and transmission coefficients for light incident on a less dense medium. One sees that the general shape of the reflection and transmission coefficients are the same as found for the case $n_1 < n_2$ discussed in Section 5.6.4, and the Brewster angle is present for the parallel case.

5.6.5.2 Critical Angle and Phase Shifts

At a larger angle than the Brewster angle, called the critical angle θ_c, the curves for the absolute value of r_p and r_s approach 1, while the transmission coefficients decrease to 0. All arguments of r_p and r_s, and t_p and t_s decrease after the critical angle. We apply the law of refraction to this case and find that the refraction angle becomes imaginary when the angle of incidence exceeds the critical angle given by

$$\sin\theta_c = n_2/n_1, \tag{5.66}$$

where n_1 is the medium with the larger index of refraction, from which the light is incident on the interface. The question of real and imaginary refraction angles is studied in FileFig 5.4.

FileFig 5.3 (M3FRN2S)

Fresnel's formulas for the case $n_1 > n_2$. Graphs are shown of the absolute value and the argument of r_p, r_s, and t_p, t_s, depending on the angle of incidence θ. The choice of $n_1 = 1$ and $n_2 = 1.5$ presents the case for light incident on a glasslike material. The Brewster angle appears again for r_p and for both r_p and

r_s the critical angle appears. The absolute value of the transmission coefficients increases before the critical angle and decreases thereafter.

M3FRN2S

Amplitudes

Fresnel's formulas as function of angle of incidence for first medium 1, second medium 2, and $n1 > n2$.

1. **Reflection coefficients**
   Absolute value and imaginary parts for p-case (parallel) and s-case (perpendicular).

$$rp(\theta) := \frac{\left(\frac{n2}{n1}\right)\cdot\cos\left(2\cdot\frac{\pi}{360}\cdot\theta\right) - \sqrt{1-\left[\left(\frac{n1}{n2}\right)\cdot\sin\left(2\cdot\frac{\pi}{360}\cdot\theta\right)\right]^2}}{\left(\frac{n2}{n1}\right)\cdot\cos\left(2\cdot\frac{\pi}{360}\cdot 0\right) + \sqrt{1-\left[\left(\frac{n1}{n2}\right)\cdot\sin\left(2\cdot\frac{\pi}{360}\cdot\theta\right)\right]^2}}$$

$$rs(\theta) := \frac{\cos\left(2\cdot\frac{\pi}{360}\cdot\theta\right) - \left(\frac{n2}{n1}\right)\cdot\sqrt{1-\left[\left(\frac{n1}{n2}\right)\cdot\sin\left(2\cdot\frac{\pi}{360}\cdot\theta\right)\right]^2}}{\cos\left(2\cdot\frac{\pi}{360}\cdot\theta\right) + \left(\frac{n2}{n1}\right)\cdot\sqrt{1-\left[\left(\frac{n1}{n2}\right)\cdot\sin\left(2\cdot\frac{\pi}{360}\cdot\theta\right)\right]^2}}$$

$$\theta \equiv \ldots 1, \ldots 2 \ldots 90 \qquad n1 \equiv 1.5 \qquad n2 \equiv 1.$$

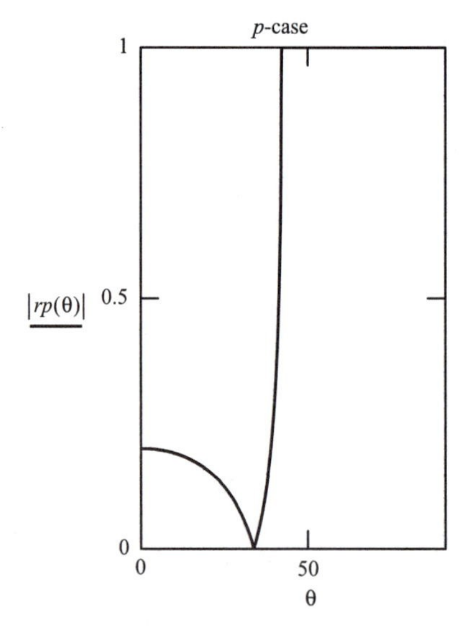

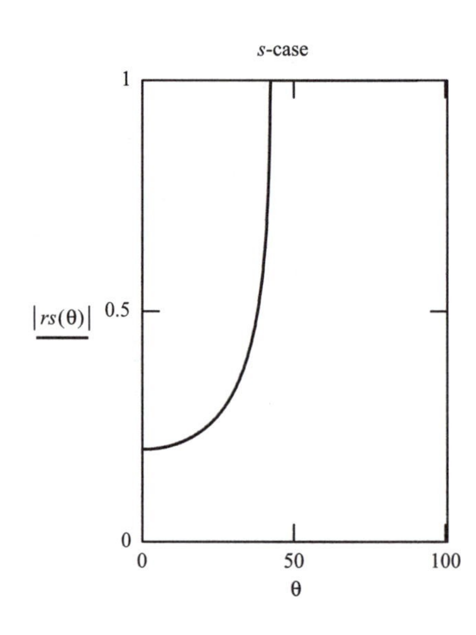

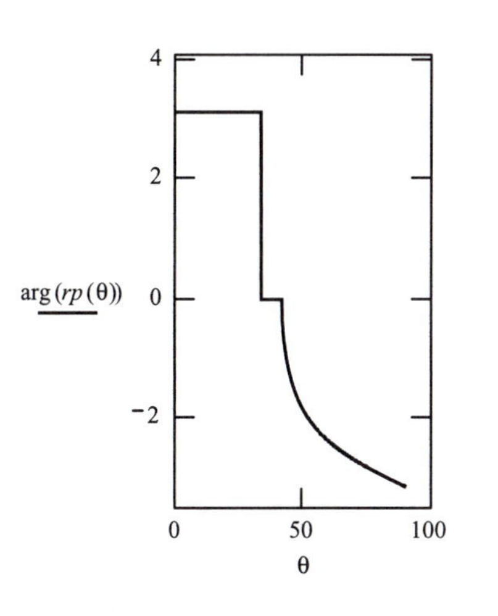

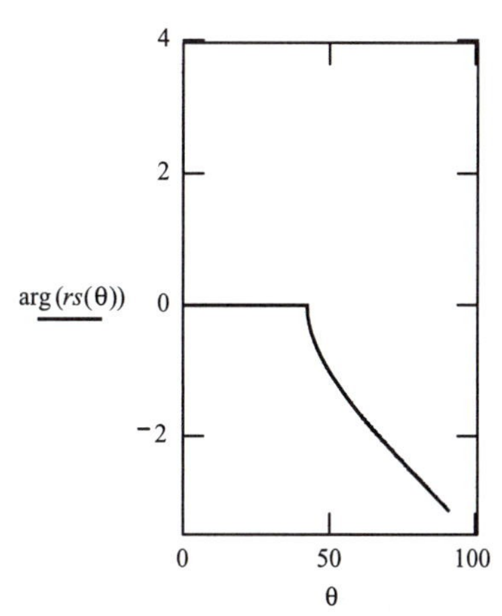

2. Transmission coefficient
Absolute value and imaginary part for p- and s-cases.

$$tp(\theta) := \frac{2 \cdot \cos\left(2 \cdot \frac{\pi}{360} \cdot \theta\right)}{\left(\frac{n2}{n1}\right) \cdot \cos\left(2 \cdot \frac{\pi}{360} \cdot \theta\right) + \sqrt{1 - \left[\left(\frac{n1}{n2}\right) \cdot \sin\left(2 \cdot \frac{\pi}{360} \cdot \theta\right)\right]^2}}$$

$$ts(\theta) := \frac{2 \cdot \cos\left(2 \cdot \frac{\pi}{360} \cdot \theta\right)}{\cos\left(2 \cdot \frac{\pi}{360} \cdot \theta\right) + \left(\frac{n2}{n1}\right) \cdot \sqrt{1 - \left[\left(\frac{n1}{n2}\right) \cdot \sin\left(2 \cdot \frac{\pi}{360} \cdot \theta\right)\right]^2}}$$

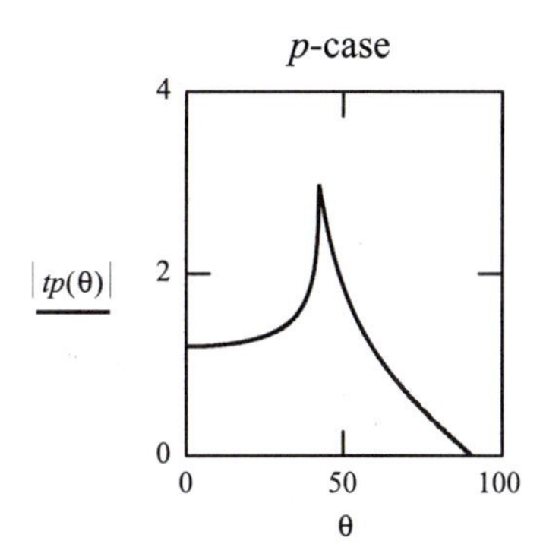

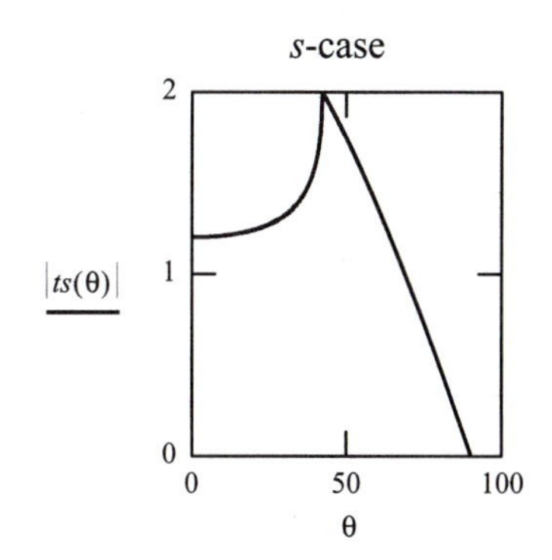

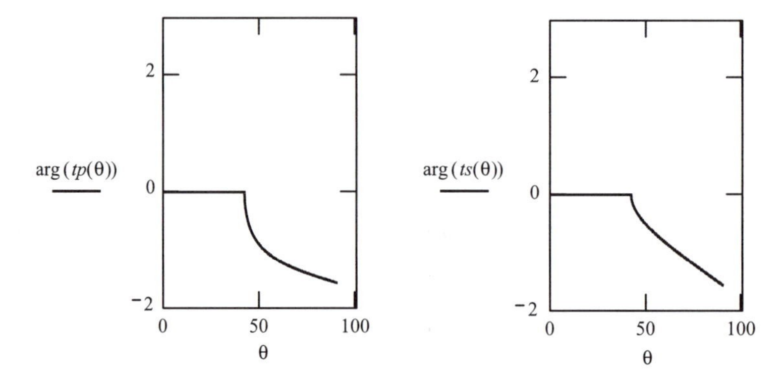

Application 5.3.

1. Study reflection and transmission coefficients by changing from $n_1 > n_2$ to $n_1 \approx n_2$ and $n_1 \approx n_2$ and compare with FileFig 5.2. Observe the different shape of t_s and t_p as they appear in FileFigs.3 and 4. This is an indication that t_s^2 and t_p^2 are not to be confused with the transmitted intensity.
2. Plot on the same graph, for $n = 1.5$ and two different indices of refraction, the reflection coefficients and look at the differences of the curves for larger and smaller differences of the two indices.
3. Change the indices of refraction and observe how the critical angle is changing.

FileFig 5.4 (M4SNELL)

The law of refraction $n_1 \sin\theta_1 = n_2 \sin\theta_2$ is plotted as $\theta_2(\theta_1) = \operatorname{asin}\left[\frac{n_1 \sin(\theta_1)}{n_2}\right]$: (1) a graph of $\theta_2(\theta_1)$ for $n_1 < n_2$; (2) a graph of $\theta_2(\theta_1)$ for $n_1 > n_2$. The graph ends at the critical angle; and (3) real and imaginary parts, separately for $n_1 > n_2$. One extends from zero to the critical angle, the other from the critical angle to $90°$.

M4SNELL is only on the CD.

5.6.6 Reflected and Transmitted Intensities

In contrast to what was done in the application of the boundary conditions, we now calculate the energy flow. At the boundary, we equated the sum of the fields of incident and reflected amplitude on one side with the transmitted amplitude on the other side. Now we want to find out how much of the incident energy is reflected and transmitted. From Eq. (5.19) we have for the Poynting vector $\mathbf{S}$ for a vacuum

$$\mathbf{S} = \varepsilon_0 c^2 \mathbf{E} \times \mathbf{B} = (1/\mu_0)\mathbf{E} \times \mathbf{B}. \tag{5.67}$$

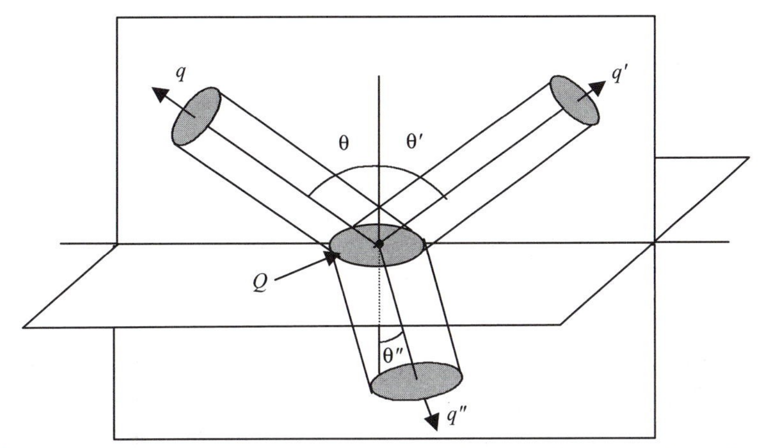

FIGURE 5.4 Areas q, q', and q'' are cross-sections of the power flow of incident, reflected, and refracted light. Q is the common area at the interface.

For a medium with refractive index n we have with Eq. (5.22) the absolute value of S,

$$S = (1/\mu_0)E_0(n/c)E_0 \tag{5.68}$$

and obtain the time average, similar to Eq. (5.21),

$$\langle S \rangle = (1/2)(n/c\mu_0)E_0^2. \tag{5.69}$$

Let us call the incident energy per area S_i, the reflected S_r, and the refracted S_t. In Figure 5.4 we have indicated the areas through which the energy flows. The area q corresponds to the incident energy and q' and q'' to the reflected and refracted energy, respectively. At the interface we have to use the same area Q for incident, reflected, and refracted energy and have to multiply S_i, S_r, and S_t by a cosine factor corresponding to the projections of q, q', q'' onto the X–Z plane. We then have

$$\begin{aligned}&(1/2)(n_1/\mu_0 c)E_{io}^2 \cos\theta \\ &\quad = 1/2(n_1/\mu_0 c)E_{ro}^2 \cos\theta + 1/2(n_2/\mu_0 c)E_{to}^2 \cos\theta''.\end{aligned} \tag{5.70}$$

5.6.6.1 The Case $n_1 < n_2$

We call the *reflectance* $R = E_{ro}^2/E_{io}^2 = r^2$, where r stands for both the parallel and perpendicular cases. Using conservation of energy,

$$R + T = 1, \tag{5.71}$$

we obtain from Eq. (5.70) for the *transmittance* T,

$$T = 1 - r^2 = (n_2 \cos\theta''/n_1 \cos\theta)t^2, \tag{5.72}$$

where E_{to}^2/E_{io}^2 and t stands again for both cases.

We call $(n^2 \cos\theta''/n_1 \cos\theta) = \alpha$ and have, writing the parallel and perpendicular cases separately,

$$R_{\parallel} = r_{\parallel}^2 \quad \text{and} \quad T_{\parallel} = \alpha t_{\parallel}^2 \tag{5.73}$$

$$R_{\perp} = r_{\perp}^2 \quad \text{and} \quad T_{\perp} = \alpha t_{\perp}^2. \tag{5.74}$$

Energy conservation holds for both cases and by using

$$T = 1 - R \tag{5.75}$$

one can avoid using the factor α. One can calculate $R_{\parallel}$ and $R_{\perp}$ from Fresnel's formulas, using the amplitude reflection coefficients, and then use $1-R$ to obtain T. In FileFig 5.5 we have plotted R_p, R_s, $T_p = 1 - R_p$ and $T_s = 1 - R_s$ for the case where $n_1 < n_2$, from 0 to 90°.

FileFig 5.5 (M5FRINTN2L)

Graphs of R_p, R_s, $T_p = 1 - R_p$, and $T_s = 1 - R_s$ depending on the angle of incidence for the case where $n_1 < n_2$.

M5FRINTN2L

Intensities

Fresnel's formulas as function of angle of incidence for $n1 < n2$ for $Rp = rp^2$, $Rs = rs^2$, and $Tp = 1 - R_p$, $Ts = 1 - Rs$.

1. Amplitude reflection coefficients

$$rp(\theta) := \frac{\left(\frac{n2}{n1}\right) \cdot \cos\left(2 \cdot \frac{\pi}{360} \cdot \theta\right) - \sqrt{1 - \left[\left(\frac{n1}{n2}\right) \cdot \sin\left(2 \cdot \frac{\pi}{360} \cdot \theta\right)\right]^2}}{\left(\frac{n2}{n1}\right) \cdot \cos\left(2 \cdot \frac{\pi}{360} \cdot \theta\right) + \sqrt{1 - \left[\left(\frac{n1}{n2}\right) \cdot \sin\left(2 \cdot \frac{\pi}{360} \cdot \theta\right)\right]^2}}$$

$$rs(\theta) := \frac{\cos\left(2 \cdot \frac{\pi}{360} \cdot \theta\right) - \left(\frac{n2}{n1}\right) \cdot \sqrt{1 - \left[\left(\frac{n1}{n2}\right) \cdot \sin\left(2 \cdot \frac{\pi}{360} \cdot \theta\right)\right]^2}}{\cos\left(2 \cdot \frac{\pi}{360} \cdot \theta\right) + \left(\frac{n2}{n1}\right) \cdot \sqrt{1 - \left[\left(\frac{n1}{n2}\right) \cdot \sin\left(2 \cdot \frac{\pi}{360} \cdot \theta\right)\right]^2}}$$

$$n1 \equiv 1, \qquad n2 \equiv 1.5 \qquad 0 \equiv 0, .4..90.$$

2. Reflection: intensities

$$Rp(\theta) := rp(\theta)^2 \qquad Rs(\theta) := rs(\theta)^2.$$

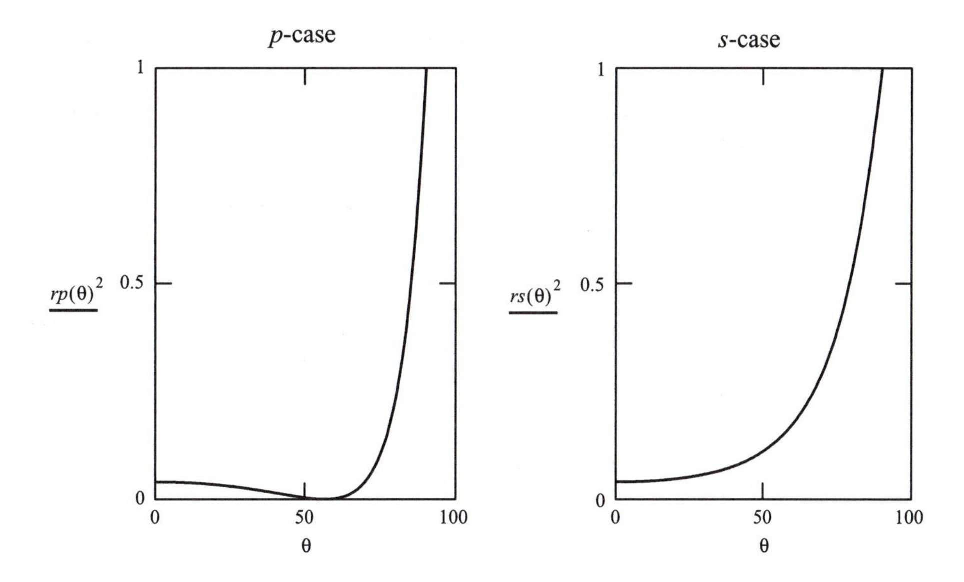

3. Transmission: intensities

$$Tp(\theta) := 1 - Rp(\theta) \qquad Ts(\theta) := 1 - Rs(\theta).$$

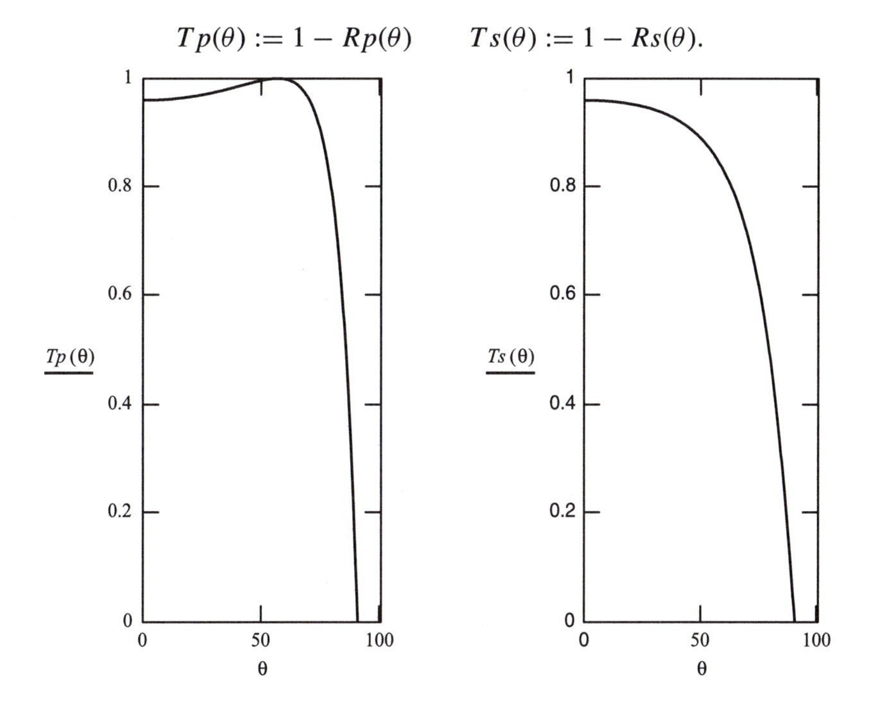

Application 5.5.

1. Make graphs of r_p^2, r_s^2, t_p^2 and t_s^2 for the case where $n_1 < n_2$. Find which quantities, just squared, are useful for calculation of corresponding intensities, and which are not. Compare the graphs of the intensities and find out if R_p

or R_s are equal to r_p^2 or r_s^2, and similarly if T_p or T_s are equal to t_p^2 or t_s^2, respectively.

2. Make a graph of the factor $\alpha = (n_2 \cos\theta''/n_1 \cos\theta)$ for $n_1 < n_2$ and $n_1 > n_2$ and compare. To do this one has to use the law of refraction to substitute for $\cos\theta''$.

5.6.6.2 $n_1 > n_2$, Critical Angle and Total Reflection

From Eqs. (5.59) to (5.62) one sees that the square root in all expressions may become negative when θ is beyond the critical angle. We may rewrite the formulas by taking out of the square root the factor $\sqrt{-1} = i$ and have all reflection and transmission coefficients be complex functions. Let us look, as an example, at the perpendicular case where we have for r,

$$r_s = \frac{\cos\theta - (n_2/n_1)i\sqrt{[(n_1/n_2)\sin\theta]^2 - 1)}}{\cos\theta + (n_2/n_1)i\sqrt{[(n_1/n_2)\sin\theta]^2 - 1)}} \tag{5.76}$$

which we may write with the abbreviations $a = \cos\theta$ and $b = n_2/n_1\sqrt{[(n_1/n_2)\sin\theta]^2 - 1}$ as

$$r_\perp = (a - ib)/(a + ib). \tag{5.77}$$

The complex conjugate is

$$r_\perp^* = (a + ib)/(a - ib). \tag{5.78}$$

To get the reflectance R, one has to take rr^*, where r^* is the complex conjugate of r and $R = r_\perp r_\perp^* = 1$, and it follows that $T = 0$. For all angles of incidence equal to or larger than the critical angle we have total reflection; that is, $R_\parallel = R_\perp = 1$ and $T_\parallel = T_\perp = 0$. In FileFig 5.6 we have plotted $R_\parallel$, $R_\perp$, $T_\parallel$, and $T_\perp$ for $n_1 = 1.5$, $n_2 = 1$ and θ from 0 to 90°.

FileFig 5.6 (M6FRINTN2S)

Graphs of R_p, R_s, $T_p = 1 - R_p$, and $T_s = 1 - R_s$ depending on the angle of incidence for the case where $n_1 > n_2$.

M6FRINTN2S

Intensities

Fresnel's formulas as function of angle of incidence for $n1 > n2$ for $Rp = rp^2$, $Rs = rs^2$, and $Tp = 1 - Rp$, $Ts = 1 - Rs$.

1. Amplitude reflection coefficients

$$rp(\theta) := \frac{\left(\frac{n2}{n1}\right) \cdot \cos\left(2 \cdot \frac{\pi}{360} \cdot \theta\right) - \sqrt{1 - \left[\left(\frac{n1}{n2}\right) \cdot \sin\left(2 \cdot \frac{\pi}{360} \cdot \theta\right)\right]^2}}{\left(\frac{n2}{n1}\right) \cdot \cos\left(2 \cdot \frac{\pi}{360} \cdot \theta\right) + \sqrt{1 - \left[\left(\frac{n1}{n2}\right) \cdot \sin\left(2 \cdot \frac{\pi}{360} \cdot \theta\right)\right]^2}}$$

$$rs(\theta) := \frac{\cos\left(2 \cdot \frac{\pi}{360} \cdot \theta\right) - \left(\frac{n2}{n1}\right) \cdot \sqrt{1 - \left[\left(\frac{n1}{n2}\right) \cdot \sin\left(2 \cdot \frac{\pi}{360} \cdot \theta\right)\right]^2}}{\cos\left(2 \cdot \frac{\pi}{360} \cdot \theta\right) + \left(\frac{n2}{n1}\right) \cdot \sqrt{1 - \left[\left(\frac{n1}{n2}\right) \cdot \sin\left(2 \cdot \frac{\pi}{360} \cdot \theta\right)\right]^2}}$$

$$n1 \equiv 1.5 \qquad n2 \equiv 1 \qquad \theta \equiv 0, \ldots, 01 \ldots 90.$$

2. Reflection: intensities

$$Rp(\theta) := rp(\theta) \cdot \overline{rp(\theta)} \qquad Rs(\theta) := rs(\theta) \cdot \overline{rs(\theta)}.$$

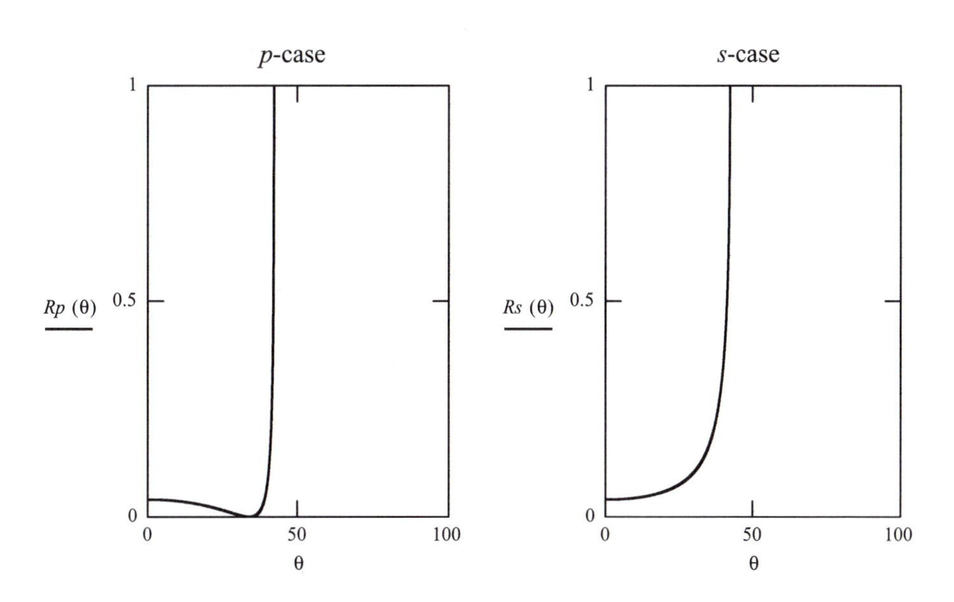

3. Transmission: intensities

$$Tp(\theta) := 1 - Rp(\theta) \qquad Ts(\theta) := 1 - Rs(\theta).$$

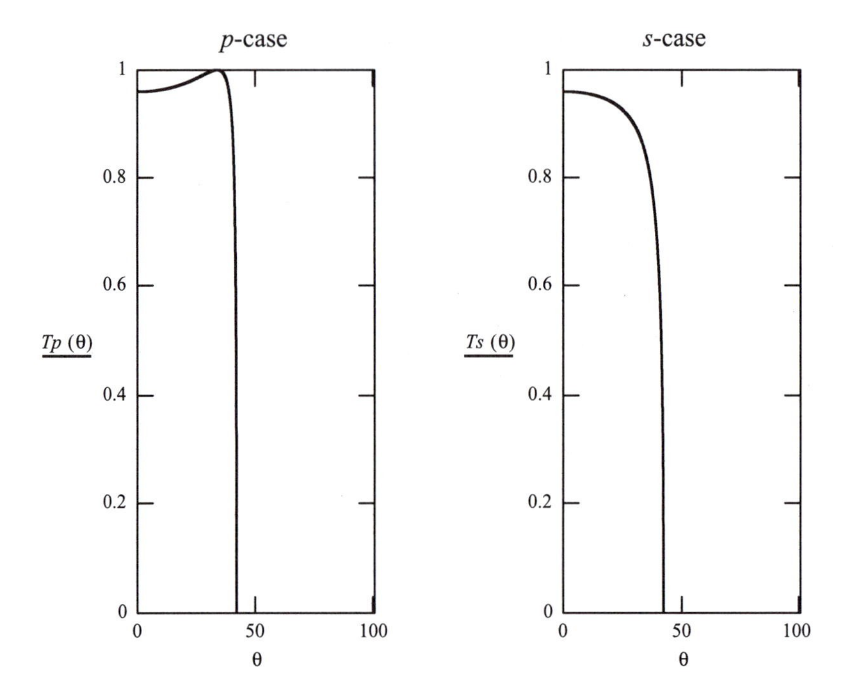

Application 5.6.

1. Make graphs of r_p^2, r_s^2, t_p^2 and t_s^2 for the case where $n_1 < n_2$, compare to the graphs of the intensities, and find out which quantities, just squared, are useful for calculation of corresponding intensities and which are not.
2. Make a graph of the factor $\alpha = (n_2 \cos\theta'')/(n_1 \cos\theta)$ for $n_1 < n_2$ and $n_1 > n_2$, compare to FileFig 5.5.

5.6.7 Total Reflection and Evanescent Wave

We look at the transmitted amplitude for the case where the angle of incidence is larger than the critical angle. We have

$$E_t = E_{to} \exp i\{k_2(\sin\theta_t X - \cos\theta_t Y)\} \exp(-i\omega t), \tag{5.79}$$

where E_{to} may be calculated from Fresnel's formulas. Using the law of refraction as $k_1 \sin\theta_i = k_2 \sin\theta_t$ we may rewrite E_t depending on the angle of incidence:

$$E_t = E_{to} \exp i\{k_1 \sin\theta_i X - k_2 Y \sqrt{1 - (k_1 \sin\theta_i / k_2)^2}\} \exp(-i\omega t). \tag{5.80}$$

The square root may be written as

$$i\left\{\sqrt{n_1 \sin\theta_i / n_2)^2 - 1}\right\} \tag{5.81}$$

and one gets

$$E_t = E_{to} \exp i\left\{k_1 \sin\theta_i X - k_2 Y i\left\{\sqrt{n_1 \sin\theta_i / n_2)^2 - 1}\right\}\right\} \exp(-i\omega t). \tag{5.82}$$

This wave is called the *evanescent wave*, traveling in the medium with the lower index of refraction in the $-Y$ direction. It is composed of a traveling wave and an attenuation factor. The attenuation factor is

$$y = A \exp(Y k_2 \sqrt{\{(n_1/n_2)^2 (\sin\theta_i)^2 - 1\}}, \tag{5.83}$$

where $k_2 = (2\pi/\lambda)n_2$.

In FileFig 5.7 we show graphs of the attenuation factor depending on different angles of incidence. One observes the rapid decrease of the magnitude depending on penetration depth $-Y_t$.

FileFig 5.7 (M7FREVA)

Graph of the attenuation factor of the amplitude of the evanescent wave for

(a) $n_1 = 1.5$, $n_2 = 1$, and critical angle $\theta_c = 41.81$, and $\theta_1 = \theta_c + 2$;

(b) $nn_1 = 3.4$, $nn_2 = 1$, and critical angle $\theta_c = 17.105$ and $\theta_2 = \theta_c + 2$ and $\lambda = 0.0005mm$, depending on coordinate $-Y$.

M7FREVA

Penetration into the Less Dense Medium at Total Reflection

Exponential factor for decrease of amplitude into the less dense medium with $-Y$ for two different refractive indices $n1$ and $nn1$ and $n2 = nn2$. θc is the critical angle. The value a is used to "be off" the critical angle. First we set

$$a \equiv 2 \qquad n1 \equiv 1.5 \qquad n2 \equiv 1 \qquad \lambda := .0005 \qquad nn1 \equiv 3.4 \qquad nn2 \equiv 1.$$

$$z := \operatorname{asin}\left(\frac{n2}{n1}\right) \qquad zz := \operatorname{asin}\left(\frac{nn2}{nn1}\right) \qquad Y := -0.00005, -.0001., -.001$$

$$\theta 1c := z \cdot \frac{360}{2 \cdot \pi} \qquad \theta 2c := zz \cdot \frac{360}{1 \cdot \pi}$$

$$\theta 1c = 41.81 \qquad \theta 2c = 17.105$$

$$\theta 1 := \theta 1c + a \qquad \theta 2 := \theta 2c + a$$

$$k2 := 2 \cdot \frac{\pi}{\lambda} \cdot n2 \qquad A := 1 \qquad kk2 := 2 \cdot \frac{\pi}{\lambda} \cdot nn2$$

$$y1(Y) := A \cdot e^{Y \cdot k2 \cdot \sqrt{\left(n1 \cdot \frac{\sin\left(\frac{2\cdot\pi}{360}\cdot\theta 1\right)}{n2}\right)^2 - 1}} \qquad y2(Y) := A \cdot e^{Y \cdot kk2 \cdot \sqrt{\left(nn1 \cdot \frac{\sin\left(\frac{2\cdot\pi}{360}\cdot\theta 2\right)}{nn2}\right)^2 - 1}}.$$

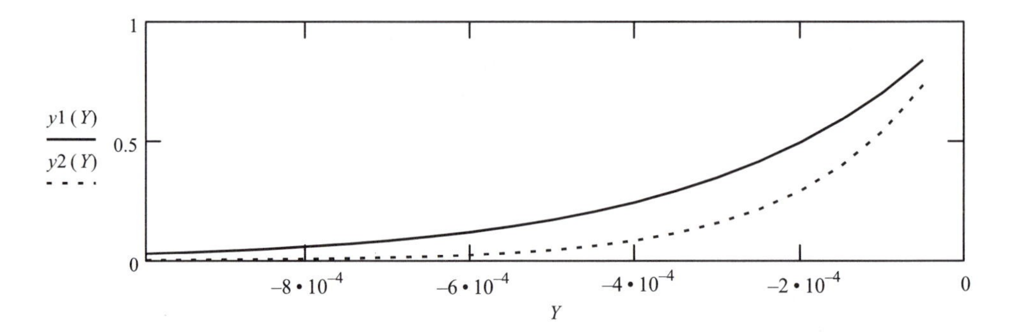

To study different angles, make refractive indices the same for both and change a to values larger than 2.

Application 5.7.

1. Study the attenuation factor for a fixed angle of incidence. Make a graph for three indices of refraction.
2. Study the attenuation factor depending on the angle of incidence. Make a graph for three angles of incidence for a large and a small difference of the refractive indices n_1 and n_2.

5.7 POLARIZED LIGHT

5.7.1 Introduction

In contrast to solutions of the scalar wave equation the solutions of Maxwell's equations are vector waves. In the discussion of Fresnel's formulas, we considered the two components of the electrical field vector, parallel and perpendicular to the plane of incidence. The two electrical field vectors vibrate in directions perpendicular to each other and each vector presents linear polarized light.Linearly polarized light may be produced by reflection under the Brewster angle at the surface of a dielectric material or when light is reflected on wire gratings with a wavelength larger than the periodicity constant.

The superposition of two linear polarized light vectors will result in linearly polarized light, but only if there is no phase difference between the two vibrations. A phase difference may be produced by using total internal reflection and will result in elliptically or circularly polarized light. The incident light is reflected at a denser medium and the two components, the p-component and the s-component, have a fixed phase angle between them, which is assumed to be zero. After reflection, each of the two reflected components "picks up" a different phase angle and the superposition results in a phase angle between the two components.

There are dielectric materials with different refractive indices in different directions of the material. Plastic films, produced by stretching, may transmit partially polarized light. Some of the large molecules of the material are oriented in the direction of the stress, and the refractive index is different in the parallel and perpendicular directions.

Uniaxial crystals have different orientations of molecules with respect to the axis and in perpendicular layers; see below.

5.7.2 Ordinary and Extraordinary Indices of Refraction

Optical materials may have a different refractive index in one direction than in another direction. These materials are called birefringent. Examples are quartz and calcite, both uniaxial crystals. These crystals are composed of layers of atoms, which are arranged in planes, and the planes are stacked in a pile. The atoms are symmetrically positioned in each plane and the normal of the planes is the symmetry axis (Figure 5.5). We use an X, Y, Z coordinate system. The symmetry axis is Z and the X and Y axes are in the plane and perpendicular to each other.

The refractive index along the Z-axis is different from the index along the X- and Y-axes. The refractive index along the Z-axis is called the extraordinary index n_e and along the X and Y-axes ordinary index n_0.

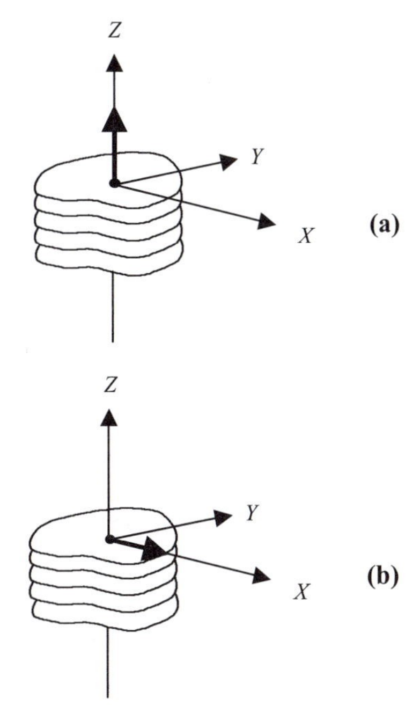

FIGURE 5.5 Propagation with respect to the optic axis (Z), shown by black arrows: (a) wave propagating parallel to the Z-axis. Possible directions of the oscillating electric fields are along the X- and Y-axes; (b) wave propagating perpendicular to the Z-axis, for example in the direction of the X-axis. Possible directions of the oscillating electric fields are along the Y- and Z- axes.

For quartz we have $n_e = 1.553$ and $n_0 = 1.544$ (positive crystal);
For calcite we have $n_e = 1.486$ and $n_0 = 1.658$ (negative crystal).
The velocity of light in the medium is calculated from $v = c/n$. For quartz the velocity along the Z-axis is smaller than along the X- and Y-axes and Z is called the *slow axis*. Crystals where the optical axis is the slow axis are called *positive crystals*. For calcite the velocity along the Z-axis is faster than along the X- and Y-axes and Z is called the *fast axis*. Crystals where the optical axis is the fast axis are called *negative crystals*.

5.7.3 Phase Difference Between Waves Moving in the Direction of or Perpendicular to the Optical Axis

The refractive index is related to the polarization of the atoms in the direction of the oscillating **E**-vector. As a result, the velocity of propagation is determined by the direction of vibration of the **E**-vector that is perpendicular to the direction of propagation. In Figure 5.6a we show waves propagating in the Z direction, but vibrating in the X and Y directions, which are in the plane of the layers perpendicular to the Z direction. The ordinary index determines the velocity of propagation, which is the same for both. In Figure 5.6b we show two waves propagating in the X direction, but one oscillates in the Y direction and the other in the Z direction. The velocity of the wave vibrating in the Y direction is determined by the ordinary index n_0, whereas the velocity of the wave vibrating in the Z direction is determined by the extraordinary index n_e. These two waves propagate with different velocities in the X direction and therefore will develop a phase difference.

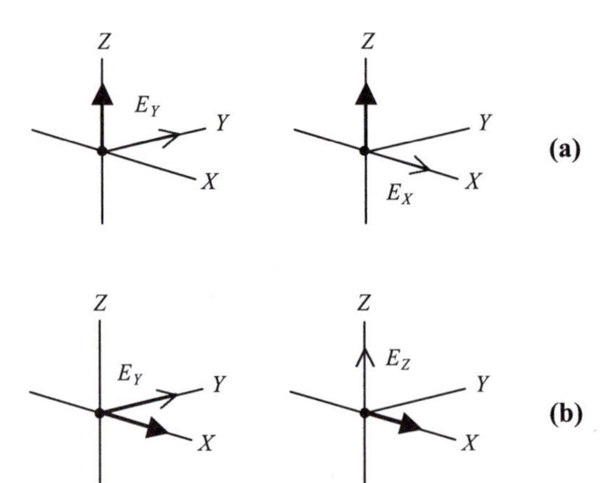

FIGURE 5.6 (a) Propagation parallel to the optical axis, vibrations perpendicular to the optical axis; (b) propagation perpendicular to the optical axis in X direction, vibrations perpendicular and parallel to the optical axis.

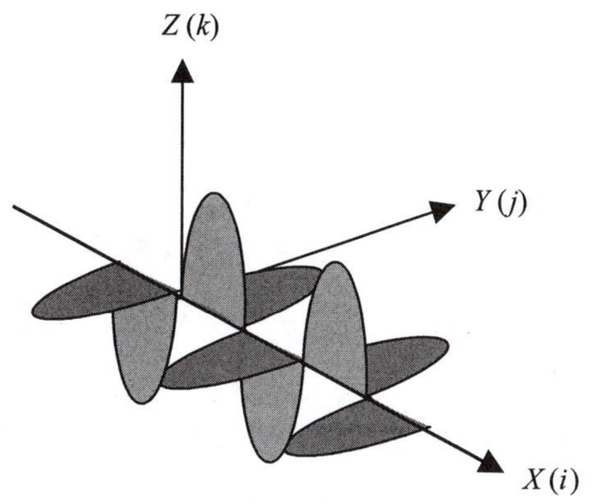

FIGURE 5.7 Waves with electrical vectors E_Y and E_Z propagating in the X direction.

For two waves, traveling in the X direction and having the same refractive index in the Y and Z directions (Figure 5.7) we have

$$E_Y = \mathbf{j}Ae^{i(k_1 X - \omega t)} \tag{5.84}$$
$$E_Z = \mathbf{k}Ae^{i(k_1 X - \omega t)}, \tag{5.85}$$

taking equal amplitudes of the electrical field vectors.

Assuming that the material has a refractive index n_1 for the wave vibrating in the Y direction and index n_2 for the wave vibrating in the Z direction with corresponding wave vectors k_1 and k_2, we write

$$E_Y = \mathbf{j}Ae^{i(k_1 X - \omega t)} \tag{5.86}$$
$$E_Z = \mathbf{k}Ae^{i(k_2 X - \omega t)}. \tag{5.87}$$

Using

$$\phi_X = (k_2 - k_1)X \tag{5.88}$$

we have

$$E_Y = \mathbf{j}A \exp i(k_1 X - \omega t) \tag{5.89}$$
$$E_Z = \mathbf{k}A \exp i(k_1 X - \omega t + \phi X). \tag{5.90}$$

In Figure 5.8 we show an example of two waves with a phase difference of ϕ.

5.7.4 Half-Wave Plate, Phase Shift of π

We consider the case where $\phi_x = \pi$ and have from Eq. (5.88) for the corresponding distance $X = \pi/(k_2 - k_1)$. At this distance there is a phase difference of π between the E_Y and the E_Z components, compared to the E_Y and the E_Z components at $X = 0$ (Figure 5.9). We apply this to the case of quartz using n_e for the Z component and n_0 for the Y component and $k_1 = 2\pi n_0/\lambda$, $k_2 = 2\pi n_e/\lambda$ and have

$$X = \pi/(2\pi n_e/\lambda - 2\pi n_0/\lambda) = (\lambda/2)/(n_e - n_0). \tag{5.91}$$

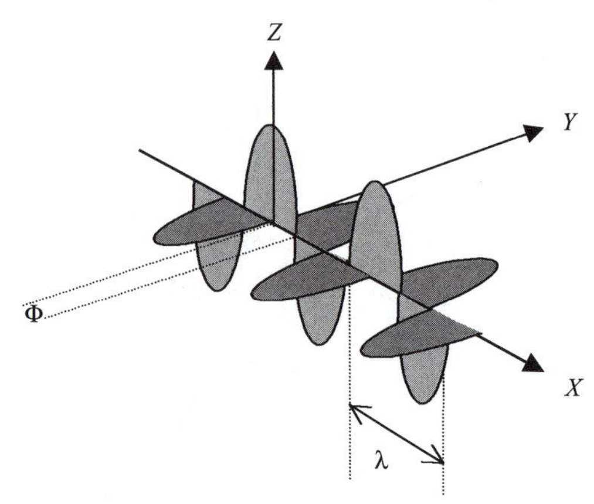

FIGURE 5.8 Waves vibrating in the E_Y and E_Z directions with a phase difference of ϕ. The waves are drawn at a time instant where $e^{i\omega t} = 1$.

This distance X is very small, but any odd integer of X will have the same effect. Therefore one may use a plate of thickness

$$L_h = (\lambda/2)(1 + 2m)/(n_e - n_0), \tag{5.92}$$

where m is an integer. Since n_e is larger than n_0 we have a positive value for d, and therefore quartz has been marked above as a positive crystal.

The case of calcite is reversed. We have n_e smaller than n_0 so d is a negative value. Consequently calcite is called a negative crystal.

In FileFigs.8 we first look at the plane $X = 0$. In the first graph the Y and Z components are plotted as functions of X and in the second graph the Z component is plotted against the Y component. The third and fourth graphs show what happens in the plane $X = L_h$, after a phase shift of π. In the third graph the Y and Z components are plotted as functions of X and the phase change of the Z component is shown. In the fourth graph the Z component is plotted against the Y component. The direction of the resulting vibration of the two waves is shifted by 90° from the second to the fourth graph (and not by π (or 180°).

FileFig 5.8 (M8POLIN)

Graphs of the superposition of the E_Y and E_Z components before entering the plate, where the phase angle $\phi_X = 0$, and at the plate $X = L$ with phase angle $\phi_X = \pi$.

M8POLIN is only on the CD.

Application 5.8. Make graphs for $\phi_X = -\pi(-180)$ and compare with $\phi_X = 0$ and $\phi_X = (180)$ and with Figure 5.9.

outside the plate k_1 is k_2

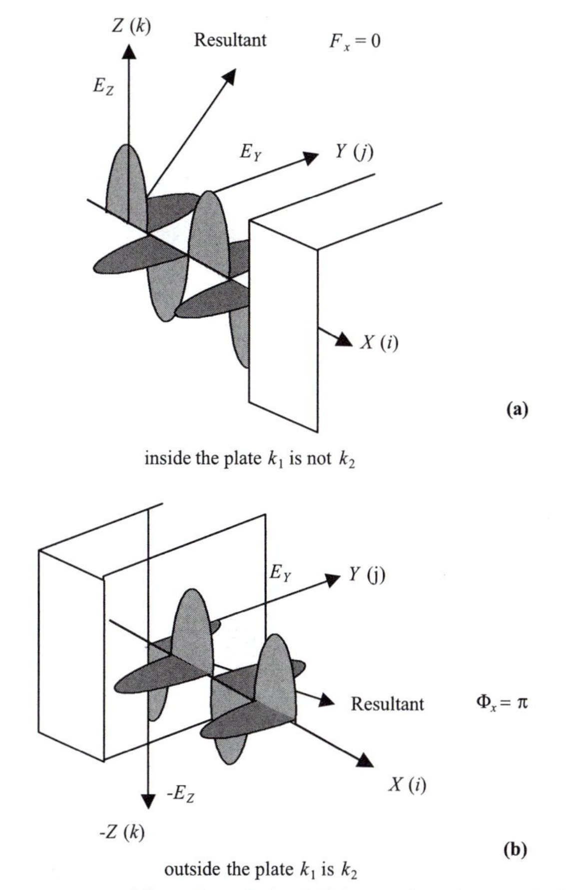

FIGURE 5.9 Two comonents of linearly polarized light passing throug a half-wave plate: (a) incident, (b) emerging.

5.7.5 Quarter Wave Plate, Phase Shift $\pi/2$

We now consider the case where $\phi_X = \pi/2$ and have the distance $X = (\pi/2)/(k_2 - k_1)$. There is a phase difference of $\pi/2$ between the E_Y and E_Z components at this distance, compared to the E_Y and E_Z components at $X = 0$ (Figure 5.10). We apply this to the case of quartz using n_e for the Z component, and n_0 for the Y component, and $k_1 = 2\pi n_0/\lambda$, $k_2 = 2\pi n_e/\lambda$, and have

$$X = (\pi/2)/(2\pi n_e/\lambda - 2\pi n_0/\lambda) = (\lambda/4)/(n_e - n_0). \tag{5.93}$$

This distance is very small, but any odd integer of X will have the same effect. Therefore one may use

$$L_q = (\lambda/4)(1 + 4m)/(n_e - n_0), \tag{5.94}$$

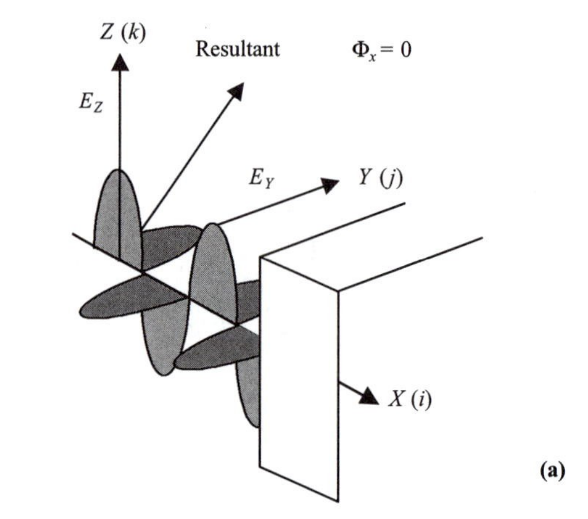

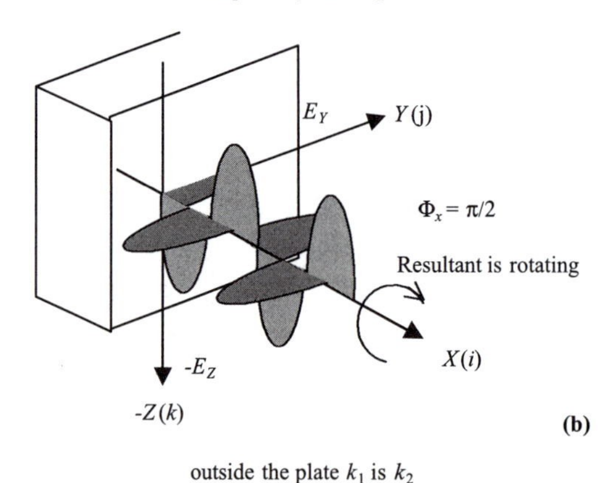

FIGURE 5.10 Phase relation of the two components of polarized light: (a) before entering; and (b) emerging from a quarter-wave plate.

where m is an integer. Since n_e is larger than n_0 we have a positive value for d, and therefore quartz has been marked above as a positive crystal, and calcite is called a negative crystal.

For the quarter-wave plate we have, with Eqs. (5.89) and (5.90),

$$E_Y = \mathbf{j}A \exp i(k_1 L_q - \omega t) \tag{5.95}$$

$$E_Z = \mathbf{k}A \exp i(k_1 L_q - \omega t + \pi/2). \tag{5.96}$$

To make a graph of the superposition of the E_Y and E_Z components, we take the real parts of the fields as the values at the Y- and Z- axis of Eqs. (5.95) and (5.96)

$$E_Y = \cos(k_1 L_q - \omega t) \tag{5.97}$$

$$E_Z = \cos(k_1 L_q - \omega t + \pi/2). \tag{5.98}$$

Or converting Eq. (5.98),

$$E_Y = \cos(k_1 L_q - \omega t) \tag{5.99}$$

$$E_Z = -\sin(k_1 L_q - \omega t). \tag{5.100}$$

We may write Eqs. (5.99) and (5.100), for a certain time interval, as $E_Y = \cos(-2\pi x_1/360)$ and $E_Z = \cos(-2\pi x_1/360 + \pi/2)$. The time interval corresponds to a certain distance in the direction of propagation and to a certain angle interval $x_1/360$. In FileFig 5.9 we show four graphs, corresponding to intervals of angles from 1° to 90°, 1° to 160°, 1° to 235°, and 1° to 315°. Looking onto the paper, in the direction of the source, we see that the resulting vibration describes a circle. The circle develops for positive $\phi_X = +\pi/2$ in a counterclockwise direction and the light is called left polarized. Considering $E_Y = \cos(-2\pi x_1/360)$ and $E_Z = \cos(-2\pi x_1/360 - \pi/2)$, for negative $\phi_X = -\pi/2$, the circle develops in the clockwise direction and the light is called right polarized.

FileFig 5.9 (M9POELIP)

Graphs of the superposition of the E_Y and E_Z components with positive and negative phase angle ϕ_X. Four graphs for four different time spans are shown.

M9POELIP

Circular and Elliptically Polarized Light

Graphs for circular and elliptically polarized light turning "left or right." Four graphs are shown, extending from 0 to 90, 0 to 160, 0 to 235, and 0 to 315 degrees. The angle ranges (x) correspond to chosen time ranges. Left and right polarized light is described by positive or negative $\Phi = \pi/2$ in one component: Positive Φ: we have $y = Ey = A\cos(-x)$, $yy = Ez = A\cos(-x + \Phi) = -A\sin(-x)$; negative Φ: we have $y = Ey = A\cos(-x)$, $yy = Ez = A\cos(-x - \Phi) = A\sin(-x)$. We write for $Ez = bA\sin(x)$. When looking in the direction of the incoming light, $b = -1$ is for "left" polarized light (counterclockwise), $b = 1$ for "right" polarized light (clockwise).

$$x1 := 1, 2 \ldots 90 \qquad x2 := 1, 2 \ldots 160$$

$$x3 := 1, 2 \ldots 235 \qquad b \equiv -1 \qquad x4 := 1, 2 \ldots 315$$

$$y1(x1) := \cos\left(-2 \cdot \pi \cdot \frac{x1}{360}\right) \qquad y2(x2) := \cos\left(-2 \cdot \pi \cdot \frac{x2}{360}\right)$$

$$y3(x3) := \cos\left(-2 \cdot \pi \cdot \frac{x3}{360}\right) \qquad y4(x4) := \cos\left(-2 \cdot \pi \cdot \frac{x4}{360}\right)$$

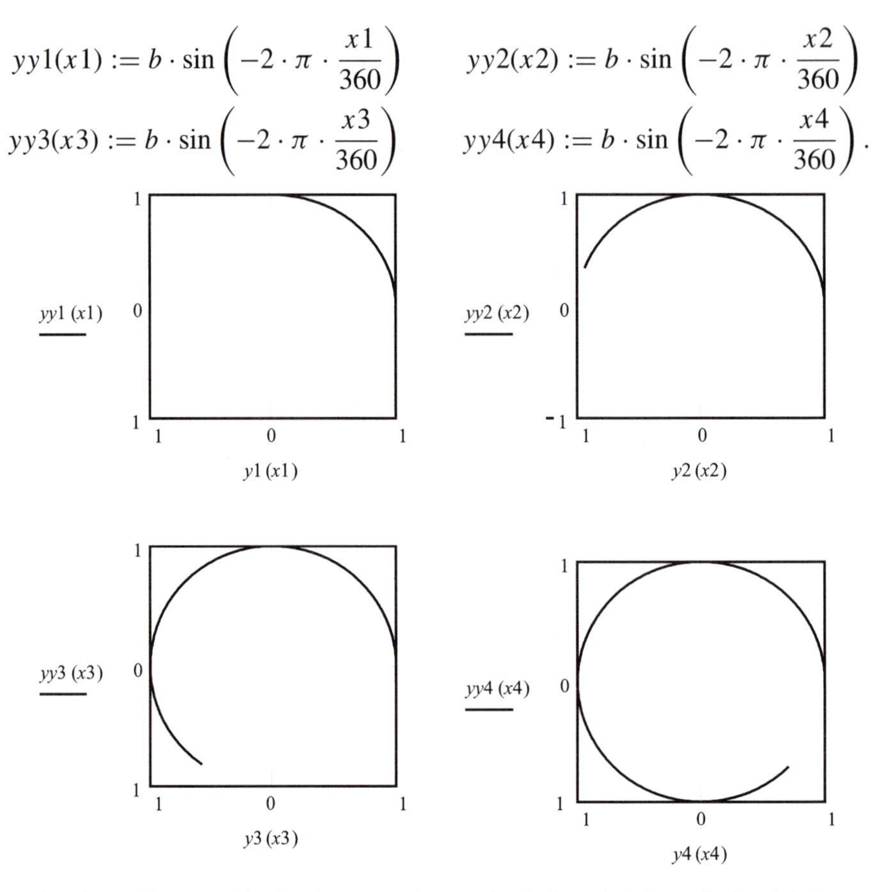

Application 5.9. Modify the four graphs to plot left and right polarized light on the same graph.

5.7.6 Crossed Polarizers

The experimental setup of crossed polarizers is a sequence of a horizontal (X direction) and a vertical (Y direction) polarizer. Any light passin the first polarizer will not pass the second. Therefore no light passes the crossed polarizers configuration.

5.7.6.1 Half Wave Plate Between Crossed Polarizers

We next discuss the case where a half-wave plate is placed between the polarizer and analyzer of the crossed polarizers configuration. We assume that the half wave plate is oriented with its optical axis (Z) at $45°$ degrees to the horizontal direction of the polarizer (Figure 5.11). The incident light is first horizontally polarized by the polarizer and then incident on the half wave plate. The horizontal polarized light is split up in a Z component along the axis of the half-wave plate

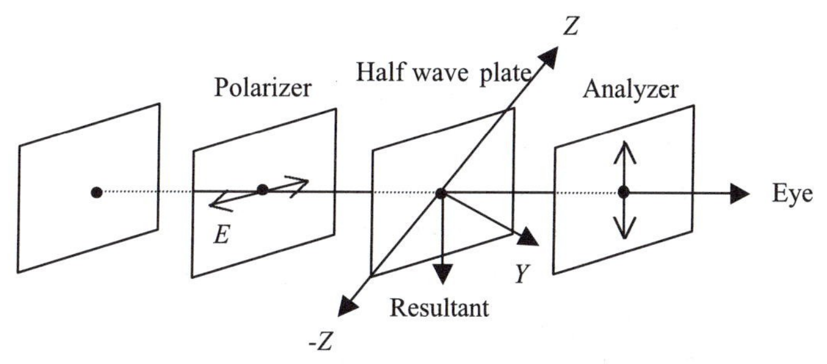

FIGURE 5.11 Half-wave plate between crossed polarizers. The horizontal polarized light is represented by the two perpendicular vectors E_Y and E_Z. The half-wave plate turns E_Z into $-E_Z$. The resultant of E_Y and $-E_Z$ is polarized in the vertical direction and may pass the analyzer.

and a Y component perpendicular to it. The Z component leaves the plate with a phase shift of π, oscillating in the Z direction, while the Y component remains oscillating in the Y direction. The resultant of the two components leaving the half-wave plate is polarized in the vertical direction, and will pass the vertical polarizer. In this setup all the light passing the horizontal polarizer will also pass the vertical polarizer, sometimes called the analyzer.

5.7.6.2 Quarter-Wave Plate Between Crossed Polarizers

A quarter-wave plate is placed between crossed polarizers with its axis at 45° (Figure 5.12). The light passing the first polarizer, incident on the quarter-wave plate, is decomposed into two components. One oscillates parallel and the other perpendicular to the axis of the quarter wave plate. The resultant, leaving the quarter-wave plate, is not stationary. It rotates around the direction of propagation X and light passes the second polarizer, also called the analyzer. Rotation of the quarter wave plate around a range of angles will not affect this. However, the axis of the quarter-wave plate should not be parallel to the linear polarizers.

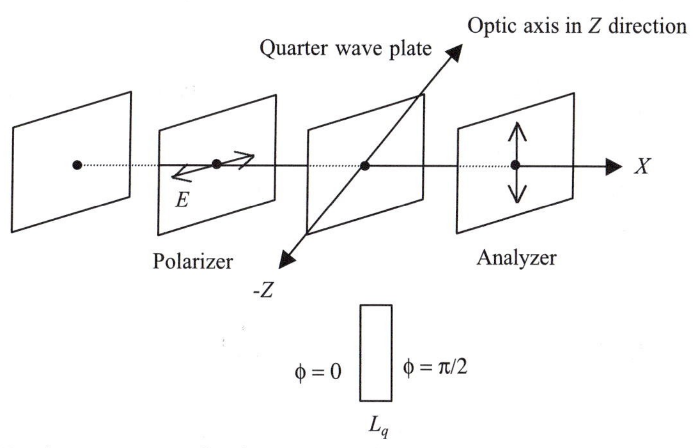

FIGURE 5.12 A quarter-wave plate between polarizers.

5.7.7 General Phase Shift

5.7.7.1 Half- and Quarter-Wave Plates

We have discussed the generation of phase differences of π and $\pi/2$ by a half-wave plate and a quarter-wave plate. Another way to produce a phase difference between the two components of the electrical field vector is by internal total reflection. In the case of $n_1 > n_2$, we have complex reflection coefficients in the region of total reflection. The range of the phase shifts is between plus or minus π, different for the p- and s-cases. We also have a change in the magnitude of reflection. If we superimpose the vectors of the total internal reflected light for the p- and s-cases, we would, in general, obtain elliptical polarized light.

5.7.7.2 Linear, Circular, and Elliptical Polarized Light

We now examine elliptically polarized light. We consider the plane $X = L$ and the corresponding phase difference of ϕ_X, for angles between 0 and 360°. We refer to Eqs. (5.89) and (5.90):

$$E_Y = \mathbf{j}A \exp i(k_1 L - \omega t) \tag{5.89}$$

$$E_Z = \mathbf{k}A \exp i(k_1 L - \omega t + \phi_X). \tag{5.90}$$

By only using the real part of the Y and Z components and substituting $\alpha = (k_1 L - \omega t)$, we have

$$E_Y = A \cos\alpha \tag{5.101}$$

$$E_Z = A\cos(\alpha + \phi) = A[\cos\alpha\cos\phi - \sin\alpha\sin\phi]. \tag{5.102}$$

Eliminating α, the equation of an ellipse is obtained:

$$E_Z^2 - 2E_Y E_Z \cos\phi + E_Y^2 = A^2 \sin\phi^2. \tag{5.103}$$

In FileFig 5.10 we show that one may write Eq. (5.103) in matrix form and that we have for $\phi = 0$, linearly polarized light,

$$E_Y = E_Z, \tag{5.104}$$

for $\phi = \pi/4$, elliptically polarized light, and have the equation of an ellipse,

$$E_Z^2/(1 - 1/\sqrt{2}) + E_Y^2/(1 + 1/\sqrt{2}) = 1, \tag{5.105}$$

and for $\phi = \pi/2$, circular polarized light, and have the equation of a circle,

$$E_Z^2 + E_Y^2 = 1. \tag{5.106}$$

FileFig 5.10 (M10POELIPSES)

The general equation of the ellipse is shown in vector-matrix notation, using the eigenvalue method. The equations are obtained for $\phi = 0$ (linear polarized

light) for $\phi = \pi/4$, (elliptically polarized light), and $\phi = \pi/2$ (circular polarized light).

M10POELIPSES is only on the CD.

Application 5.10. Derive the equations for $\phi = 3\pi/4$, $\phi = 5\pi/2$, $\pi = 3\pi/2$, and $\phi = 7\pi/4$ and compare with results of FileFig 5.11.

In Appendix A5.2 we show that the rotation of the coordinate system may be equivalent to a transformation to principal axes. In FileFig 5.11 we show graphs of one component plotted against the other for

$$\begin{array}{llll} \phi = 0, & \phi = \pi/4, & \phi = \pi/2, & \phi = 3\pi/4, \\ \phi = \pi, & \phi = 5\pi/4, & \phi = 3\pi/2, & \phi = 7\pi/4, \; . \\ \phi = 2\pi & & & \end{array} \tag{5.107}$$

One sees that the two components of linearly polarized light, vibrating along perpendicular directions, result in linear polarized light when the phase difference is $\phi_X = 0, \pi$, and 2π. The resulting vibration takes place along a line tilted by 45° for $\phi_X = 0, 2\pi$, and tilted by 135° for $\phi_X = \pi$. For $\phi_X = \pi/2$ we have left circular polarized light and for $\phi_X = 3\pi/2$, equivalent to $-\pi/2$, right circular polarized light. The ellipse is left turning, when the ϕ_X values are first larger and then smaller than $\phi_X = \pi/2$, and "right turning" for ϕ_X values first larger and then smaller than $\phi_X = 3\pi/2$. The large axis of the ellipse is always oriented in the same direction as the axis of the "closest" linear polarized light.

FileFig 5.11 (M11POELIPLIS)

Graphs are shown of the equation of the ellipse, that is, Eq. (5.103) for $\phi = 0$, $\phi = \pi/4$, $\phi = \pi/2$, $\phi = 3\pi/4$, $\phi = \pi$, $\phi = 5\pi/4$, $\phi = 3\pi/2$, $\phi = 7\pi/4$, and $\phi = 2\pi$.

M11POELIPLIS

Elliptical Polarized Light

Similarly to that discussed in FileFig 5.9 we plot $\cos(-2\pi x/360)$ on the z-axis and $\cos(-2\pi x/360 + \Phi)$ on the y-axis.

$$x \equiv 1, 2 \cdots 360 \qquad \phi 1 := 0$$

$$y1(x) := \cos\left(-2 \cdot \pi \cdot \frac{x}{360}\right) \qquad yy1(x) := \cos\left(-2 \cdot \pi \cdot \frac{x}{360} + \phi 1\right)$$

$$\phi 2 := \frac{\pi}{4}$$

$y2(x) := \cos\left(-2 \cdot \pi \cdot \frac{x}{360}\right)$ $yy2(x) := \cos\left(-2 \cdot \pi \cdot \frac{x}{360} + \phi 2\right)$

$\phi 3 := \frac{\pi}{2}$

$y3(x) := \cos\left(-2 \cdot \pi \cdot \frac{x}{360}\right)$ $yy3(x) := \cos\left(-2 \cdot \pi \cdot \frac{x}{360} + \phi 3\right)$

$\phi 4 := 3 \cdot \frac{\pi}{4}$

$y4(x) := \cos\left(-2 \cdot \pi \cdot \frac{x}{360}\right)$ $yy4(x) := \cos\left(-2 \cdot \pi \cdot \frac{x}{360} + \phi 4\right)$

$\phi 5 := \pi$

$y5(x) := \cos\left(-2 \cdot \pi \cdot \frac{x}{360}\right)$ $yy5(x) := \cos\left(-2 \cdot \pi \cdot \frac{x}{360} + \phi 5\right)$

$\phi 6 := \frac{5 \cdot \pi}{4}\pi$

$y6(x) := \cos\left(-2 \cdot \pi \cdot \frac{x}{360}\right)$ $yy6(x) := \cos\left(-2 \cdot \pi \cdot \frac{x}{360} + \phi 6\right)$

$\phi 7 := \frac{3 \cdot \pi}{2}\pi$

$y7(x) := \cos\left(-2 \cdot \pi \cdot \frac{x}{360}\right)$ $yy7(x) := \cos\left(-2 \cdot \pi \cdot \frac{x}{360} + \phi 7\right)$

$\phi 8 := 7 \cdot \frac{\pi}{4}\pi$

$y8(x) := \cos\left(-2 \cdot \pi \cdot \frac{x}{360}\right)$ $yy8(x) := \cos\left(-2 \cdot \pi \cdot \frac{x}{360} + \phi 8\right)$

APPENDIX 5.1

A5.1.1 Wave Equation Obtained from Maxwell's Equation

$$\begin{aligned} \nabla \times \mathbf{E} &= -\partial \mathbf{B}/\partial t \\ c^2 \nabla \times \mathbf{B} &= +\partial \mathbf{E}/\partial t \\ \nabla \cdot \mathbf{E} &= 0 \\ \nabla \cdot \mathbf{B} &= 0. \end{aligned} \tag{A5.1}$$

From the first equation of A5.1 we have by taking the cross product with ∇

$$\nabla \times \nabla \mathbf{E} = -\partial/\partial t \nabla \times \mathbf{B} \tag{A5.2}$$

and using the identity

$$\nabla \times \nabla \times \mathbf{E} = \nabla(\nabla \cdot \mathbf{E}) - \nabla^2 \mathbf{E} \tag{A5.3}$$

we get

$$\nabla(\nabla \cdot \mathbf{E}) - \nabla^2 \mathbf{E} = -\partial/\partial t \nabla \times \mathbf{B}. \tag{A5.4}$$

Inserting the second equation of Eq. (A5.1), we obtain

$$\nabla(\nabla \cdot \mathbf{E}) - \nabla^2\mathbf{E} = -(1/c^2)\partial/\partial t(\partial\mathbf{E}/\partial t). \tag{A5.5}$$

With $\nabla \cdot \mathbf{E} = 0$ of the third equation of (A5.1) we have the vector wave equation for the electrical field vector **E**,

$$\partial^2\mathbf{E}/\partial x^2 + \partial^2\mathbf{E}/\partial y^2 + \partial^2\mathbf{E}/\partial z^2 = (1/c^2)\partial^2\mathbf{E}/\partial t^2. \tag{5.3}$$

Using a similar formalism we derive the vector wave equation for the magnetic field vector **B**,

$$\partial^2\mathbf{B}/\partial x^2 + \partial^2\mathbf{B}/\partial y^2 + \partial^2\mathbf{B}/\partial z^2 = (1/c^2)\partial^2\mathbf{B}/\partial t^2. \tag{5.4}$$

A5.1.2 The Operations ∇ and ∇^2

Cartesian coordinates (x, y, z)

$$\nabla = \mathbf{i}\partial/\partial x + \mathbf{j}\partial/\partial y + \mathbf{k}\partial/\partial z$$
$$\nabla^2 = \partial^2/\partial x^2 + \partial^2/\partial y^2 + \partial^2/\partial z^2$$

Spherical coordinates (r, θ, ϕ)

$$\nabla = \mathbf{i}\partial/\partial r + \mathbf{j}(1/r)\partial/\partial\theta + \mathbf{k}(1/r\sin\phi)\partial/\partial\phi$$
$$\nabla^2 = (1/r^2)\partial/\partial r(r^2\partial/\partial r) + [1/(r^2\sin\theta)]\partial/\partial\theta(\sin\theta\partial/\partial\theta) + [1/(r^2\sin^2\theta)]\partial^2\partial\phi^2$$

APPENDIX 5.2

A5.2.1 Rotation of the Coordinate System as a Principal Axis Transformation and Equivalence to the Solution of the Eigenvalue Problem

FileFig A5.12 (MA2ROTMAS)

1. Rotation matrices and their multiplication.
2. Demonstrated that the equation of the ellipse is obtained without cross terms when introducing $\phi = \pi/4$.
   - **a.** Introduction of $\phi = \pi/4$ into the general equation of the ellipse
   - **b.** Introduction of $\phi = \pi/4$ as rotation angle

   If the value of $\phi = \pi/4$ is not known, it may be determined from the transformation making the matrix of the general equation of the ellipse diagonal.

MA2ROTMAS is only on the CD.

APPENDIX 5.3

A5.3.1 Phase Difference Between Internally Reflected Components

We have mentioned above that elliptically polarized light may be produced by total internal reflection. The incident light is reflected at a denser medium and the two components, the p and s components, have a fixed phase angle between them which is assumed to be zero. The reflected light is the superposition of the two reflected components, each "picking up" a different phase angle upon reflection. The difference between the two "new" phase angles after reflection is the phase angle between the two components and may be calculated as

$$\tan \Delta/2 = (\sin\theta)^2/[\cos\theta\sqrt{((\sin\theta)^2 - (n_2/n_1)^2)}]. \tag{A5.6}$$

We can get a graph of the angle Δ from the complex reflection coefficients r_p and r_s. We just have to take the argument of r_p/r_s. This is done in FileFigA3.

FileFig A5.13 (MA3DIFINTRO)

A graph is shown of the difference between the arguments of the reflection coefficients for internal total reflection.

MA3DIFINTRO is only on the CD.

Application A5.13. Observe the change of the difference angle depending on the refractive index.

APPENDIX 5.4

A5.4.1 Jones Vectors and Jones Matrices

We have presented the two mutually perpendicular components propagating in the X direction of the electrical field vectors. The phase angle ϕ between them is

$$E_Y = \mathbf{j}A\exp i(k_1X - \omega t) \tag{A5.7}$$

$$E_Z = \mathbf{k}A\exp i(k_1X - \omega t + \phi). \tag{A5.8}$$

One may want to write this in vector notation as

$$\begin{pmatrix} E_Y \\ E_Z \end{pmatrix} = A\exp i(k_1X - \omega t)\begin{pmatrix} 1 \\ e^{i\phi} \end{pmatrix}. \tag{A5.9}$$

Disregarding the common factor $A \exp i(k_1 X - \omega t)$ we may describe polarized light by such vectors and have

$$\begin{pmatrix} 1 \\ 0 \end{pmatrix} \text{horizontal} \qquad \begin{pmatrix} 0 \\ 1 \end{pmatrix} \text{vertical} \tag{A5.10}$$

$$1/\sqrt{2}\begin{pmatrix} 1 \\ 0 \end{pmatrix} \text{+45 degrees} \qquad 1/\sqrt{2}\begin{pmatrix} 0 \\ 1 \end{pmatrix} \text{-45 degrees} \tag{A5.11}$$

$$1/\sqrt{2}\begin{pmatrix} 1 \\ -i \end{pmatrix} \text{right circular} \qquad 1/\sqrt{2}\begin{pmatrix} 0 \\ i \end{pmatrix} \text{left circular} \tag{A5.12}$$

All Jones vectors are listed in FileFig 5.14.

A5.4.2 Jones Matrices

We have discussed above the transformation between coordinate systems using the rotation matrix and found, for example, that the rotation of 45° degrees is expressed as

$$EE_Y = E_Y - E_Z \tag{A5.13}$$

$$EE_Z = E_Y + E_Z. \tag{A5.14}$$

In matrix formulation we have

$$\begin{pmatrix} EE_Y \\ EE_Z \end{pmatrix} = \begin{pmatrix} 1 & -1 \\ 1 & 1 \end{pmatrix} \begin{pmatrix} E_Y \\ E_Z \end{pmatrix} \tag{A5.15}$$

In a similar martix representation we can obtain other operations on the two components of the electrical field vectors (FileFig 5.14).

A5.4.3 Applications

We discuss two applications of the Jones vectors and Jones matrices.

Half-Wave Plate Between Crossed Polarizers

We start off with linear polarized light and apply the half-wave plate with 45° orientation, disregarding the normalization factor. Then we do the multiplication

$$\begin{pmatrix} 1 & 1 \\ 1 & 1 \end{pmatrix} \begin{pmatrix} 1 \\ 0 \end{pmatrix} = \begin{pmatrix} 1 \\ 1 \end{pmatrix}. \tag{A5.16}$$

The result is 45° polarized light. Then we apply the vertical linear polarizer and obtain vertically polarized light.

$$\begin{pmatrix} 0 & 0 \\ 0 & 1 \end{pmatrix} \begin{pmatrix} 1 \\ 1 \end{pmatrix} = \begin{pmatrix} 0 \\ 1 \end{pmatrix}. \tag{A5.17}$$

Quarter-Wave Plate Between Crossed Polarizers

We start off with linear polarized light and apply the quarter-wave plate as the right circular polarizer, disregarding the normalization factor.

$$\begin{pmatrix} 1 & i \\ -i & 1 \end{pmatrix} \begin{pmatrix} 1 \\ 0 \end{pmatrix} = \begin{pmatrix} 1 \\ -i \end{pmatrix}. \tag{A5.18}$$

The result is circular polarized light. Then we apply the vertical polarizer

$$\begin{pmatrix} 0 & 0 \\ 0 & 1 \end{pmatrix} \begin{pmatrix} 1 \\ -i \end{pmatrix} = \begin{pmatrix} 0 \\ -i \end{pmatrix} \tag{A5.19}$$

and obtain right circular polarized light, passing the vertical polarizer.

FileFig A5.14 (MA4JONES)

Vector formulation of Jones vectors for linear and circular polarized light. Matrix formulation of Jones matrices for linear polarizer, circular polarizer, and half-wave and quarter-wave plates.

MA4JONES is only on the CD.

Application A5.14.

1. Derive Jones vectors for linear (−45°), left circular, and right circular polarized light.
2. Derive Jones matrices for the half-wave and quarter-wave plates.
3. Apply Jones matrices to Jones vectors for:
   - **a.** Matrices of horizontal and vertical polarizers, and the half-wave plate to each horizontal, vertical, and 45° Jones vector. Comment on the results.
   - **b.** Let light first be polarized horizontally, then pass a 45° polarizer, and a −45° polarizer. Comment on the results.
   - **c.** What is the resulting matrix of a quarter-wave plate and then a half-wave plate? What is the resulting operation?
   - **d.** What is the resulting matrix of a half-wave plate and then a quarter-wave plate? What is the resulting operation?
   - **e.** What is the resulting matrix of a half-wave plate, then a quarter-wave plate, and then another half-wave plate? What is the resulting operation?

C H A P T E R

Maxwell II. Modes and Mode Propagation

6.1 INTRODUCTION

In the chapter on interference we discussed the resonance mode of a Fabry–Perot. We found that all light was transmitted for a specific wavelength λ and separation D of the Fabry–Perot plates. The condition for generation of the modes was $D = m\lambda/2$, where m was an integer.

The modes may be considered as a standing wave. A standing wave can be described by the superposition of two traveling waves moving in opposite directions. The standing waves are characterized by $m + 1$ nonmoving nodes, including both nodes on the Fabry–Perot plates. Between the nodes, the amplitudes oscillate from maximum to minimum. A rectangular box of dimensions a, b, c with reflecting walls (Figure 6.1) has three pairs of parallel plates in the x, y, and z directions. Each pair may be considered as a Fabry–Perot. The walls are assumed to be perfectly reflecting, and the boundary conditions for the mode formation are required to have nodes at the walls. Using plane wave solutions, the modes are represented by sine functions. For the x direction, we have at length a of the box that $\sin(2\pi a/\lambda_x) = 0$ or $2\pi a/\lambda_x = \pi$. A similar result is found for the y and z directions and we have three standing wave conditions

$$a = n_1\lambda_x/2, \qquad b = n_2\lambda_y/2, \qquad c = n_3\lambda_z/2. \tag{6.1}$$

Since the n_is are integers, and a, b, c are constants, the possible values of the wavelengths are restricted. Using wave vectors, Eqs. (6.1) may be written as

$$k_x = \pi n_1/a \qquad k_y = \pi n_2/b \qquad k_z = \pi n_3/c. \tag{6.2}$$

These are the wave vectors for the three components of a general mode of the box. The corresponding wave vector is obtained by vector addition of the three

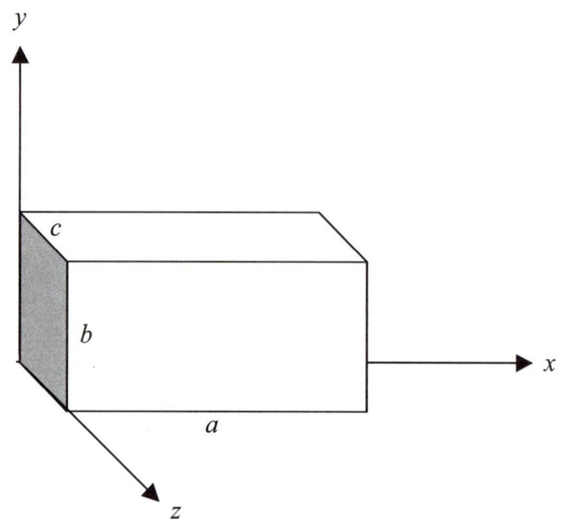

FIGURE 6.1 Coordinates for the discussion of the modes of a box with reflecting walls and dimensions a, b, c.

components of Eq. (6.2). For the square of the scalar value one gets

$$k^2 = \pi^2\{(n_1/a)^2 + (n_2/b)^2 + (n_3/c)^2\}. \tag{6.3}$$

The corresponding wave is a product of three standing waves in the three directions x, y, and z. The standing waves have fixed nodes, and the number of nodes is related to the values of n_i as $(1 + n_i)$ for $i = 1, 2, 3$. Therefore, the number of nodes may be used for the characterization of the modes. In FileFig 6.1 the first two graphs show the one-dimension standing waves for the x and y directions for the case of $n_x = 2$ and $n_y = 2$. The third and fourth graphs are the contour, and surface plots of the amplitude for $M_{i,k}$ with $i = 2$ and $k = 2$ and the fifth and sixth graph, show the intensity. In the third through sixth graphs we see six node lines.

FileFig 6.1 (N1RECBOX)

Graphs of sine functions in one and two dimensions with nodes depending on the integers n_1, n_2. The modes are numbered by Mik, where i gives the number of nodes in the x direction and k in the y direction. Note that the node lines are the same for the graphs of amplitude and intensity.

N1RECBOX

Modes of the Rectangular Box in Two Dimensions

Standing sine waves in x and y directions. Mode number constants. x direction $n1$ and a; y direction $n2$ and b. The wave in each direction is shown as well as contour and surface plots. The square is also shown as surface plot.

$$i := 0 \ldots N \qquad j := 0 \ldots N$$

$$x_i := (-40) + 2.001 \cdot i \qquad y_j := -40 + 2.00001 \cdot j$$

$$\lambda 1 := 2 \cdot \frac{a}{n1} \qquad \lambda 2 := 2 \cdot \frac{b}{n2}$$

$$y1(x) := \sin\left(2 \cdot \pi \cdot \frac{x}{\lambda 1}\right) \qquad y2(y) := \sin\left(2 \cdot \pi \cdot \frac{y}{\lambda 2}\right).$$

1. One dimension

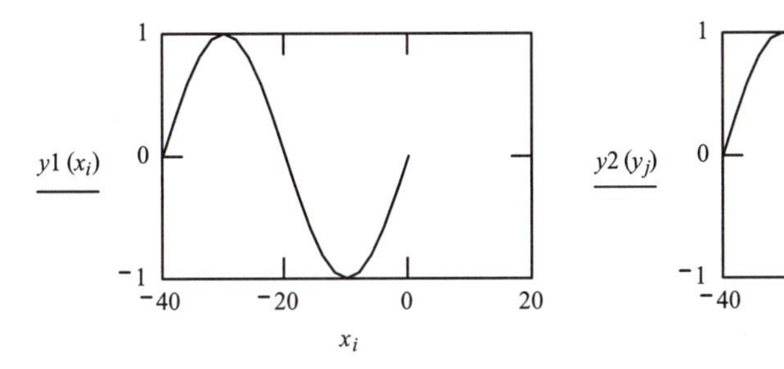

2. Amplitude, 2 D

$$M11_{i,j} := y1(x_i) \cdot y2(y_j) \quad n1 \equiv 2 \quad a \equiv 40 \qquad n2 \equiv 2 \quad b \equiv 40 \quad N \equiv 20.$$

3. Intensity, 2D

$$MM11_{i,j} := (y1(x_i) \cdot y2(y_j))^2.$$

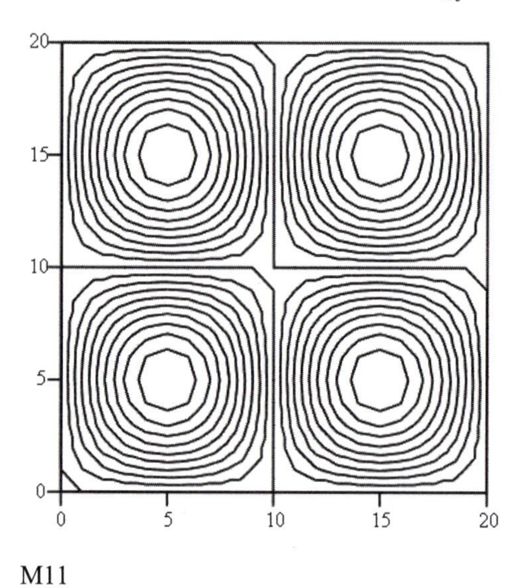

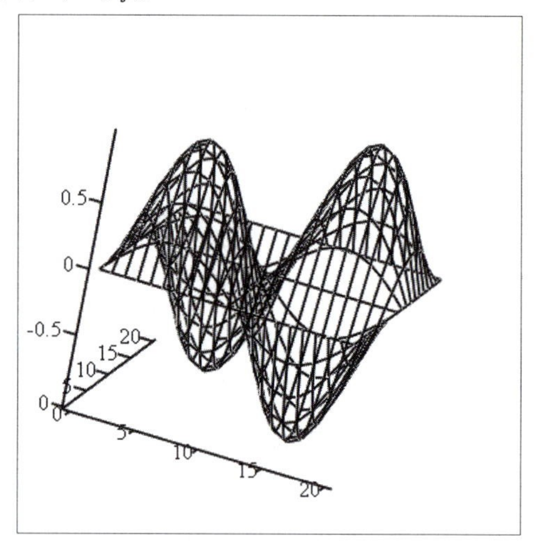

M11

M11

Application 6.1.

1. Print out M00, M01, M10, M11, and M22.
2. Make graphs of M00, M01, M10, and M11, using a cosine function for x. The corresponding boundary condition would be an "open end" of the box.
3. Make graphs of M00, M01, M10, and M11, using cosine functions for x and y.

6.2 STRATIFIED MEDIA

In the chapter on geometrical optics we discussed image formation with spherical mirrors. We also discussed astronomical telescopes composed of an objective lens and magnifiers. Lenses always reflect some of the incident light, according to Fresnel's formulas. To reduce the reflection, an antireflection coating was developed. A thin film of a specific material was vacuum-deposited on the lens surface, preventing most of the incident light of a specific wavelength from being reflected. In the same chapter we also discussed laser cavities. The mirrors of a laser cavity need to have high reflectivity only for a limited wavelength range, but the reflectivity should be close to one in order to obtain a high gain.

In this section we show that stacks of dielectric layers may have very high reflectance or transmittance. Applications are antireflection coatings of lenses and mirrors for laser cavities with extremely high reflectivity. We study the

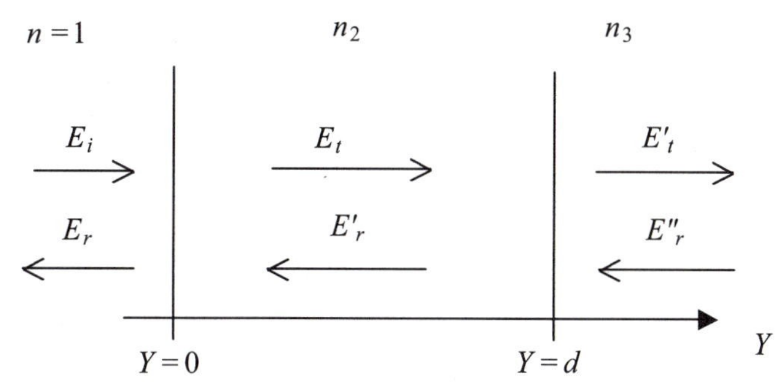

FIGURE 6.2 Three media of index of refraction $n_1 = 1, n_2$, and n_3. In each medium we consider a forward and a backward traveling wave.

reflection and transmission of light incident on stacks of dielectric layers. All layers have the thickness $d = (1/4)(\lambda/n) = n/4\lambda$, but some may have different refractive indices from others.

6.2.1 Two Interfaces at Distance d

When calculating Fresnel's formulas, we equated the fields on both sides of a boundary and obtained a set of linear equations for the amplitudes of the reflection and the transmission coefficients. This method is called the boundary value method and is now applied to two interfaces at distance d. In the chapter on interference we studied the plane parallel plate and summed up all reflected and transmitted waves. This method is called the *summation method* and is less rigorous than the boundary value method. We consider three media and assume normal incident light. In each medium we assume forward and backward traveling waves (Figure 6.2). We do not show the time dependence of the waves, giving us these equations for incident, reflected, and transmitted waves at the two interfaces:

1. before the first interface

$$E_i = A_1 \exp ik_1(+Y) \qquad \text{incident} \tag{6.4}$$

$$E_r = A'_1 \exp ik_1(-Y) \qquad \text{reflected,} \tag{6.5}$$

2. after the first interface

$$E_t = A_2 \exp ik_2(+Y) \qquad \text{transmitted after 1} \tag{6.6}$$

$$E'_r = A'_2 \exp ik_2(-Y) \qquad \text{reflected back to 1,} \tag{6.7}$$

3. after the second interface

$$E_t = A_3 \exp ik_3(+Y) \qquad \text{transmitted after 2} \tag{6.8}$$

$$E''_r = A'_3 \exp ik_3(-Y) \qquad \text{reflected back to 2.} \tag{6.9}$$

We take the electrical field vector perpendicular to the plane of incidence, which is the case of perpendicular polarization discussed in Chapter 5. We first take into account the electrical field vectors at boundary 1, where $Y = 0$. The sum of Eqs. (6.4) and (6.5) is equal to the sum of Eqs. (6.6) and (6.7). At the boundary for $Y = d$ the sum of Eqs. (6.6) and (6.7) is equal to the sum of Eqs. (6.8) and (6.9). We obtain the following equations

$$A_1 + A_1' = A_2 + A_2' \tag{6.10}$$

$$A_2 e^{ikd} + A_2' e^{-ikd} = A_3 e^{ik'd} + A_3' e^{-ik'd}, \tag{6.11}$$

where $k = 2\pi n_2/\lambda$, $k' = 2\pi n_3/\lambda$, and d is the distance between the two interfaces. In a similar manner we obtain a second set of two equations for the B-fields. This is similar to the calculations of Fresnel's formulas in Chapter 5, applying the corresponding boundary conditions for the B-field at $Y = 0$ and $Y = d$.

$$-A_1 + A_1' = -n_2 A_2 + n_2 A_2' \tag{6.12}$$

$$-n_2 A_2 e^{ikd} + n_2 A_2' e^{-ikd} = -n_3 A_3 e^{ik'd} + n_3 A_3' e^{-ik'd}. \tag{6.13}$$

In order to derive a vector-matrix formulation, we abbreviate the sum of the amplitudes before the first boundary (Eq. (6.10)), by $E_{1,d=0}$. Since Eq. (6.11) was derived using the magnetic field vector we call the amplitude $B_{1,d=0}$. In a similar way we use $E_{2,d=0}$ and $B_{2,d=0}$ for the sum of the amplitudes after the first boundary. For the amplitudes before and after the second boundary, we use $E_{2,d=d}$ and $B_{2,d=d}$, and $E_{3,d=d}$ and $B_{3,d=d}$, respectively. We may now write

$$E_{1,d=0} = A_1 + A_1' = A_2 + A_2' = E_{2,d=0} \tag{6.14}$$

$$B_{1,d=0} = -A_1 + A_1' = -n_2 A_2 + n_2 A_2' = B_{2,d=0} \tag{6.15}$$

$$E_{2,d=d} = A_2 e^{ikd} + A_2' e^{-ikd} = A_3 e^{ik'd} + A_3' e^{-ik'd} = E_{3,d=d} \tag{6.16}$$

$$\begin{aligned} B_{2,d=d} &= -n_2 A_2 e^{ikd} + n_2 A_2' e^{-ikd} \\ &= -n_3 A_3 e^{ik'd} + n_3 A_3' e^{-ik'd} = B_{3,d=d}. \end{aligned} \tag{6.17}$$

Equations (6.14) and (6.15) may be written in matrix notation as

$$\mathbf{Field}_{1,d=0} = \begin{pmatrix} E_{1,d=0} \\ B_{1,d=0} \end{pmatrix} = \begin{pmatrix} 1 & 1 \\ -1 & 1 \end{pmatrix} \begin{pmatrix} A_1 \\ A_1' \end{pmatrix} = \mathbf{M_{1,0}A_1} \quad \text{(6.14) and (6.15)}$$

$$\mathbf{Field}_{2,d=0} = \begin{pmatrix} E_{2,d=0} \\ B_{2,d=0} \end{pmatrix} = \begin{pmatrix} 1 & 1 \\ -n_2 & +n_2 \end{pmatrix} \begin{pmatrix} A_2 \\ A_2' \end{pmatrix} = \mathbf{M_{2,0}A_2} \quad \text{(6.14) and (6.15)}$$

and have

$$\begin{aligned} &\mathbf{M_{2,0}A_2} = \mathbf{M_{1,0}A_1} \text{ or} \\ &\quad \mathbf{A_2} = \mathbf{M_{2,0}^{-1}M_{1,0}A_1} = \mathbf{M_{2,0}^{-1}Field_{1,d=0}}. \end{aligned} \tag{6.18}$$

Equations (6.16) and (6.17) may be written in matrix notation as

$$\mathbf{Field}_{2,d=d} = \begin{pmatrix} e^{ikd} & e^{-ikd} \\ -n_2 e^{ikd} & +n_2 e^{-ikd} \end{pmatrix} \begin{pmatrix} A_2 \\ A_2' \end{pmatrix} = \mathbf{M_{2,d}A_2} \quad \text{(6.16) and (6.17)}$$

$$\mathbf{Field}_{3,d=d} = \begin{pmatrix} e^{ik'd} & e^{-ik'd} \\ -n_3 e^{ik'd} & n_3 e^{-ik'd} \end{pmatrix} \begin{pmatrix} A_3 \\ A_3' \end{pmatrix} = \mathbf{M_{3,d}A_3} \quad (6.16) \text{ and } (6.17)$$

and we have

$$\mathbf{Field}_{3,d=d} = \mathbf{M_{3,d}A_3} = \mathbf{M_{2,d}A_2}. \tag{6.19}$$

We want to express the vector $\mathbf{Field_{3,d=d}}$ as the product of matrices times the vector $\mathbf{Field}_{1,d=0}$, and have

$$\mathbf{Field}_{3,d=d} = \mathbf{M_{2,d}A_2} = \mathbf{M_{2,d}M_{2,0}^{-1}E_{1,d=0}}. \tag{6.20}$$

The manipulation of the matrices is shown in FileFig 6.2.

FileFig 6.2 (N2SYMATR)

Demonstration of the matrix manipulations and multiplication of $M_{2,d}M_{2,0}^{-1}$ and the resulting matrix M_2.

$$\mathbf{M_2} = \begin{pmatrix} \cos(kd) & -i\sin(kd)/n_2 \\ -n_2 i\sin(kd) & \cos(kd) \end{pmatrix}. \tag{6.21}$$

N2SYMATR is only on the CD.

The final result is

$$\begin{pmatrix} E_{3,d=d} \\ B_{3,d=d} \end{pmatrix} = \begin{pmatrix} \cos(kd) & -i\sin(kd)/n_2 \\ -n_2 i\sin(kd) & \cos(kd) \end{pmatrix} \begin{pmatrix} E_{1,d=0} \\ B_{1,d=0} \end{pmatrix}. \tag{6.22}$$

Using this matrix approach, we discuss some applications.

6.2.2 Plate of Thickness $d = (\lambda/2n_2)$

We apply Eq. (6.22) to a plate of thickness $d = (\lambda/2n_2)q$, where q is an integer. The thickness of the plate is a multiple of half a wavelength divided by the refractive index n_2 of the material of the plate. Using $k = 2\pi n_2/\lambda$ and $d = (\lambda/2n_2)q$ we have for the product $kd = q\pi$. At the boundary we have for the exponentials $e^{-ik'd} = e^{-ikd}$. Both are 1 for even q. Equation (6.22) is now

$$\begin{pmatrix} E_{3,d=d} \\ B_{3,d=d} \end{pmatrix} = \begin{pmatrix} 1 & 0 \\ 0 & 1 \end{pmatrix} \begin{pmatrix} E_{1,d=0} \\ B_{1,d=0} \end{pmatrix}. \tag{6.23}$$

The matrix M is the unit matrix and the result is that the medium of thickness d has no effect on the transmitted fields. The same result is obtained when q is odd. A similar result has been obtained in Chapter 2 for the plane parallel plate. All incident power of the particular wavelength λ will be transmitted when the plate has the thickness $d = (\lambda/2n_2)q$.

6.2.3 Plate of Thickness d and Index n_2

We apply Eq. (6.22) to a plate of thickness d with refractive index n_2. We also assume that there is no backwards traveling wave in medium 3 and that the refractive indices of the first and third media are assumed to be 1. We then have

$$\begin{pmatrix} A_3 e^{ik'd} \\ -A_3 e^{ik'd} \end{pmatrix} = \begin{pmatrix} \cos(kd) & -i\sin(kd)/n_2 \\ -n_2 i \sin(kd) & \cos(kd) \end{pmatrix} \begin{pmatrix} A_1 + A_1' \\ -A_1 + A_1' \end{pmatrix}. \quad (6.24)$$

In FileFig 6.3 we calculate the transmitted intensity of a plane parallel plate. Calling $x = A_1'/A_1$ and $y = A_3/A_1$ and observing that $e^{ik'd} = 1$, we have $R = xx^*$ for the reflected intensity and $T = yy^*$ for the transmitted intensity. Equation (6.24) is a system of two linear equations in x and y and is solved for x and y. The result for T is

$$T = 1/[1 + \{(n^2 - 1)^2/4n^2\}(\sin(kd)^2]. \quad (6.25)$$

Here we can use T for the transmitted intensity because we have assumed that the refractive index in the first medium is $n_1 = 1$. This is the same result one obtains with the summation method for the case of normal incidence, discussed in Chapter 2.

FileFig 6.3 (N3SYMATPL)

Calculation of the transmitted intenstiy T of a plane parallel plate of thickness d with indices outside the plate equal to 1. We use $x = A_1'/A_1$ and $y = A_3/A_1$ and have $T = yy^*$ for the transmitted intensity and $R = xx^*$ for the reflected intensity.

N3SYMATPL is only on the CD.

Application 6.3.

1. Calculate the reflected intensity R.
2. Make graphs for T and R in the wavelength range from 1 to 20 microns. Use for d values equal to $kd = q\pi$, one for q even and one for q odd.
3. Make graphs for T for two different refractive indices between 1.1 and 4 in the wavelength range from 1 to 20 microns. Use for the thickness d values not equal to $kd = q\pi$, q even or odd.

6.2.4 Antireflection Coating

Antireflection coating may be found on camera lenses. A thin dielectric film of refractive index n_2 is vacuum-deposited on the surface of a lens of refractive index n_3. We assume for the film a thickness $\lambda/(4n_2)$. The light is incident from a medium with index n_1 and the product kd in Eq. (6.22) is $[(2\pi n_2/\lambda)(\lambda/4n_2)] = \pi/2$. One gets

$$\begin{pmatrix} A_3 e^{ik'd} \\ -n_3 A_3 e^{ik'd} \end{pmatrix} = \begin{pmatrix} 0 & -i/n_2 \\ -n_2 i & 0 \end{pmatrix} \begin{pmatrix} A_1 + A_1' \\ -n_1 A_1 + n_1 A_1' \end{pmatrix}. \tag{6.26}$$

Because $e^{ik'd} = i$ one has

$$iA_3 = (-i/n_2)(-n_1 A_1 + n_1 A_1') \tag{6.27}$$

$$-n_3 i A_3 = (-i n_2)(A_1 + A_1'). \tag{6.28}$$

To get to the condition for no reflection, we set the ratio $A_1'/A_1 = 0$ and have

$$A_3/A_1 = n_1/n_2 = n_2/n_3 \tag{6.29}$$

or

$$n_1 n_3 = n_2^2. \tag{6.30}$$

We have the result: the antireflection film of a thickness of a quarter-wavelength must have a refractive index equal to the square root of the product of the refractive index on both sides. Let us consider a glass lens of index $n_3 = 1.5$. The incident light travels in the medium with $n_1 = 1$ and $n_3 = 1.5$ being the index of the lens. The square root of the product is 1.22. A material with exactly that refractive index is hard to find, but material with approximately that value is used for antireflection coating.

In FileFig 6.4 we calculate from Eqs. (6.27) and (6.28) the reflection coefficient $r = A_1'/A_1$ and the transmission coefficient $t = A_3/A_1$ and investigate the reflected and transmitted intensities. Using $rr^* = R$ and tt^*, the sum $R + tt^*$ is not 1. However, if we use the correct expression for T (see Chapter 5), we show that $(n_3/n_1)tt^*$ is equal to T and that one has $R + T = 1$. The graph in FileFig 6.4 shows a plot of the antireflection coating depending on the refractive index n_2.

FileFig 6.4a (N4SYMULANTa)

Antireflection coating. The reflected amplitude $r = A_1'/A_1$ and the transmitted amplitude $t = A_3/A_1$ as solutions of Eq. (6.26). The products rr^* and tt^* are also calculated.

FileFig 6.4b (N4SYMULANTb)

Numerical calculations.

1. Special case of zero thickness for demonstration of reflected and transmitted intensities.
2. Antireflection coating. Graph of reflected intensity depending on the refractive index of the coating.

N4SYMULANTa and N4SYMULANTb are only on the CD.

Application 6.5.

1. Using the correct expression for T (see chapter 5), derive formulas for the transmitted and reflected intensities for one interface with refractive indices $n1$ and $n2 \neq 1$. Show that $T + R = 1$.
2. Find the refractive index for an antireflection coating material for silicon ($n = 3.4$).
3. Use polyethylene ($n = 1.5$) as a coating of silicon and calculate the percentage of reflected amplitude and intensity.

6.2.5 Multiple Layer Filters with Alternating High and Low Refractive Index

High-reflecting dielectric mirrors are composed of a large number of dielectric layers with alternating high and low refractive indices. We extend Eq. (6.22) to $f - 1$ layers of equal thickness and obtain

$$\begin{pmatrix} E_{f,d=d} \\ B_{f,d=d} \end{pmatrix} = M_{f-1}M_{f-2},\ldots,\begin{pmatrix} E_{1,d=0} \\ B_{1,d=0} \end{pmatrix}. \tag{6.31}$$

For a rigorous derivation see Born and Wolf (1964; p.66).

We apply Eq. (6.31) to a sequence of double layers, consisting of a high and a low refractive index, assuming that the refractive index outside is 1. Assuming N double layers the sequence of the refractive indices is then

$$(1)(n_H n_L),\ldots,(n_H n_L)(1). \tag{6.32}$$

The product of the matrices is

$$(M_H M_L),\ldots,(M_H M_L). \tag{6.33}$$

The thickness of one layer is assumed to be one-quarter of a wavelength and we have the product $kd = [(2\pi n_2/\lambda)(\lambda/4n_2)]$, resulting in $\sin(kd) = 1$ and $\cos(kd) = 0$. For one double layer we get for the product of M_H times M_L

$$\begin{pmatrix} -n_L/n_H & 0 \\ 0 & -n_H/n_L \end{pmatrix} \tag{6.34}$$

and for N double layers

$$\begin{pmatrix} (-n_L/n_H)^N & 0 \\ 0 & (-n_H/n_L)^N \end{pmatrix}. \tag{6.35}$$

In FileFig 6.5a we calculate the reflected intensity R symbolically and in FileFig 6.5b we calculate it numerically. One finds values close to 1 for $N = 10$, 20, 40 or more. Such multiple layer mirrors, consisting of a large number of thin films of alternating high and low refractive indices, are used in laser cavities to reduce losses.

FileFig 6.5a (N5STRASYMa)

Symbolic calculation of the reflected intensity for multilayer reflection coatings.

FileFig 6.5b (N5STRANUMb)

Numerical calculation for $nH = 2.5$, $nL = 1.5$, and $N = 20$.

N5STRASYMa and N5STRASYMb are only on the CD.

Application 6.7. Calculate the reflected intensity (R) and transmitted intensity ($T = 1 - R$) for $N = 10, 20, 40, 100$ for chosen values of n_L and n_H.

6.3 GUIDED WAVES BY TOTAL INTERNAL REFLECTION THROUGH A PLANAR WAVEGUIDE

6.3.1 Traveling Waves

When laser light was applied to telecommunication, one looked at the possibilities of light traveling through some type of guide. Travel through the open air resulted in too many losses. Long wavelength electromagnetic waves travel through metal cables, but microwaves may travel inside a rectangular waveguide. These waveguides have parallel reflecting metal surfaces, and the wavelength of a traveling mode is characterized by the dimensions of the rectangular cross-section. As the first step for propagation of modes of laser light one considered a dielectric film of refractive index n_1 with refractive indices n_2 and n_3 equal to 1 on the outside. Later, dielectric fibers were used for guiding the light, discussed in the next section.

In Chapter 2 we discussed the modes of a Fabry–Perot. The incident light wave was traveling perpendicular to the planes, and a relation between the wavelength and the distance between the planes characterized the modes. We now look at very similar modes, formed inside a plane parallel dielectric film but traveling parallel to the boundaries of the plate.

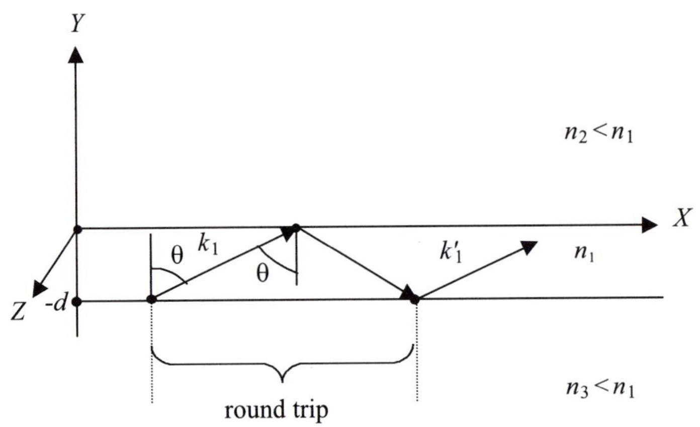

FIGURE 6.3 Thin film of refractive index n_1 larger than the indices n_2 and n_3 of the outside media. Propagation is in the X direction and mode formation in the Y direction. There are exponential decreasing solutions in the outside medial of indices n_2 and n_3.

We consider three layers of dielectric materials extending in the X to Z direction and stacked up in the Y direction. We assume that the layer in the middle has thickness d and refractive index n_1. Above and below are two other dielectric materials with refractive indices n_2 and n_3, both smaller than n_1 (Figure 6.3). A wave in the layer with refractive index n_1 is totally reflected on the two interfaces, above and below, and travels effectively in the X direction.

When treating the plane parallel plate (see Chapter 2) we summed up all the light reflected and transmitted at the different interfaces and found resonance conditions corresponding to modes. Summing up all the reflected and transmitted light is called the summation method in contrast to the boundary value method, used in Section 6.2 to describe the modes for multilayer dielectric material. To treat the problem of waves traveling with internal total reflection in a dielectric layer, we apply another method, the traveling wave method. For comparison, the boundary value method is given in the appendix.

We assume a wave is traveling in the X direction in medium (1) between the two media 2 and 3. We assume that the angle to the normal θ is larger than the critical angle, in order to have total reflection. A ray is launched from a point inside the plate and after reflection on each interface, the pattern repeats. If the component of the wave traveling in the Y direction has the same wave vector after one period of travel in the X direction, we have a traveling mode. This is only possible for discrete values of the angle θ, the angle corresponding to the direction of the k vector and the normal (Figure 6.3). We use complex notation for presenting the wave traveling between internal reflections in the X, Y plane

$$u_1 = e^{i(2\pi X n_1/\lambda)} e^{i(2\pi Y n_1/\lambda)} e^{-i\omega t}, \tag{6.36}$$

where λ is the wavelength in free space, ω the frequency, t the time, and n_1 the refractive index in medium (1). Equation (6.36) is a function of X and Y

and satisfies the scalar wave equation depending on X, Y, and t. We rewrite Eq. (6.36) using the wave vector notation

$$u_1 = e^{i(k_{1X}X)}e^{i(k_{1Y}Y)}e^{-i\omega t}, \tag{6.37}$$

and have for the components of the wave vectors $(k_{1X})^2 + (k_{1Y})^2 = k_1^2$, with $k_1 = n_1 k$ and $k = 2\pi/\lambda$.

As we know from our discussion of total internal reflection, there is an extension of the internally reflected wave into the optically denser medium, called the evanescent wave. To describe these waves for the two media with refractive indices n_2 and n_3, we write for the wave in (1), similarly to what we wrote for the wave in medium (2),

$$u_2 = e^{i(k_{2X}X)}e^{i(k_{2Y}Y)}e^{-i\omega t}, \tag{6.38}$$

with $(k_{2X})^2 + (k_{2Y})^2 = k_2^2$, $k_2 = n_2 k$, and $k = 2\pi/\lambda$, and similarly for medium (3)

$$u_3 = e^{i(k_{3X}X)}e^{i(k_{3Y}Y)}e^{-i\omega t} \tag{6.39}$$

with $(k_{3X})^2 + (k_{3Y})^2 = k_3^2$, $k_3 = n_3 k$, and $k = 2\pi/\lambda$.

6.3.2 Restrictive Conditions for Mode Propagation

We recall that we derived a restriction on certain parameters when discussing the modes in the Fabry–Perot. For the traveling mode in a waveguide we have a similar, but more complicated situation. The k vectors of the three waves have a real value for the Y component in medium (1) and an imaginary value in media (2) and (3) because of total reflection. We have for the k vectors:

1. Components of the k vector in the Y direction,

$$k_{1Y} \text{ real, and } \tilde{k}_{2Y} \text{ and } \tilde{k}_{3Y} \text{ imaginary; and} \tag{6.40}$$

2. Components of the k vector in the X direction

For the three waves traveling in the X direction, the X component of the k vector has the same real value. We use the notation

$$k_{1X} = k_{2X} = k_{3X} = \beta. \tag{6.41}$$

The value of β is restricted to $k_1 \geq \beta \geq k_2$ and $k_1 \geq \beta \geq k_3$, which follows from the relations

$$k_2^2 - \beta^2 = k_{2Y}^2, \text{ but since } \tilde{k}_{2Y} \text{ is imaginary, } \beta \text{ must be larger than } k_2$$
$$k_1^2 - \beta^2 = k_{1Y}^2, \text{ but since } k_{1Y} \text{ is real, } \beta \text{ must be smaller than } k_1$$
$$K_3^2 - \beta^2 = k_{3Y}^2, \text{ but since } \tilde{k}_{3Y} \text{ is imaginary,} \beta \text{ must be larger than } k_3.$$

6.3.3 Phase Condition for Mode Formation

We define the phase change upon reflection on medium 2, $(1 \to 2)$, as $\Phi_{1,2}$ and for reflection on medium 3, $(1 \to 3)$, as $\Phi_{1,3}$. Considering a round trip in the Y direction, we have for the phase shift: on medium 2: $(2\pi d/\lambda_1)\Phi_{1,2}$, and on medium 3: $2\pi d/\lambda_1 + \Phi_{1,3}$. The sum must have values of $2\pi m$ which is written

$$2\pi m = 2[2\pi d/\lambda_1] + \Phi_{1,2} + \Phi_{1,3} \text{ with } m = 0, 1, 2, 3. \tag{6.42}$$

This is the resonance condition for the mode, involving the k values in the Y direction. The phase values are calculated from Fresnel's formulas depending on the Y components of k. The resonance conditions are different for the s-polarization (TE) modes and the p-polarization (TM) modes, and are discussed separately in Sections 6.3.4 and 6.3.5.

6.3.4 (TE) Modes or s-Polarization

The reflection coefficient on the interfaces 1,2 and 1,3 are obtained from Fresnel's formulas (see Chapter 5),

$$r_s = \frac{\left(n_1 \cos\theta - n_2\sqrt{[1 - [(n_1/n_2)\sin\theta]^2]}\right)}{\left(n_1 \cos\theta + n_2\sqrt{[1 - [(n_1/n_2)\sin\theta]^2]}\right)}. \tag{6.43}$$

We multiply by $2\pi/\lambda$ and introduce the definitions of the k vectors. Since the values of k in media 1 and media 3 are imaginary, we write $\tilde{k}_{2Y}$ and $\tilde{k}_{3Y}$ and get complex numbers for the two reflection coefficients

$$\tilde{r}_{s1,2} = (k_{1Y} - \tilde{k}_{2Y})/(k_{1Y} + \tilde{k}_{2Y}) \tag{6.44}$$

$$\tilde{r}_{s1,3} = (k_{1Y} - \tilde{k}_{3Y})/(k_{1Y} + \tilde{k}_{3Y}). \tag{6.45}$$

For the phase angle we have

$$\Phi_{s1,2} = \tan^{-1}(-\tilde{k}_{2Y}/k_{1Y}) - \tan^{-1}(\tilde{k}_{2Y}/k_{1Y}) = -2\tan^{-1}(\tilde{k}_{2Y}/k_{1Y}) \tag{6.46}$$

$$\Phi_{s1,3} = -2\tan^{-1}(\tilde{k}_{3Y}/k_{1Y}). \tag{6.47}$$

The resonance condition (Eq. (6.42)) may now be written as

$$2k_{1Y}d = -2\tan^{-1}(\tilde{k}_{2Y}/k_{1Y}) - 2\tan^{-1}(\tilde{k}_{3Y}/k_{1Y}) + \pi m, \tag{6.48}$$

or using the formula $\tan^{-1} A + \tan^{-1} B = \tan^{-1}(A + B)/(1 - AB)$ we may write

$$\tan k_{1Y}d = [k_{1Y}(\tilde{k}_{2Y} + \tilde{k}_{3Y})]/(k_{1Y}^2 - \tilde{k}_{2Y}\tilde{k}_{3Y}). \tag{6.49}$$

For the numerical calculation we prefer to write the condition of Eq. (6.49) as

$$2\pi n_1 \cos\theta d/\lambda = -\operatorname{atan}\left(\sqrt{[n_1^2 \sin^2\theta - n_2^2]}/(n_1\cos\theta)\right) \tag{6.50}$$

$$- \operatorname{atan}\left(\sqrt{[n_1^2 \sin^2\theta - n_3^2]}/(n_1\cos\theta)\right) + m\pi,\ m = 1, 2, 3.$$

In FileFig 6.6 we calculate the condition of Eq. (6.50) using

$$\begin{aligned} ys(\theta) &= ys2(\theta) - ys3(\theta) + m\pi \quad &\text{for } m = 1\\ yys(\theta) &= ys2(\theta) - ys3(\theta) + m\pi \quad &\text{for } m = 2\\ yyys(\theta) &= ys2(\theta) - ys3(\theta) + m\pi \quad &\text{for } m = 3 \end{aligned}$$

and at the crossover point of the graph the resonance condition is fulfilled. The crossover point indicates the angle θ for the mode with mode number m. The characterization of the mode depends on the refractive indices, thickness d, and wavelength λ.

FileFig 6.6 (N6PLSPS)

Resonance condition for s-polarization (TE). The crossover point indicates the angle θ of the mode with mode number m depending on the refraction indices, thickness d, and wavelength λ. The lowest number of the mode corresponds to the lowest curve.

N6PLSPS

Wave Traveling with Total Internal Reflection Through a Planar Waveguide

Resonance condition of s-polarization. Global definition of $n1$, $n2$, $n3$, d, and λ above the graph.

$$\theta := 0, 1 \ldots 90$$

$$y(\theta) := 2 \cdot \pi \cdot n1 \cdot \frac{d}{\lambda} \cdot \cos\left(2 \cdot \pi \frac{\theta}{360}\right)$$

$$ys1(\theta) := -\operatorname{atan}(zs1(\theta)) \qquad zs1(\theta) := \frac{\left(\sqrt{n1^2 \cdot \sin\left(2 \cdot \pi \frac{\theta}{360}\right)^2 - n2^2}\right)}{n1 \cdot \cos\left(2 \cdot \pi \frac{\theta}{360}\right)}$$

$$ys3(\theta) := -\operatorname{atan}(zs3(\theta)) \qquad zs3(\theta) := \frac{\left(\sqrt{n1^2 \cdot \sin\left(2 \cdot \pi \frac{\theta}{360}\right)^2 - n3^2}\right)}{n1 \cdot \cos\left(2 \cdot \pi \frac{\theta}{360}\right)}.$$

ys is for $m = 1$, yys for $m = 2$, and $yyys$ for $m = 3$. For these parameters the angle θ of the first three possible modes is determined:

$$\begin{aligned} ys(\theta) &:= -ys1(\theta) - ys3(\theta) + \pi\\ yys(\theta) &:= -ys1(\theta) - ys3(\theta) + \pi \cdot 2\\ yyys(\theta) &:= -ys1(\theta) - ys3(\theta) + \pi \cdot 3. \end{aligned}$$

Global definition

$$\theta c := \operatorname{asin}\left(\frac{n2}{n1}\right) \qquad \theta\theta c := 360 \cdot \frac{\theta c}{2 \cdot \pi} \qquad \theta\theta c = 41.81$$

$$n1 \equiv 1.5 \qquad n2 \equiv 1 \qquad n3 \equiv 1 \qquad d \equiv 18 \qquad \lambda \equiv 2.$$

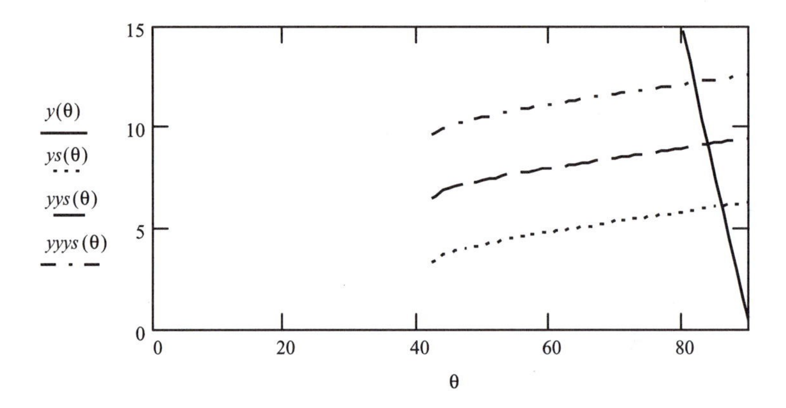

At the crossover point of y with ys, yys, or $yyys$, respectively, the resonance condition is fulfilled. The functions ys, yys, and $yyys$ are complex in the region from horizontal appearance to zero. This is shown in the next graph where the argument is plotted. The complex region has to be disregarded for the determination of the crossover point.

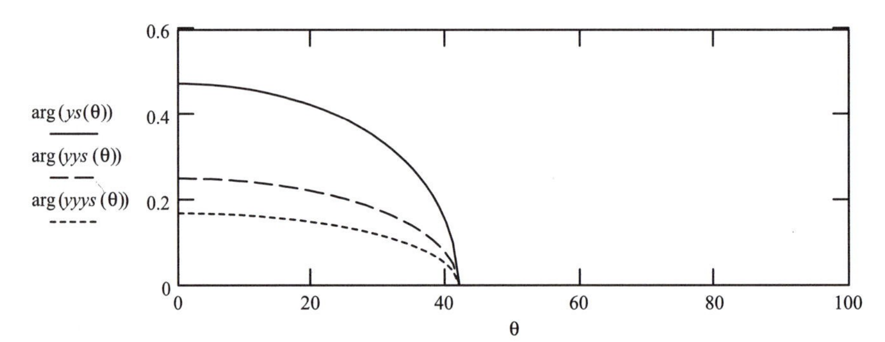

Application 6.6.

1. Change the refractive index and choose d and λ such that all three modes are possible.
2. Change the thickness d and choose n_1 and λ such that all three modes are possible.
3. Change the wavelength λ and choose the refractive index n_1 and the thickness d such that all three modes are possible.

6.3.5 (TM) Modes or p-Polarization

The coefficients of reflection on media 1 and 3 are obtained from Fresnel's formulas. First multiply by $2\pi/\lambda$ and then introduce the definition of k. One can observe that the k value in the medium above and below is imaginary. One obtains complex numbers for the two reflection coefficients similarly as for the s-case

$$\tilde{r}_{p1,2} = (n_2^2 k_{1Y} - n_1^2 \tilde{k}_{2Y})/(n_2^2 k_{1Y} + n_1^2 \tilde{k}_{2Y}) \tag{6.51}$$

$$\tilde{r}_{p1,3} = (n_3^2 k_{1Y} - n_1^2 \tilde{k}_{3Y})/(n_3^2 k_{1Y} + n_1^2 \tilde{k}_{3Y}) \tag{6.52}$$

and also for the phase changes

$$\begin{aligned}\Phi_{s1,2} &= \tan^{-1}(-n_1^2 \tilde{k}_{2Y}/n_2^2 k_{1Y}) - \tan^{-1}(n_1^2 \tilde{k}_{2Y}/n_2^2 k_1 Y)\\ &= -2\tan^{-1}(n_1^2 \tilde{k}_{2Y}/n_2^2 k_{1Y})\end{aligned} \tag{6.53}$$

and

$$\Phi_{s1,3} = -2\tan^{-1}(n_1^2 \tilde{k}_{3Y}/n_3^2 k_{1Y}). \tag{6.54}$$

The resonance condition, similar to Eq. (6.42), may now be written as

$$2k_{1Y}d = -2\tan^{-1}(n_1^2 \tilde{k}_{2Y}/n_2^2 k_{1Y}) - 2\tan^{-1}(n_1^2 \tilde{k}_{3Y}/n_3^2 k_{1Y}) + \pi m \tag{6.55}$$

or

$$\tan k_{1Y}d = (n_1^2 k_{1Y}(n_2^2 \tilde{k}_{2Y} + n_3^2 \tilde{k}_{3Y})/(n_2^2 n_3^2 k_{1Y}^2 - n_1^4 \tilde{k}2Y \tilde{k}_{3Y}). \tag{6.56}$$

For the numerical calculation we prefer to write the condition as

$$\begin{aligned}2\pi n_1 \cos\theta d/\lambda = &- \operatorname{atan}\left(n_1^2\sqrt{[n_1^2 \sin^2\theta - n_2^2]}/(n_2^2 n_1 \cos\theta)\right)\\ &- \operatorname{atan}\left(n_1^2\sqrt{n_1^2 \sin^2\theta - n_3^2})/(n_3^2 n_1 \cos\theta)\right) + \pi m, m = 1, 2, 3.\end{aligned} \tag{6.57}$$

In FileFig 6.7 we calculate the resonance condition of Eq. (6.57) using

$$\begin{aligned} yp(\theta) &= yp2(\theta) - yp3(\theta) + m\pi && \text{for } m = 1\\ yyp(\theta) &= yp2(\theta) - yp3(\theta) + m\pi && \text{for } m = 2\\ yyyp(\theta) &= yp2(\theta) - yp3(\theta) + m\pi && \text{for } m = 3\end{aligned}$$

and at the crossover point of the graph the resonance condition is fulfilled. The crossover point indicates the angle θ for the mode with mode number m, depending on the refraction indices, thickness d, and wavelength λ. Given m, d, and λ, the k vector for the traveling mode is given and if real, the mode is possible.

FileFig 6.7 (N7PLPPS)

Resonance condition for p-polarization (TE). The crossover point indicates the angle θ of the mode with mode number m, depending on the refraction indices, the thickness d, and wavelength λ. The lowest number of the mode corresponds to the lowest curve.

N7PLPPS is only on the CD.

Application 6.7.

1. Change the refractive index and choose d and λ such that all three modes are possible.
2. Change the thickness d and choose n1 and λ such that all three modes are possible.
3. Change the wavelength λ and choose the refractive index $n1$ and the thickness d such that all three modes are possible
4. Give an example for $\lambda = 0.00025$ mm, $d = 0.0005$ mm and show that we have for $m = 1$ a *cosine mode*, and for $m = 2$ a *sine mode*.

6.4 FIBER OPTICS WAVEGUIDES

6.4.1 Modes in a Dielectric Waveguide

In Section 6.3 we discussed mode propagation in a dielectric film of thickness of several wavelengths and refractive index n_2. The modes were guided by two outer dielectric media of refractive indices smaller than n_1. The mode propagation was in the X direction. The media were stacked in the Y direction and extended in the Z direction without limits. A constant refractive index was assumed.

We now consider a dielectric fiber of radius a and homogeneous refractive index n_1 in a surrounding medium of refractive index n_o (Figure 6.4). Following in general second Jackson (1975, p.364), we choose the direction of propagation in the z direction, but use refractive indices instead of dielectric constants. We start with the general wave equations

$$\partial^2\mathbf{E}/\partial x^2 + \partial^2\mathbf{E}/\partial y^2 + \partial^2\mathbf{E}/\partial z^2 = (1/c^2)\partial^2\mathbf{E}/\partial t^2 \tag{6.58}$$

$$\partial^2\mathbf{B}/\partial x^2 + \partial^2\mathbf{B}/\partial y^2 + \partial^2\mathbf{B}/\partial z^2 = (1/c^2)\partial^2\mathbf{B}/\partial t^2 \tag{6.59}$$

and call the differentiation with respect to the variables perpendicular to the direction of propagation the transverse Laplacian $\nabla_t - \partial^2/\partial z^2$.

Assuming periodic exponential solutions for the time dependence and the field in the z direction, one has the two constants $(n_1\omega/c)^2$ and k^2 after application of the second derivative. The ratio ω/c is equal to k_1 in the dielectric and k is the wave vector for the wave traveling in the Z direction. For "inside" and "outside"

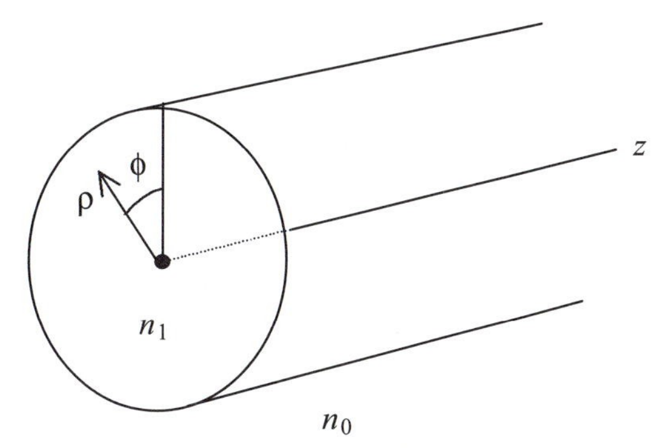

FIGURE 6.4 Coordinates for mode propagation in a fiber of radius a. The tranverse coordinates are ρ and ϕ and the modes propagate in the z-direction.

we have

$$[\nabla_t^2 + ((n_1\omega/c)^2 - k^2)]\mathbf{E} = 0 \qquad \text{“inside”} \tag{6.60}$$

$$[\nabla_t^2 + ((n_1\omega/c)^2 - k^2)]\mathbf{B} = 0 \qquad \text{“inside”} \tag{6.61}$$

$$[\nabla_t^2 + ((n_0\omega/c)^2 - k^2)]\mathbf{E} = 0 \qquad \text{“outside”} \tag{6.62}$$

$$[\nabla_t^2 + ((n_0\omega/c)^2 - k^2)]\mathbf{B} = 0 \qquad \text{“outside”.} \tag{6.63}$$

“Inside” we may use the positive constant $\gamma^2 = ((n_1\omega/c)^2 - k^2)$. “Outside” we expect an exponential decrease of the solutions and define $\beta^2 = (k^2 - (n_o\omega/c)^2$ with β real. We use cylindrical coordinates ρ and ϕ, and assume that we have no azimuthal variation, which means there is no dependence on ϕ. We have two differential equations for the two transversal components $\mathbf{E}_t$ and $\mathbf{B}_t$, which have Bessel functions as solutions

$$d^2/d\rho^2 + (1/\rho)d/d\rho + \gamma^2)E_t = 0 \qquad \text{“inside”} \tag{6.64}$$

$$d^2/d\rho^2 + (1/\rho)d/d\rho + \gamma^2)B_t = 0 \qquad \text{“inside”} \tag{6.65}$$

and

$$d^2/d\rho^2 + (1/\rho)d/d\rho - \beta^2)E_t = 0 \qquad \text{“outside”} \tag{6.66}$$

$$d^2/d\rho^2 + (1/\rho)d/d\rho - \beta^2)B_t = 0 \qquad \text{“outside”.} \tag{6.67}$$

We have $J_o(\gamma\rho)$ as solutions for E_t and B_t “inside,” and $K_o(\beta\rho)$ for E_t and B_t “outside.” From Maxwell's equations we obtain relations between the field components depending on the “transverse” coordinates ρ and ϕ and the components depending on z. The relations are divided into two groups, β_ρ and E_ϕ depending on B_z, and B_ϕ and E_ρ depending on E_z.

These relations are for “inside”:

$$B_\rho = (ik/\gamma^2)dB_z/d\rho \quad B_\phi = (in_1^2\omega/c\gamma^2)dE_z/d\rho \tag{6.68}$$

$$E_\phi = (-\omega/ck)B_\rho \quad E_\rho = (ck/n_1^2\omega)E_\phi. \tag{6.69}$$

For "outside" one has a similar set of equations.

Here we treat only the TE modes, that is for a nonvanishing B_z component. One has explicitly

$$B_z = J_o(\gamma\rho)$$
$$B_\phi = (-ik/\gamma)J_1(\gamma\rho) \qquad \text{“inside”} \tag{6.70}$$
and
$$E_\phi = (i\omega/c\gamma)J_1(\gamma\rho)$$

$$B_z = AK_o(\beta\rho)$$
$$B_\phi = (ikA/\beta)K_1(\beta\rho) \qquad \text{“outside,”} \tag{6.71}$$
and
$$E_\phi = -(i\omega A/c\beta)K_1(\beta\rho)$$

where A is a constant. Only the first two equations of Eqs. (6.70) and (6.71) are independent, and application of the boundary conditions at $\rho = a$ yields the equations:

$$AK_o(\beta a) = J_o(\gamma a) \tag{6.72}$$
$$(-A/\beta)K_1(\beta a) = (1/\gamma)J_1(\gamma a). \tag{6.73}$$

Elimination of A results in the characteristic equation for the determination of k^2, written in γ and β with both depending on k,

$$(J_1(\gamma a)/(\gamma J_o(\gamma a)) = -(K_1(\beta a)/(\beta K_o(\beta a)). \tag{6.74}$$

Since γ and β are both functions of k, we plot the right and the left sides of Eq. (6.74) on the same graph, with both depending on k. At the crossing of the curves we get the resulting value of k. This is shown in the second graph of FileFig 6.8. The "cutoff" frequency is obtained for $J_o(\gamma a) = 0$, which is $\gamma a = 2.405$ and the corresponding wavelength is $\lambda_c = \{[\sqrt{(n_1^2 - n_o^2)}a2\pi]/2.405\}$. At that wavelength β^2 is 0 and k is equal to the free space value.

FileFig 6.8 (N8CWGK)

Determination of k for dielectric circular waveguide.

N8CWGK

Dielectric Circular Waveguide, Determination of k

Check for positive values of argument for $J0$, $J1$ and $K0$, $K1$. Since $x = (\gamma a)^2$ and $y = (\beta a)^2$, we have for the square of the arguments of the Bessel functions

$$xx(k) := a^2 \cdot \left(\frac{n1^2 \cdot 4 \cdot \pi^2}{\lambda^2} - k^2\right) \qquad yy(k) := a^2 \cdot \left(k^2 - \frac{no^2 \cdot 4 \cdot \pi^2}{\lambda^2}\right)$$

and for the arguments

$$x(k) := \sqrt{xx(k)} \qquad y(k) := \sqrt{yy(k)}.$$

We try the interval $k := 2.65, 2.66 \ldots 3.8$ and make a graph.

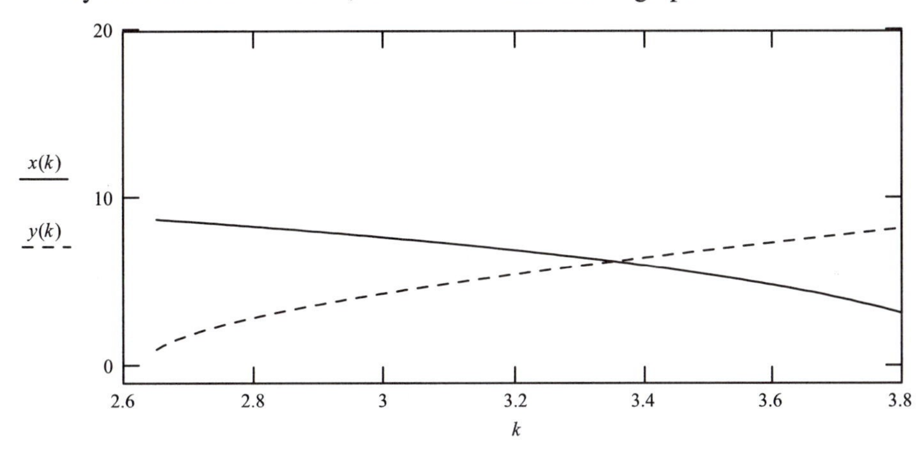

Input data: radius, wavelength, and refractive indices:

$$a \equiv 3 \qquad \lambda \equiv 2.39 \qquad n1 \equiv 1.5 \qquad no \equiv 1.$$

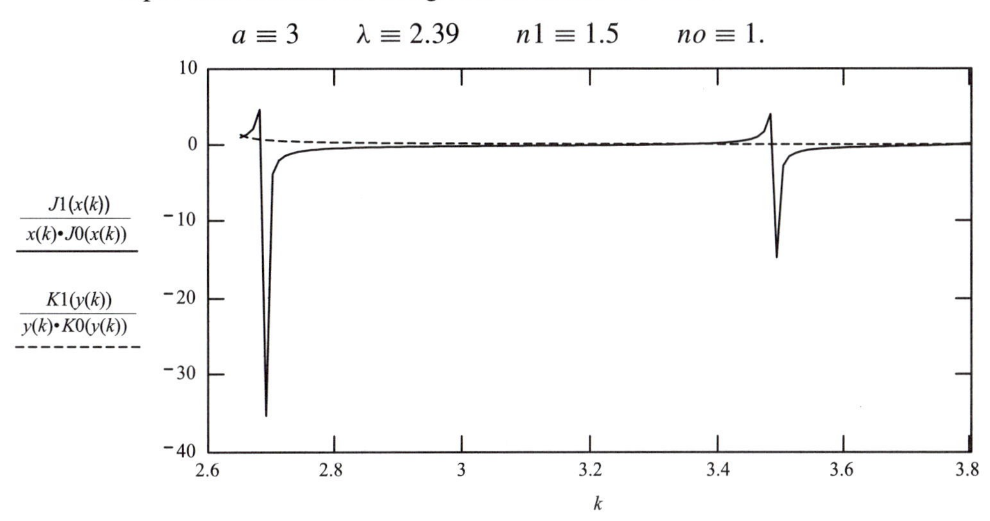

From graph: first intersection

$$kk := 2.66$$

$$\lambda\lambda := \frac{2 \cdot \pi}{kk}$$

$$\lambda\lambda = 2.362.$$

Side calculation. If $\lambda\lambda\lambda := 8$, we get

$$kkk := 2 \cdot \frac{\pi}{\lambda\lambda\lambda} \qquad \text{and} \qquad kkk = 0.785.$$

Application 6.8.

1. Change the refractive index on the outside to 1.1, 1.2, 1.3.
2. Change the radius to 2 times the wavelength and 4 times the wavelength.

APPENDIX 6.1

A6.1.1 Boundary Condition Method Applied to TE Modes of Plane Plate Waveguide

The wave travels in the X direction, but it is sufficient for mode formation to consider only the Y direction. For the fields E and B one assumes wavelike solutions in (1) and exponential decreasing solutions in (2) and (3). Application of the boundary conditions matches the fields. The objective is to calculate the characteristic equation for k_{1Y} depending on $\hat{k}_{2Y}$ and $\hat{k}_{3Y}$.

We have for the solutions of the wave equation in the media:

Medium (2)

$$E_{2Y} = Ae^{-i\hat{k}_{2Y}Y} \tag{A6.1}$$

$$B_{2Z} = \hat{k}_{2Y}Ae^{-i\hat{k}_{2Y}Y}; \tag{A6.2}$$

Medium (1)

$$E_{1Y} = BB\cos k_{1Y}Y - C\sin k_{1Y}Y \tag{A6.3}$$

$$B_{1Z} = k_{1Y}BB\sin k_{1Y}Y - k_{1Y}C\cos k_{1Y}Y; \tag{A6.4}$$

Medium (3)

$$E_{3Y} = De^{-i\hat{k}_{3Y}Y} \tag{A6.5}$$

$$B_{3Z} = -\hat{k}_{2Y}De^{-i\hat{k}_{3Y}Y}. \tag{A6.6}$$

Matching at the boundary at $Y = 0$:

$$A = BB \tag{A6.7}$$

$$\hat{k}_{2Y}A = -k_{1Y}C. \tag{A6.8}$$

Matching at the boundary at $Y = -d$:

$$BB\cos k_{1Y}d - C\sin k_{1Y}d = De^{-i\hat{k}_{3Y}d} \tag{A6.9}$$

$$-k_{1Y}B\sin k_{1Y}d - k_{1Y}C\cos k_{1Y}d = -\hat{k}_{3Y}De^{-i\hat{k}_{3Y}d}. \tag{A6.10}$$

We have four linear homogeneous equations for the coefficients A, B, C, D and therefore the determinant of the system of the four equations must be zero. The solutions of the resulting equation determine the values of θ for which modes are possible.

Coefficients of the characteristic determinant (A.11)

A	BB	C	D
1	-1	0	0
$\hat{k}_{2Y}$	0	k_{1Y}	0
0	$\cos k_{1Y}d$	$-\sin k_{1Y}d$	$-e^{-i\hat{k}_{3Y}d}$
0	$-k_{1Y}\sin k_{1Y}d$	$-k1Y\cos k_{1Y}d$	$\hat{k}_{3Y}e^{-i\hat{k}_{3Y}d}$

The exponential factors cancel out, and the determinant may be developed with respect to the first row into a sum of two 3×3 determinants

$$\begin{pmatrix} 0 & k_{1Y} & 0 \\ \cos k_{1Y}d & -\sin k_{1Y}d & -1 \\ -k_{1Y}\sin k_{1Y}d & -k_{1Y}\cos k_{1Y}d & \hat{k}_{3Y} \end{pmatrix} (-1)(-1)\text{times}. \quad \text{(A6.12)}$$

$$\begin{pmatrix} \hat{k}_{2Y} & k_{1Y} & 0 \\ 0 & -\sin k_{1Y}d & -1 \\ 0 & -k_{1Y}\cos k_{1Y}d & \hat{k}_{3Y} \end{pmatrix} = 0. \quad \text{(A6.13)}$$

Calculation of the determinants results in

$$\tan k_{1Y}d = k_{1Y}(\hat{k}_{3Y} + \hat{k}_{2Y})/(k_{1Y}^2 - \hat{k}_{2Y}\hat{k}_{3Y}), \quad \text{(A6.14)}$$

which is the condition for the TE modes (see Eq. (6.49)). Similarly one has for the TM modes

$$\tan k_{1Yd} = n_1^2 k_{1Y}(n_2^2\hat{k}_{3Y} + n_3^2\hat{k}_{2Y})/(n_2^2 n_3^2 k_{1Y}^2 - n_1^4\hat{k}_{2Y}\hat{k}_{3Y}). \quad \text{(A6.15)}$$

The boundary value method gives us the same result with less effort. One only has to assume that the k values in the media (2) and (3) are imaginary and apply the boundary conditions to the solutions in the three sections.

FileFig A6.9 (NA1PLTE)

Calculation of the mode conditions for the TE case, which is Eq. (6.49) for the planar waveguide.

NA1PLTE is only on the CD.

Application A6.9. Derive in a similar way the condition for the TM case, which is the p-polarization.

CHAPTER

Blackbody Radiation, Atomic Emission, and Lasers

7.1 INTRODUCTION

At the end of the nineteenth century electromagnetic theory and thermodynamics were well developed and there was the question of reunification of these theories. Blackbody radiation is the emission of electromagnetic radiation from a closed cavity at temperature T through a small hole (Figure 7.1). It was found that the frequency distribution of the emitted radiation was dependent upon the temperature T. M. Planck formulated the famous blackbody radiation law in 1900. He used electromagnetic theory and thermodynamics, but needed an assumption that marked the beginning of quantum theory. Planck's law involved a fundamental physical constant, now called Planck's constant. The emission of light from the blackbody was interpreted as quantum emission of light. In the first half of the last century, the quantum emission of atoms and molecules was studied and in

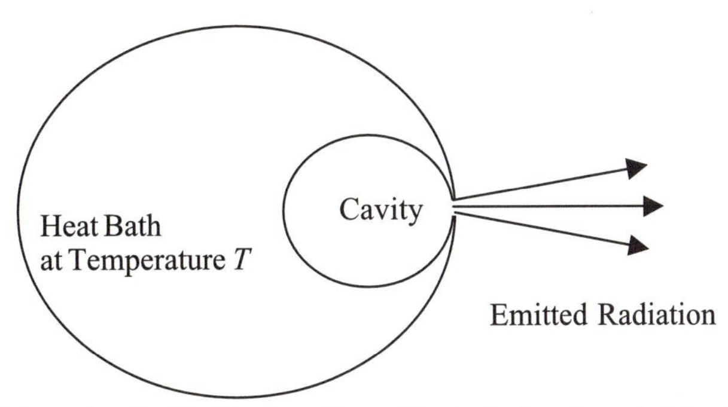

FIGURE 7.1 Schematic of a blackbody radiator. The body is surrounded by the heat bath of temperature T. Electromagnetic radiation is emitted and the frequency distribution depends on temperature T.

the second half of the century the laser was developed. ("Laser" stands for light amplification by stimulated emission of radiation).

7.2 BLACKBODY RADIATON

7.2.1 The Rayleigh–Jeans Law

An example of blackbody radiation is a wire heated by electricity. First it gets red and when the temperature increases it becomes white. Increasing the temperature, one finds that the maximum of the frequency distribution of the emitted electromagnetic radiation shifts to a shorter wavelength. In Figure 7.1 we show a schematic of a blackbody. Radiation is emitted at thermal equilibrium and the energy is drawn from the heat bath at temperature T. The first, but insufficient, analysis of blackbody radiation was done by Rayleigh and Jeans, analyzing the modes of a rectangular box in a heat bath (Figure 7.2). Oscillators were assumed to be at the walls of the box, emitting and absorbing light. The modes of the box are standing waves and in Figure 7.3 we show one standing wave in one direction as we have in a Fabry–Perot. The number of standing waves in three dimensions was analyzed and the number of modes determined with respect to the same energy which is the same frequency or wavelength. It was assumed that in thermal equilibrium, each mode carried the energy kT, where k is Boltzmann's constant. The energy density per frequency interval $du/d\nu$ was equal to the energy dE per volume V and frequency interval $d\nu$. The energy density per frequency interval $du/d\nu$ was then calculated to be

$$du/d\nu = (1/V)dE/d\nu = 8\pi kT\nu^2/c^3. \tag{7.1}$$

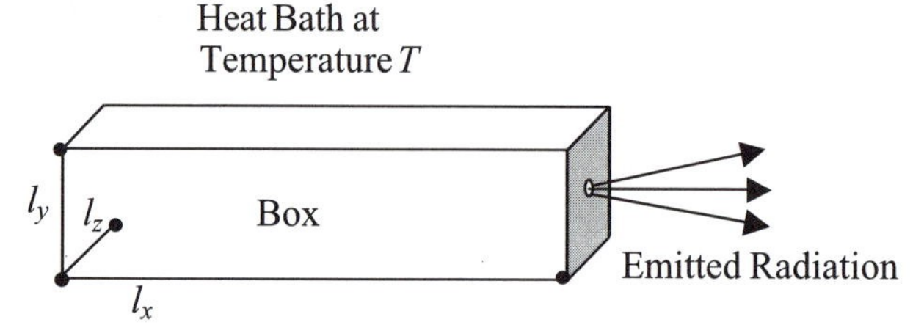

FIGURE 7.2 A box as cavity in the heat bath of temperature T; see Figure 7.1.

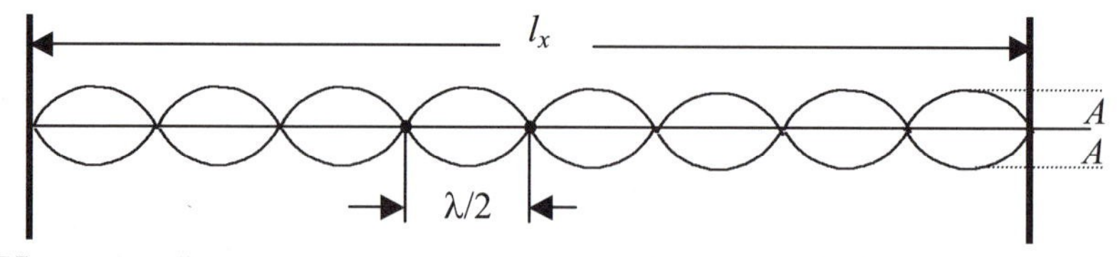

FIGURE 7.3 Standing wave pattern in one dimension. The length is l_x, the magnitude of the wave is A, and the wavelength λ.

C H A P T E R 

Blackbody Radiation, Atomic Emission, and Lasers

7.1 INTRODUCTION

At the end of the nineteenth century electromagnetic theory and thermodynamics were well developed and there was the question of reunification of these theories. Blackbody radiation is the emission of electromagnetic radiation from a closed cavity at temperature T through a small hole (Figure 7.1). It was found that the frequency distribution of the emitted radiation was dependent upon the temperature T. M. Planck formulated the famous blackbody radiation law in 1900. He used electromagnetic theory and thermodynamics, but needed an assumption that marked the beginning of quantum theory. Planck's law involved a fundamental physical constant, now called Planck's constant. The emission of light from the blackbody was interpreted as quantum emission of light. In the first half of the last century, the quantum emission of atoms and molecules was studied and in

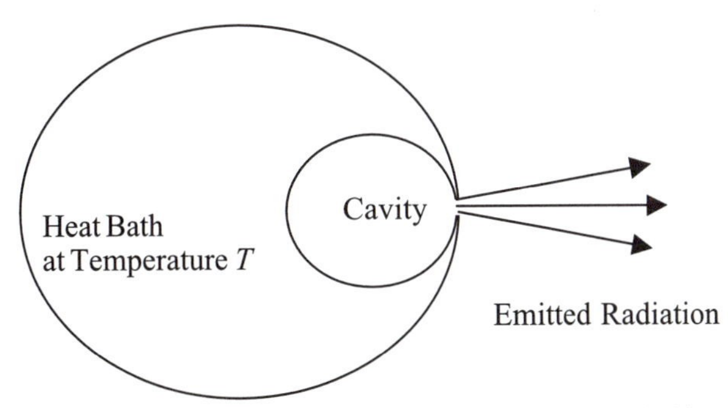

FIGURE 7.1 Schematic of a blackbody radiator. The body is surrounded by the heat bath of temperature T. Electromagnetic radiation is emitted and the frequency distribution depends on temperature T.

the second half of the century the laser was developed. ("Laser" stands for light amplification by stimulated emission of radiation).

7.2 BLACKBODY RADIATON

7.2.1 The Rayleigh–Jeans Law

An example of blackbody radiation is a wire heated by electricity. First it gets red and when the temperature increases it becomes white. Increasing the temperature, one finds that the maximum of the frequency distribution of the emitted electromagnetic radiation shifts to a shorter wavelength. In Figure 7.1 we show a schematic of a blackbody. Radiation is emitted at thermal equilibrium and the energy is drawn from the heat bath at temperature T. The first, but insufficient, analysis of blackbody radiation was done by Rayleigh and Jeans, analyzing the modes of a rectangular box in a heat bath (Figure 7.2). Oscillators were assumed to be at the walls of the box, emitting and absorbing light. The modes of the box are standing waves and in Figure 7.3 we show one standing wave in one direction as we have in a Fabry–Perot. The number of standing waves in three dimensions was analyzed and the number of modes determined with respect to the same energy which is the same frequency or wavelength. It was assumed that in thermal equilibrium, each mode carried the energy kT, where k is Boltzmann's constant. The energy density per frequency interval $du/d\nu$ was equal to the energy dE per volume V and frequency interval $d\nu$. The energy density per frequency interval $du/d\nu$ was then calculated to be

$$du/d\nu = (1/V)dE/d\nu = 8\pi kT\nu^2/c^3. \tag{7.1}$$

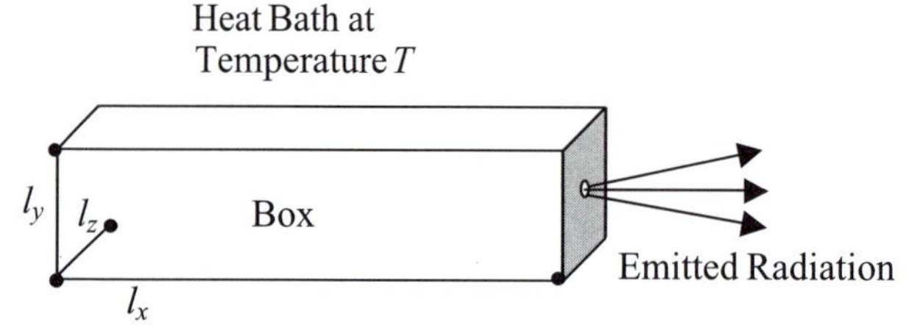

FIGURE 7.2 A box as cavity in the heat bath of temperature T; see Figure 7.1.

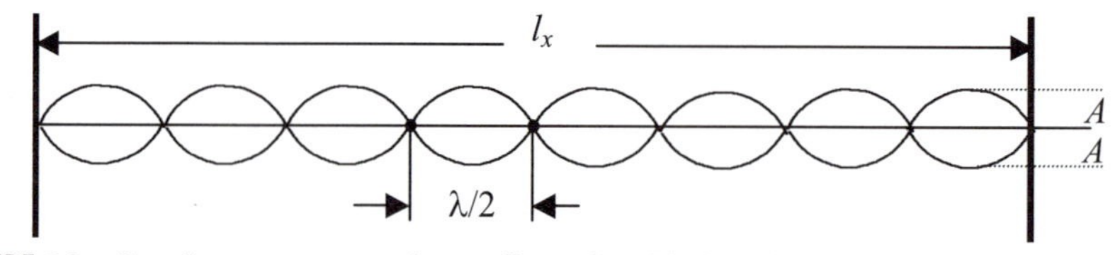

FIGURE 7.3 Standing wave pattern in one dimension. The length is l_x, the magnitude of the wave is A, and the wavelength λ.

This was the Rayleigh–Jeans law and turned out to be valid for the long wavelength region only. In the short wavelength region, the energy density increased by ν^2 and led to very high and unrealistic large energy densities, the so-called "violet catastrophe."

The graph in FileFig 7.1 shows the Rayleigh–Jeans radiation law in the visible spectral region, where one can observe the increase of the radiation density to shorter wavelengths.

FileFig 7.1 (L1RAJEANS)

Graph of the Raleigh–Jean law using units of energy density per frequency interval.

L1RAJEANS is only on the CD.

7.2.2 Planck's Law

The radiation law of blackbody radiation was discovered by Max Planck and is valid in both the short and the long wavelength regions. Planck's radiation law agreed with the Rayleigh-Jeans law for the long wavelength region. Many years later Einstein derived Planck's radiation law using the concept of transition probabilities. One assumes fictitious oscillators to be on the walls of the blackbody cavity. These oscillators are in contact with the heat bath at temperature T. When radiating, they get the energy from the heat bath and transform it into radiation energy. Einstein defined the processes of induced absorption, induced emission, and spontaneous emission for the emission and absorption of radiation by the oscillators (Figure 7.4). The probability of a transition of induced absorption between the levels numbered 1 and 2 is called W_{12} and is proportional to the

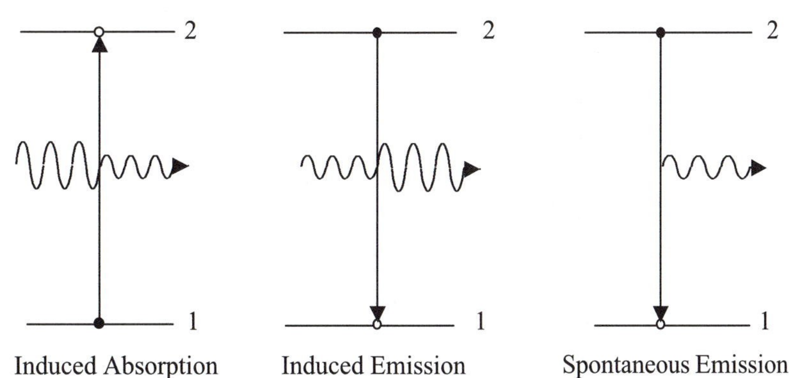

FIGURE 7.4 Schematic of induced absorption and induced and spontaneous emission.

energy density $du/d\nu$,

$$W_{12} = B_{12} du/d\nu. \tag{7.2a}$$

Similarly the probability of a transition of induced emission from level 2 to level 1 is called W_{21},

$$W_{21} = B_{21} du/d\nu. \tag{7.2b}$$

The probability of spontaneous emission is not proportional to $du/d\nu$ and is called

$$Wa_{21} = A_{21}. \tag{7.3}$$

B_{12}, B_{21}, and A_{21} are called the Einstein probability coefficients. In thermal equilibrium, there are as many transitions up as down, and for the "down" we have in addition spontaneous emission. One has

$$N_1(B_{12} du/d\nu) = N_2(B_{21} du/d\nu + A_{21}). \tag{7.4}$$

The numbers N_1 and N_2 are the occupation numbers of the two states of energy E_1 and E_2. In thermal equilibrium N_1 and N_2 follow from the Boltzmann distribution

$$N_1 = N_0 \exp -(E_1/kT) \text{ and } N_2 = N_0 \exp -(E_2/kT), \tag{7.5}$$

where N_0 is a constant. In thermal equilibrium one has $N_2 < N_1$,which means the occupation number for the lower energy states is always higher.

From Eqs. (7.4) and (7.5) one gets for the energy density per frequency interval

$$du/d\nu = A_{21}/\{B_{12} \exp[(E_2 - E_1)/kT] - B_{21}\}. \tag{7.6}$$

The constants A_{21}, B_{12}, and B_{21} are determined by considering two limiting cases. When $T \to \infty$, one has $du/d\nu \to \infty$ and it follows that $B_{12} = B_{21}$. We may write for Eq. (7.6),

$$du/d\nu = A_{21}/\{B_{12}(\exp[(h\nu)/kT] - 1)\}, \tag{7.7}$$

where $E_2 - E_1 = h\nu$ and h is Planck's constant. To consider the long wavelength region, we develop the exponential. In, the limit of long wavelengths, where $\nu \to 0$, the energy density per frequency interval (right-hand side of Eq. (7.7)) is

$$du/d\nu = A_{21}/\{B_{12}(h\nu/kT)\}. \tag{7.8}$$

This must be equal to the Rayleigh–Jeans Law, $du/d\nu = 8\pi kT\nu^2/c^3$ (see Eq. (7.1)) and one obtains

$$(A_{21}/B_{12}) = 8\pi h\nu^3/c^3. \tag{7.9}$$

Introduction of Eq. (7.9) into Eq. (7.7) gives us Planck's formula for the energy density per frequency interval

$$du/d\nu = 8\pi h\nu^3/c^3\{1/[\exp(h\nu/kT) - 1]\}. \tag{7.10}$$

Graphs of Planck's formula, depending on wavelength and frequency, are given in FileFigs.2 and 3, respectively.

7.2.3 Stefan–Boltzmann Law

The Stefan–Boltzmann law gives the integrated energy over all wavelengths or frequencies depending on the temperature T. Integration of Eq. (7.10) over the frequency results in the energy density u:

$$u = (8/15)\pi^5 (kT)^4/(hc)^3. \tag{7.11}$$

This is the energy density equal to the energy per unit volume in the cavity of the blackbody. To calculate the energy emitted from the hole of the blackbody, we introduce the radiance L_B (or brightness) of the blackbody. The power dW leaving the blackbody is calculated by the product

$$dW = L_B da \cos\theta d\Omega, \tag{7.12}$$

where da is the area from which the power is emitted, $d\Omega$ the solid angle into which it travels, and Ω the angle between the normal of the area and the center line of the solid angle (Figure 7.5). The radiance L_B is measured by placing a power meter before the opening of the blackbody, taking into account the area

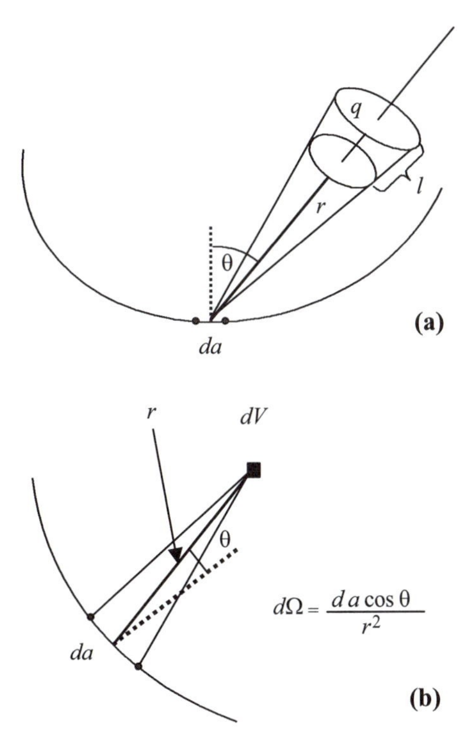

FIGURE 7.5 (a) Power emitted from the surface element da traverses the volume element $dV = ql$ in time l/c; (b) the solid angle seen from dV.

and solid angle. This can be written

$$dW/(da\cos\theta d\Omega) = L_B. \tag{7.13}$$

The radiation is emitted into a hemisphere and when integrating over the solid angle of the hemisphere we have

$$W = L_B da\pi. \tag{7.14}$$

The radiance L_B is related to the energy density u of the blackbody and we want to find the relation between these quantities. We consider a small volume $dV = l$ times q (Figure 7.5a). The power dW, transmitted in the time interval dt from the area da of the blackbody into the solid angle $d\Omega$, is then

$$dWdt = L_B da\cos d\Omega dt. \tag{7.15}$$

Using from Figure 7.5 $d\Omega = q/r^2$ and $dt = l/c$ and observing that $dWdt = dudV$ one obtains

$$dudV = (L_B da\cos\theta dV)/(r^2c). \tag{7.16}$$

Integration over the area times the solid angle, seen from dV (Figure 7.5b), results in

$$u = (L_B/c)[\int da(\cos\theta/r^2)] = (L_B/c)\int d\Omega_V. \tag{7.17}$$

The integral is over the surrounding sphere and is 4π. Using Eqs. (7.11) and (7.14) one has

$$L_B \cdot \pi = \frac{2c}{15}\pi\frac{(\pi kT)^4}{(hc)^3}. \tag{7.18}$$

This is the Stefan–Boltzmann law telling us that the "total emission" is proportional to T^4. The constant $\sigma = (2c/15)(\pi k)^4/(hc)^3$ has the value $5.670310^{-8}\mathrm{W}/(m^2\mathrm{K}^4)$.

In FileFig 7.4 graphs shown of the Stefan–Boltzmann law in units of power/area.

7.2.4 Wien's Law

The wavelength at maximum emission of the blackbody may be calculated by rewriting Eq. (7.10) as a function of the wavelength and setting it to zero. One gets

$$\lambda_m T = 2.8910^{-3} mK. \tag{7.19}$$

This is called Wien's displacement law.

Graphs of Wien's displacement law are shown in FileFig 7.5.

7.2.5 Files of Planck's, Stefan–Boltzmann's, and Wien's Laws. Radiance, Area, and Solid Angle

Using the radiance L_B as discussed in Section 7.2.3, we can rewrite Planck's law in terms of the radiance per wavelength or per frequency interval as

$$dL_B/d\lambda = (C_1/\lambda^5)/(\exp(C_2/\lambda T)) - 1)[W/\{m^3 sr\}] \quad (7.20)$$
$$dL_B/d\nu = (C_3\nu^3)/(\exp(C_4\nu/T)) - 1)[W/\{1/sm^2 sr\}], \quad (7.21)$$

where

$$\begin{aligned} C_1 &= 2hc^2 = 1.17610^{-16} W m^2 \\ C_2 &= hc/k = 1.43210^{-2} m K \\ C_3 &= 2h/c^2 = 1.4710^{-50} W s^4/m^2 \\ C_4 &= h/k = 4.7810^{-11} s K. \end{aligned} \quad (7.22)$$

FileFig 7.2 (L2BBLS)

Graph of blackbody radiation depending on wavelength. The calculation of the radiance of a particular wavelength range and calculation of the corresponding radiant energy by multiplication with area times solid angle.

L2BBLS

1. Blackbody radiation. Graph of $dL/d\lambda$

$c2 := 1.43 \cdot 10^{-2}$ $\quad c1 := 1.18 \cdot 10^{-16}$ $\quad T := 1000.$

Planck's law depending on wavelength is

$$x := 3 \cdot 10^{-5}, 2.99 \cdot 10^{-5} \ldots 10^{-7}$$
$$f(x) := \frac{c1}{x^5 \left[e^{\left(\frac{c2}{x \cdot T}\right)} - 1\right]}$$

x in meters.

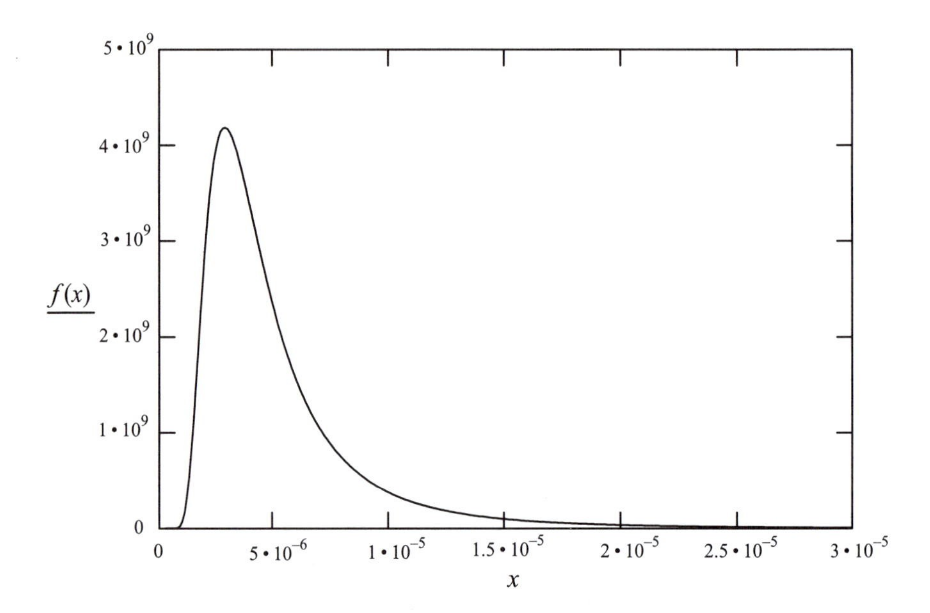

2. Integration over the wavelength range from $3*10^{-6}$ to $3*10^{-5}$ meters to obtain the radiance

$$R := \int_{3\cdot 10^{-6}}^{3\cdot 10^{-5}} f(x)dx.$$

Radiance

$$R = 1.316 \cdot 10^4.$$

3. Multiplication with area times solid angle to obtain the radiant energy Area A, Solid angle SA: $A := 1$, $SA := 4$; radiant energy RR: $RR := A \cdot SA \cdot R$; $RR = 5.263 \cdot 10^3$ watts. RR has the same value as the corresponding value when integrating over the frequency.

Application 7.2.

1. Calculate the radiance for different wavelength intervals.
2. Calculate the radiant energy and choose an area of 5 mm^2 and a solid angle of π^4.
3. Consider the wavelength range from 1 mm to 100 mm for comparison with the Rayleigh–Jeans law. Calculate the corresponding frequencies and derive from Planck's law the corresponding energy density for this frequency interval. Do the same calculation for the Rayleigh–Jeans law and give the difference in the numerical values.

FileFig 7.3 (L3BBFS)

Graph of blackbody radiation depending on the frequency. The calculation of the radiance of a particular frequency range and calculation of the corresponding radiant energy by multiplication with area times solid angle.

L3BBFS is only on the CD.

Application 7.3.

1. Calculate the radiance for different frequency intervals.
2. Calculate the radiant energy, and choose the area times solid angle such that the radiant energy is the same as you calculated in Application FF2.
3. Numerical calculation of the Stefan–Boltzmann law. Calculate, using the same units, the integrated radiation from Planck's law for a chosen temperature T and compare with the Stefan–Boltzmann law (FileFig 7.2).

FileFig 7.4 (L4STEFS)

The Stefan–Boltzmann law is plotted using linear and logarithmic scales.

L4STEFS is only on the CD.

FileFig 7.5 (L5WIENS)

Wien's law is plotted for two ranges of the temperature.

L5WIENS is only on the CD.

7.3 ATOMIC EMISSION

7.3.1 Introduction

The operation of a laser needs an "active medium" between the two mirrors of the laser cavity. Energy is "pumped" from the outside of the cavity into this medium and produces atoms or molecules in excited states. Here we discuss only excited energy states of atoms in the gas phase and consider the hydrogen atom and atoms with hydrogenlike spectra. The energy states are labeled by letters with subscripts and superscripts and some of the notations have their origin in the "old days" of spectroscopy. At that time, for example, one used for the characterization of some spectral lines: s for "sharp," and d for "diffuse." The letters *s* and *d* are still in use for the characterization of the angular momentum. There is some truth to these notations, since we know that the s-state has less degeneracy than the d-state.

7.3.2 Bohr's Model and the One Electron Atom

In Bohr's model, an electron with a negative charge circulates around the positive charge in the center of its orbits, determined by the *principal quantum number* n. The energy of such an orbit is given by

$$E = (-2\pi^2 K^2 e^4 m)/(n^2 h^2), \tag{7.23}$$

where K is the constant of Coulombs' law, e the electron charge, m the electron mass, and h Planck's constant. The principal quantum number n has integer numbers 1, 2.3 .... Radiation is emitted when the electron changes its orbits and the energy of the emitted photon is given as

$$\nu_{ni,nf} h = [2\pi^2 K^2 e^4 m)/(h^2)][1/n_i^2 - 1/n_f^2], \tag{7.24}$$

where n_i is the quantum number of the initial orbit and n_f is the quantum number of the final orbit. In Figure 7.6 we show the series of lines originating from the state $n_f = 1$, called the Lyman series, from the state $n_f = 2$, called the Balmer series, and from the state $n_f = 3$, called the Paschen series. The significant achievement of Bohr's derivation was that he used fundamental physical constants and could reproduce exactly the empirical constant of the expression of the Balmer series.

7.3.3 Many Electron Atoms

7.3.3.1 Principal Quantum and Angular Momentum Quantum Numbers

The Schroedinger equation and the Pauli principle are needed to understand the atomic energy schematics and transitions. As an example we look at an atom with Z electrons and a nucleus with a positive charge of Ze. A list of some lower energy states is shown in Figure 7.7. For such an atom, we have energy levels labeled by the principal quantum number n and the angular momentum quantum number l. The Schroedinger equation tells us that for each n there are $n - 1$ different possible states of the angular momentum.

7.3.3.2 Magnetic Quantum Number and Degeneracy

To each state labeled by the angular quantum number l, there are $2l+1$ substrates. They only have different energy values if the atom is in a magnetic field. The corresponding quantum number is called the magnetic quantum number m. If the magnetic field is zero, all states have the same energy, and therefore the state is m fold degenerate.

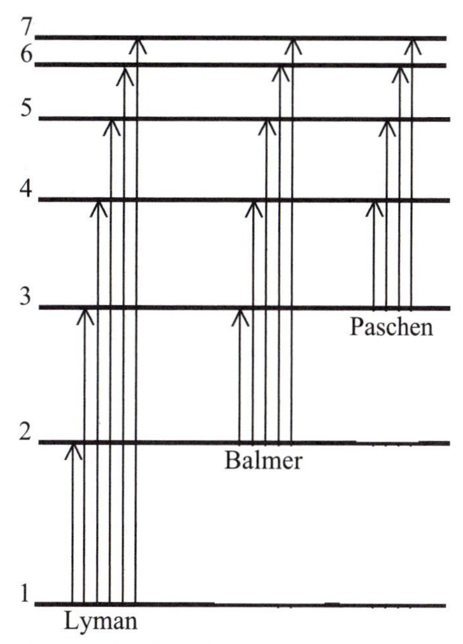

FIGURE 7.6 Diagram of energy levels and transitions of Bohr's model. The electrons change from the state labeled n_i to n_f and emit light: Lyman series: $n_f = 1, n_i = 2, 3, 4 \ldots$; Balmer series: $n_f = 2, n_i = 3, 4, 5 \ldots$; Paschen series: $n_f = 3, n_i = 4, 5, 6$. The energy difference between $n = 1$ and $n = \infty$ is the dissociation energy.

7.3.3.3 Spin States

Each electron has an angular momentum with respect to its own axis called the spin, described by the spin quantum number s. In a magnetic field, the projection has the values $1/2$ and $-1/2$.

7.3.3.4 Pauli Principle and Occupation Rule

Each nondegenerate state of the atom has a different set of quantum numbers n, l, m, s. For each n there are $n - 1$ values of l. For each l there are $2l + 1$ values of m and for each m there are two values of s (Figure 7.7).

7.3.3.5 Buildup Principle of Atoms and Labels for Energy Levels

The energy schematic of an atom with Z electrons may be obtained, in first approximation, by using the "buildup principle." There are special notations for the principal and angular quantum numbers.

The principal quantum number $n = 1, 2, 3 \ldots$ is labeled $K, L, M, \ldots$.

The angular quantum number $l = 0, 1, 2, 3 \ldots$ is labeled s, p, d, f.

Following the buildup principle, the electrons first occupy the lowest n, then the next, and so on. For each n, electrons fill all possible states labeled by l, m, and s.

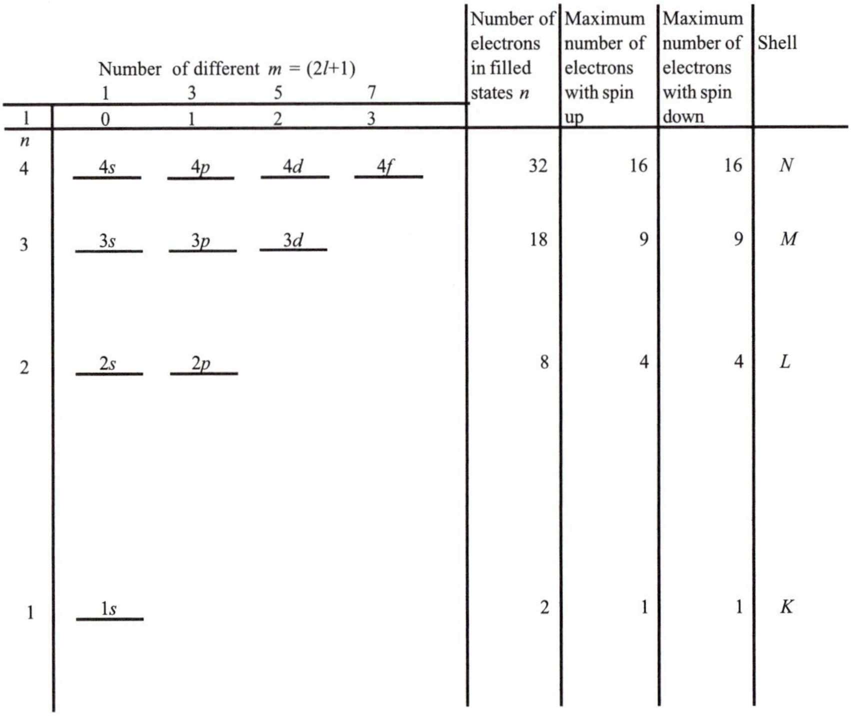

	Number of different $m = (2l+1)$				Number of electrons in filled states n	Maximum number of electrons with spin up	Maximum number of electrons with spin down	Shell
	1	3	5	7				
l	0	1	2	3				
n								
4	$4s$	$4p$	$4d$	$4f$	32	16	16	N
3	$3s$	$3p$	$3d$		18	9	9	M
2	$2s$	$2p$			8	4	4	L
1	$1s$				2	1	1	K

FIGURE 7.7 Energy levels and quantum numbers for a few states of the hydrogenlike atom. The states are shown for each n and labeled by s, p, d, f. Spin states are also indicated.

In Figure 7.8 the number of electrons occupying different states is listed for the elements from $Z = 1$ to $Z = 25$. There are irregularities, explained by quantum mechanics. An example is $Z = 19$, where the $4s$ state has lower energy than the $3d$ state.

7.3.3.6 Transitions Between Energy States

Photons may be emitted when the atom changes from a higher energy state to a lower one. In an atom like the sodium atom, there is just one "outer" electron. All other electrons are in "closed" shells. This outer most electron is an "s" electron (Figure 7.9), and therefore the spectrum has a similarity to the spectrum of the hydrogen atom. The columns in Figure 7.9 are labeled with capital letters, referring to compound states of all electrons. Since the K and L shells are closed, they do not contribute and the angular momentum of all electrons is the same as the single electron in the M shell.

There are selection rules restricting the energy levels between which transitions are possible. The general rule is that the angular momentum has to change by plus or minus 1. Transitions are only allowed between the levels of the columns labeled by s, p, and d (Figure 7.9).

		Shell	K	L		M			N
		n	1	2		3			4
Z	Element	l	s	s	p	s	p	d	s
1	H	Hydrogen	1						
2	He	Helium	2						
3	Li	Lithium	2	1					
4	Be	Beryllium	2	2					
5	Be	Boron	2	2	1				
6	C	Carbon	2	2	2				
7	N	Nitrogen	2	2	3				
8	O	Oxygen	2	2	4				
9	F	Fluorine	2	2	5				
10	Ne	Neon	2	2	6				
11	Na	Sodium	2	2	6	1			
12	Mg	Magnesium	2	2	6	2			
13	Al	Aluminum	2	2	6	2	1		
14	Si	Silicon	2	2	6	2	2		
15	P	Phosphorus	2	2	6	2	3		
16	S	Sulfur	2	2	6	2	4		
17	Cl	Chlorine	2	2	6	2	5		
18	Ar	Argon	2	2	6	2	6		
19	K	Potassium	2	2	6	2	6		1
20	Ca	Calcium	2	2	6	2	6		2
21	Sc	Scandium	2	2	6	2	6	1	2
22	Ti	Titanium	2	2	6	2	6	2	2
23	V	Vanadium	2	2	6	2	6	3	2
24	Cr	Chromium	2	2	6	2	6	5	1
25	Mn	Manganese	2	2	6	2	6	5	2

FIGURE 7.8 Electronic states in atoms. Electrons having the same n value in "shells," named K, L, M. For each n we have $n - 1$ substates of angular momentum with quantum number l, named s, p, d, f. There are $2l + 1$ possible values of m, and two spin states. Therefore the maximum occupations of K and L are $K : n = 1, 2[(2\cdot 0+1)] = 2; L : n = 2, 2[(2\cdot 0+1)]+2(2\cdot 1+1)] = 8$ (the M shell is more complicated).

7.4 BANDWIDTH

7.4.1 Introduction

The atom emits light when an electron makes a transition from a higher energy state to a lower one. The emitted light is not monochromatic, since the emission process last only for a short time. Therefore we have a wavetrain of limited length. Such wavetrains are described by the superposition of a large number of monochromatic waves having a certain frequency distribution (see Chapter 4). The frequency spectrum shows a maximum at the transition frequency and the bandwidth of the frequency distribution is related to the time of the emission process.

We follow the book *Lasers* by F. K. Kneubühl and M. W. Sigrist, B. G. Teubner, Stuttgart, 1988, and use some of their numerical values in the examples.

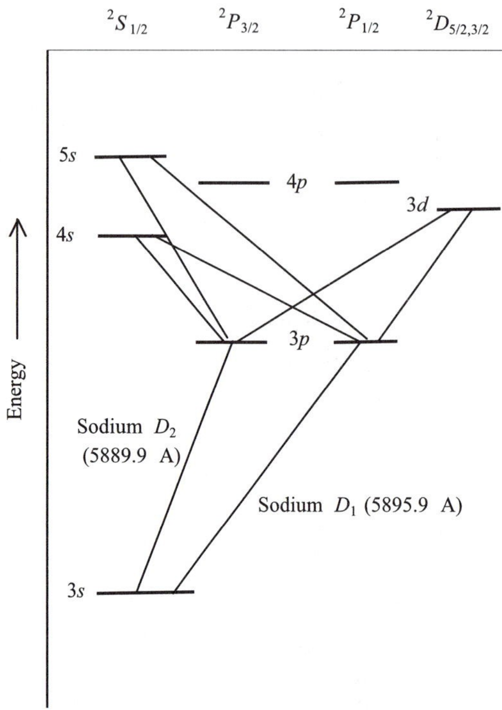

FIGURE 7.9 Energy schematic for sodium. The energy level of 3*s*, that is, the ground state of the outermost electron, has been chosen as 0. The two well-known "Sodium *D*" lines are indicated.

7.4.2 Classical Model, Lorentzian Line Shape, and Homogeneous Broadening

Light is emitted from an atom when an electron leaves an excited state E_2 and occupies a state with lower energy E_1. One has for the energy of the process

$$E_2 - E_1 = h\nu. \tag{7.25}$$

Before the electron can make the transition, it has to be placed into the upper state. This process is called population inversion (see Section 5.2). The electron remains in the upper state for a limited time. On average the life-time of an excited state is 10^{-8} sec. However, longer lifetimes corresponding to metastable states play an important role in laser action. The emitted light is described by a wavetrain with decreasing amplitude (Figure 7.10), and may be described in first approximation as

$$A = A_0 e^{-i\omega_0 t} e^{-t/\tau}. \tag{7.26}$$

The lifetime τ is defined as the time in which the initial amplitude A_0 drops to a value of $A_0/e = A_0/2.718$. The frequency distribution of the wavetrain now

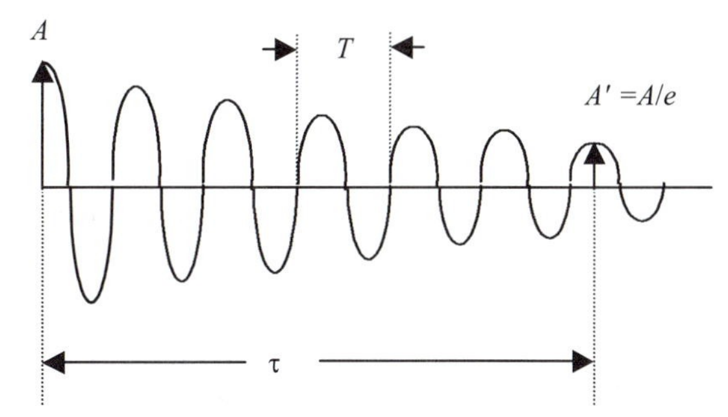

FIGURE 7.10 Wavetrain decreasing over the time τ to the value $A/e = A/2.71$.

discussed by application of a time-dependent Fourier transformation,

$$y(\omega) = (A_0/\sqrt{2\pi}) \int_0^{\infty} e^{i\omega_0 t} e^{-t/2\tau} e^{-i\omega t} dt. \tag{7.27}$$

The integral may be calculated analytically and one gets

$$y(\omega) = (A_0/\sqrt{2\pi})[-1/\{i(\omega_0 - \omega) - 1/2\tau\}]. \tag{7.28}$$

The intensity is obtained as

$$I(\omega) = y(\omega)y(\omega)^* = \Psi_0[1/\{(\omega_0 - \omega)^2 + (1/2\tau)^2\}]. \tag{7.29}$$

The constant ψ_0 is the total intensity and depends on the lifetime τ. From Eq. (7.29) one defines the profile of the line or the *Lorentzian line shape* as

$$gl(\omega) = 2[(1/2\tau)/\{(\omega_0 - \omega)^2 + (1/2\tau)^2\}]. \tag{7.30}$$

A graph of $gl(\omega)$ is shown in in Figure 7.11 with ω_0 at its center and a fixed value of τ. The bandwidth $\Delta_\omega = (\omega - \omega_0)$ at half-height is obtained from

$$(1/2)gl(\omega = \omega_0) = gl(\omega = \omega_0 + \Delta\omega(/2) \tag{7.31}$$

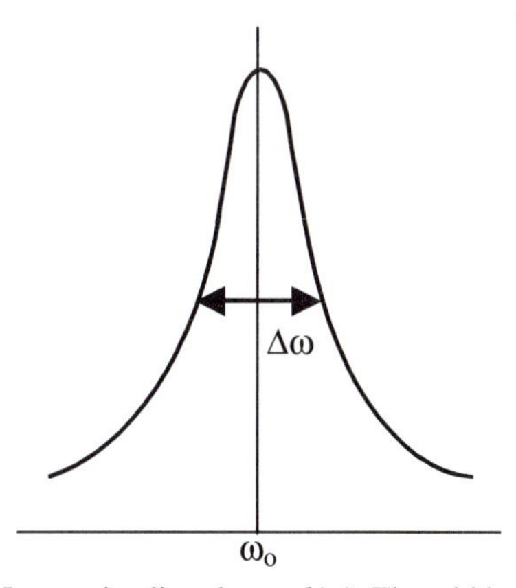

FIGURE 7.11 Graph of the Lorentzian line shape $gl(\omega)$. The width at half-height $\Delta\omega$ is equal to $1/\tau$.

and one gets

$$\Delta\omega = 1/\tau. \qquad (7.32)$$

Therefore, the bandwidth is related to the lifetime of the atomic emission process, assuming that one has a waveform as given in Eq. (7.26). Introduction of Eq. (7.32) into Eq. (7.30) results in

$$gl(\omega) = 2[(\Delta\omega/2)/\{(\omega - \omega_0)^2 + (\Delta\omega/2)^2\}]. \qquad (7.33)$$

For an oscillator the quality factor is $Q = \omega_0/\Delta\omega$. This expression is similar to the resolving power discussed for the Fabry–Perot in Chapter 2 and the grating in Chapter 3. In FileFig 7.6 we show an example of the band shape of Eq. (7.33), where the lifetime is chosen to be $\tau = 1000$ in order to show a graph in the chosen frequency region.

FileFig 7.6 (L6BANDS)

Lorentzian line shape spectrum with angular resonance frequency ω_0 and lifetime τ.

L6BANDS

Lorentzian Line Shape

Frequency interval $m := 11$.

$$\omega 0 := \frac{49}{(2^m - 1)} \qquad \omega := 1\frac{1}{(2^m - 1)}, \frac{2}{(2^m - 1)} \cdots 1.$$

To make a graph the lifetime is chosen such that the Lorentzian line shape can be demonstrated.

$$\tau := 1000$$

$$gl(\omega) := 2\frac{\frac{1}{(2\cdot\tau)}}{\frac{1}{(2\cdot\tau)^2} + (\omega - \omega 0)^2} \qquad Q := \tau\cdot\omega 0$$

$$Q := 23.937.$$

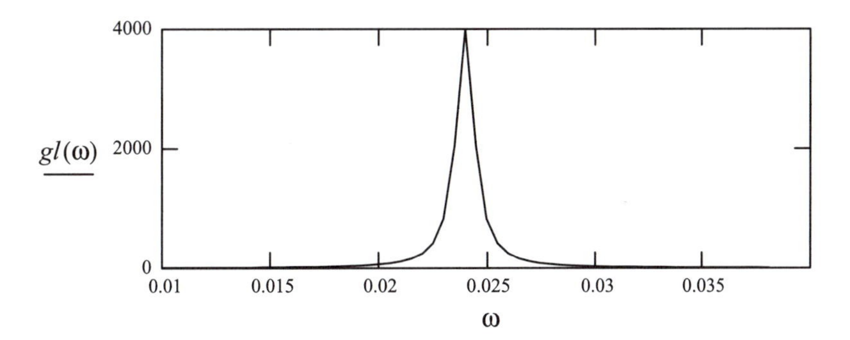

Application 7.6.

1. For the Lorentzian line shape, change the lifetime τ to more realistic values in the range of 10^{-3} to 10^{-8} and find a corresponding frequency range.
2. Change the lifetime and find the bandwidth. Calculate Q as $\omega_0/\Delta\omega$ and compare with the data from a graph.

7.4.3 Natural Emission Line Width, Quantum Mechanical Model

In Section 7.4.2 we saw how the width of the Lorentzian line shape depends on the lifetime τ of the electron in the upper state. The uncertainty principle of quantum mechanics relates the lifetime τ to the width ΔE of the corresponding energy state, in which the electron is placed. The uncertainty principle may be expressed as

$$\Delta E\tau = h/2\pi. \tag{7.34}$$

In Figure 7.12 we indicate three possible transitions from the higher to the lower state. One corresponds to the transition from center-to-center of the two bands, and the others to the transition between two different energies of the bands, corresponding to lower and higher frequencies. Quantum mechanics shows that the center-to-center transition has a higher transition probability than the other possible transitions. The shape of the emission line shows these differences. The Einstein coefficient of spontaneous emission A_{21} describes a transition probability, and is related to the lifetime of the emission process as

$$A_{21} = 1/\tau. \tag{7.35}$$

7.4.4 Doppler Broadening (Inhomogeneous)

The classical Doppler effect may be observed when a train passes with a blowing horn. The signal seems to have a higher frequency when the train is traveling towards us and a lower frequency when traveling away from us. The atoms in a

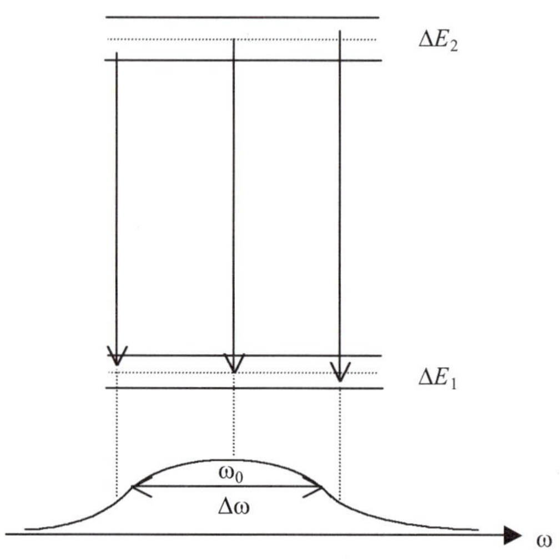

FIGURE 7.12 The upper energy level has the width ΔE_2, the lower ΔE_1. Transitions of higher and lower frequencies are indicated, with respect to the peak frequency of the Lorentzian line shape.

gas also move while emitting light. The frequency shift, related to their particular velocity v, is given as

$$\nu - \nu_0 = \nu_0(v/c). \tag{7.36}$$

The distribution of the velocities depends on the temperature and follows Maxwell's velocity distribution law. One gets

$$\begin{aligned} I(\nu) &= I_{\nu o} \exp(-E_k/kT) = I_{\nu o} \exp(-mv^2/2kT) \\ &= I_{\nu o} \exp\{(-mc^2/2kT)((\nu - \nu_0)^2/(\nu_0)^2)\}, \end{aligned} \tag{7.37}$$

where I_0 is a constant, m the mass of the oscillator, c the speed of light, k the Boltzmann constant, and T the absolute temperature. As we did in Section 7.4.2 for the Lorentzian line shape, we define the *Doppler line shape* as $I(\nu) = I_0 gd(\nu)$ and get

$$gd(\nu) = [(\sqrt{\pi}(ln\ 2)(2/\pi\,\Delta\nu)] \exp\{-(\nu - \nu_0)^2(ln\ 2)/(\Delta\nu/2)^2\}. \tag{7.38}$$

The line shape of Eq. (7.38) has a Gaussian profile which is different from the Lorentzian, schematically shown in Figure 7.13a.

The halfwidth at half height is calculated to be

$$\Delta\nu = 2\nu_0\sqrt{[(2kT/mc^2)(ln\ 2)]}. \tag{7.39}$$

The Gaussian line shape may be looked at as the envelope of a large number of Lorentzian line shapes. Each oscillator moves with a different velocity in a different direction and the peak frequencies of the emitted light have a Gauss distribution, (shown in schematically Figure 7.13b). When discussing an exam-

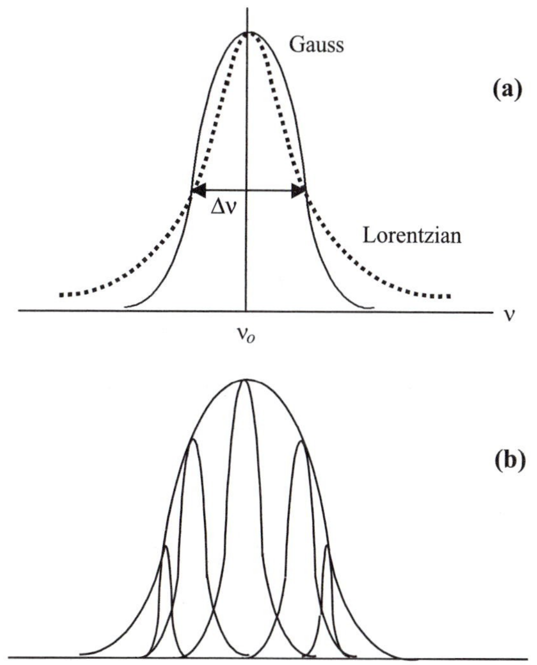

FIGURE 7.13 (a) Comparison of Lorentzian and Gaussian line shapes of approximately same line width at halfweight; (b) Gaussian line shape is the envelope of Lorentzian line shapes emitted statistically at different velocities in different directions.

ple of the Ne–He laser below, we give numerical values for the halfwidth of Eq. (7.39).

7.5 LASERS

7.5.1 Introduction

We discussed blackbody radiation in Section 7.2, atomic emission in Section 7.3, the Fabry–Perot in Chapter 2, and modes in Chapter 6. We now discuss the following simplified model for a two-level laser. We consider a Fabry–Perot filled with an active medium of atomic oscillators. First we need *population inversion*. The oscillators will be excited into the upper state 2 by using the light of frequency ν_P from the outside. We need the excited states to remain excited for sometime, for example, 10^{-3}sec. Then we need *stimulated emission* to transfer the energy of the excited oscillators to the modes of the Fabry–Perot. To have laser light emitted from the Fabry–Perot, one must couple out part of the light of the modes. This is done by making one of the mirrors of the Fabry–Perot not totally reflective, such as when there is a small hole in the mirror.

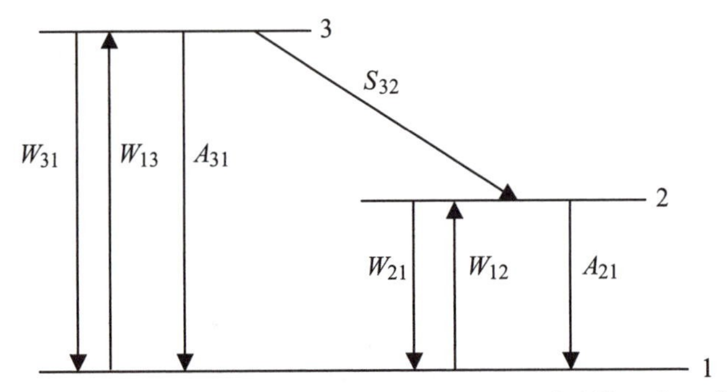

FIGURE 7.14 Schematic of a three-level laser laser. S_{23} is the probability of a radiationless transition, W_{ik} and W_{ki} are probabilities of induced absorption and emission, and A_{ik} is the probability of spontaneous emission.

7.5.2 Population Inversion

7.5.2.1 Two-Level System with Stimulated and Spontaneous Transitions

When discussing the blackbody radiation law, we used for the occupation of the different energy levels in thermal equilibrium the Boltzmann distribution $Ne^{-E/kT}$, where k is Boltzmann's constant. This distribution law tells us that, at temperature T, the lower states are more occupied than the higher states. For laser action we need just the reverse, which is more electrons in a higher energy state. This is called population inversion and is in contrast to the population of the electrons in thermal equilibrium.

We assume that we have to deal with oscillators having just two levels, as shown in Figure 7.14. We call the upper level E_2 and the lower level E_1. The change in time of the number of photons N_1 of E_1 and N_2 of E_2, the rate equations, is

$$dN_1/dt = -N_1 B_{12} u(\nu) + N_2 B_{21} u(\nu) + N_2 A_{21} \tag{7.40}$$

$$dN_2/dt = +N_1 B_{12} u(\nu) - N_2 B_{21} u(\nu) - N_2 A_{21}, \tag{7.41}$$

where $u(\nu)$ is the radiation density. The coefficient B_{12} is the Einstein coefficient of stimulated absorption. The coefficient B_{21} is the Einstein coefficient of stimulated emission and A_{21} is the coefficient of spontaneous emission. We have used these coefficients when deriving Planck's radiation law, (see Section 7.2), and used the relation

$$A_{21}/B_{12} = 8\pi h(\nu/c)^3 \tag{7.42}$$

as well as $B_{12} = B_{21}$. This relation has to be slightly changed for atomic oscillators. We have to take into account that the atomic energy levels may be degenerate, which means there are several energy levels having the same energy. For example, if there is a threefold degeneracy, we have to use the weight $g = 3$

for the transition. Therefore we have to use

$$g_1 B_{12} = g_2 B_{21}. \tag{7.43}$$

7.5.2.2 Changes in the Upper Level Considering Stimulated Transitions

A necessary condition for laser action is population inversion which as stated before, means there must be more photons in the upper state than in the lower state. Using only the stimulated emission and absorption processes, we have for the number of photons in the upperstate.

$$N_1 B_{12} u(\nu), \tag{7.44}$$

and the number in the lowerstate

$$N_2 B_{21} u(\nu). \tag{7.45}$$

The change in the number of photons in time is then

$$dn_p/dt = N_2 B_{21} u(\nu) - N_1 B_{12} u(\nu). \tag{7.46}$$

With Eq. (7.43) we write

$$dn_p/dt = [N_2 - N_1(g_2/g_1)] B_{21} u(\nu). \tag{7.47}$$

To have laser action one needs

$$N_2 - N_1 \frac{g_2}{g_1} > 0 \tag{7.48}$$

and one has the condition for laser action, $N_2/g_2 > N_1/g_1$.

7.5.3 Stimulated Emission, Spontaneous Emission, and the Amplification Factor

To have the condition $N_2/g_2 > N_1/g_1$ fulfilled, one has to make the stimulated emission larger than the spontaneous emission. The stimulated emission transfers the photons from the upper energy state E_2 into the modes of the Fabry–Perot. The bandwidth for the emission process was discussed above and we assume that we have a Lorentzian line shape. The modes of the Fabry–Perot are much narrower and close to delta functions (see Figure 7.15). The probability of stimulated emissions per unit time is calculated by considering Eq. (7.46), written for a number of n photons as

$$dn/dt = [N_2 - (g_2/g_1)N_1] B_{21} u(\nu). \tag{7.49}$$

The radiation density is $u(\nu) = h\nu n$. Taking into account the line shape of the emission process, (see Eq. (7.30)), we have to multiply the coefficient of stimulated emission B_{21} by $u(v)gl(v)$, that is, $B_{21}u(\nu)gl(\nu)$, and get

$$du(\nu)/dt = [N_2 - (g_2/g_1)N_1] h\nu gl(\nu) B_{21} u(\nu). \tag{7.50}$$

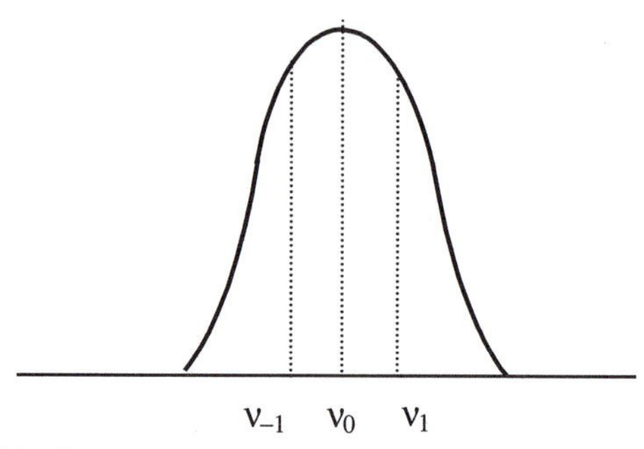

FIGURE 7.15 The width of the modes of the cavity at frequencies ν_{-1}, ν_0, and ν_1 are almost delta functions compared to the bandwidth of the Lorentzian-shaped emission line.

We can express the change of $u(\nu)$ with respect to the time dt by considering the length dz in which the light travels in the time dt This gives us $dt = dz/c'$. Here we use the speed of light c', modified in the medium of the laser cavity. The change of the radiation density over the interval dz is then

$$du(\nu)/dz = [N_2 - (g_2/g_1)N_1](h\nu/c')gl(\nu)B_{21}u(\nu, z). \tag{7.51}$$

Using $A_{21} = 1/\tau$, and $A_{21}/B_{21} = 8\pi h(\nu/c')^3$, we have

$$du(\nu)/dz = \{(c'^2/8\pi\nu^2\tau)[N_2 - (g_2/g_1)N_1]gl(\nu)\}u(\nu, z). \tag{7.52}$$

The expression in the curly braces is called the amplification factor $\varepsilon(\nu)$:

$$\varepsilon(\nu) = \{(c'^2/8\pi\nu^2\tau)[N_2 - (g_2/g_1)N_1]gl(\nu)\}. \tag{7.53}$$

The gain of the beam $\varepsilon(\nu)$ is "per length" and an example for the Ruby laser is calculated in FileFig 7.7.

FileFig 7.7 (L7RUBYS)

Gain calculation for the example of the Ruby laser. (See also Kneubühl and Sigmst, 1988.)

L7RUBYS is only on the CD.

7.5.4 The Fabry–Perot Cavity, Losses, and Threshold Condition

The Fabry–Perot cavity has two plane parallel mirrors with reflectivities R_1 and R_2, and the modes are standing waves. These standing waves may be considered as two traveling waves, moving in opposite directions. If R_1 and R_2 have high reflectivity, the traveling waves will pass forward and backward through the volume filled with the oscillators in the excited state E_2. By stimulated emission, energy will be picked up and the intensity of each mode will increase. This process will come to an end when the increase in energy is set off by the losses. Large distances between the two mirrors of the Fabry–Perot correspond to a high order of longitudinal modes and these high-order modes correspond to a very narrow bandwidth, schematically shown in Figure 7.15. The intensity depending on the length z of the light traveling in the active medium of the Fabry–Perot is related to the radiation density as $I(z) = c'u(\nu, z)$. From Eq. (7.50) and (7.51) we have

$$dI(z)/dz = \varepsilon(\nu)I(\nu, z). \tag{7.54}$$

Integration gives us

$$I(z) = I(0)e^{\varepsilon(\nu)z}. \tag{7.55}$$

We see the exponential increase of the light traversing the active medium; energy is transferred to the modes of the Fabry–Perot.

In order to use some of the energy one has to couple it out of the cavity. This may be achieved by using a mirror with a small hole for one of the two mirrors of the Fabry–Perot. The length of one round trip is $z = 2L$ and the losses are taken into account by the factor $(\alpha(\nu))$ in the exponent of Eq. (7.55). We multiply Eq. (7.55) with the reflectivities R_1R_2 to account for the reflection losses of one round trip and get

$$I(2z) = I(0)e^{[\varepsilon(\nu)-\alpha(\nu)]2L}(R_1R_2). \tag{7.56}$$

The thereshold condition is then obtained from Eq. (7.56) when

$$e^{[\varepsilon(\nu)-\alpha(\nu)]2L}R_1R_2 = 1 \tag{7.57}$$

or $[\varepsilon(\nu) - \alpha(\nu)] = (1/2L)ln(1/R_1R_2)$. The threshold gain is

$$\varepsilon_T(\nu) = \alpha(\nu) + (1/2L)ln(1/R_1R_2) \tag{7.58}$$

We rewrite Eq. (7.58) by calling $\sigma = [N_2 - N_1(g_2/g_1)]$ and insert $\varepsilon_T(\nu)$ in Eq. (7.53) and get

$$\sigma_T = [\alpha(\nu) + (1/2L)ln(1/R_1R_2)](8\pi\nu^2\tau/c'^2)(1/gl(\nu)). \tag{7.59}$$

This is the famous threshold condition by Schawlow and Townes (Kneubühl and Sigrist, 1988, p.38). In FileFig 7.8 we present a calculation with numerical values for the He–Ne laser.

FileFig 7.8 (L8HENES)

Gain calculations for the example of the Ne–He laser. We have used for m in mc^2 the mass of the proton times 20 for Ne. The Lorentzian line width has been replaced by the Doppler line width. The line shape $gd(\nu)$ is approximately 1 divided by $(2\nu_0\sqrt{(2kT/mc^2)\ln 2})$ (see Eq. (7.39).)

L8HENES is only on the CD.

7.5.5 Simplified Example of a Three-Level Laser

We consider an energy schematic of the oscillators shown in Figure 7.14 with an upper state 3, a lower state 2, and the ground state 1. We have for transitions between 3 and 1 induced absorption, induced emission, and spontaneous emission, as discussed in Section 7.5.2. The transition probabilities are called W_{31}, W_{13}, and A_{31}, respectively. A similar description holds for the transitions from 2 to 1. The transitions from 3 to 2 are now special. They are radiationless transitions and their probability is called S_{32}. The occupation number of the oscillators in the states 1, 2, and 3 are called N_1, N_2, N_3 and the total number N_0 is assumed to be a constant.

$$N_0 = N_1 + N_2 + N_3. \tag{7.60}$$

The change in time of the number of oscillators in state 3 is

$$dN_3/dt = W_{13}N_1 - (W_{31} + A_{31} + S_{32})N_3 \tag{7.61}$$

and in state 2

$$dN_2/dt = W_{12}N_1 - (W_{21} + A_{21})N_2 + S_{32}N_3. \tag{7.62}$$

For the steady state, the time derivatives are zero. We assume that $A_{31} \ll W_{31}$, which gives us for the ratio N_2/N_1 of the oscillators

$$N_2/N_1 = \{(W_{13}S_{32})/(W_{13} + S_{32}) + W_{12}\}/(A_{21} + W_{21}). \tag{7.63}$$

From the discussion of blackbody radiation, $W_{13} = W_{31}$ and $W_{12} = W_{21}$, and we assume that the radiationless transition probability S_{32} is much larger than the transition probability W_{13} (the probability to empty, state 3). We then have from Eq. (7.63)

$$N_2/N_1 = \{(W_{13} + W_{12})/(A_{21} + W_{12})\}. \tag{7.64}$$

Using Eqs. (7.60) and (7.63) we may get after some calculations

$$(N_2 - N_1)/N_0 = \{(W_{13} - A_{21})/(W_{13} + A_{21} + 2W_{12})\}. \tag{7.65}$$

The threshold condition for operation is obtained when, in the steady state, there are as many oscillators in state 1 as in state 2, $N_1 = N_2$. From Eq. (7.65) one has $W_{13} = A_{21}$. The minimum power for operation is calculated from the condition that the power corresponding to induced absorption of level 3 is equal to the power corresponding to the spontaneous emission of state 2. In FileFig 7.9 we calculate $P = NA_{21}h\nu$ for a process involving a metastable state and spontaneous emission.

FileFig 7.9 (L9MINPOWS)

Model calculation of minimum power for operation of a laser, $P = NA_{21}h\nu$:

1. for a metastable state, and
2. for spontaneous emission.

L9MINPOWS is only on the CD.

Application 7.9. How would the numbers change if S_{32} must be taken into account?

Take $S_{32} = 10W_{13}$ or $S_{32} = 1000W_{13}$.

7.6 CONFOCAL CAVITY, GAUSSIAN BEAM, AND MODES

7.6.1 Paraxial Wave Equation and Beam Parameters

Laser light travels forwards and backwards between the two mirrors of a resonance cavity. The first Ne–He laser used a Fabry–Perot cavity of a length of one meter between the two plane parallel mirrors and was extremely difficult to align. Later it was found that the confocal cavity was much easier to align, and was also very efficient from the point of view of diffraction losses. In Chapter 1, we discussed several types of resonators from the perspective of geometrical optics, and in Chapter 2 the Fabry–Perot as a resonance interferometer.

In this section we discuss the confocal cavity with radii of curvature $R_m = d$, where d is the distance between the two mirrors placed at $z = \pm d/2$. Here the origin of the coordinate system is at $z = 0$ in the middle between the mirrors. The laser beam in the cavity is approximated by a wave traveling in the z direction and having a bell-shaped profile in the transversal (x, y) direction (Figure 7.16). For the mathematical presentation of these propagating modes, we use Cartesian coordinates and rectangular mirrors. We follow H. Kogelnick and

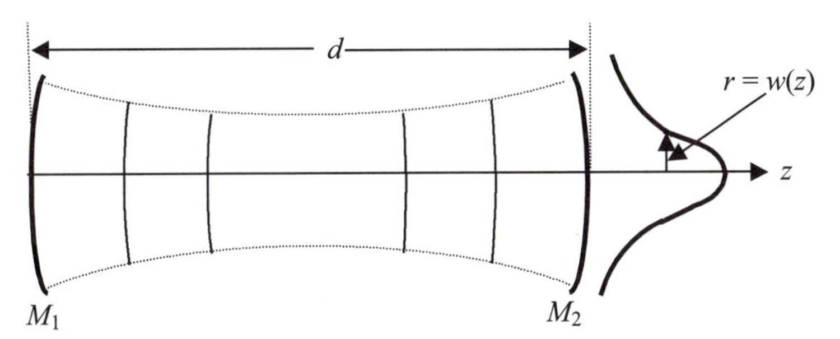

FIGURE 7.16 Waist of a Gaussian beam in a confocal cavity depending on the z coordinate.

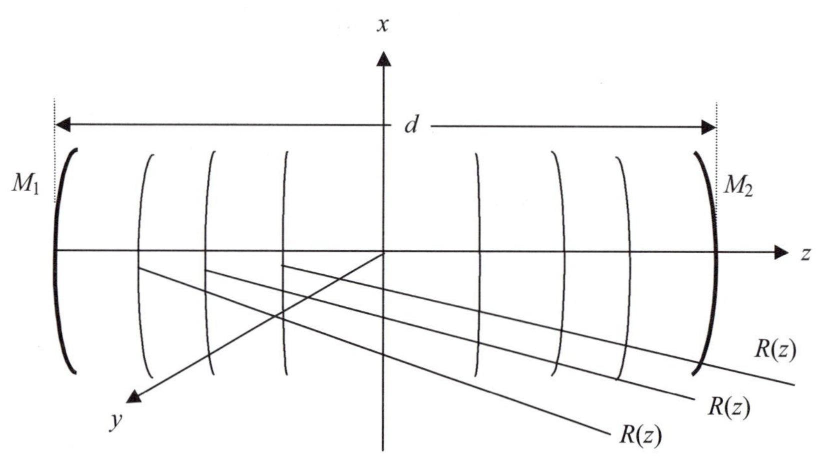

FIGURE 7.17 (a) Coordinates for discussion of the radius of curvature of the wavefront of a mode in a confocal cavity; (b) radius of curvature of the wavefront is indicated for different values of the z coordinate.

T. Li and start with the scalar wave equation

$$\delta^2 u/\delta x^2 + \delta^2 u/\delta y^2 + \delta^2 u/\delta z^2 + k^2 u = 0, \tag{7.66}$$

where $k = 2\pi/\lambda$ and try to find a solution of the wave equation of the form

$$u = \psi(x, y, z)\exp(-ikz). \tag{7.67}$$

Inserting Eq. (7.67) into Eq. (7.66) and assuming for the calculation that the second derivative $\delta^2\psi/\delta z^2$ may be neglected, we further consider the paraxial wave equation

$$\delta^2\psi/\delta x^2 + \delta^2\psi/\delta y^2 - 2ik\delta\psi/\delta z = 0. \tag{7.68}$$

Solutions of Eq. (7.68) describe a Gaussian beam profile in the transverse direction, depending on the distance r from the axis, where $r^2 = x^2 + y^2$ (Figure 7.17). A solution of Eq. (7.68) is

$$\psi = \exp\{-i[P(z) + k(x^2 + y^2)/2q(z)]\} \tag{7.69}$$

with

$$\delta q(z)/\delta z = 1 \tag{7.70}$$

and

$$\delta P(z)/\delta z = -i/q(z), \tag{7.71}$$

where the two functions $q(z)$ and $P(z)$ are not independent of each other. We are mainly interested in the solution for $q(z)$ and can solve for $P(z)$ if we need it, which is a phase factor.

We write the solution of Eq. (7.70) as

$$q(z) = iz_R + z, \tag{7.72}$$

where z_R is a constant. We present the function $q(z)$ as a combination of the radius of curvature of the wavefront $R(z)$ and the beam waist $w(z)^2$ as

$$1/q(z) = 1/(iz_R + z) = 1/R(z) - i\lambda/(\pi w(z)^2). \tag{7.73}$$

From Eqs. (7.72) and 7.73, after separation of real and imaginary parts, we have

$$w(z)^2 = (\lambda/\pi)\{z_R + z^2/z_R\} \tag{7.74}$$

and

$$R(z) = z + z_R^2/z. \tag{7.75}$$

The function $w(z)^2$ is the beam waist, which is the width of the beam depending on z (Figure 7.16). $R(z)$ is the curvature of the wavefront of the beam depending on z (Figure 7.17).

7.6.2 Fundamental Mode in Confocal Cavity

The confocal cavity was discussed in the chapter on geometrical optics. It is a stable cavity with radii of curvature of the mirrors equal to the length d of the cavity. The fundamental mode of the solution of the paraxial wave equation (see Eq. (7.68)), is the same for rectangular mirrors of a cavity for which Cartesian coordinates are used and for circular mirrors for which cylindrical coordinates are used.

We show that the radius of curvature of the wavefront of the Gaussian beam in the confocal cavity matches the curvature of the mirrors at distance $z = d/2$ and $-d/2$, counting z from the middle at 0.

7.6.2.1 Beam Waist

The beam waist is indicated in Figure 7.16. Inserting Eq. (7.73), into Eq (7.69) and taking the real part one has $(\exp -kr^2\lambda/2\pi w(z)^2)$. With $q = k\lambda/2\pi$ and setting $q = 1$ for simplicity we get $(\exp -r^2/w(z)^2)$. This factor decreases in the transversal direction and is 1 for $r = 0$ and $1/e$ for $r = w(z)$. The beam is attenuated from its value at $r = 0$ at the axis to $1/e$ at distance r from the axis.

7.6.2.2 Wavefront of Beam at Center and at Mirror

Wavefront at Center

The wavefront is plain when $R(z) = \infty$. If we choose this value at $z = 0$, one gets from Eq. (7.73)

$$1/q(0) = -i\lambda/(\pi w(0)^2) \tag{7.76}$$

From Eq (7.74) we have for the total waist in the middle $w_0^2 = w(0)^2 = (\lambda/\pi)z_R$.

Wavefront at Mirrors

For $z = z_R$ we have

$$R(z = z_R) = 2z_R \quad \text{and} \quad w(z = z_R)^2 = (\lambda/\pi)^2 z_R = 2w_0^2. \tag{7.77}$$

We choose $2z_R = d$ equal to the distance d between the two mirrors, which is also the radius of curvature of the spherical mirrors in the confocal cavity. The radius of curvature of the wavefront at $z = d/2$ matches the radius of curvature of the mirrors.

Confocal Cavity

For the confocal cavity the radius of curvature of the wavefront, when approaching the mirror, is equal to the radius of curvature of the mirror (see Figure 7.17). The beam waist is $2w_0^2$ at the point $z = d/2$, which is twice as large as at its minimum at $z = 0$. In FileFig 7.10 we have plotted $w(z)$ and $R(z)$ over the range $z = -100$ to 100, which is the distance between the mirrors equal to the length $d = 2z_R = 200$. One observes the beam waist at $z = 0$ and one obtains for the radius of the wavefront at $z = 100$, that is, $R(100)$, the value 200, corresponding to the radius of curvature of the mirror $R_m = d$.

The half angle of the opening of the beam in the far field approximation is the angle θ_{ap} of the asymptote of $w(z)$, which passes through $z = 0$. It is obtained as

$$\lim_{z\to\infty} w(z)/z = w_0/z_R = \lambda/\pi w_0. \tag{7.78}$$

The asymptote $y = zw_0/z_R$ is also indicated in FileFig 7.10.

FileFig 7.10 (L10WRS)

Graphs are shown for the radius of the wavefront $R(z)$, beam waist $w(z)$, and the asymptote $y(z)$.

L10WRS

Radius of Curvature and Beam Waist

1. Radius of curvature
   Beam waist is normalized to 1; that is, we plot $(w(z) = w0SQR(1+(z/zR)^2)$ and set $w0 = .1$ in cm, and $zR = \pi(\omega 0)2/\lambda - .01\pi/\lambda$, λ in cm. Radium of curvature $R(z) = z + (zR)^2/z$

   $$R(z) := \left| z + \frac{zR^2}{z} \right| \qquad zR := 100$$

   $$z := -100, -99.99 \ldots 100.$$

$Rm = 2zR$. At $z = 1/2$ of distance of mirrors, that is, for distance 200 at 100, the radius of curvature must be equal to the distance of the mirrors.

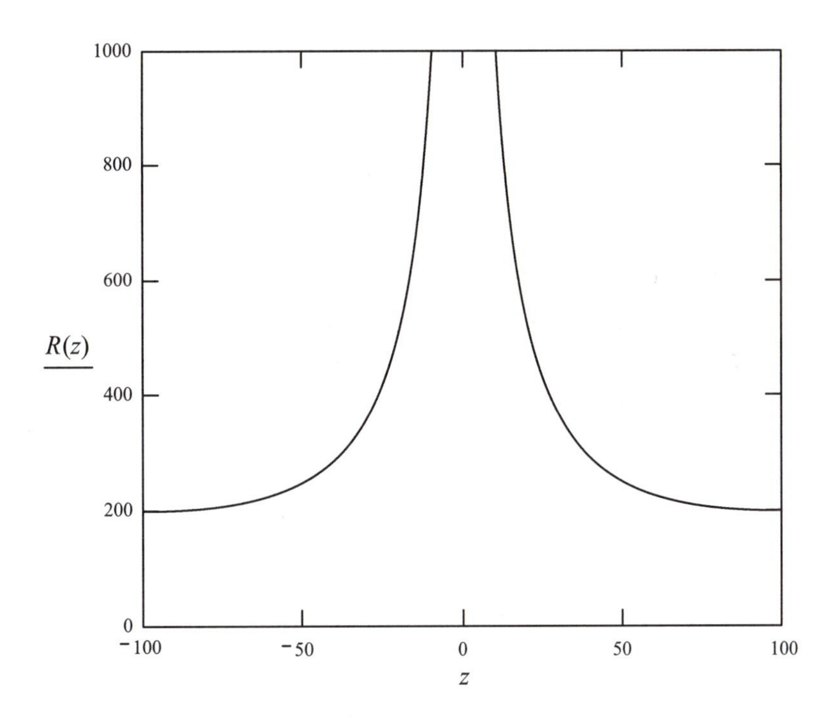

2. **Beam waist**

   Plots of two branches of the beam waist and the asymptote to $w(z)$; that is, $y = z/zR$.

   If z is in cm, we have set for $w0 = .1$, $\lambda = 3.14*0.01/zR$ in cm (about 3 microns for $zR = 100$).

$$w(z) := .1 \cdot \sqrt{1 + (\frac{z}{zR})^2}$$

and for the asymptote $yy(z) := -y(z)$,

$$ww(z) := -.1 \cdot \sqrt{1 + \left(\frac{z}{zR}\right)^2}.$$

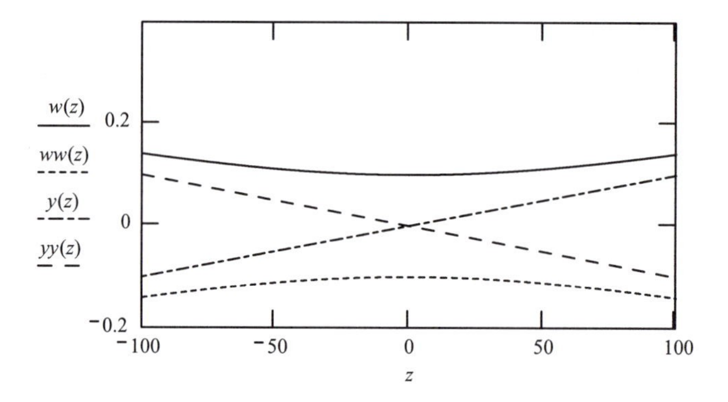

Application 7.10.

1. Repeat the calculations for $\lambda = 10$ microns, $z_R = 100$ cm.
2. Repeat the calculations for $\lambda = 3$ microns, $z_R = 160$ cm.
3. Repeat the calculations for $\lambda = 10$ microns, $z_R = 160$ cm.

7.6.3 Diffraction Losses and Fresnel Number

The diffraction losses of a mirror of diameter $2a$ at distance d are characterized by the Fresnel number F. We assume parallel light incident on an obstacle of diameter $2a$ and have for the diffraction angle $\theta = \lambda/2a$ (see Figure 7.18). The geometrical shadow AG at distance d is $4a^2\pi$ and the total area AT, enlarged by diffraction, is $4(a + \theta d)^2\pi$. The Fresnel number is defined as $F = AG/(AT - AG)$, which is

$$\begin{aligned} F &= 4a^2\pi/\{(4a^2 + 8ad(\lambda/2a) + \ldots)\pi - 4a^2\pi\} \\ &= 4a^2\pi/(8ad\lambda\pi/2a) = a^2/d\lambda, \end{aligned} \tag{7.79}$$

where a term in θ^2 was neglected in the denominator.

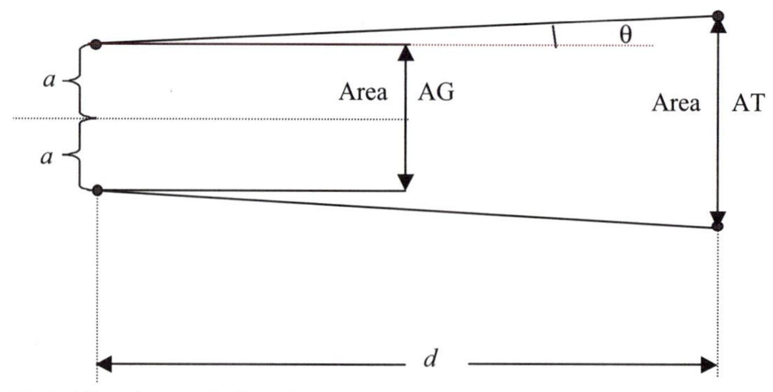

FIGURE 7.18 Diffraction angle for mirror. The geometrical shadow has the area AG. The total area of the light is AT. The Fresnel number is defined as $F = (AG)/(AT - AG)$.

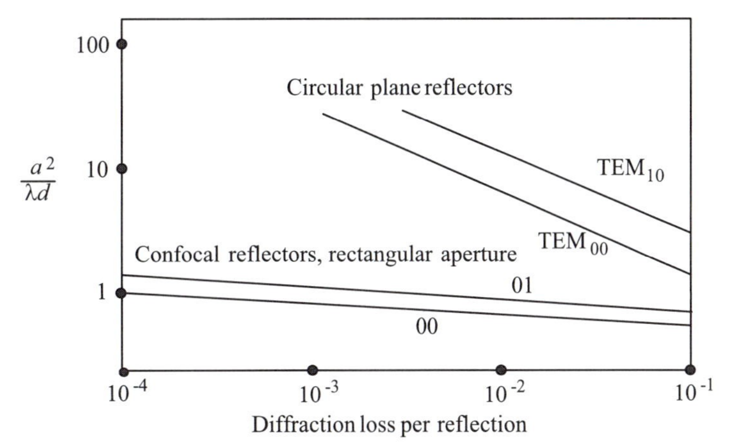

FIGURE 7.19 Fresnel numbers for circular plain reflectors and confocal reflectors. The confocal cavity is about two orders of magnitude better than a Fabry–Perot cavity.

A confocal resonator has considerably fewer losses for Fresnel numbers larger than 1 compared to Fabry–Perot resonators. For example, at $F = 1$ the confocal resonator does more than 300 times better than the Fabry–Perot (Figure 7.19).

7.6.4 Higher Modes in the Confocal Cavity

We have discussed in Section 7.2 the fundamental mode in a confocal cavity. The fundamental mode is the same for a cavity using rectangular mirrors or spherical mirrors. The higher modes need for their description Cartesian coordinates for rectangular mirrors and cylindical coordinates for round mirrors. In both cases we have TEM modes characterized by three mode numbers, the first two for the transversal modes and the last for the longitudinal modes.

7.6.4.1 Confocal Cavity and Rectangular-Shaped Mirrors (Cartesian Coordinates)

For the higher modes of cavities with rectangular-shaped mirrors, using Cartesian coordinates (x, y, z), the solution of the paraxial wave equation (see Eq. (7.68)), is

$$\psi = H_m(\sqrt{2}x/w) \cdot H_n(\sqrt{2}y/w) \cdot \exp\{-i(P(z) + k(x^2 + y^2)/2q(z))\}, \quad (7.80)$$

where $H_m(\sqrt{2}x/w)$ and $H_n(\sqrt{2}y/w)$ are Hermitian polynomials, each a solution of a differential equation written in the variable u as

$$d^2 H_m/\delta u^2 - 2u \cdot dH_m/du + 2mH_m = 0. \quad (7.81)$$

The indices m and n are the transverse mode numbers. The Hermitian polynomials for m equals 0 to 3 are

$$H_0(u) = 1$$
$$H_1(u) = u \quad (7.82)$$

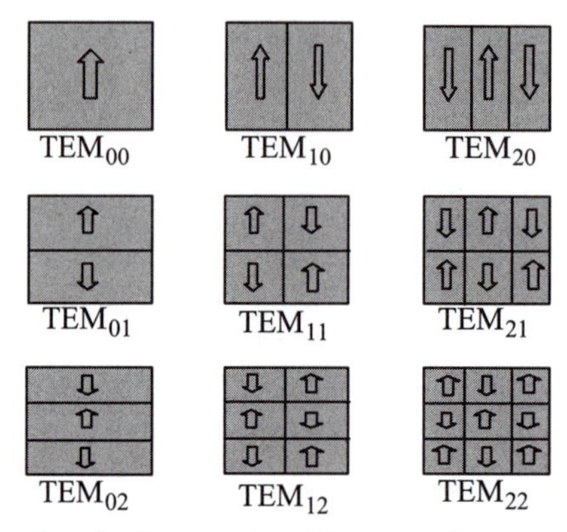

FIGURE 7.20 Schematic of modes for a cavity with rectangular mirrors.

$$H_2(u) = 4u^2 - 2$$
$$H_3(u) = 8u^3 - 12u.$$

The fundamental mode, discussed above, has the indices $m = 0$ and $n = 0$.

The higher modes are labeled by $\mathrm{TEM}_{\mathrm{mnq}}$, where n and m refer to the Hermitian polynomials (Eqs. (7.81) and (7.82)). They are transversal mode numbers and correspond to the number of vertical or horizontal zero-intensity lines in the transversal pattern. A schematic representation is given in Figure 7.20. The longitudinal mode is characterized by the large number q. Since the phase has to be the same after one round trip, one obtains the following resonance condition,

$$2L/\lambda_{nmq} = 1/2(m + n + 1) + q. \tag{7.83}$$

There are degenerate modes if $m + n + 2q = m^* + n^* + 2q^*$. For the mode separation one has $v_{00\ q+1} - \nu_{00q} = c'/2L$, where c' is the phase velocity in the medium of the cavity and is independent of the Fresnel number.

In FileFig 7.11 we show surface graphs of the higher-order modes in the transversal direction from (0,0) to (2,2).

FileFig 7.11 (L11MOCONFCS)

Modes for confocal cavity and rectangular mirrors. Surface plots of Hermitian polynomials with exponential factor for indices 0, 1, and 2, and scaling factors X and Y.

L11MOCONFCS

Cartesian Coordinates for Rectangular Mirrors in the Confocal Resonator

Field distribution as contour plot. The mode numbers m and n are for hermitian polynomials. The constant in the exponential is simulated by X. Small X

corresponds to small waist width.

$$N := 40 \qquad i := 0 \ldots N \qquad j := 0 \ldots N$$
$$x_i := (-20) + 1.00 \cdot i \qquad y_j := -20 + 1.00 \cdot j$$
$$H0(x) := 1 \quad H0(y) := 1 \quad H1(x) := x \cdot \sqrt{\frac{2}{Y}} \quad H1(y) := y \cdot \sqrt{\frac{2}{Y}}$$
$$H2(y) := 4 \cdot \left(\sqrt{\frac{2}{Y}}\right)^2 - 2 \qquad H2(x) := 4 \cdot (x)^2 - 2$$
$$H00(x, y) := H0(x) \cdot H0(y)$$
$$H20(x, y) := H2(x) \cdot H0(y)$$
$$H02(x, y) := H0(x) \cdot H2(y)$$
$$H01(x, y) := H0(x) \cdot H1(y)$$
$$H11(x, y) := H1(x) \cdot H1(y)$$
$$H10(x, y) := H1(x) \cdot H0(y)$$
$$H21(x, y) := H2(x) \cdot H1(y)$$
$$H12(x, y) := H1(x) \cdot H2(y)$$
$$R(x, y) := (x)^2 + ((y))^2.$$

Constant X:

$$X \equiv 100 \qquad Y \equiv 100 \qquad g(x, y) := \left(e^{\frac{-R(x,y)}{X}}\right).$$

$$M00_{i,j} := \left(g(x_i, y_j) \cdot H00(x_i, y_j)\right)^2 \qquad M10_{i,j} := \left(g(x_i, y_j) \cdot H10(x_i, y_j)\right)^2$$

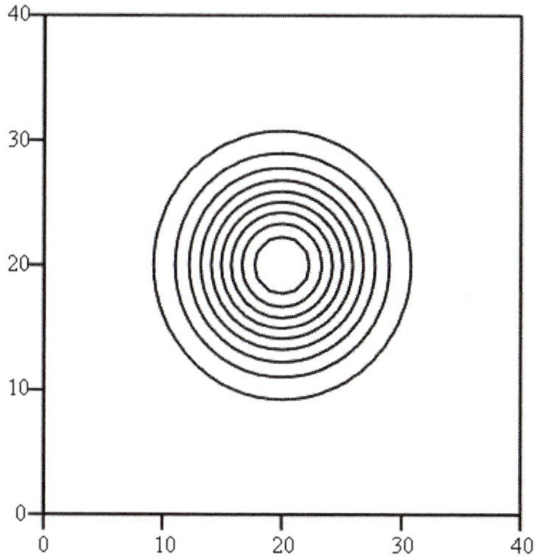

M00

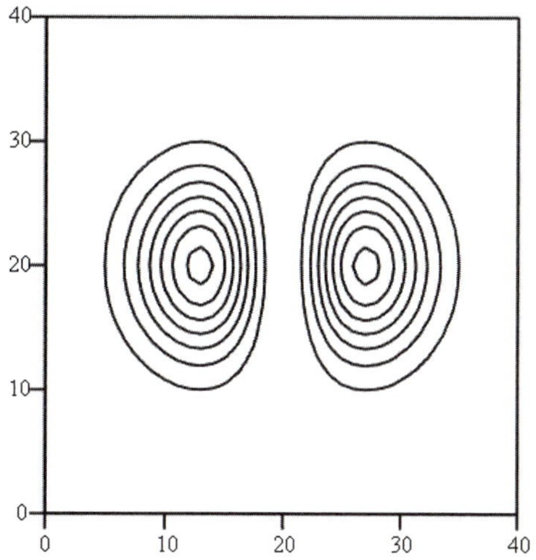

M10

$M01_{i,j} := \left(g(x_i, y_j) \cdot H01(x_i, y_j)\right)^2$

$M11_{i,j} := \left(g(x_i, y_j) \cdot H11(x_i, y_j)\right)^2$

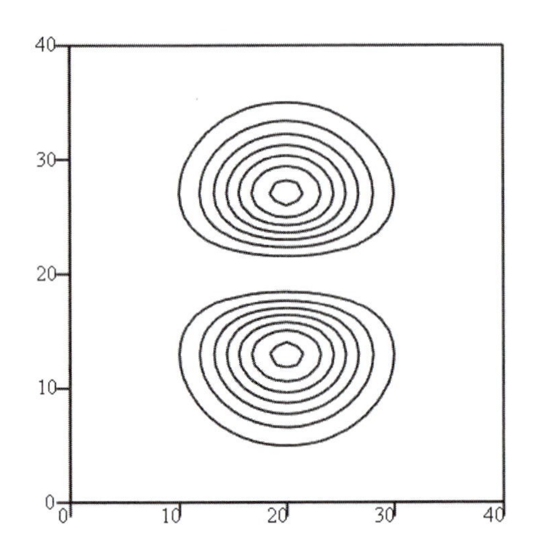

M01

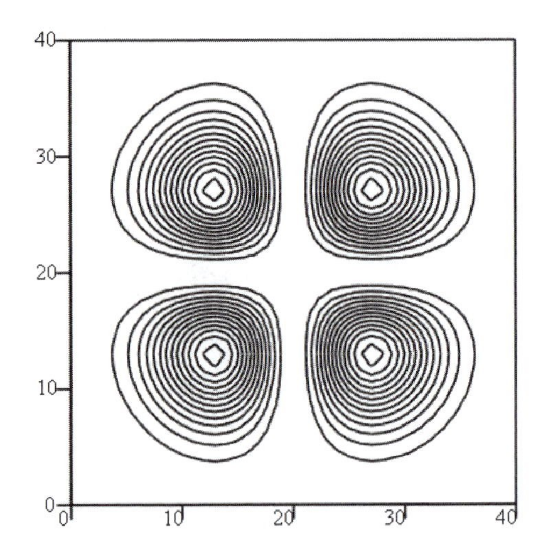

M11

$M20_{i,j} := \left(g(x_i, y_j) \cdot H20(x_i, y_j)\right)^2$

$M21_{i,j} := \left(g(x_i, y_j) \cdot H21(x_i, y_j)\right)^2$

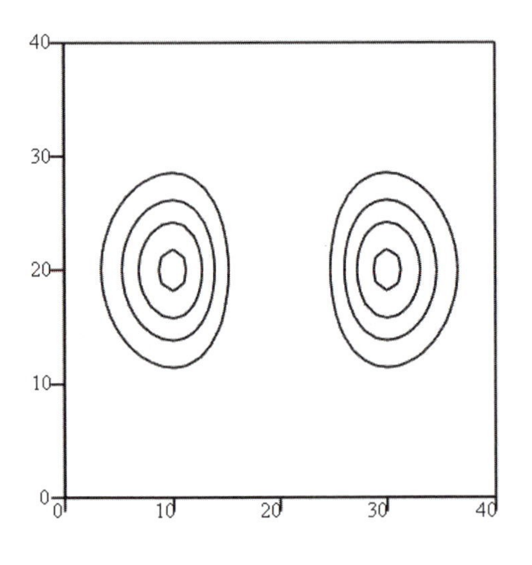

M20

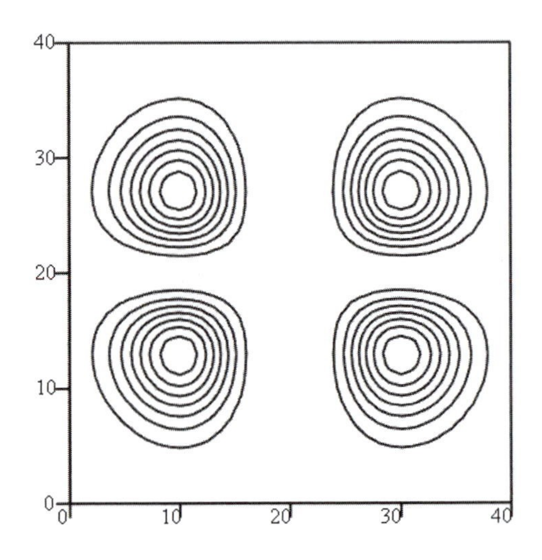

M21

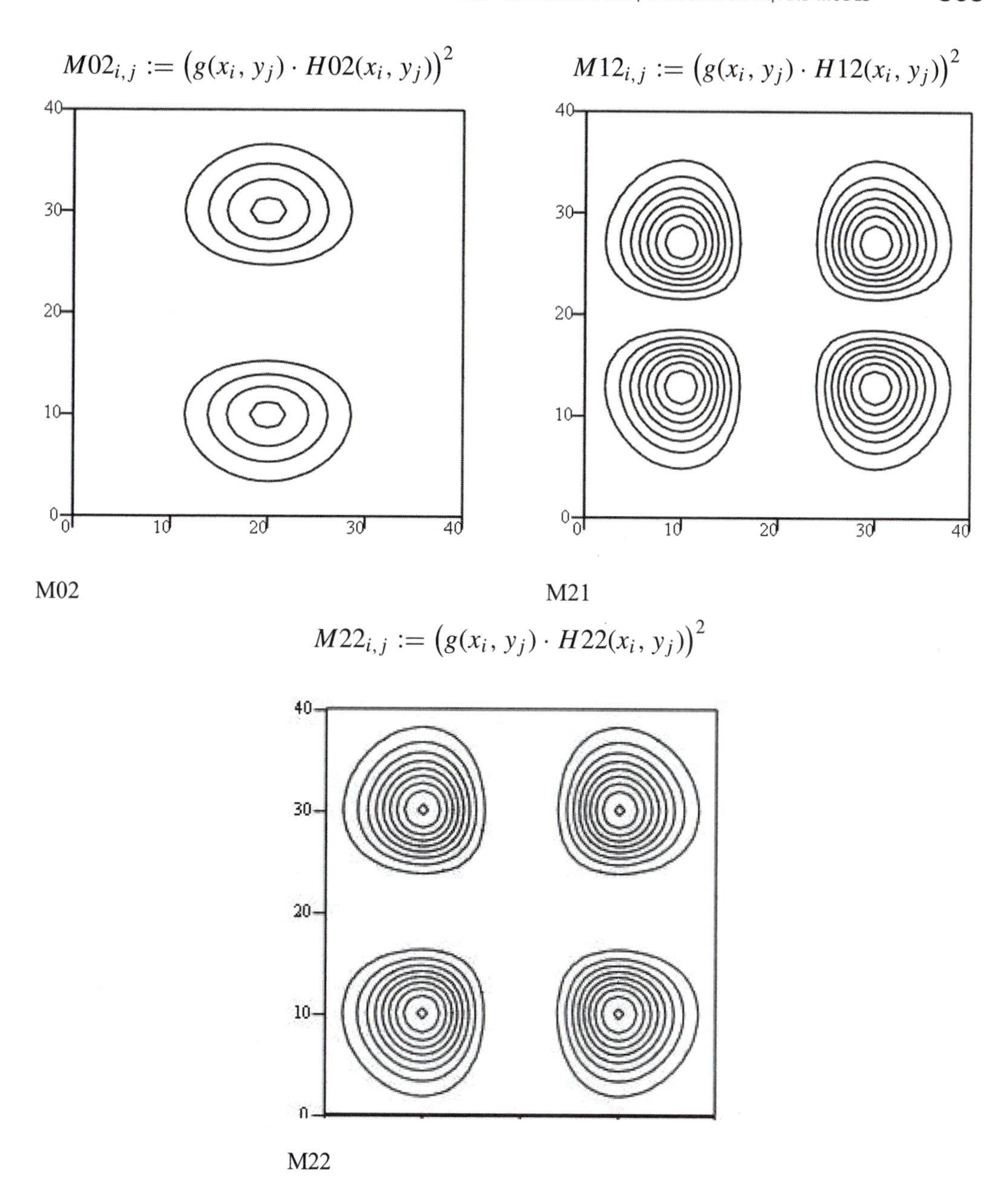

M02

M21

M22

Application 7.11.

1. Compare with Figure 7.20 and the number of zero-intensity lines with the mode numbers.
2. Extend the modes using $H3(x)$ and $H3(y)$ and complete all graphs up to indices 0, 1, 2, and 3.
3. Convert surface to contour plots.

7.6.4.2 Confocal Cavity and Circular Mirrors (Cylindrical Coordinates)

For circular mirrors, one uses cylindrical coordinates (r, ϕ, z). After rewriting the wave equation in these coordinates, one has the solutions

$$\psi = (\sqrt{2}r/w)^l \cdot L^l_p(2r^2/w^2) \cdot \exp\{-i(P(z) + kr^2/2q(z))\}, \tag{7.84}$$

where $Lp^l(2r^2/w^2)$ is a generalized Laguerre polynomial. The $L^l_p(2r^2/w^2)$ are solutions of the differential equation written in the variable u,

$$ud^2L^l_p/du^2 + (l + 1 - u)dL^l_p/du + pL^l_p = 0. \tag{7.85}$$

The index p is the radial mode number and l the angular mode number. For $p = 0$ to 2 we have for the polynomials

$$L^l_0(u) = 1 \tag{7.86}$$

$$L^l_1(u) = l + 1 - u$$

$$L^l_2(u) = (1/2)(l + 1)(l + 2) - (l + 2)u + (1/2)(u^2). \tag{7.87}$$

The fundamental mode has the numbers $l = 0$ and $p = 0$. The modes in confocal cavities with round mirrors are labeled by TEM_{lpq}, where l is the angular mode number corresponding to the number of angular zero-intensity lines in the transversal pattern (see the schematic presentation in Figure 7.21). The radial mode number correspond to the number of zero-intensity rings. The longitudinal mode is characterized by the large number q. The resonance condition, which is the condition that the phase is the same after one round trip, is

$$2L/(\lambda_{lpq} = (1/2)(l + 2p + 1) + q. \tag{7.88}$$

There are degenerate modes if $l+2p+2q = l^*+2p^*+2q^*$. The mode separation is again independent of the Fresnel number and one has again $\nu_{00\ q+1} - \nu_{00q} = c'/2L$.

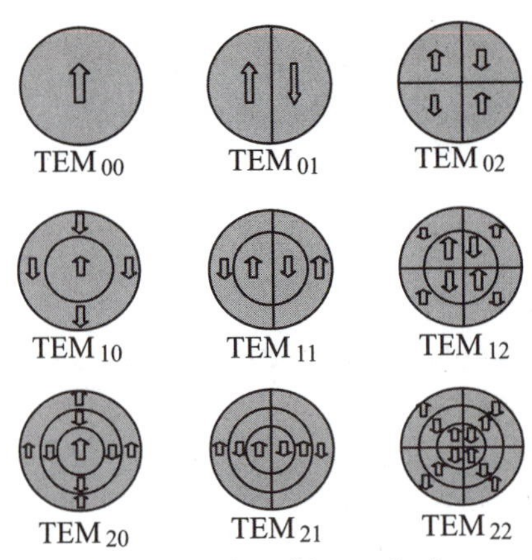

FIGURE 7.21 Schematic of modes for a cavity with round mirrors.

In FileFig 7.12a we show graphs of the first four modes from (0,0) to (1,1). In FileFig 7.12b we show graphs of the next five modes from (1,2) to (2,2). Contour plots were chosen for better reference to Figure 7.21, and identification of the zero-intensity lines and mode numbers.

FileFig 7.12 (L12MOCY1to4S)(L12MOCY5to9S)

Modes for confocal cavity and circular mirrors. Contour plots of generalized Laguerre polynomial for indices l and p equal 0, 1, and 2. Scaling factor is X.

L12MOCY1to4S

Cylindrical Coordinates for Circular Mirrors in Confocal Resonator

Field distribution as contour plot for graph 00, 10, 01, and 11. The $L(l, p)$ functions are written out for 00 to 22. The constant in the exponential is X:

$$i := 0 \ldots N \qquad j := 0 \ldots N \qquad N = 40$$

$$x_i := (-2) + .10001 \cdot i \qquad y_j := -2 + .10001 \cdot j$$

$$R(x, y) := (x)^2 + (y)^2 \qquad \beta(x, y) := \left(a \tan\left(\frac{x}{y}\right)\right)^2 \qquad q(x, y) := \left[e^{\frac{-(R(x,y))}{x}}\right].$$

Constant $X : X \equiv 3$. The L's are given below.

$$u(x, y) := 4 \cdot \frac{R(x, y)}{X} \qquad g(x, y) := \cos(0 \cdot \beta(x, y))$$

$$L00(x, y) := 1 \qquad L01(x, y) := 1 - u(x, y)$$

$$L10(x, y) := 1 \qquad L11(x, y) := 2 - u(x, y)$$

$$M00_{i,j} := (\cos(0 \cdot \beta(x_i, y_j)) \cdot q(x_i, y_j) \cdot L00(x_i, y_j))^2$$

$$M10_{i,j} := (\cos(0 \cdot \beta(x_i, y_j)) \cdot q(x_i, y_j) \cdot L01(x_i, y_j))^2.$$

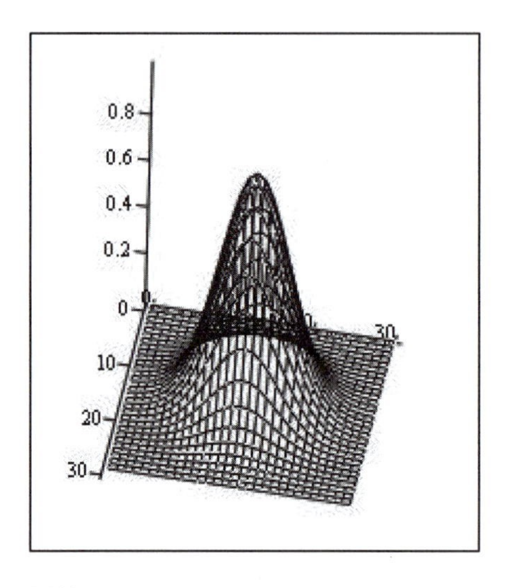

M00

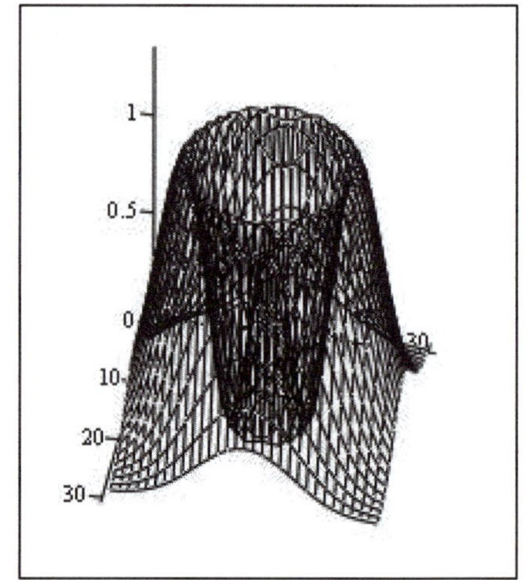

M10

$$M01_{i,j} := (\cos(1 \cdot \beta(x_i, y_j)) \cdot q(x_i, y_j) \cdot L10(x_i, y_j))^2$$
$$M11_{i,j} := (\cos(1 \cdot \beta(x_i, y_j)) \cdot q(x_i, y_j) \cdot L11(x_i, y_j))^2.$$

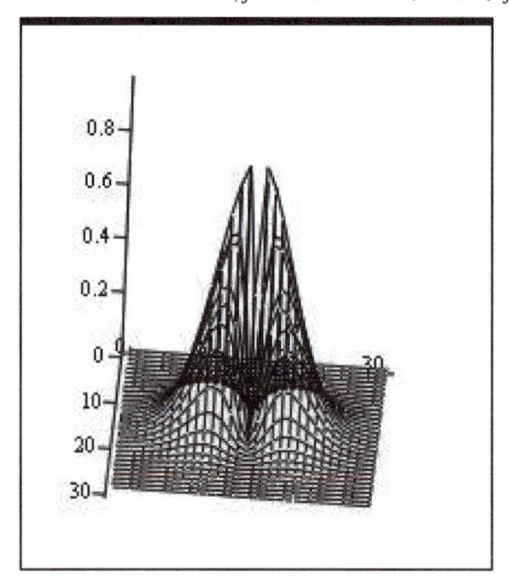

M01

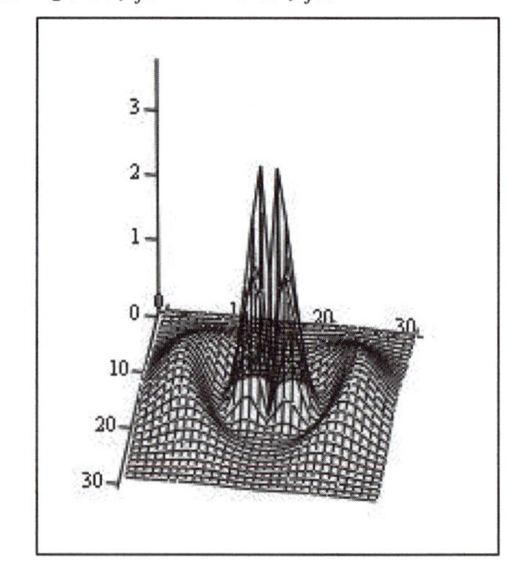

M11

L12MOCY5to9S

Cylindrical Coordinates for Circular Mirrors in Confocal Resonator

Field distribution as contour plot for graph 02 to 20. The $L(l, p)$ functions are written out for 00 to 22. The constant in the exponential is X:

$$i := 0 \ldots N \qquad j := 0 \ldots N \qquad N = 40$$

$$x_i := (-2) + .10001 \cdot i \qquad y_j := -2 + .10001 \cdot j$$

$$R(x, y) := (x)^2 + (y)^2$$

$$\beta(x, y) := \left(a \tan\left(\frac{x}{y}\right)\right)^2$$

$$q(x, y) := \left[e^{\frac{-(R(x,y))}{x}}\right]$$

Constant X : $X \equiv 2$. There h stands for l and p runs from 0 to 2.

$$Lh2(x, y) := [1/2(h+1)(h+2) - (h+2)u(x, y)] + (1/2)u(x, y)^2$$

$$u(x, Y) := 4 \cdot \frac{R(x, y)}{X} \qquad g(x, y) := \cos(0 \cdot \beta(x, y))$$

$$L02(x, y) := 1 - 2 \cdot u(x, y) + \frac{1}{2} \cdot u(x, y)^2$$

$$L22_{(}x, y) := 6 - 4 \cdot u(x, y) + \frac{1}{2} \cdot u(x, y)^2$$

$$L12(x, y) := 3 - 3 \cdot u(x, y) + \frac{1}{2} \cdot u(x, y)^2$$

$$L21(x, y) := 3 - u(x, y)$$

$$L20(x, y) := 1.$$

$$M02_{i,j} := (\cos(2 \cdot \beta(x_i, y_j)) \cdot q(x_i, y_j) \cdot L00(x_i, y_j))^2$$

$$M20_{i,j} := (\cos(0 \cdot \beta(x_i, y_j)) \cdot q(x_i, y_j) \cdot L02(x_i, y_j))^2.$$

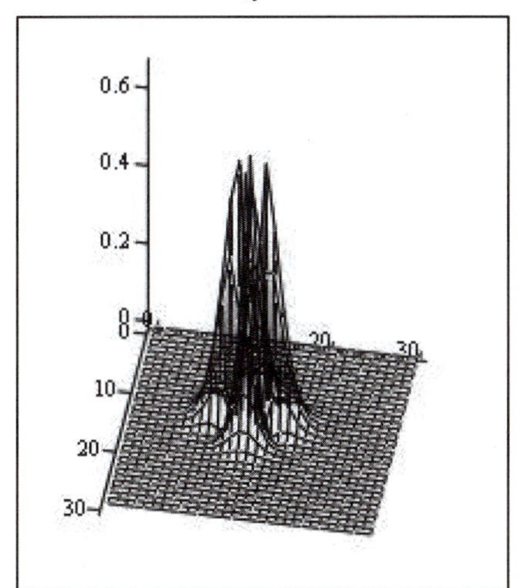

M02

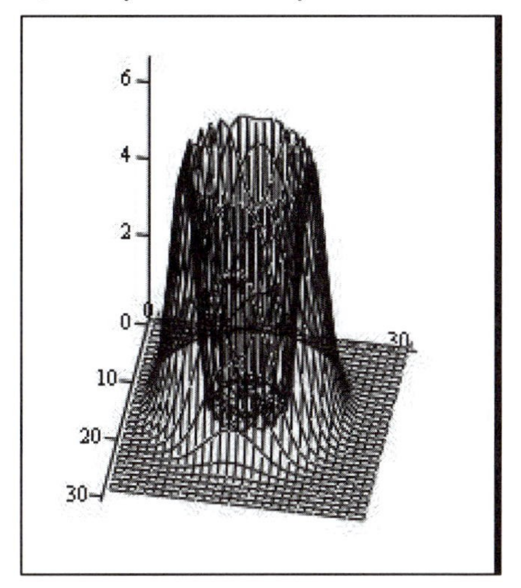

M20

$$M12_{i,j} := (\cos(2 \cdot \beta(x_i, y_j)) \cdot q(x_i, y_j) \cdot L21(x_i, y_j))^2$$

$$M21_{i,j} := (\cos(1 \cdot \beta(x_i, y_j)) \cdot q(x_i, y_j) \cdot L12(x_i, y_j))^2.$$

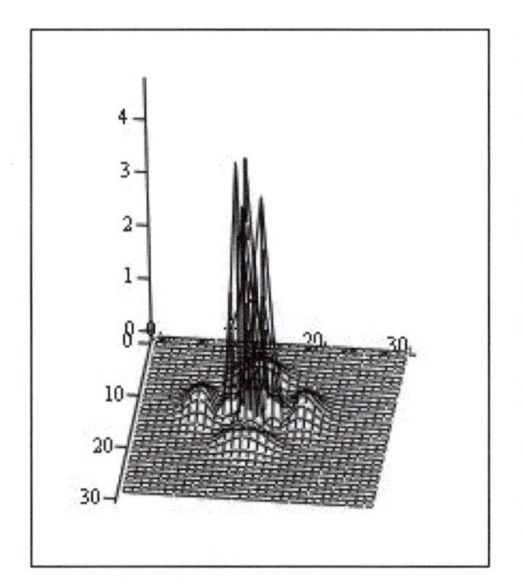

M12

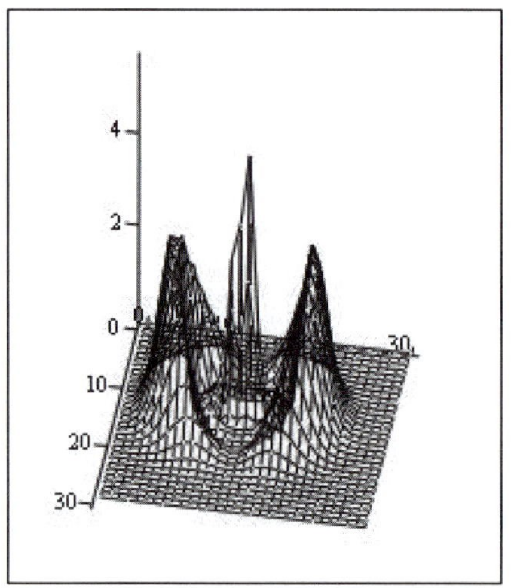

M21

$$M22_{i,j} := (\cos(2 \cdot \beta(x_i, y_j)) \cdot q(x_i, y_j) \cdot L22(x_i, y_j))^2.$$

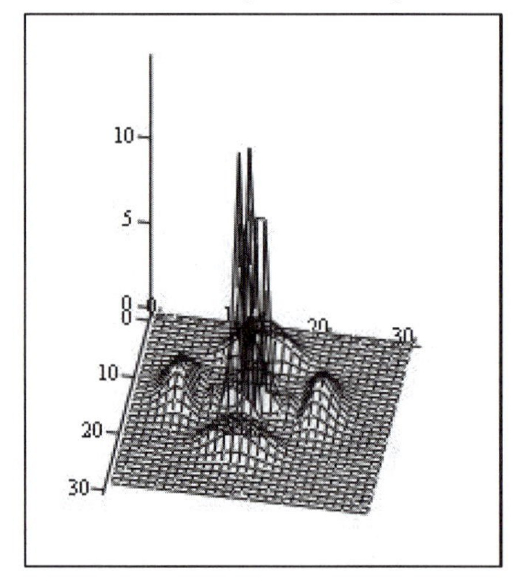

M22

Application 7.12.

1. Compare with Figure 7.21 and the number of zero intensity lines with the mode numbers.
2. Convert surface to contour plots and try to identify the modes.

C H A P T E R

Optical Constants

8.1 INTRODUCTION

In the chapters on geometrical optics, interference, and electromagnetic theory we have sometimes used the refractive index $n = c/v$. When light enters a dielectric medium, it interacts with the atoms and changes its speed from c in vacuum to v in the medium. The medium is called *isotropic* when the speed of light is the same in all directions. The refractive index may be obtained from the real part of the dielectric constant in Maxwell's equations. In the case where there are losses in the medium, light will be absorbed, and one uses the complex dielectric constant in Maxwell's equations.

In this chapter we first look at the dielectric constants in Maxwell's equations and then use a simple model for the analytical representation. As a result we get the index of refraction depending on the frequency of the light and model parameters. The model we use is a damped forced oscillator. The incident light drives these oscillators, representing the material, and loses some of its intensity. The losses of the electromagnetic wave are described by a complex refractive index $(n - iK)$. In books on solid state physics the complex refractive index is often called $n + i\kappa$ (and not $n - iK$). We have to relate n and K or κ, called *optical constants*, to the parameters of our model.

8.2 OPTICAL CONSTANTS OF DIELECTRICS

8.2.1 The Wave Equation, Electrical Polarizability, and Refractive Index

We write Maxwell's equations for an isotropic and nonmagnetic material without free charges, which means we assume $\nabla \cdot \mathbf{P} = 0$ and $\rho = 0$ and have

$$\begin{aligned} \nabla \times \mathbf{E} &= -\partial \mathbf{B}/\partial t \\ c^2 \nabla \times \mathbf{B} &= \partial \mathbf{E}/\partial t + \mathbf{j}/\varepsilon_0 \\ \nabla \cdot \mathbf{E} &= \mathbf{0} \\ \nabla \cdot \mathbf{B} &= \mathbf{0}, \end{aligned} \tag{8.1}$$

where $\mathbf{E}$ is the electrical field vector, $\mathbf{B}$ the magnetic field vector, $\mathbf{j}$ the current density vector ρ the charge density, and $\varepsilon_0 = 8.854 \times 10^{-12}$ F/m, the permittivity of vacuum. We now study the effect of an outside electrical field on the bound charges in an isotropic material. The outside electrical field is assumed to be a harmonic wave and will act on the bound charges and make them vibrate. For the current density N of such vibrating charges we have

$$\mathbf{j} = N e \mathbf{v}, \tag{8.2}$$

where N is the number of charges per unit volume, e the charge of the electron, and v the velocity vector. Since we assume an isotropic medium, we only have to take into account one direction of vibration, which we call y, and one direction of motion, which we call x. With $v = dx/dt$ we have

$$j_y = Ne\, dx/dt. \tag{8.3}$$

The induced dipoles are ex, and $P = N(ex)$ is called the polarization. The current density is then

$$j_y = dP_y/dt. \tag{8.4}$$

The wave equation for vibration in the y direction and propagation in the x direction is

$$\partial^2 E_y/\partial x^2 - (1/c^2)\partial^2 E_y/\partial t^2 = [1/(\varepsilon_0 c^2)]\partial^2 P_y/\partial t^2, \tag{8.5}$$

where the right side of Eq. (8.5) is called the *source* term.

In the first approximation, we set P_y proportional to the incident electrical field and write

$$P_y = \varepsilon_0 N \alpha E_y, \tag{8.6}$$

where α is the atomic polarizability, a constant characteristic for the material. We introduce into the wave equation a trial solution $E_y = A\cos(kx - \omega t)$, and get

$$-k^2 E_y + (1/c^2)\omega^2 E_y = (1/\varepsilon_0 c^2)(-\omega^2 \alpha E_y) \tag{8.7}$$

or

$$k^2 = (\omega^2/c^2)(1 + N\alpha). \tag{8.8}$$

We associate the velocity v with the the phase velocity ω/k in the medium and obtain for n

$$n^2 = c^2/v^2 = c^2k^2/\omega^2 = (1 + N\alpha). \tag{8.9}$$

We have obtained a relation between the optical constant n and the material constant α, the atomic polarizability.

8.2.2 Oscillator Model and the Wave Equation

8.2.2.1 Less Dense Medium

To study the dependence of the refractive index on frequencies and losses, we relate the refractive index to the parameters of an oscillator model.

The polarization vector P of the medium is defined as the number of electrical dipoles per unit volume. The induced electrical dipole moment is eE_y, which we now call (eu), where u is the displacement of an electron in an atom. The number of dipoles per unit volume is N and we consider only one component of vibration y and have

$$P = Neu. \tag{8.10}$$

We describe the displacement u of the charges by the vibrations of a damped oscillator,

$$md^2u/dt^2 + m\gamma du/dt + m\omega_0^2 u = 0, \tag{8.11}$$

where u is the displacement of the charge from its equilibrium position, m is the mass, f the force, and γ the damping constant. The frequency without the damping term is $\omega_0'2 = f/m$, and the resonance frequency for the damped oscillator is $\omega_0 \neq \omega_0'$. The electromagnetic wave of the light drives these oscillators. The forced damped oscillator equation is

$$md^2u/dt2 + m\gamma du/dt + m\omega_o^2 u = eE_o e^{-i\omega t}. \tag{8.12}$$

We introduce the trial solution $Ae^{-i\omega t}$ and obtain

$$u(t) = [eE(t)/m]/[(\omega_o^2 - \omega^2) - i\gamma\omega], \tag{8.13}$$

where $E(t) = E_0 e^{-i\omega t}$. The driving electromagnetic wave produces polarization

$$P(t) = Neu(t) = [Ne^2\varepsilon_o E(t)]/[m\varepsilon_o(\omega_o^2 - \omega^2 - i\gamma\omega)] = \chi^*\varepsilon_0 E(t), \tag{8.14}$$

where ε_0 is the permittivity of free space. As a result of the imaginary damping term in $u(t)$ one has a complex susceptibility χ, indicated by a star. In Eq. (8.14) we have related the polarization $P(t)$ and the electrical susceptibility χ to the parameters of our model and the electrical field $E(t)$ of the light.

From the wave equation, we have another expression of $P(t)$ (see Eq. (8.6)) and introducing $P(t) = \varepsilon_0 N\alpha E(t)$ into Eq. (8.14) we get

$$\alpha = (1/N)\omega_p^2/(\omega_0^2 - \omega^2 - i\gamma\omega), \tag{8.15}$$

where α is the *atomic polarizability* and $\omega_p^2 = Ne^2/m\varepsilon_0$ the *plasma frequency*. One should not confuse this with ω_0^2, the square of the angular frequency of the oscillator model, representing the dielectric. We have obtained a relation between the material constant of the atomic polarizability α and the parameters of our oscillator model. For the square of the refractive index, (see eq. (8.9)) we get

$$(n^*)^2 = 1 + \omega_p^2/(\omega_0^2 - \omega^2 - i\gamma\omega). \tag{8.16}$$

Since $(n^*)^2$ is a complex number, we marked it with a star. It is customary to call the real part of the complex refractive index n, and the imaginary part K. The imaginary part K may be called the *extinction index*. We have $n^* = n - iK$ and $(n^*)^2$ is

$$(n - iK)^2 = 1 + \omega_p^2/((\omega_0^2 - \omega^2) - i\gamma\omega). \tag{8.17}$$

For the real and imaginary parts we obtain

$$n^2 - K^2 = \varepsilon' = 1 + \omega_p^2(\omega_0^2 - \omega^2)/((\omega_0^2 - \omega^2)^2 + (\gamma\omega)^2) \tag{8.18}$$

and

$$-2nK = \varepsilon'' = \omega_p^2\gamma\omega/((\omega_0^2 - \omega^2)^2 + (\gamma\omega)^2), \tag{8.19}$$

where ε' is the real part of the dielectric constant and ε'' is the imaginary part. For optically thin media, such as gases, one has n close to 1 and K is small and has the approximation

$$n = 1 + (\omega_p^2/2)(\omega_0^2 - \omega^2)/((\omega_0^2 - \omega^2)^2 + (\gamma\omega)^2) \tag{8.20}$$

and

$$K = -(\omega_p^2/2)\gamma\omega/((\omega_0^2 - \omega^2)^2 + (\gamma\omega)^2). \tag{8.21}$$

The damping term of the damped oscillator equation appears in the imaginary part of n^*, indicating the losses present in the description of our model. When the damping term is zero we get

$$n^2 = 1 + (\omega_p^2)/(\omega_0^2 - \omega^2) \tag{8.22}$$

or

$$n(\omega) = \sqrt{(1 + [(\omega_p^2)/(\omega_0^2 - \omega^2)]}, \tag{8.23}$$

where ω_0 is the resonance frequency of the case without losses. The refractive index depends on the frequency and the model parameters. In Figure 8.1a the dependence of n on the frequency is shown schematically. When $\gamma = 0$ we

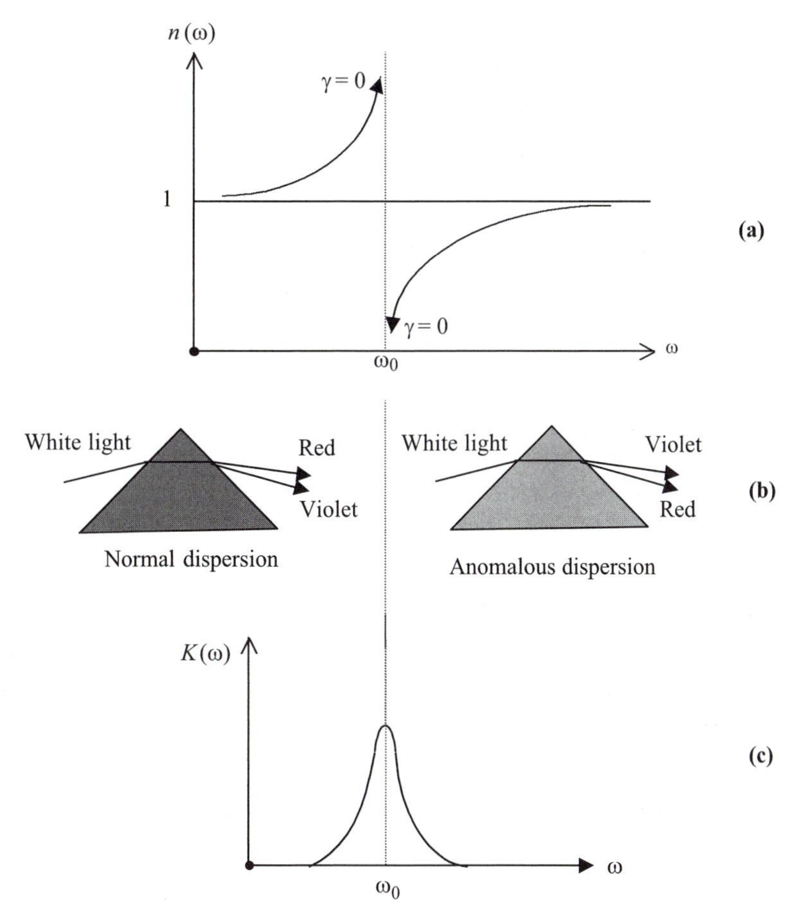

FIGURE 8.1 (a) Dependence of n on frequency. For no damping we have a singularity; (b) normal and anomalous dispersion; (c) dependence of K on frequency. The maximum is not at infinity if $\gamma \neq 0$.

have singularities at the resonance frequency, and for $\gamma \neq 0$ these singularities are avoided. On the left side of Figure 8.1a the refractive index increases with higher frequencies. This region is called normal dispersion, shown on a prism on the left in Figure 8.1b. The reverse case is on the right side of Figure 8.1a, called anomalous dispersion and shown on a prism on the right in Figure 8.1b. In Figure 8.1c we show an absorption curve, which is the dependence of K on the frequency.

8.2.2.2 Dense Medium

So far, in the preceding discussion, it was assumed that the local field E_0 at the site of the oscillators is the same as the applied field. This is only true if the density of the oscillators in the medium is low, as it would be for a gas. For a dense distribution of the oscillators in a solid, the surrounding area is also electrically polarized and has an effect on the oscillator under consideration. The local field at the oscillator must be corrected. This is called the Lorentz correction and the effective field at the site of the charges is

$$E + (1/3\varepsilon_0)P. \tag{8.24}$$

Using $P = \varepsilon_0 N\alpha E$ we have

$$P_y = \varepsilon_0 N\alpha\{E_y + (P_y/3\varepsilon_0)\}. \tag{8.25}$$

Similar to the low density case, using Eq. (8.9), the square of the refractive index for the denser medium is

$$n^2 = 1 + N\alpha/(1 - (N\alpha/3). \tag{8.26}$$

This equation may also be written as

$$3(n^2 - 1)/(n^2 + 2) = N\alpha \tag{8.27}$$

and is called the Clausius–Mossotti equation, or with the parameters of our model,

$$n^2 = 1 + 3\varepsilon_0\omega_p^2/\{3(\omega_0^2 - \omega^2 - i\gamma\omega) - \omega_p^2\}. \tag{8.28}$$

We have in a solid that the interaction between the oscillators becomes very strong. The oscillation frequencies are modified and the damping constants become large. In addition, in crystals the periodicity must be taken into account.

8.3 DETERMINATION OF OPTICAL CONSTANTS

8.3.1 Fresnel's Formulas and Reflection Coefficients

The determination of the two parts n and K of the complex refractive index may be accomplished by using reflection measurements. In Chapter 5 we found that Fresnel's formulas relate the reflection coefficients of the p- and s-polarization cases to the real index of refraction. In a similar way to that of Chapter 5, one can show that the complex reflection coefficients are related to complex refractive indices through Fresnel's formulas. Replacing n_2 by $n_2^* = n_2 - iK_2$ we have for the reflection coefficients

$$r_\parallel^* = \frac{(n_2 - iK_2)\cos\theta - n_1\cos\theta''}{(n_2 - iK_2)\cos\theta + n_1\cos\theta''} \tag{8.29}$$

$$r_\perp^* = \frac{n_1\cos\theta - (n_2 - iK_2)\cos\theta''}{n_1\cos\theta + (n_2 - iK_2)\cos\theta''}. \tag{8.30}$$

In order to represent $r_\parallel$ and $r_\perp$ depending on the angle of incidence. We need the law of refraction in complex terms (see FileFig 8.4 (M4SNELL) of Chapter 5). The law of refraction is

$$n_1\sin\theta = (n_2 - iK_2)\sin\theta''. \tag{8.31}$$

The angle θ'' must be a complex quantity since the left side of Eq. (8.31) is real. Introduction of Eq. (8.31) into Eqs. (8.29) and (8.30) gives us

$$r_{\parallel}^{*} = \frac{(n_2 - iK_2)\cos\theta - n_1\sqrt{1 - \{(n_1 \sin\theta)/(n_2 - iK_2)\}^2}}{(n_2 - iK_2)\cos\theta + n_1\sqrt{1 - \{(n_1 \sin\theta)/(n_2 - iK_2)\}^2}} \tag{8.32}$$

$$r_{\perp}^{*} = \frac{n_1\cos\theta - (n_2 - iK_2)\sqrt{1 - \{(n_1 \sin\theta)/(n_2 - iK_2)\}^2}}{n_1\cos\theta + (n_2 - iK_2)\sqrt{1 - \{(n_1 \sin\theta)/(n_2 - iK_2)\}^2}}. \tag{8.33}$$

In FileFig 8.1 we have graphs of the absolute value and the argument for reflected amplitudes of the parallel (zrp) and perpendicular (zrs) cases depending on n and K. For $K \neq 0$ we see from the first graph that for the parallel case the minimum corresponding to the Brewster angle is not zero. The angle at the minimum is called the *principal angle*. The second graph shows the phase jump at the Brewster angle, which is now a smooth transition. In the application of FileFig 8.1, Section 3, we plot the parallel (zrp) and perpendicular (zrs) cases for several different values of n and K on the same graph.

FileFig 8.1 (O1FRNKPSS)

Graphs are shown for reflected amplitudes of the parallel (zrp) and perpendicular (zrs) cases depending on n and K for $n_1 = 1$. For both cases the absolute value of the reflected amplitudes and the phase angle are plotted for $n1 = 1, n2 = 1.5$, and $K = 2$.

O1FRNKPSS is only on the CD.

Application 8.1.

1. Make a graph and find the values of the principal angle for one value of n and three values of K.
2. Make a graph and find the values of the principal angle for one value of K and three values of n.
3. Observe that the curve of $arg(zrp(\theta))$ is not continuously approaching the curve of $arg(zrp0(\theta))$ when K is going from 0.01 to zero. What is the remaining phase difference?

8.3.2 Ratios of the Amplitude Reflection Coefficients

Equations (8.32) and (8.33) give the dependence of the reflection coefficients on n and K. We have the general problem that we cannot represent n and K as functions of r_s and r_p. This is only possible for approximations, (see Appendix 8.1), and other methods have to be considered.

One method is to apply reflection measurements at several angles of incidence to determine the absolute values of the ratio r_s/r_p. The ratio r_s/r_p is calculated similarly to that done by Born and Wolf (1964, p.617).

From Chapter 5 the ratio $r_\perp/r_\parallel$ from Fresnel's formulas is

$$r_\perp/r_\parallel = [(n_1 \cos\theta - n_2 \cos\theta'')/(n_1 \cos\theta + n_2 \cos\theta'')]/nonumber \quad (8.34)$$

$$[(n_2 \cos\theta - n_1 \cos\theta'')/(n_2 \cos\theta + n_1 \cos\theta'')]. \quad (8.35)$$

Using the law of refraction and the trigonometric formula for the sum of angles

$$\cos(a \pm b) = \cos a \cos b \pm \sin a \sin b \quad (8.36)$$

we get

$$\mid r_s/r_p \mid = \mid \cos(\theta - \theta'')/\cos(\theta + \theta'') \mid . \quad (8.37)$$

In FileFig 8.2 we show graphs of the real parts of r_p and r_s for various values of n and K. The third graph shows the ratio r_p/r_s and the fourth r_s/r_p. From these two graphs it appears that the ratio r_p/r_s is much more useful for the determination of optical constants than the ratios r_s/r_p, because they are smooth and do not show a resonance, related to the appearance of the Brewster angle. The optical constants may be obtained by measuring values of $|r_p/r_s|$ for two different angles of θ, and solving the two equations for the unknowns n and K.

FileFig 8.2 (O2FRSOPS)

Graphs are shown for the real part of the ratios r_s/r_p and r_p/r_s, calculated with the expressions used in FileFig 8.1, for $n1 = 1$, $n2 = 1.5$, $nn2 = 1.5$, $K = 0.1$, and $K = 0.01$, $KK = 0.5$, $KKK = 2$.

O2FRSOPS is only on the CD.

8.3.3 Oscillator Expressions

8.3.3.1 One Oscillator

To fit experimental data of a narrow frequency range, in which we have a resonance feature, one may use for $n + iK$ a similar expression derived in Eq. (8.23) but extended to four parameters,

$$n + iK = \sqrt{A + S/[1 - (\nu/\nu_0)^2 - \gamma(\nu/\nu_0)]}, \quad (8.38)$$

where A is a general constant, S the oscillator strength, γ the damping constant, and $\nu_0 = \omega_0/2\pi$ the resonance frequency. An example is shown in Figure 8.2 and a calculation given in FileFig 8.3.

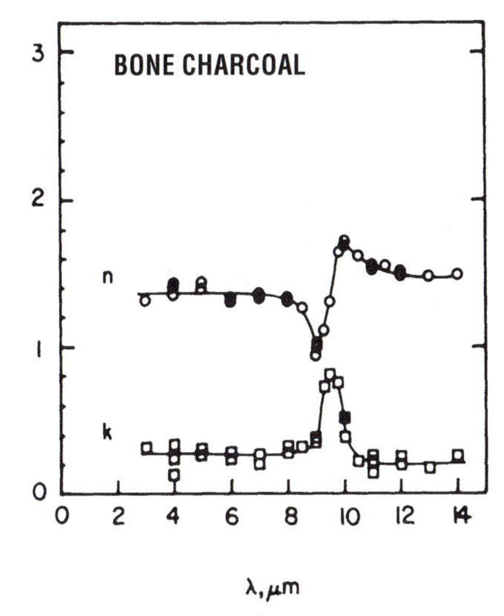

FIGURE 8.2 Optical constants of bone charcoal powder. Resonance of vibrations of *Ca*, *P*, and *O* atoms against each other. (From Tomasecli et al., infrared optical constants of black powders determined from measurements, *Applied Optics*, 1981, 20, 3961–3967.)

FileFig 8.3 (O3OSTINS)

Graphs are shown of Eq. (8.38) for $A = 20$, $S1 = 0.09$, $\gamma 1 = 0.002$, and $\nu_0 1 = 1050\ \text{cm}^{-1}$. An analytical approximation for n close to 1 and small K is also presented.

O3OSTINS is only on the CD.

Application 8.3.

1. Change γ and observe the change in resonance wavenumber and the height of the imaginary part.
2. Change S and observe the change in resonance wavenumber and the height of the imaginary part.
3. Modify the graphs by plotting in addition a graph using $S2 = 0.19$, $\gamma 2 = 0.03$, and $\nu_0 2 = 1150\ \text{cm}^{-1}$. Change parameters and study the effect on n and K.
4. Modify the graphs of the thin medium by also plotting a graph using different values of $a2$ and $c2$. Change $a1$, $a2$ and $c1$ and cc and compare the effect on n and K.

8.3.3.2 Many Oscillator Terms

The dependence of the electrical polarization on the frequency over a large range suggests that one uses more than one oscillator term in the formula for the

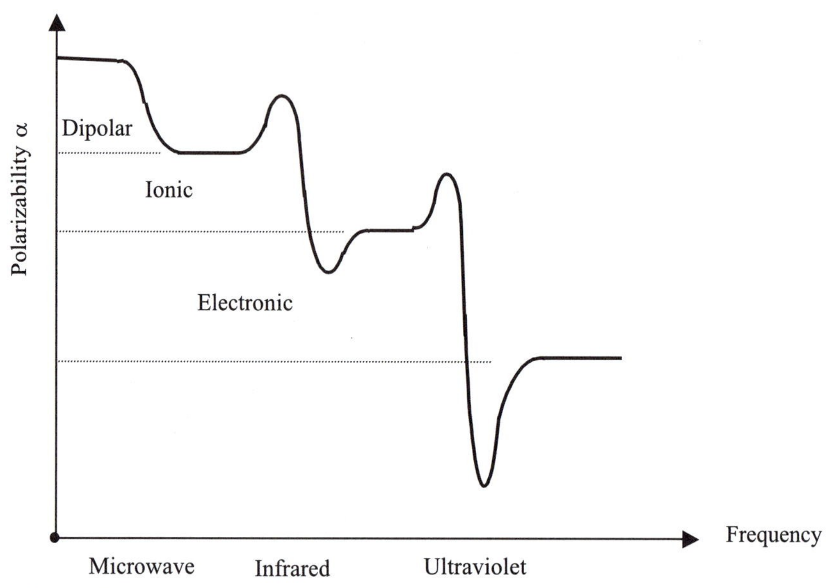

FIGURE 8.3 The dependence of the polarizability on frequency for the microwave, infrared, and ultraviolet regions.

representation of n and K (schematically shown in Figure 8.3). Having measured n and K over a large range of frequencies, the experimental data are fit to formulas such as

$$n^2 - K^2 = 1 + \sum_j f_j \omega_p^2 (\omega_{oj}^2 - \omega^2)/((\omega_{oj}^2 - \omega^2)^2 + (\gamma j \omega)^2) \quad (8.39)$$

$$2nK = \sum_j \omega_p^2 f_j \gamma_j \omega/((\omega_{oj}^2 - \omega^2)^2 + (\gamma j \omega)^2), \quad (8.40)$$

where f_j, γ_j, and ω_{oj} are empirical constants. The constants in these expressions are determined by a "best fit" calculation over a large range of frequencies with respect to the measured values of n and K

8.3.4 Sellmeier Formula

Similarly to what has been discussed for the oscillator expression, one may fit experimental data to represent the dependence of n and K on the wavelength by using a polynomial approach of the type

$$n^2 = c_1 + c_2\lambda^4/(\lambda^2 - c_3) + \sum_{j=1}^{j=N} a_j\lambda^2/(\lambda^2 - b_j). \quad (8.41)$$

This is called a Sellmeier-type equation and has been used, for example, to fit the data for potassium bromide in the spectral region from .2 to 42 microns using 11 empirical constants.[1] An example is given in FileFig 8.4.

When fitting experimental data one has to keep in mind that n and K are not independent. They are related by the Kramer–Kroning model.[2] In some spectral regions, for example in the x-ray region, one obtains the K value from absorption measurements and calculates the corresponding n value with the Kramer–Kroning model.

FileFig 8.4 (04SELMRS)

A graph of a Sellmeier expression $n(\lambda)$ for fused quartz is shown for the range of λ from 4000 to 8000 Angstrom using parameters c_i with $i = 1$ to 3.

04SELMRS

Graph for Demonstration of the Sellmeier Presentation of the Refractive Index

For fused quartz we have

$$c1 := 1.448 \qquad c2 := 3.3 \cdot 10^5 \qquad c3 := 1.23 \cdot 10^{11}$$

$$\lambda := 4000, 4001 \ldots 8000$$

$$n(\lambda) := c1 + \frac{c2}{\lambda^2} + \frac{c3}{\lambda^4}.$$

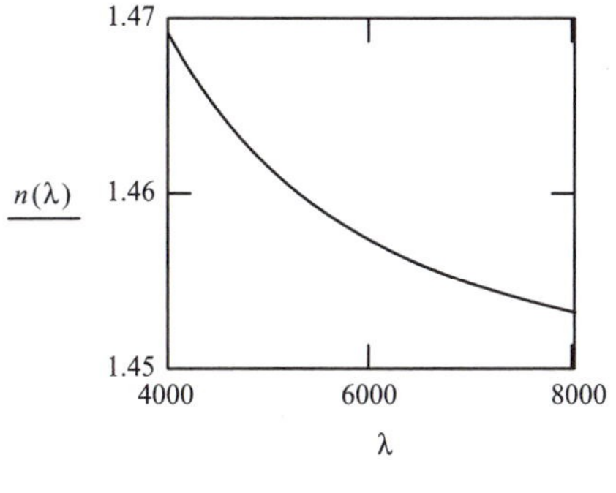

Application 8.4. Determination (backward) of the constants $c1$, $c2$, and $c3$. Read from the graph three values of λ_i and the corresponding value of $n(\lambda_i)$. Consider $c1$ to $c3$ as unknown and formulate a system of linear equations. One would be $n(\lambda_i) = c1 + c2/\lambda_i^2 + c3/\lambda_i^4$. Solve the system of these inhomogeneous

[1] E. D. Palik, *Handbook of Optical Constants of Solids II*, Academic Press, New York, 1991.

[2] Charles Kittel, *Introduction to Solid State Physics*, John Wiley & Sons, New York, 1967.

linear equations with one of the available computer programs. Make an estimate of the error.

8.4 OPTICAL CONSTANTS OF METALS

8.4.1 Drude Model

In Section 8.2 we discussed the optical constants of dielectrics and in Section 8.3 their determination. We now discuss metals and how their optical constants are also described by a complex refractive index. The determination of the n and K values for metals is similar to what has been discussed before, but the model representing the material is different.

Metals show high reflectivity in the visible and infrared spectral regions and their attenuation increases with lower frequencies. The values of real and imaginary parts of the refractive index of metals are in the lower frequency region which is much higher than one usually finds for dielectrics.

The interaction of the electromagnetic wave with the metal is described by the Drude model. The electrons are assumed to move almost freely in the metal and there is no restoring force to make the electrons vibrate, as discussed for the dielectric. For an isotropic medium with free conducting charges we write Maxwell's equations as

$$\begin{aligned} \nabla \times \mathbf{E} &= -\partial \mathbf{B}/\partial t \\ c^2 \nabla \times \mathbf{B} &= \partial \mathbf{E}/\partial t + \mathbf{j}/\varepsilon_0 \\ \nabla \cdot \mathbf{E} &= \mathbf{0} \\ \nabla \cdot \mathbf{B} &= \mathbf{0}. \end{aligned} \tag{8.42}$$

This is similar to Eq. (8.1). For the current density vector $\mathbf{j}$ we take

$$\mathbf{j} = Nev_j, \tag{8.43}$$

where v_j is called the drift velocity. The electrical field of the light and the current density in the material are related by the wave equation. We assume $\mathbf{E}$ and $\mathbf{j}$ are vibrating in the y direction and propagating in the x direction and from Eq. (8.42) we get for the wave equation

$$\partial^2 E_y/\partial x^2 - (1/c^2)\partial^2 E_y/\partial t^2 = [1/(c^2\varepsilon_0)]\partial j_y/\partial t. \tag{8.44}$$

As in Section 8.2, we now find an expression for j_y in terms of the parameters of the damped oscillator model and the vibrating electrical field E_0. The differential equation of the model is now without the force term,

$$md^2u/dt^2 + m\gamma du/dt = eE_0 e^{-i\omega t}. \tag{8.45}$$

The general solution of Eq. (8.45) is the sum of the solutions of the homogeneous and inhomogeneous equations. For the homogeneous equation,

$$md^2u/dt^2 + m\gamma du/dt = 0. \tag{8.46}$$

Using the trial solution $u = u_0 e^{-t/\tau}$, we get for $\gamma = \tau^{-1}$. Typically one has a value of 10^{-13} for τ and we neglect this solution.

The inhomogeneous equation may be rewritten, using $v = du/dt$, as

$$mdv/dt + m\gamma v = eE_0 e^{-i\omega t}, \tag{8.47}$$

Using the trial solution $v = v_0 e^{-i\omega t}$ one obtains

$$v_0 = E_0(e/m)/(\gamma - i\omega). \tag{8.48}$$

The current density $j_0 = Nev_0$ can be expressed as

$$j_0 = (\tau E_0)(Ne^2/m)/(1 - i\omega\tau) = (\sigma E_0)/(1 - i\omega\tau), \tag{8.49}$$

where we used $j_0 = Nev_0$, $\gamma = \tau^{-1}$, and the static conductivity

$$\sigma = \tau Ne^2/m. \tag{8.50}$$

Equation (8.49) relates the current density j_0 of our model to the electrical field E_0 of the light, vibrating with angular frequency ω.

We now turn to the wave equation, (see Eq. (8.44)), using the trial solutions

$$E_y = E_0 e^{i(kx-\omega t)} \quad \text{and} \quad j_y = j_0 e^{i(kx-\omega t)} \tag{8.51}$$

and introducing j_0 from our model, Eq. (8.49), we obtain for the complex wave vector k^*

$$(k^*)^2 = 1/c^2\{\omega^2 + (i\sigma\omega/\varepsilon_0)/(1 - i\omega\tau)\} \tag{8.52}$$

and the complex refractive index $n^* = k^*c^2/\omega^2$,

$$(n^*)^2 = 1 + i\sigma/\{\omega\varepsilon_0(1 - i\omega\tau)\} = 1 - \sigma/\{\omega\varepsilon_0(i + \omega\tau)\}. \tag{8.53}$$

Equation (8.53) relates the refractive index to the static conductivity of the metal σ, the frequency of light ω and the relaxation time τ, which is a parameter of our model and is related to the metal. In FileFig 8.5 we show graphs over a large frequency region of the real and imaginary parts of Eq. (8.53).

8.4.2 Low Frequency Region

For low frequencies, that is, when $\omega\tau \ll 1$, we may neglect $\omega\tau$ with respect to i (see Eq. (8.53)), and get

$$(n^*)^2 = 1 + i\sigma/\omega\varepsilon_0. \tag{8.54}$$

Since ω is small, which means $i\sigma/\omega\varepsilon_0$ is large compared to 1, we write

$$n - iK = (\sqrt{i})(\sqrt{\sigma/\omega\varepsilon_0}). \tag{8.55}$$

Using the identity $\sqrt{i} = (1+i)/\sqrt{2}$ one has

$$| n |=| K |= (\sqrt{\sigma/2\omega\varepsilon_0}); \tag{8.56}$$

that is, n and K have the same value and the wave is strongly attenuated in the medium. To find the frequency limit of Eq. (8.56), we have from $\omega\tau \ll 1$, that $\omega << 1/\tau$, with $\tau = m\sigma/Ne^2$. For copper we have $\sigma = 5.76 \ 10^7$ (ohm meter)$^{-1}$, $N = 8.5\ 10^{28}$ meter^{-3}, $e = 1.6\ 10^{-19}$ coulomb, $m = 9.11\ 10^{-31}$ kgm, and obtain for $1/\tau = 4.1\ 10^{13}$ 1/sec. Therefore this approximation is valid for angular frequencies smaller than $10^{11} - 10^{12}$ Hz, which are lower frequencies than the far infrared. This approximation is plotted in FileFig 8.5 as the last graph. The light is strongly attenuated when entering metals in this spectral region and is therefore highly reflected (see also FileFig 8.7).

8.4.3 High Frequency Region

For high frequencies we have $\omega\tau \gg 1$, and from Eq. (8.53) one has

$$n^2 = 1 - \sigma/\omega^2\tau\varepsilon_0. \tag{8.57}$$

Using Eq. (8.50) and the plasma frequency $\omega_p = Ne^2/m\varepsilon_0$ we have

$$n^2 = 1 - \omega_p^2/\omega^2. \tag{8.58}$$

The plasma frequency $\omega_p = Ne^2/m\varepsilon_0$ has the value $1.6\ 10^{16}$ 1/sec when using $N = 8.5\ 10^{28}$ meter^{-3}, $e = 1.6\ 10^{-19}$ coulomb, $\varepsilon_0 = 8.85\ 10^{-12}$ farad-meter^{-1}, and $m = 9.11\ 10^{-31}$ kg. Therefore the approximation of Eq. (8.58) is valid for angular frequencies larger than 10^{18}, corresponding to the x ray region. This is also plotted in FileFig 8.5 on the third graph. In this region the refractive index n is real and less than 1 and light is penetrating the metal without being attenuated.

We see that the plasma frequency divides the frequency range into two parts. One for high frequencies when n is smaller than 1 and one for low frequencies when n is complex (see Eq. (8.54)).

FileFig 8.5 (O5METALS)

Graphs of the real and imaginary part of $n^* = n + iK$ of copper, for the general case and for the high and low frequency approximations.

O5METALS

Calculation of n and k for Copper Using the Drude Model

Calculation of real and imaginary parts. Expression for low and high frequencies depending on angular frequency.

1. General Expression

$$c := 3 \cdot 10^8 \text{m/s} \qquad \sigma := 6 \cdot 10^7\ (\text{OHMm})^{-1} \qquad \varepsilon o := 8.85 \cdot 10^{-12}\ \ \text{C}^2/\text{Nm}$$

$$\tau := \frac{1}{4.1 \cdot 10^{11}}\ \ \text{sec} \qquad i := \sqrt{-1}$$

$$\omega := 10^{11}, (2 \cdot 10)^{11} \ldots 10^{18}.$$

Angular frequency for 1 mm wavelength is 2π*300*10^9; see below. The general expression for $n - ik = zm(\omega)$

$$zm(\omega) := \sqrt{1 + \left(\frac{i \cdot \sigma}{\varepsilon o \cdot \omega}\right) \cdot \frac{1}{1 - i \cdot \omega \cdot \tau}}.$$

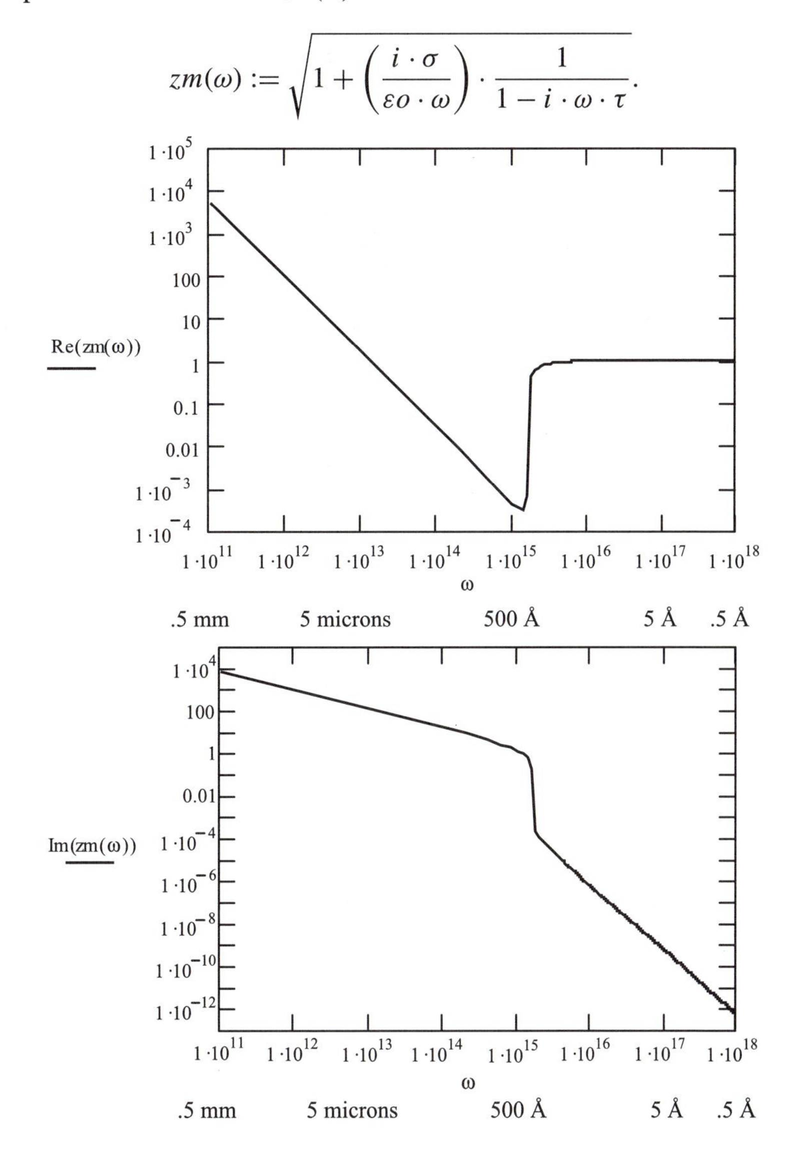

2. High frequency limit

$$\mathrm{nh}(\omega) := 1 - \frac{\sigma}{\varepsilon o \cdot \omega^2 \cdot \tau}.$$

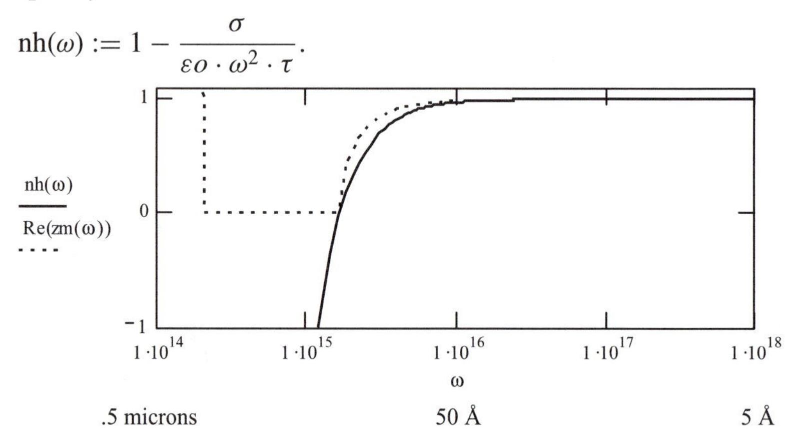

3. Low frequency limit

$$nl(\omega) := \sqrt{\frac{\sigma}{2 \cdot \varepsilon o \cdot \omega}}.$$

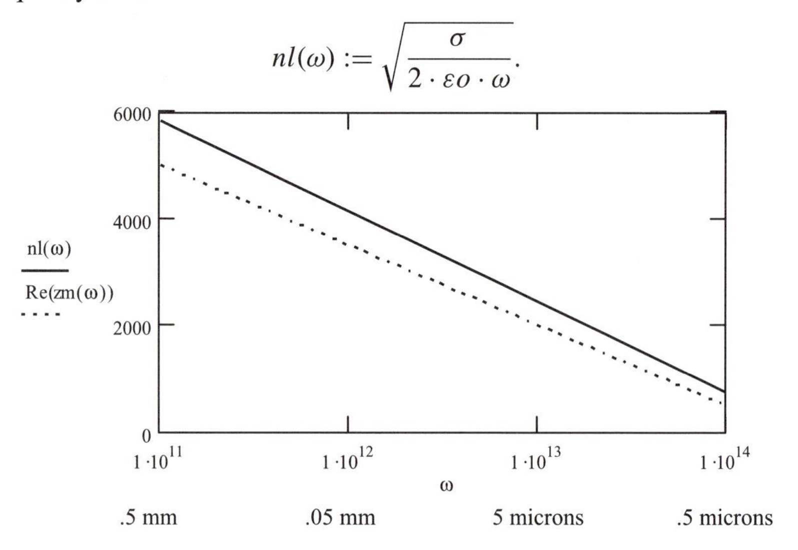

frequency	frequency	angular frequency
$3 \cdot 10^{11}$ is 1 mm	$1 \cdot 10^{11}$ is 3 mm	$1 \cdot 10^{11}$ is $3/(2 \cdot \pi)$ mm
$3 \cdot 10^{14}$ is 1 micron	$1 \cdot 10^{14}$ is 3 micron	$1 \cdot 10^{14}$ is 0.477 micron
$3 \cdot 10^{17}$ is 1 mm is 10 Å	$1 \cdot 10^{17}$ is 3 mm is 30 Å	$1 \cdot 10^{17}$ is 0.477 mm is 40 Å

Application 8.5.

1. Check your computer program manual and get familiar with physical unit systems, MKS, and cgs.
2. Modify the calculations for gold, $4.5 \cdot 10^7$ [1/Ohm m].
3. Modify the calculations for silver, $6.3 \cdot 10^7$ [1/Ohm m].

4. Modify the calculations for nickel, $1.5 \cdot 10^7$ [1/Ohm m].
5. Modify the calculations for lead, $0.5 \cdot 10^7$ [1/Ohm m].
6. Compare the graphs for the long wavelength region. At what frequency have the absolute values of n and K about the same value? At what frequency have the optical constants of all metals about the same value?

8.4.4 Skin Depth

In the low frequency region, where the light is strongly attenuated in the metal, the light may only penetrate into the metal a small distance, called the *skin depth*.

An incident wave $u_0 \exp -i(\omega t + kX)$ may be written with complex k^* as

$$u = u_0 \exp -i\omega t \exp -i\{(n - iK)\omega X/c\}. \tag{8.59}$$

We write Eq. (8.59) in real and complex factors as

$$u = u_0 \exp -(\omega K X/c) \exp -i\omega\{t - nX/c\}. \tag{8.60}$$

One defines the *penetration depth* as the length at which the wave is attenuated to $1/e$, that is,

$$X_l = l = c/\omega K = \sqrt{[(2c^2\varepsilon_0)/(\omega\sigma)]}. \tag{8.61}$$

In most cases one is interested in the intensity and has

$$X_l = \sqrt{[c^2\varepsilon_0/2\omega\sigma]} \tag{8.62}$$

or with a good conductor such as copper, with $\sigma = 5.76\ 10^7$ (ohm meter)$^{-1}$ one gets with ω in sec^{-1},

$$X_l = .17\{\sqrt{[(1sec^{-1}/\omega)]}\} \text{ meter}. \tag{8.63}$$

In FileFig 8.6 the first graph shows the penetration depth for the intensity of frequencies in the range of 10^{14}, the visible and near infrared. The second graph shows the penetration depth for the intensity of frequencies in the long wavelength range.

FileFig 8.6 (O6SKINS)

Graph of the skin depth of the intensity for copper in the wavelength range from 10^{-3} to 10^{-7}.

O6SKINS

Skin Depth

1. Skin depth (in meters) for intensity depending on frequency

$$\varepsilon o := 8.85 \cdot 10^{-12} \qquad C^2/Nm \qquad c := 3 \cdot 10^8 \qquad \text{m/s}$$

$$\sigma := 6 \cdot 10^7 \qquad (\text{Am})^{-1}$$

$$\omega := 10^{10}, (10)^{11} \ldots 10^{14} \qquad i := \sqrt{-1}$$

$$l(\omega) := \sqrt{\frac{\varepsilon o \cdot c^2}{2 \cdot \omega \cdot \sigma}} \qquad \text{in meters.}$$

l(ω)

8·10⁻⁸, 6·10⁻⁸, 4·10⁻⁸, 2·10⁻⁸, 0

2·10¹³, 4·10¹³, 6·10¹³, 8·10¹³, 1·10¹⁴

ω

$8 \cdot 10^{-8}$
$6 \cdot 10^{-8}$
$l(\omega)$
$4 \cdot 10^{-8}$
$2 \cdot 10^{-8}$
0
$2 \cdot 10^{13}$ $4 \cdot 10^{13}$ $6 \cdot 10^{13}$ $8 \cdot 10^{13}$ $1 \cdot 10^{14}$
ω

2. Skin depth (in meters) for intensity depending on wavelength
   (For checking: for 1 mm wavelength angular frequency is 2π*300*10 ∧ 9.)

$$l(\lambda) := \sqrt{\frac{\varepsilon o \cdot c \cdot \lambda}{4\pi \cdot \sigma}}.$$

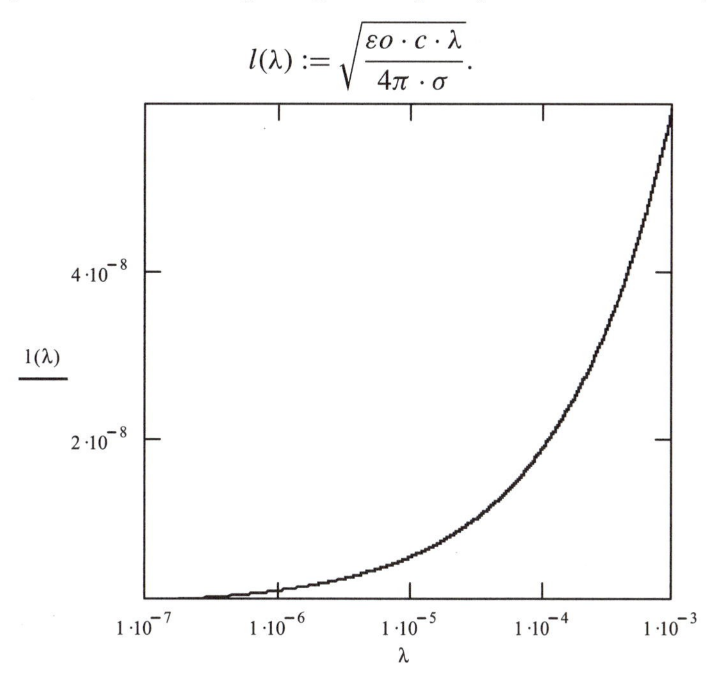

$1 \cdot 10^9$ meter is 1 nm $=$ .001 microns $= 10\text{Å}$.

Application 8.6.

1. Derive the penetration depth for the intensity, Eq. (8.62).
2. Check your Mathcad manually and get familiar with physical unit systems, MKS, and cgs.
3. Modify the calculations for gold, $4.5 \cdot 10^7$ [1/Ohm m].
4. Modify the calculations for silver, $6.3 \cdot 10^7$ [1/Ohm m].
5. Modify the calculations for nickel, $1.5 \cdot 10^7$ [1/Ohm m].
6. Modify the calculations for lead, $0.5 \cdot 10^7$ [1/Ohm m].

8.4.5 Reflectance at Normal Incidence and Reflection Coefficients with Absorption

We have seen in Section 8.3 that the $r_{\|}$ component is not zero at the *principal angle*, which means the angle corresponding to the Brewster's angle for the case of the lossless dielectric. The reflectance R is equal to the square of the reflection coefficients of Fresnel's formulas, (Eq. (8.32) and (8.33)) and is valid for both dielectric and metals if the corresponding values of n and K are used. For normal incidence, when the parallel and perpendicular components are the same we have for the reflectance R,

$$R = (1 - n)^2 + K^2/[(1 + n)^2 + K^2]. \tag{8.64}$$

In FileFig 8.7 we show for normal incidence the dependence of the reflectance R on K. When $K = 0$ we go back to Fresnels formulas, depending only on the real part of the refractive index, as discussed in Chapter 5 for lossless dielectrics. When K is large, we have high reflectance R as observed for metals.

FileFig 8.7 (O7REFNKS)

Graph of the reflectance at normal incidence depending on K.

O7REFNKS is only on the CD.

8.4.6 Elliptically Polarized Light

We mentioned in Chapter 5 that elliptically polarized light may be produced when light is totally internally reflected in a dielectric medium. Reflection on metal surfaces shows a similar phenomenon. The reflection coefficients r_p and r_s may have arguments depending on the angle of incidence. Since each component picks up a different change in the argument, the difference of the arguments of the reflected components is not the same and corresponds to the angle ϕ of elliptically polarized light, as discussed in Chapter 5.

In FileFig 8.8 we have plotted the difference of the phase angles after reflection, depending on the angle of incidence.

FileFig 8.8 (O8ARDELS)

Graph of the difference Δ of the arguments of zrp and rzs depending on specific values of n and K.

O8ARDELS is only on the CD.

Application 8.8.

1. Change the optical constants and plot a graph of Δ depending on a range of values n for fixed K and three values of θ, for example, 35°, 45°, 55°.
2. Change the optical constants and plot a graph of Δ depending on a range of K for fixed value of n and three values of θ, for example, 35°, 45°, 55°.

APPENDIX 8.1

A8.1.1 Analytical Expressions and Approximations for the Determinaton of n and K

As discussed in Section 3.2 there are approximate methods to represent n and K analytically. We show that an approximation is used to get formulas for n and K depending on the ratio of r_s to r_p, the phase difference Δ of r_s and r_p after reflection, and the angle of incidence θ. This method is called *ellipsometry* because one measures the absolute value of the ratio r_s to r_p and the phase difference Δ of r_s and r_p; see Section 8.4.6. Using the law of refraction one may rewrite Fresnel's formulas for ratios r_s and r_p as

$$r_s = (-1)\sin(\theta - \theta'')/sin(\theta + \theta'') \tag{A8.1}$$

$$r_p = \tan(\theta - \theta'')/\tan(\theta + \theta''), \tag{A8.2}$$

and obtain for r_s/r_p

$$r_s/r_p = -\{\cos(\theta - \theta'')/\cos(\theta + \theta'')\}. \tag{A8.3}$$

The ratio r_s/r_p may be complex, and therefore we call P the absolute value of the ratio r_s/r_p, and Δ the argument.

$$Pe^{-i\Delta} = r_s/r_p = -\cos(\theta - \theta'')/\cos(\theta + \theta''). \tag{A8.4}$$

Using the formula for the sum and difference of angles in a cosine function one may write

$$(1 - Pe^{-i\Delta})/(1 + Pe^{-i\Delta}) = -(\cos\theta\cos\theta'')/(\sin\theta\sin\theta'') \tag{A8.5}$$

and with the complex law of refraction

$$(1 - Pe^{-i\Delta})/(1 + Pe^{-i\Delta}) = (-1)\sqrt{\{n'^{*2} - (\sin\theta)^2\}}/(\sin\theta)(\tan\theta). \tag{A8.6}$$

By using $P = \tan\psi$ one can write for the left side of Eq. (A8.6),

$$\begin{aligned}&(1 - \tan\psi e^{-i\Delta})/(1 + \tan\psi e^{-i\Delta})\\ &= (\cos 2\psi + i\sin\Delta\sin 2\psi)/(1 + \cos\Delta\sin 2\psi)\end{aligned} \tag{A8.7}$$

and one has for Eq. (A8.7)

$$\begin{aligned}&\sqrt{\{n'^{*2} - (\sin\theta)^2\}}\\ &= -(\sin\theta)(\tan\theta)(\cos 2\psi + i\sin\Delta\sin 2\psi)/(1 + \cos\Delta\sin 2\psi).\end{aligned} \tag{A8.8}$$

We obtain for n' and K' the complex expression

$$\begin{aligned}&n' + iK'\\ &= \sqrt{[(\sin\theta)^2 + \{(\sin\theta)(tan\theta)(\cos 2\psi + i\sin\Delta\sin 2\psi)/(1 + \cos\Delta\sin 2\psi)\}^2]}.\end{aligned} \tag{A8.9}$$

In order to get to n' and K' we have to determine the real and imaginary parts of the right side of Eq. (A8.9). As done in Born and Wolf (1964, p. 619), one can make an approximation by neglecting in the square root of Eq. (A8.8) the term $(\sin\theta)^2$ with respect to n'^{*2}, and obtain explicit expressions for n' and K':

$$n' = \{(\sin\theta)(\tan\theta)(\cos 2\psi)\}/\{1 + \cos\Delta \sin 2\psi\} \qquad \text{(A8.10)}$$

$$K' = \{(\sin\theta)(\tan\theta)(\sin\Delta)(\sin 2\psi)\}/\{1 + \cos\Delta \sin 2\psi\}. \qquad \text{(A8.11)}$$

We show in FileFig 8.9, graphs of P, Δ, and ψ depending on the angle of incidence θ. These graphs are for specific values of n and K. A comparison of the exact and approximate calculation, again for specific values of n and K, is shown in FileFig 8.10. In praxis one often uses iteration for the determination of n' and K' and uses more than two input data for a best fit calculation.

FileFig A8.9 (OA1DELTAFfS)

For $zp = r_p \exp(i\delta_p)$ and $zs = r_s \exp(i\delta_s)$, graphs are shown for $P = \tan\psi$ with $P = r_s/r_p$, Δ (difference of the arguments of r_s and r_p), and of $atan(zs/zp)$.

OA1DELTAfS is only on the CD.

Application A8.9.

1. Change the optical constants and plot a graph of P depending on a range of values n for fixed K and three values of θ, for example, 35°, 45°, 55°.
2. Change the optical constants and plot a graph of $atan(zp/zs)$ depending values of on a range of K for fixed value of n and three values of θ, for example, 35°, 45°, 55°.

FileFig A8.10 (OA2METPDS)

Graphs are shown for $z = n + iK$ depending on ψ because one has $P = \tan\psi$, $P = r_s/r_p$ and Δ is the difference of the arguments of r_s and r_p. Curves of the exact expressions are compared with the approximations.

OA2METPDS is only on the CD.

Application A8.10.

1. Study the approximation for different values of P and fixed value of Δ.
2. Study the approximation for different values of Δ and fixed value of P.

C H A P T E R

Fourier Transformation and FT-Spectroscopy

9.1 FOURIER TRANSFORMATION

9.1.1 Introduction

In this chapter we present some basic properties of Fourier transformation and applications of Fourier transform spectroscopy. A simple example of an application of Fourier transformation is the determination of the frequencies one needs to compose a function $f(x)$, presenting a rectangular pulse. Such a pulse may be generated by superposition of many monochromatic waves with many different wavelengths and amplitudes. The input data to Fourier transformation are the space coordinates of $f(x)$. The result of Fourier transformation is the frequency spectrum corresponding to the different wavelengths used to compose $f(x)$. A more complicated application is the analysis of the interferogram obtained from incident light traversing an absorbing material. The Fourier transformation of the interferogram will calculate the absorption spectrum of the material.

The discussion uses a considerable number of examples and not much of the mathematical theory of Fourier transformations. Most important is numerical Fourier transformation, available in most computational computer programs and applied in Fourier transform spectroscopy.

9.1.2 The Fourier Integrals

The integrals we used in the far field approximation of the Kirchhoff–Fresnel diffraction theory are Fourier integrals. The integral transforms the "input function" (in our case the aperture function), into the "output function" (in our case the diffraction pattern). For the example of the diffraction at a slit, the aperture function is the "(double) step function" $S(y)$ and is transformed into the diffraction pattern $G(\nu)$. The step function is defined as

$$S(y) = \begin{cases} 0 & \text{for } \infty \text{ to } d/2 \\ 1 & \text{for } d/2 \text{ to } -d/2 \\ 0 & \text{for } -d/2 \text{ to } -\infty, \end{cases} \tag{9.1}$$

and the Fourier integral transforms $S(y)$ into the Fourier transform $G(\nu)$.

Using slightly different coordinates from those in Chapter 3, we have that the Fourier transform of $S(y)$ is $G(\nu)$.

$$G(\nu) = \int_{-\infty}^{\infty} S(y) e^{-i2\pi\nu y} dy. \tag{9.2}$$

From Fourier transform theory we find that the inverse relationship is also true and we have the Fourier transform of (note the plus sign in the exponent)

$$S(y) = \int_{-\infty}^{\infty} G(\nu) e^{i2\pi\nu y} d\nu. \tag{9.3}$$

Both variables ν and y have continuous values from $-\infty$ to $+\infty$. The variable y is the variable in the space domain and the variable ν is the variable in the frequency domain and has the dimension of 1/space coordinate. (In infrared spectroscopy one uses as the unit cm^{-1}). Both $G(\nu)$ and $S(y)$ are not normalized. When $G(\nu)$ is symmetric with respect to zero, the integral may be written as

$$S(y) = 2 \int_{0}^{\infty} G(\nu) \cos(2\pi\nu y) d\nu. \tag{9.4}$$

We now discuss some analytical Fourier transformations.

9.1.3 Examples of Fourier Transformations Using Analytical Functions

We present two examples to calculate the Fourier transformation, and the Fourier transformation of the Fourier transformation. We show that the latter is indeed the original function.

9.1.3.1 Gauss Function

We consider

$$S(y) = \exp(-a^2 y^2/2). \tag{9.5}$$

We insert it into the integral of Eq. (9.2) and use the integral formula

$$\int_{0}^{\infty} \exp(-c^2 x^2) \cos(bx) dx = (\sqrt{\pi}/2c) \exp(-b^2/4c^2) \tag{9.6}$$

and obtain

$$G(\nu) = (\sqrt{2\pi}/a) \exp(-2\pi^2 \nu^2/a^2). \tag{9.7}$$

If we take the function $G(\nu) = (\sqrt{2\pi}/a)\exp(-2\pi^2\nu^2/a^2)$ and insert it into Eq. (9.3) and use the same integral formula, we get back the original function $S(y)$:

$$S(y) = \exp(-a^2y^2/2). \tag{9.8}$$

This example shows that the Fourier transformation of the Fourier transformation reproduces the original function.

9.1.3.2 The Functions $1/(1+x^2)$ and $\pi e^{-2\pi\nu}$

If we use

$$S(y) = 1/(1+y^2) \tag{9.9}$$

and apply the integral formula

$$\int_0^\infty \{(\cos ax)/(1+x^2)\}dx = (\pi/2)e^{-a}, \tag{9.10}$$

we obtain

$$G(\nu) = \pi e^{-2\pi\nu}, \tag{9.11}$$

and for the Fourier transform of $G(\nu)$ we find the original function by using the integral formula

$$\int_0^\infty e^{-ax}\cos mx dx = a/(a^2+m^2). \tag{9.12}$$

As result we find that the Fourier transform of $1/(1+y^2)$ is $\pi e^{-2\pi\nu}$, and the Fourier transform of $\pi e^{-2\pi\nu}$ is $1/(1+y^2)$:

$$e^{-2\pi\nu} \leftrightarrow 1/(1+y^2). \tag{9.13}$$

These two examples are exceptions to the point of view that one may calculate analytically the Fourier transformation and that one can get analytically the inverse Fourier transformation. A simple example when this is not the case is the Fourier transformation of the step function $S(y)$ of Eq. (9.1). The Fourier transformation is a function of the type $(\sin a\nu)/a\nu$, and its Fourier transform may not be calculated analytically because the result of the Fourier transformation of $(\sin a\nu)/a\nu$ is a discontinuous function. However, using numerical methods, one may perform the Fourier Transformation and the inverse Fourier transformation.

9.1.4 Numerical Fourier Transformation

9.1.4.1 Fast Fourier Transformation

For numerical calculations of Fourier transformations we use the fast Fourier transformation program, available in most computational computer programs.

Real Fourier Transformation (fft)

This program (fft) is used for real input data and works with 2^n input and 2^{n-1} output points. For a real Fourier transformation we have, from Eq. (9.4),

$$S(y) = 2 \int_0^\infty G(\nu) \cos(2\pi \nu y) d\nu. \tag{9.14}$$

This integral may be written as

$$\int_{-\infty}^0 G(\nu) \cos(2\pi \nu y) d\nu + \int_0^\infty G(\nu) \cos(2\pi \nu y) d\nu. \tag{9.15}$$

The first term over the negative part of ν is the "mirror" image around 0 of the second term over the positive part. Negative frequencies are a formality in Fourier transformations. They may be eliminated in order to correlate to observable results.

The input data of the Fast Fourier transformation is arranged in such a way that the negative part of the Fourier transformation follows the positive part. Let us assume we have a total of 128 points. The positive part is from point 1 to 64, and the negative part follows as a "mirror image" from 65 to 128. The frequency content of the negative part is the same as that of the positive part. The fast Fourier transformation therefore considers only one part, analyzes it, and plots the determined frequencies for only 1/2 of the total points (in our example for 64 points). The inverse transformation $(ifft)$ works backward. It has 64 input points, but takes care of the imaginary part and again ends up with 128 output points. The Fourier transform program of Mathcad numbers $2^6 = 64$ points from 0 to 63, and $2^7 = 128$ points from 0 to 127.

We demonstrate in FileFig 9.1 the real Fourier transformation of a single-sided step function with 256 points. For comparison in FileFig 9.2, we demonstrate the real Fourier transformation of a double-sided step function with 256 points. Both show the same transformation with 128 points, and the inverse transformation for both is the original function.

FileFig 9.1 (F1FTSTEPS)

Real: The original function is a one-sided step function, 256 points. The transform is a single-sided $\sin z/z$ function shown for 128 points. The inverse transformation reproduces the original function with 128 points. The imaginary part is zero for the original, appears in the transform, and is zero again for the inverse transformation.

F1FTSTEPS

Fourier Transform of a Single-Sided Step Function of Width 0 to d

The real FT is used. Orginal function

$$i := 0 \ldots 255$$
$$x_i := (\Phi(i) - \Phi(i - d)).$$

Global definition of d:

$$d \equiv 20.$$

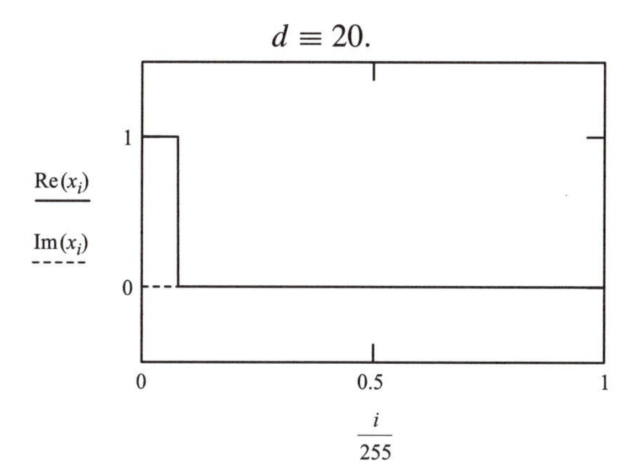

Fourier transform

$$c := fft(x)$$
$$N := \text{last}(c) \qquad N = 128$$
$$j := 0 \ldots N.$$

The first zero of FT is at $1/2d$.

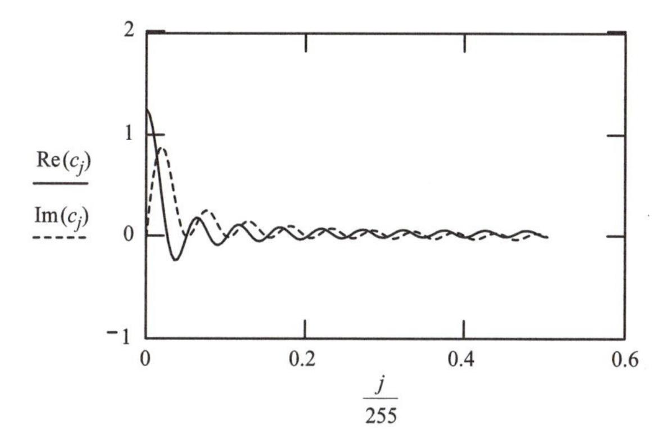

Fourier transform (inverse) of Fourier transform

$$y := fft(c)$$
$$N := \text{last}(c) \qquad N = 128$$

$$j := 0 \ldots N.$$
$$\frac{1}{2 \cdot d} = 0.025.$$

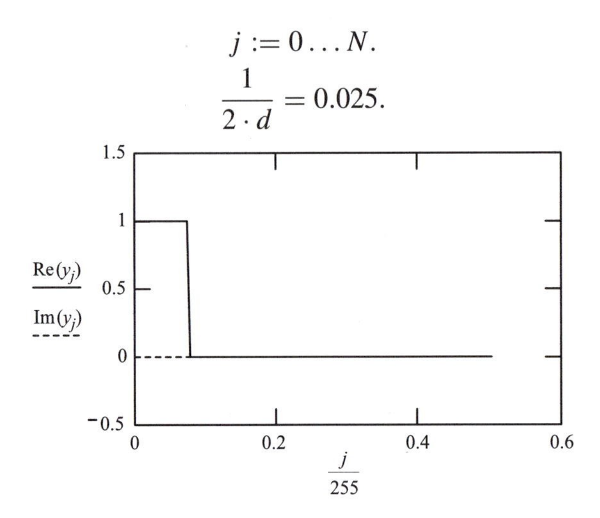

FileFig 9.2 (F2FTSTEPDS)

Real Fourier transformation: The Original Function is a double-sided step function, 256 points. The Fourier transform is a single-sided $\sin z/z$ function shown for 128 points. The inverse transformation reproduces the original function. The imaginary part is zero for the original, appears in the transform, and is zero again for the inverse transformation.

F2FTSTEPDS

Fourier Transform of a Double-Sided Step Function of Width 0 to d

The real FT is used. Orginal function

$$i := 0 \ldots 255$$
$$x_i := [\Phi(i) - \Phi(i - d)] + \Phi(i - 255 + d).$$

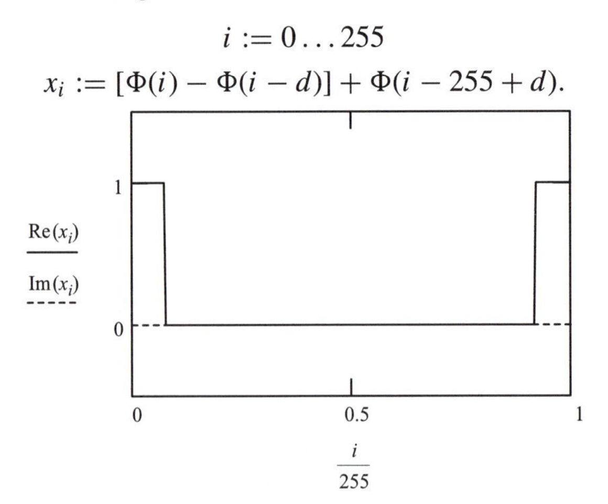

Fourier transform

$$c := fft(x)$$

$$N := \text{last}(c) \qquad N = 128$$

$$j := 0 \ldots N.$$

Global definition of d: $d \equiv 20$. The first zero of FT is at $1/2d$.

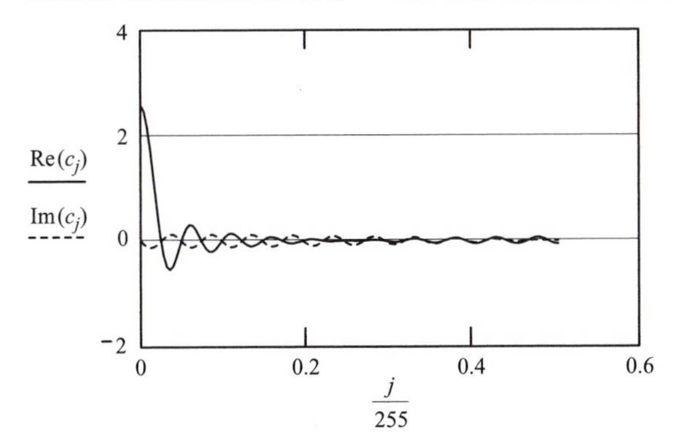

Fourier transform (inverse) of Fourier transform

$$y := fft(c)$$

$$N := \text{last}(z) \qquad N2 = 255$$

$$k := 0 \ldots N2$$

$$\frac{1}{2 \cdot d} = 0.025.$$

Re(z_i)
Im(z_i)
1
0
0
100
200
300
i

Complex Fourier Transformation ($cfft$)

We saw in FileFigs 9.1 and 9.2 that the real Fourier transformation of the step function, which is real, has a nonzero imaginary part. The imaginary part of the transformations is different for the original single- and double-sided step functions. To learn more about real and complex Fourier transformations, we apply the complex Fourier transform in FileFig 9.3 to the single-sided step function and in FileFig 9.4 to the double-sided step function.

The complex Fourier transformation program ($cfft$) works with 2^n input and 2^n output points. The inverse transformation ($icfft$) works backwards with 2^n input points and 2^n output points. The imaginary part of the transformation is again different for the single- and double-sided original step functions. The inverse transformation reproduces the original function.

FileFig 9.3 (F3FTSTEPC1S)

Complex Fourier transformation: the original function is a one-sided step function, 256 points. The complex transformation is a double-sided $\sin z/z$ function shown for all 256 points. The second part is a mirror image of the first part. The inverse transformation reproduces the original function. The imaginary part is zero for the original, appears in the transform, and is zero again for the inverse transform.

F3FTSTEPC1S

Fourier Transform of a Single-Sided Step Function of Width 0 to d

The complex FT is used. Orginal function

$$i := 0 \ldots 255$$
$$x_i := [\Phi(i) - \Phi(i-d)].$$

Global definition of d: $d \equiv 20$.

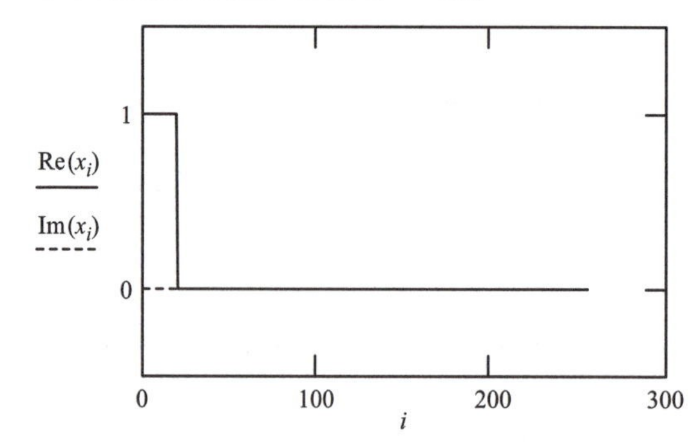

Fourier transform

$$c := cfft(x) \qquad N := \text{last}(c)$$
$$N = 255 \qquad j := 0 \ldots N.$$

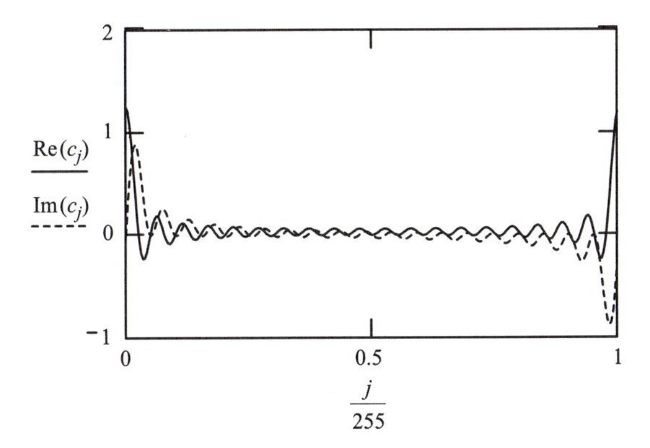

Fourier transform (inverse) of Fourier transform

$$y := fft(c) \qquad N := \text{last}(z)$$
$$N2 = 255 \qquad k := 0 \ldots N2.$$

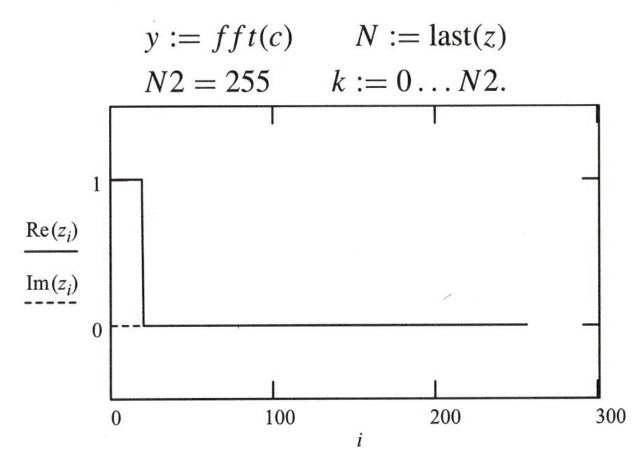

FileFig 9.4 (F4FTSTEPOSS)

Complex Fourier transformation; the original function is a double-sided step function, 256 points. The complex transformation is a double-sided $\sin x/x$ function shown for all 256 points. The second part is a mirror image of the first part. The inverse transformation reproduces the original function. The imaginary part is zero for the original, appears in the transform, and is zero again for the inverse transform.

F4FTSTEPOSS

Fourier Transform of a Double-Sided Step Function of Width 0 to d

The complex FT is used. Original function

$$i := 0 \ldots 255$$
$$x_i := [\Phi(i) - \Phi(i - d)] + \Phi(i - 255_d).$$

Global definition of d: $d \equiv 20$.

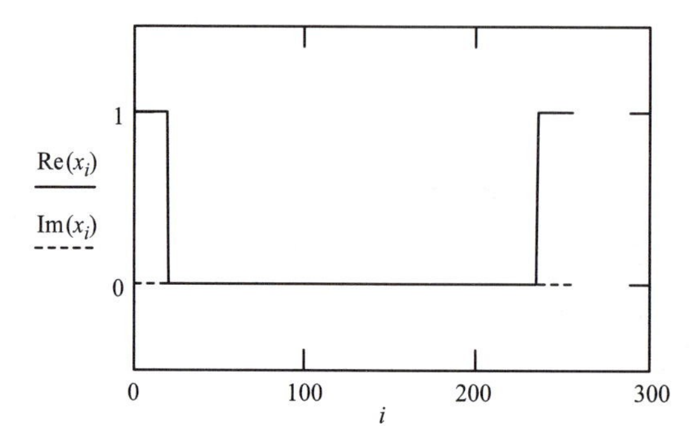

Fourier transform

$$c := cfft(x) \qquad N := \text{last}(c)$$

$$N = 255 \qquad j := 0 \ldots N.$$

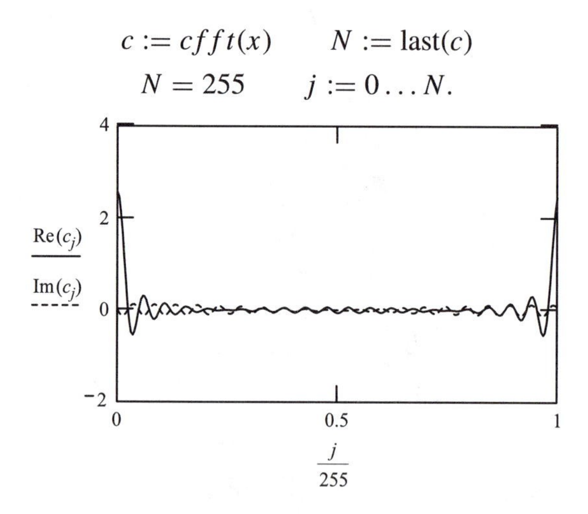

Fourier transform (inverse) of Fourier transform

$$z := icfft(c) \qquad N2 := \text{last}(z)$$

$$N2 = 255 \qquad k := 0 \ldots N2.$$

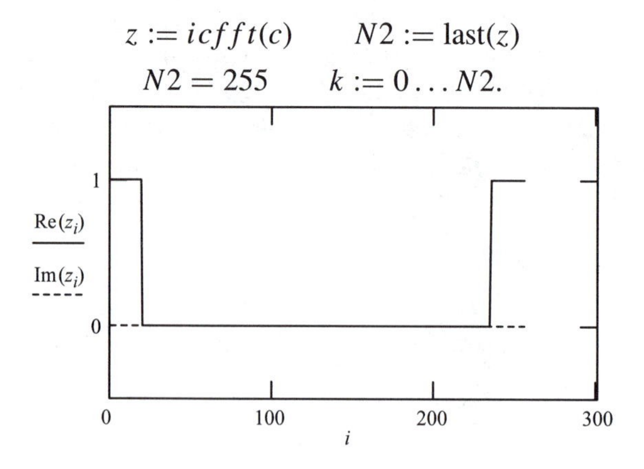

In FileFigs.5 and 6 we compare real and complex Fourier transformations for the $\sin x/x$ function, and observe the difference in the second graphs of the FileFigs.5 and 6.

FileFig 9.5 (F5FTSINCRS

Real Fourier transformation: the original function is a one sided $\sin x/x$ function, 256 points. The real transformation is a single-sided step function shown for 128 points. The real inverse transformation reproduces the original function. The imaginary part is zero for the original, appears in the transform, and is zero again for the inverse transform.

F5FTSINCRS is only on the CD.

FileFig 9.6 (F6FTSINCCS)

Complex Fourier transformation: the original function is a one-sided $\sin x/x$ function, 256 points. The complex transform is a double-sided step function shown for 256 points. The complex inverse transformation reproduces the original function. The imaginary part is zero for the original, appears in the transform, and is zero again for the inverse transform.

F6FTSINCCS is only on the CD.

9.1.4.2 General Fourier Transformation

The fast Fourier transformation needs 2^n input points. In the case where we have a different number of input points we have to use the complex Fourier transformation and its inverse. In FileFig 9.7 we show the complex Fourier transformation for the step function using 184 points, and in FileFig 9.8 for the $\sin z/z$ function.

FileFig 9.7 (F7FTSTEP183S)

The original function is a step function. The number of points is 184. The complex Fourier transformation results in a double-sided $\sin z/z$ function. The inverse complex transformation reproduces the original function. The imaginary part is zero for the original, appears in the transform, and is zero again for the inverse transform.

F7FTSTEP183S is only on the CD.

FileFig 9.8 (F8FTSINC183S)

The original function is a $\sin z/z$ function. The number of points is 184. The complex Fourier transformation results in a double-sided step function. The inverse complex transformation reproduces the original function. The imaginary part is zero for the original, appears in the transform, and is zero again for the inverse transform.

F8FTSINC183S is only on the CD.

A comparison of the fast Fourier transformation with the general Fourier transformation is given in FileFigs.9 and 10. The complex Fourier transformation of a Gauss functions with different a values is given in FileFig 9.9 for 256 points, and in FileFig 9.10 for 326 points.

FileFig 9.9 (F9FTGAUSS)

The original function is the Gauss function with values of 50 and 100 for a. The number of points is 256. The complex Fourier transformation again results in a Gauss function, however, with a narrower shape. The inverse transformation reproduces the original function. The imaginary part is zero for the original, appears in the transform, and is zero again for the inverse transform.

F9FTGAUSS is only on the CD.

FileFig 9.10 (F10FTGAUSGS)

The original function is the Gauss function with values of 50 and 100 for a. The number of points is 326. The complex Fourier transformation again results in a Gauss function, however, with a narrower shape. The inverse transformation reproduces the original function. The imaginary part is zero for the original, appears in the transform, and is zero again for the inverse transform.

F10FTGAUSGS is only on the CD.

9.1.5 Fourier Transformation of a Product of Two Functions and the Convolution Integral

The Fourier transformation $S(y)$ of the function $G(\nu)$ is

$$S(y) = \int_{-\infty}^{\infty} G(\nu) \exp 2\pi \nu y d\nu. \tag{9.16}$$

We now discuss the case where $G(\nu)$ is the product of two functions $g_1(\nu)$ and $g_2(\nu)$,

$$G(\nu) = g_1(\nu)g_2(\nu). \tag{9.17}$$

In the cases where one wants to keep the two functions separate, one can represent the Fourier transformation of $G(\nu)$ by a convolution integral. The convolution integral theorem tells us that we may write the Fourier transform of the product $g_1(\nu)g_2(\nu)$ as the convolution integral of the Fourier transformations of $g_1(\nu)$ and $g_2(\nu)$. The Fourier transformation of $g_1(\nu)$ is

$$s_1(y) = \int_{-\infty}^{\infty} g_1(\nu) \exp 2\pi\nu y d\nu \tag{9.18}$$

and of $g_2(\nu)$

$$s_2(y) = \int_{-\infty}^{\infty} g_2(\nu) \exp 2\pi\nu y d\nu. \tag{9.19}$$

The Fourier transform $S(y)$ of the product $G(\nu) = g_1(\nu)g_2(\nu)$ is calculated from the convolution integral

$$S(y) = \int_{-\infty}^{\infty} s_1(\tau)s_2(y-\tau)d\tau, \tag{9.20}$$

where $s_1(y)$ and $s_2(y)$ are the Fourier transforms of $g_1(\nu)$ and $g_2(\nu)$ (see Eqs. (9.18) and (9.19)).

A simple example can be discussed by splitting the Gauss function

$$G(\nu) = \exp -(a^2\nu^2)/2 \tag{9.21}$$

into two, as

$$\exp - (3/4)(a^2\nu^2)/2 \tag{9.22}$$

$$\exp - (1/4)(a^2\nu^2)/2. \tag{9.23}$$

We calculate the Fourier transform of Eqs. (9.22) and (9.23) as

$$s_1(y) = \int_{-\infty}^{\infty} \{\exp -(3/4)(a^2\nu^2)/2\} \cos 2\pi\nu y d\nu \tag{9.24}$$

and

$$s_2(y) = \int_{-\infty}^{\infty} \{\exp -(1/4)(a^2\nu^2)/2\} \cos 2\pi\nu y d\nu \tag{9.25}$$

and using the formula of Eq. (9.6), one gets

$$s_1(y) = (\sqrt{2\pi}/a\sqrt{3}) \exp -(8\pi^2y^2)/(3a^2) \tag{9.26}$$

$$s_2(y) = (\sqrt{2\pi}/a) \exp -(8\pi^2y^2)/(a^2). \tag{9.27}$$

The Fourier transform of $G(\nu)$ (i.e., $S(y)$) is now obtained from the convolution integral

$$S(y) = \int_{-\infty}^{\infty} s_1(\tau)s_2(y-\tau)d\tau \tag{9.28}$$

or

$$S(y) = 8\pi/(\sqrt{3})(a^2) \int_{-\infty}^{\infty} \exp\{-(8\pi^2\tau^2)/(3a^2) - (8\pi^2/a^2)(y-\tau)^2\}d\tau. \tag{9.29}$$

Using the integral formula

$$\int_{-\infty}^{\infty} \exp -t^2 dt = \{\sqrt{\pi}\} \tag{9.30}$$

and introducing a complementary term $+(3y/2)^2 - (3y/2)^2$ in the exponent of the integral, one may write $[2\tau - (3y/2)]^2$ and obtain for Eq. (9.30),

$$S(y) = (\{\sqrt{2\pi}\}/a)\exp(-2\pi^2 y^2/a^2). \tag{9.31}$$

If we had calculated the Fourier transformation of $G(\nu) = \exp -(a^2\nu^2)/2$, we would have obtained the same result:

$$S(y) = (\sqrt{2\pi/a})\exp(-2\pi^2 y^2/a^2). \tag{9.32}$$

In FileFig 9.11 we show the convolution integral for the case where the two factors are a step and a $\sin cx$ function.

FileFig 9.11 (F11CONVOS)

Convolution integral of the step function and the $\sin x/x$ function.

F11CONVOS is only on the CD.

Application 9.11. Calculate analytically the integral and show the effect of the convolution process of the step function on the $\sin x/x$ function.

9.2 FOURIER TRANSFORM SPECTROSCOPY

9.2.1 Interferogram and Fourier Transformation. Superposition of Cosine Waves

We next discuss the superposition of plane waves in order to understand the formulas of Fourier transform spectroscopy. We consider a plane wave solution of the scalar wave equation

$$u(x,t) = A\cos 2\pi(x/\lambda - t/T + \phi). \tag{9.33}$$

In Chapter 2 we discussed the dependence on x and t and on the phase factor ϕ. We write the two cosine waves with path difference $x_2 = x_1 - \delta$ as

$$u_1(x,t) = A\cos 2\pi(x_1/\lambda - t/T) \tag{9.34}$$

$$u_2(x,t) = A\cos 2\pi[(x_1 - \delta)/\lambda - t/T]. \tag{9.35}$$

Using the formula

$$\cos\alpha + \cos\beta = 2[\cos(\alpha+\beta)/2][\cos(\alpha-\beta)/2], \tag{9.36}$$

we may write for the superposition $u = u_1 + u_2$,

$$\begin{aligned} u &= u_1 + u_2 \\ &= 2A[\cos(2\pi(\delta/2)/\lambda)]\{\cos[2\pi(x_1/\lambda - t/T) - 2\pi(\delta/2)/\lambda]\} \end{aligned} \tag{9.37}$$

and obtain for the intensity

$$I = \{2A[\cos(2\pi(\delta/2)/\lambda)]\}^2.\{\cos[2\pi(x_1/\lambda - t/T) - 2\pi(\delta/2)/\lambda]\}^2. \tag{9.38}$$

Only the second factor depends on the time. Since it oscillates very quickly; we can assume that only the average value is detected and results in a constant. Therefore we have

$$I = \{2A\cos(2\pi(\delta/2)/\lambda)\}^2 \quad \text{(times a constant)}. \tag{9.39}$$

We see that the intensity of the superposition of the two harmonic waves depends only on the optical path difference δ.

9.2.2 Michelson Interferometer and Interferograms

The Michelson interferometer (see Figure 9.1) has been discussed in Chapter 2, and we assume that an ideal beam splitter reflects and transmits 50% of the incident light. The incident beam is divided at the beam splitter and each beam is reflected at one of the two mirrors M_1 and M_2. Part of the reflected beam from M_1 is transmitted, and part of the beam coming from M_2 is reflected. These two parts are superimposed and travel to the detector.

The optical path difference is introduced by the displacement of the mirror M_2. If the mirror is displaced by $n\delta/2$, the optical path difference is $n\delta$ for each wavelength component of the incident beam. The center position is at $\delta = 0$. There we have constructed interference for all wavelengths. For constructed interference in the direction of the detector, for a specific wavelength no light travels back to the source. For destructive interference in the direction of the detector, the light travels back to the source.

Since only the space dependence of the harmonic wave is of importance we write the amplitude of the input wave as

$$u_1(x) = A\cos 2\pi(x_1/\lambda). \tag{9.40}$$

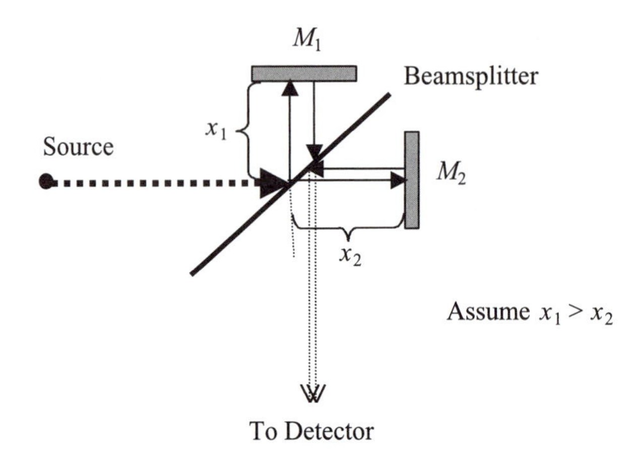

FIGURE 9.1 Schematic of Michelson interfometer. The optical path difference $x_2 - x_1 = \delta/2$ is produced by displacement of mirror $M2$.

The superposition of the two waves emerging from the beam splitter is (see Eq. (9.38))

$$u(\delta) = B\cos(2\pi(\delta/2)/\lambda), \tag{9.41}$$

where B is a constant. For the intensity we obtain

$$I(\delta) = \{B\cos(2\pi(\delta/2)/\lambda)\}^2, \tag{9.42}$$

where B^2 represents the effective intensity at the detector for $\delta = 0$.(Note the factor $\frac{1}{2}$ in the argument which comes from $\delta/2$ in Eq. (9.39)).

So far we have assumed for the input wave (Eq. (9.40)) only one wavelength or one frequency. The case of interest for Fourier transform spectroscopy is the determination of the frequency spectrum when the incident wave contains many frequencies. We assume that we have p waves of amplitude A_i and (discrete) wavelength λ_i which as frequencies we write as $\nu_i = 1/\lambda_i$ in cm^{-1}. The input of the sum of waves (amplitudes) is

$$u = \sum_{i=1}^{p} A_i \cos(2\pi\nu_i x_1) \tag{9.43}$$

and for the intensity of the output at the detector one has

$$I(y) = \sum_{i=1}^{p} \{B_i \cos(2\pi\nu_i y/2)\}^2, \tag{9.44}$$

where y is the optical path difference produced by the two mirrors. Each of the incident waves in the sum of Eq. (9.43) will separately produce a superposition pattern and all intensity patterns will be superimposed (see Eq. (9.44)). We apply to each term the formula

$$\cos^2 \alpha = (1/2)(1 + \cos 2\alpha) \tag{9.45}$$

and obtain

$$I(y) = \sum_{j=1}^{p} 2C_j^2\{1 + \cos(2\pi \nu_j y)\}. \tag{9.46}$$

Note that the factor 1/2 in the argument of Eq. (9.44) has now disappeared. Although we need a discontinuous representation for the realistic case of Fourier transform spectroscopy, we use integral presentation for the manipulations to obtain the interferogram function.

9.2.3 The Fourier Transform Integral

In Fourier transform spectroscopy we observe the intensity pattern at the detector, depending on a "length" coordinate, and calculate the frequency pattern using Fourier transformation.

We now consider a continuous frequency distribution from $\nu = 0$ to $\upsilon = \infty$ and write Eq. (9.46) as an integral over $d\nu$,

$$J(y) = 2\int_0^\infty G(\nu)\{1 + \cos(2\pi \nu y)\}d\nu, \tag{9.47}$$

where we have replaced C_j^2 by the continuous function $G(\nu)$ and the sum by an integral. For the specific case of $y = 0$ we get

$$J(0) = 4 \int_0^\infty G(\nu)d\nu. \tag{9.48}$$

Fourier integrals need for ν and y the range from $-\infty$ to ∞. We therefore formally extend the frequency range by assuming that $G(\nu)$ is symmetric around $\nu = 0$ and therefore that $G(-\nu) = G(\nu)$. For Eq. (9.48) we then may write

$$J(0) = 2 \int_{-\infty}^{\infty} G(\nu)d\nu \tag{9.49}$$

and write for the cos-function in Eq. (9.47) the corresponding exponential, including negative frequencies, and get

$$J(y) = \int_{-\infty}^{\infty} G(\nu)\{1 + e^{i(2\pi \nu y)}\}d\nu. \tag{9.50}$$

Introduction of Eq. (9.49) into Eq. (9.50) gives us

$$J(y) = (1/2)J(0) + \int_{-\infty}^{\infty} G(\nu)\{e^{i(2\pi \nu y)}\}d\nu. \tag{9.51}$$

The function $J(y)$ contains the observed data.

We define the interferogram function $S(y)$ as

$$S(y) = J(y) - (1/2)J(0) = \int_{-\infty}^{\infty} G(\nu)\{e^{i(2\pi\nu y)}\}d\nu. \tag{9.52}$$

The subtraction of 1/2 $J(0)$ from $J(y)$ produces negative values for $S(y)$ and makes it a Fourier integral of the type we have discussed above. If a double-sided interferogram is obtained, y runs from negative to positive values related to the position of the mirror M_2 of the Michelson interferometer. Using the fast Fourier transformation, we can only consider 2^n points. The input function starts with the positive part of $S(y)$ and the negative part follows as a mirror image. Therefore, the information we are interested in is only contained in the single-sided interferogram.

The integral $S(y)$ is one of the pair of Fourier transform integrals

$$S(y) = \int_{-\infty}^{\infty} G(\nu)\{e^{i(2\pi\nu y)}\}d\nu. \tag{9.53}$$

The other is

$$G(\nu) = \int_{-\infty}^{\infty} (S(y)\{e^{-i(2\pi\nu y)}\}dy. \tag{9.54}$$

(Note the negative sign in the exponent.) The integral of Eq. (9.54) is the one we need to calculate the frequency spectrum of the observed data, presented by $S(y)$.

We have assumed that both variables ν and y have continuous values from $-\infty$ to $+\infty$. When $G(\nu)$ and $S(y)$ are symmetrical with respect to zero, the imaginary part disappears and one has to deal only with one side of the integral. The Fourier transformation pairs are then

$$S(y) = 2\int_{0}^{\infty} G(\nu)\{\cos(2\pi\nu y)\}d\nu \tag{9.55}$$

$$G(\nu) = 2\int_{0}^{\infty}(S(y)\{\cos(2\pi\nu y)\}dy. \tag{9.56}$$

9.2.4 Discrete Length and Frequency Coordinates

9.2.4.1 Discrete Length Coordinates

The Fourier integral of Eq. (9.56) is the Fourier transformation of the interferogram function. For numerical calculations it is written as

$$G(\nu_i) = \sum_{k=1}^{n} S(y_k)\cos(2\pi\nu_i y_k). \tag{9.57}$$

Since the coordinates ν_i and y_k each have n discrete values, the Fourier transformation is equivalent to the solution of a system of n linear equations.

The input data of the interferogram function $S(y_k)$ are the measured data. The mirror M_2 of the Michelson interferometer (Figure 9.1) is moved at equal steps from one position to the next, and the values of the interferogram are recorded at the detector. The zero path length position is the location of the central maximum. The input values of the interferogram function $S(y_k)$ are calculated by subtraction of half the value of the central maximum (Eq. (9.52)).

In the cosine functions of Eq. (9.57) we have a set of values of y_k and as a consequence we have a repetition of possible values of ν_i. Let us choose for the length interval $y_k = k$, which means we have 128 points from $k = 1$ to 128, and consider

$$\cos[2\pi(\nu_i k)]. \tag{9.58}$$

When k is 128, and ν_i is 1/128 we have $\nu_i k = 1$, which means all values of the cosine are in one cycle for $k = 1$ to 128 and $\nu_i = 1/128$. When the product $(\nu_i k)$ is larger than 1, we have repetition. Let us assume the value $\nu_i = 133/128 = 1 + 5/128$,

$$\begin{aligned} \cos[2\pi(133/128)k] &= \cos[2\pi(1 + 5/128)k] \\ &= \cos[2\pi(5/128)k). \end{aligned} \tag{9.59}$$

Because of the periodicity of the cos-function, the frequency $\nu_i = 133/128$ is equivalent to $\nu_i = 5/128$. When using $k = 1$ to 128 points in the space coordinate domain, after application of the Fourier transformation, all that we can get are 128 frequencies in the frequency domain. The repetition at 2π is shown in FileFig 9.12. In addition, the cos-functions of Eq. (9.57) have "mirror" symmetry at the center of the interval (at $i = 64$), and at both ends (shown in FileFig 9.13). In FileFig 9.14 graphs are shown of amplitudes and intensities of the superposition of cosine waves. These superpositions are simulations of interferograms depending on the optical path difference.

FileFig 9.12 (F12FTDISC1S)

Graphs are shown of $\cos(2\pi\nu_i k)$ for the space coordinate from 1 to 128 and frequencies 1/128, 2/128, 3/128, 64/128, 127/128, 128/128, 129/128, and 130/128.

F12FTDISC1S is only on the CD.

Application 9.12. Consider a total of $n = 32$ points in the space domain and show on a graph the first five frequencies. Then find the functions that have the same appearance around frequency numbers 32 and 64.

FileFig 9.13 (F13FTDISC2S)

Graphs are shown of $\cos(2\pi\nu_i k)$ for the space coordinate from 1 to 128 and frequencies 1/128, 2/128, 63/128, 64/128, 65/128. A section of $k = 20$ to 80 of $\cos(2\pi\nu_i k)$ for 63/128, 64/128, and 65/128 shows that the frequency of 63/128 appears similar to 65/128 and both have a larger cycle than 64/128.

F13FTDISC2S is only on the CD.

Application 9.13. Make graphs of $\cos(2\pi\nu_i k)$ for higher and lower frequencies than 64/128 and show that they appear similar to graphs of lower and higher frequency numbers.

FileFig 9.14 (F14MICHOPS)

Simulations of interferograms for the Michelson interferometer. Graphs of amplitudes and intensities of cosine functions depending on optical path difference.

F14MICHOPS is only on the CD.

9.2.4.2 Sampling

We have seen in FileFig 9.13 that the cos-function of Eq. (9.57) has a symmetrical appearance around the frequency position 64/128, when using 128 points in space and having exactly two space points per cycle. For this special example two points per cycle have been used for the characterization of the frequencies. In general, we have from information theory the sampling theorem.

Theorem 9.1. *Two points per cycle are needed for sampling a continuous periodic function in order not to lose information.*

In our example, the continuous functions are cos-functions. The cos-function with the highest frequency is sampled by using two points per cycle. For example, one is at the maximum and one is at the minimum. Lower frequencies are also correctly sampled, but higher frequencies present a problem. They appear as lower frequencies and need to be removed by a filter. When talking about wavelengths instead of frequencies, we similarly have that the sampling interval is equal to one-half of the wavelength of the shortest wavelength (highest frequency). Calling the sampling interval l, we have

$$l = 1/(2\nu_M), \tag{9.60}$$

where ν_M is the highest frequency of the spectrum under consideration. In Fourier transform spectroscopy it is assumed that all higher frequencies are removed.

9.2.5 Folding of the Fourier Transform Spectrum

What happens when the choice of the sampling interval is done incorrectly? How is the spectrum affected? We show that higher frequencies are folded around the highest frequency, corresponding to the correct sampling interval, and appear as lower frequencies in the spectrum. We use for the space domain the coordinate points $1/256, 2/256, \ldots, 1$ and consider the sum of cos-functions with frequencies 65, 85, and 105.

1. The sampling interval is 1/256. We have

$$y1_i = \cos(2\pi \cdot 65(i/256) + \cos(2\pi \cdot 85(i/256) + \cos(2\pi \cdot 105(i/256), \tag{9.61}$$

where i runs from 1 to 256.

2. The sampling interval is 2/256. We consider

$$y2_i = \cos(2\pi \cdot 65(2i/256) + \cos(2\pi \cdot 85(2i/256) + \cos 2\pi 105(2i/256), \tag{9.62}$$

where i runs from 1 to 256.

3. The sampling interval is 4/256. We have

$$y4_i = \cos(2\pi \cdot 65(4i/256) + \cos(2\pi \cdot 85(4i/256) + \cos(2\pi \cdot 105(4i/256), \tag{9.63}$$

where i runs from 1 to 256.

In FileFig 9.15 we compare the appearance of the spectrum for these three sampling intervals 1/256, 2/256 and 4/256.

FileFig 9.15 (F15FOLDS)

Folding of the spectrum:

1. Sampling interval 1/256 for the function $y1$. The three frequencies to be investigated are 65, 85, and 105. The highest frequency is 128. All frequencies appear at the right place in the Fourier transformation spectrum.

2. Sampling interval 2/256 for function $y2$. We now use a sampling interval twice as large. Consequently the highest frequency is now 64 and the three frequencies 65, 85, and 105 are all higher than 64. The spectrum appears folded. In the Fourier transformation the three frequencies appear at 45, 85, and 125. Since the frequency spectrum is shown over 128 points, we may look at the function $y2$ as it would have frequencies $2 \cdot 65$, $2 \cdot 85$ and $2 \cdot 105$. It is sampled with the sampling interval 1/256 (as we did under (a), where the highest frequency is 128. We then have to look for the frequencies 130, 170, and 210, which all exceed the frequency interval from 1 to 128. We saw in

FileFigs 9.12 and 13 that higher frequencies are folded back into the spectrum around the highest frequency (which is 128 because we sample $y2$ now like $y1$). We have to subtract from 130, 170, and 210 the highest frequency 128 and find 2, 42, and 82. These values have to be traced back from 128 because of the folding. We get 126, 86, and 46, which are the frequencies we should find in the graph of the Fourier transformation. However, we find 125, 85, 45 because we have not taken into account that the Fourier transformation program starts at 0 and not at 1.

3. **Sampling interval 4/256 for function $y4$.** We now use a sampling interval 4/256 to sample the same function with 4/256. The highest frequency is now 32. In the Fourier transformation the three frequencies appear at 5, 85, and 90. Since the frequency spectrum is shown over 128 points, we look at $y4$ as it would have frequencies $4 \cdot 65$, $4 \cdot 85$, and $4 \cdot 105$. It is sampled with the sampling interval 1/256, as $y1$ was. We have to look for the frequencies 260, 340, and 420, which all exceed the frequency interval from 1 to 128. We saw in FileFigs.12 and 13 that higher frequencies are folded back into the spectrum around the highest frequency (which is 128 because we sample $y4$ now like $y1$). We have to subtract from 260, 340, and 420 the highest frequency 128, and get 132, 212, and 293. Then we have to trace back from 128, because they all exceed 128 and get 4, 84, and 165 and have to fold again at 0 into forward (i.e., 4, 84, and 165). The first two are now in the right position, but 165 exceeds 128. We have to subtract 128 from 165 which is 37 and trace it back from 128, that is, we get to 91. We finally get for the position of the three frequencies 4, 84, and 91, and those are the frequencies we should find in the graph of the Fourier transformation. However, we find 5, 85, and 90. The first two are folded twice, the last three times. We have not taken into account that the Fourier transformation program starts at 0 and not at 1.

F15FOLDS

Folding of the Spectrum

For the sampling interval 1.255, hightest frequency is 128; the frequencies are at 65, and 105, all below 127.

1. Sample interval $i/255$

$$i := 0 \ldots 255$$

$$y1_i := \cos\left(2 \cdot \pi \cdot 65 \cdot \frac{i}{255}\right) + \cos\left(2 \cdot \pi \cdot 85 \cdot \frac{i}{255}\right) + \cos\left(2 \cdot \pi \cdot 105 \cdot \frac{i}{255}\right).$$

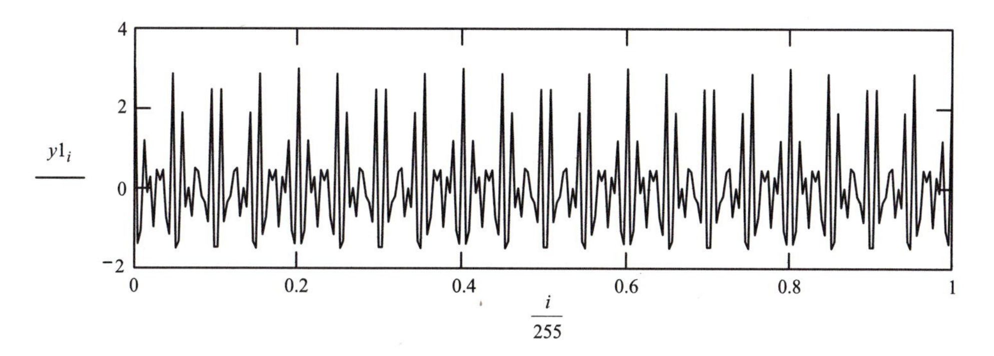

$$c := fft(y1)$$

$$N := \text{last}(c) \qquad N = 128$$

$$j := 0 \ldots 128$$

Frequency peaks are at 65, 85, 105.

2. Sample interval $2i/255$

For the sampling interval 2/255, highest frequency is 64: the original frequencies are at 65, 85, 105: they are all larger 64 and appear folded.

$$y2_i := \cos\left(2 \cdot \pi \cdot 65 \frac{i}{255}\right) + \cos\left(2 \cdot \pi \cdot 85 \cdot \frac{i}{255}\right) + \cos\left(2 \cdot \pi \cdot 105 \frac{i}{255}\right).$$

$$c := fft(y2)$$

$$N := \text{last}(cc) \qquad N = 128$$

$$j := 0 \ldots 128$$

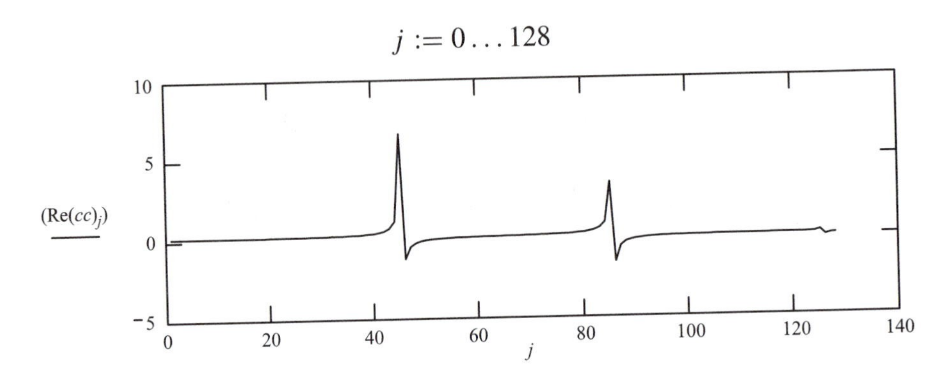

Frequency peaks appear at 45, 85, 125.

3. Sample interval $4i/255$
For the sampling interval 4/255, highest frequency 32, the frequencies are higher than 1 times 32 and 2 times 32.

$$y4_i := \cos\left(2 \cdot \pi \cdot 65 \frac{4 \cdot i}{255}\right) + \cos\left(2 \cdot \pi \cdot 85 \cdot \frac{4 \cdot i}{255}\right) + \cos\left(2 \cdot \pi \cdot 105 \frac{4 \cdot i}{255}\right).$$

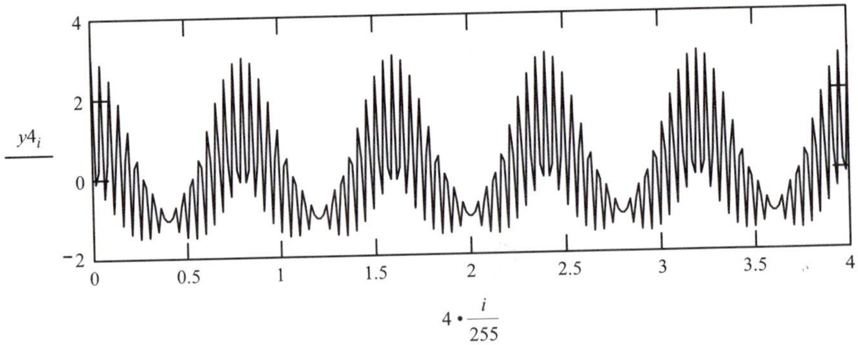

$$ccc := fft(y4)$$

$$N := \text{last}(ccc) \qquad N = 128$$

$$j := 1 \ldots 128$$

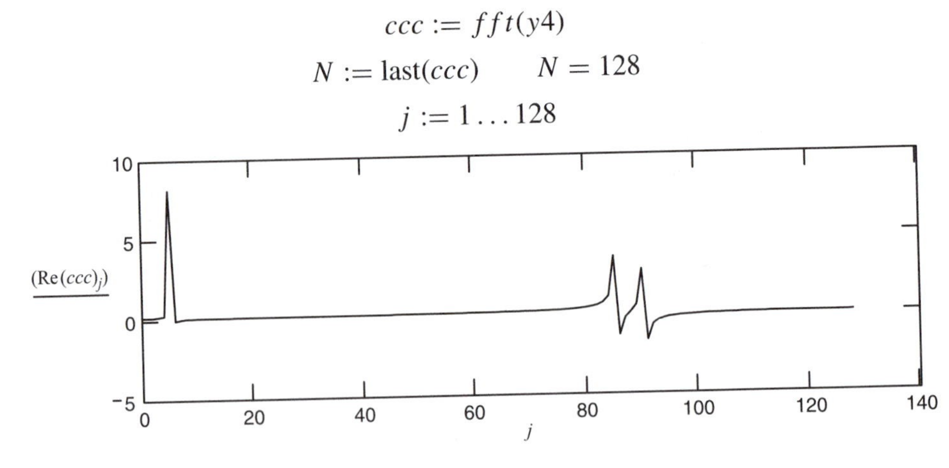

Freguency peaks appear

$$65 \text{ at } \rightarrow 125$$
$$85 \text{ at } \rightarrow 185$$
$$105 \text{ at } \rightarrow 45.$$

Application 9.15. Use the three frequencies 15, 34, and 97. Determine the appearance in the Fourier transformation as discussed for the original set (65, 85, 105) of frequencies.

9.2.6 High Resolution Spectroscopy

In high resolution spectroscopy one is often interested in investigating a narrow bandwidth of frequencies for high resolution. The sampling interval l in the length domain is related to the highest frequency ν_M in the 1/length domain. When increasing the number of points N in the length domain, one also increases the number N of points in the frequency domain. Since the sampling interval determines the frequency interval from 0 to ν_M, a higher number of points in the length domain results in a higher number of points in the frequency domain. One gets higher resolution in the interval from 0 to ν_M. The resolution is inverse proportional to the total length L of the interferogram. The length may be expressed as $L = N\,l$, and one has

$$Nl = L = 1/(2(\nu_M/N)) \tag{9.64}$$

and obtains for ν_M/N

$$\nu_M/N = \Delta\nu = 1/2L. \tag{9.65}$$

A high resolution interferometer uses a large optical path difference L in order to make $\Delta\nu$ small. If we take, for example, $L = 50$ cm, we have for $\Delta\nu = (1/100)$ cm^{-1} and for the resolving power $R = \nu_M/\Delta\nu$ at 100 microns (equal to 100 cm−1)

$$R = \nu_M/\Delta\nu = 100/(1/100) = 10,000. \tag{9.66}$$

The number of points in length space is equal to the number of frequency intervals in frequency space, which is

$$N = 100/(1/100) = 10,000 \tag{9.67}$$

and is also equal to R. To record 10,000 points, assuming about 3 sec for each point, would take 9 hours. The information we obtain in this way is a spectrum from 1 to 100 cm^{-1} with resolution of 1/100 cm^{-1}.

In the case where one is only interested in the study of a section of the spectrum one may use folding of the spectrum. Let us assume that we are interested in the section from 2/3 of 100 cm^{-1} to 100 cm^{-1}. In that case we use a bandpass

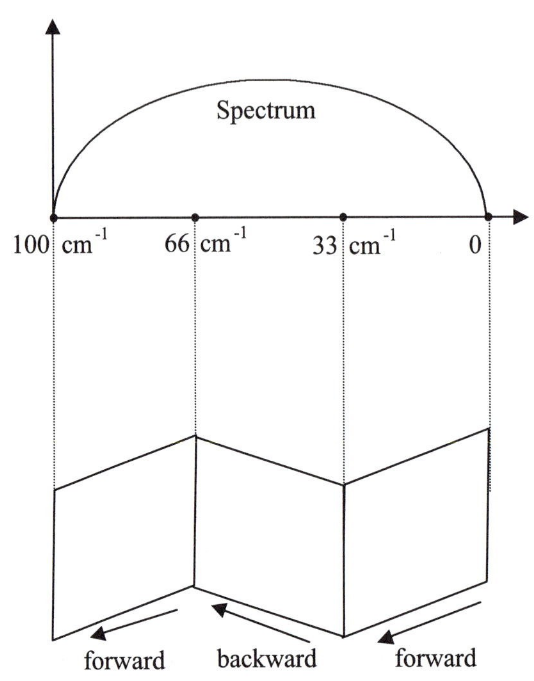

FIGURE 9.2 We assume that a bandpass filter eliminates the spectrum from 1 to 66 cm^{-1}. The spectrum is folded by using a sampling interval of 3 times l, where l is the sampling interval, corresponding to the highest frequency 100 cm^{-1}; that is, $1 = 1/2 \cdot 100$ cm^{-1} = 0.005 cm = 50 microns. The frequency scale for the sections changes the direction to the highest value for even or odd factors of folding.

filter and eliminate all other frequencies, and fold the spectrum three times by choosing a sample interval three times larger (see Figure 9.2). This is similar to what we discussed above in FileFig 9.15, where we studied the folding of a sum of cos-functions sampled with intervals 1/256, 2/256, and 4/256.

We have $\nu_M/3$ for the width of the spectral sections to be studied with the length L of the interferogram. Consequently, the sampling interval has the value $3\,l = L/3333$, instead of $l = L/(10, 000)$. The total number of points is reduced from 10,000 to 3333, but the length of the interferogram remains the same. Therefore we obtain the same high resolution in a smaller spectral region with one-third of the points. A very important fact is that we have to use the bandpass filter we assumed to apply. The spectra of the two sections of 1 to 33 cm^{-1} and 33 cm^{-1} to 66 cm^{-1} are folded onto the spectrum of 66 cm^{-1} to 100 cm^{-1}. If they contain spectral information, the spectrum will be “messed up,” like the Fourier transforms in FileFig 9.15. We obtain the same high resolution for a smaller part of the spectrum, using a larger sampling interval and fewer points. In our case the spectrum is obtained in one-third of the time.

In Figure 9.3 we show the background spectrum, taken with a Michelson interferometer and a bandpass filter. The bandpass has to have a width of ν_M/integer.

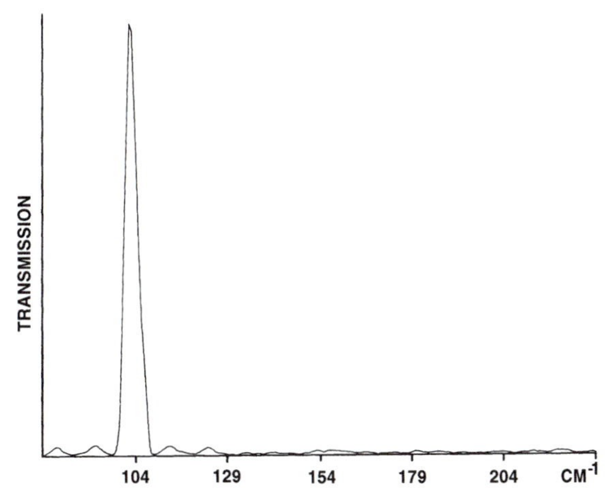

FIGURE 9.3 The background spectrum taken with a Michelson interferometer and a bandpass filter. (From J. Kachmarsky et al., Far-infrared high-resolution Fourtier transform spectrometer: applications to H_2O, NH_3, and NO_2 lines. *Applied Optics, 15*, 1976, p 708–713.)

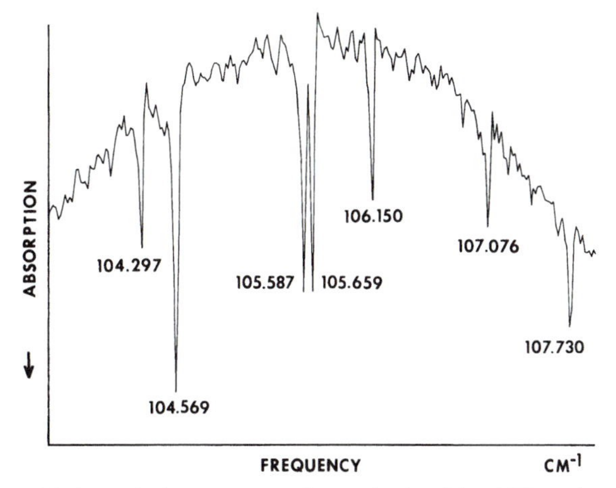

FIGURE 9.4 High-resolution spectrum of water in the 104 to 107 cm^{-1} region obtained by folding the spectrum six times. (From J. Kachmarsky et al., Far-infrared high-resolution Fourtier transform spectrometer: applications to H_2O, NH_3, and NO_2 lines. *Applied Optics, 15*, 1976, p 708–713.)

In Figure 9.4 we show a high-resolution spectrum of water in the 104 to 107 cm^{-1} region, obtained with a sampling interval six times larger and in a time span six times shorter. We state the procedure in more general terms as follows. A filter is used to eliminate all parts of the spectrum with frequencies higher than ν_M and lower than ν_m, where m stands for minimum. The value of ν_m has to be chosen in such a way that $q = \nu_M/(\nu_M - \nu_m)$ is an integer. We then obtain with N sample intervals, each calculated by using

$$l = 1/\{2(\nu_M - \nu_m)\}, \tag{9.68}$$

a spectrum of width ν_M/q and a resolution of $\Delta\nu = 1/(2L)$, where $L = Nl$. The scale has its highest value ν_M/q and runs forward for q odd and backward for q even (Figure 9.2).

9.2.7 Apodization

The Fourier integrals, as shown in Eq. (9.55) and (9.56),

$$S(y) = 2\int_0^\infty G(\nu)\{\cos(2\pi\nu y)\}d\nu \tag{9.69}$$

$$G(\nu) = 2\int_0^\infty S(y)\{\cos(2\pi\nu y)\}dy \tag{9.70}$$

have a range of integration from 0 to ∞. The discrete Fourier transform, as shown in Eq. (9.57),

$$G(\nu_j) = \sum_{k=1}^{n} S(y_k)\{\cos(2\pi\nu_j y_k)\}, \tag{9.71}$$

has a finite summation range, instead of the infinite large integration range. In the section on Fourier transformation, we reduced the integration range for the Fourier transformation of the slit from $-\infty$ to ∞ to a range from $-a$ to a. We did this because the width of the slit is $2a$, and outside the interval from $-a$ to a there is no information. We may multiply the interferogram function $S(y)$ in the integral of Eq. (9.70) with a step function $p(y)$, which is 1 in the interval from $-a$ to a and otherwise zero. We introduce into Eq. (9.70) the step function $p(y)$,

$$G(\nu) = \int_{-\infty}^{\infty} p(y)S(y)\exp(-i2\pi\nu y)dy \tag{9.72}$$

and reduce the infinite range of integration to a finite range $-a$ to a, which we may write as

$$G(\nu) = \int_{-a}^{a} S(y)\exp(-i2\pi\nu y)dy. \tag{9.73}$$

From the point of view of calculating a Fourier transformation over a limited range of space, Eq. (9.73) is similar to the discrete Fourier transformation of Eq. (9.71). In FileFig 9.17 we discuss an example of the Fourier transformation of a cos-function integrated over a finite range. We use the function $y_k = \cos(2\pi fk/255)$, with frequency $f = 31$, plotted over a range of $k = 0$ to 400 and do the multiplication with a step function p_k of width 256 points. The result of the real Fourier transformation is that we obtain a peak at $f = 31$, not infinitely narrow and with a loop extending to negative values. Since negative intensities do not appear in spectroscopy, it is desirable to find a procedure to avoid this artifact. The situation can be corrected by using in place of p_k, which is a step function, the function q_k, which is a triangular function, defined as

$$q_k = 1 - k/255. \tag{9.74}$$

The multiplication of y_k by q_k and their Fourier transformation is shown in FileFig 9.16. The function y_k decreases to 0, and remains 0 in the range of integration, which was set to zero when changing from the infinite range of integration to the finite range. Application of the Fourier transformation shows that the negative values of the frequency spectrum have disappeared. This is called *apodization*. A drawback to this process is that the resolution is reduced by a factor of about 2.

FileFig 9.16 (F16APODIS)

Fourier transformation, sin-function, and apodization:

1. the infinite range of integration is reduced to a finite range by the use of a step function. As a result, the Fourier transformation shows negative intensity values;
2. the use of the triangular function $q_k = 1 - k/255$ eliminates the negative values but reduces the resolution by a factor of 2.

F16APODIS

Fourier Transformation of Sine Function and Apodization

1. Original function

$$k := 0 \ldots 255 \qquad y_k := \left(\cos\left(2 \cdot \pi \cdot f \cdot \frac{k}{255}\right)\right) \qquad f \equiv 31.$$

2. Step function: $i := 0 \ldots 400$

$$p_i := (\Phi(i) - \Phi(i - (d))) \qquad d \equiv 255$$

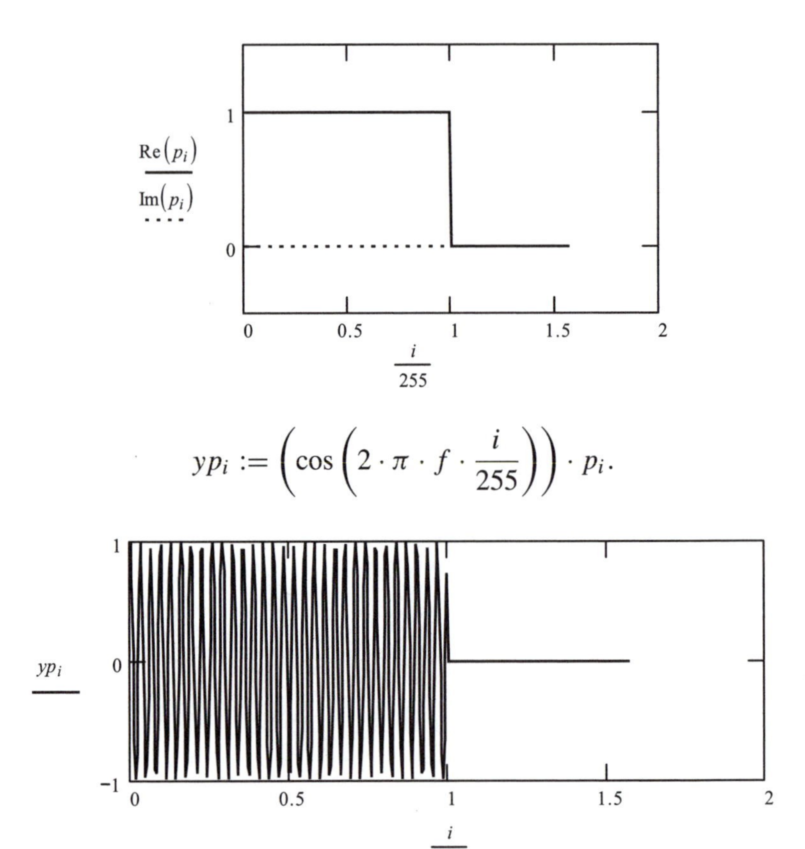

$$yp_i := \left(\cos\left(2 \cdot \pi \cdot f \cdot \frac{i}{255}\right)\right) \cdot p_i.$$

3. Fourier transformation of $y \times p$: we have to use 255 points

$$x_k := (\cos(2 \cdot \pi \cdot f \cdot \frac{k}{255})) \cdot (\Phi(k) - \Phi(k - (d))) \qquad k := 0 \ldots 255$$

$$c := fft(x)$$

$$N := \text{last}\,(c) \quad N = 128$$

$$j := 0 \ldots N.$$

5
0
$(\mathrm{Re}(c)_j)$
0 50 100 150
j
trace 1

4. Triangle function

$$ay_k := q_k \cdot y_k \qquad q_k := 1 - \frac{k}{255}.$$

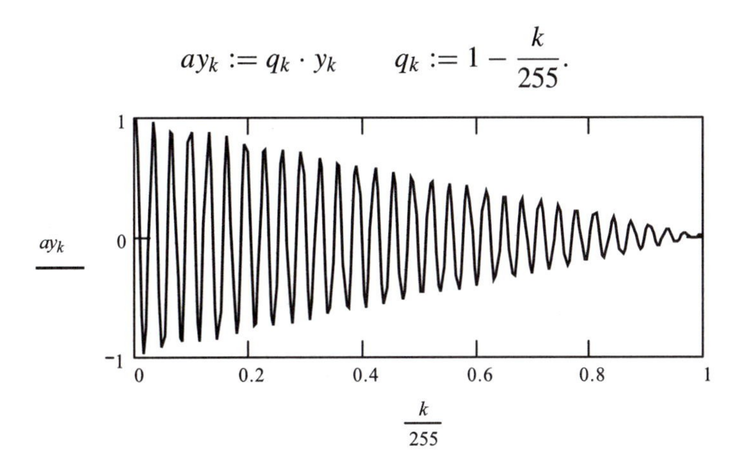

5. Fourier transformation of $y \times p$; we have to use 255 points

$$N = 128$$
$$j := 0 \ldots N$$
$$c := fft(ay)$$
$$N := \text{last}(c).$$

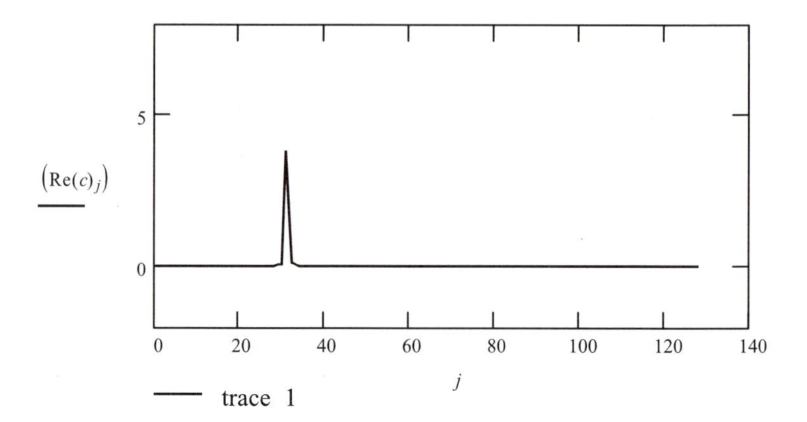

Application 9.16.

1. Use the apodization function $qq_k = (1 - k/255)^2$ and calculate the apodized function and the Fourier transform. Compare with the use of p_k and q_k.
2. Study the impact of the reduction of resolution. Repeat the procedure for a sum of five cosine functions of five different frequencies, such as 30, 35, 40, 45, and 50. Study the resolution using the apodization functions p_k, q_k, and qq_k.

APPENDIX 9.1

A9.1.1 Asymmetric Fourier Transform Spectroscopy

We have discussed the use of the Michelson interferometer for spectroscopy. The sample was assumed to be positioned either before the light was divided at the beam splitter or after the light was recombined. The reflection and transmission properties of the two arms of the interferometer were assumed to be equal and the only asymmetry was the difference in the path. To calculate the spectrum we could use the "real" Fourier transformation.

We now assume that the sample is at the surface of one of the mirrors of the Michelson interferometer. The reflection properties of the two arms of the interferometer are then different and the complex Fourier transformation must be applied. As a result of using the complex Fourier transformation one has more information available (complex numbers instead of real numbers). We may calculate not only the intensity (square of amplitude) but also obtain phase information. If in the case of reflection for a certain frequency range, the reflected amplitude and the phase change at reflection are known, the optical constants n and K may be calculated. We start with Eqs. (9.34) and (9.35), describing the two beams reflected from the two mirrors in the Michelson interferometer. We assume that the light is reflected in the fixed arm by a mirror producing the same magnitude of the amplitude and a phase shift of π. Using complex notation we have

$$u_1 = A(\exp i\pi)\exp i\{2\pi(x_1/\lambda - t/T)\}. \tag{A9.1}$$

The sample is assumed to be on the mirror in the movable arm of the interferometer, and we assume an amplitude reflection coefficient r and a phase change φ. We call y the displacement from the equal path length position and have

$$u_2 = A(r\exp i\varphi)\exp i\{2\pi(\{x_1 + y\}/\lambda - t/T)\}. \tag{A9.2}$$

Superposition of the two waves gives us

$$\begin{aligned} u &= u_1 + u_2 \\ &= A\{r\exp i\varphi + \exp i\{2\pi y/\lambda + (\pi)\}\exp i\{2\pi(x_1/\lambda - t/T)\} \end{aligned} \tag{A9.3}$$

and for uu^* we calculate

$$uu^* = A^2\{1 + r^2 + 2r\cos(\varphi - \pi - 2\pi y/\lambda)\}. \tag{A9.4}$$

Introduction of R for r^2 and for the frequency $\nu = 1/\lambda$, one has for the intensity I,

$$I(y) = A^2\{1 + R + 2r\cos(\varphi - \pi - 2\pi y\nu)\}. \tag{A9.5}$$

If we have a continuous spectrum of frequencies ν, from 0 to ∞ we have

$$I(y) = \int_0^\infty A^2\{1 + R + 2r\cos(\varphi - \pi - 2\pi y\nu)\}d\nu. \tag{A9.6}$$

For very large values of y the cosine term oscillates so fast that the average will be zero. We call

$$I(\infty) = \int_0^\infty A^2\{1 + R\}d\nu \tag{A9.7}$$

and get for Eq. (A9.6)

$$I(y) - I(\infty) = \int_0^\infty A^2\{2r\cos(\varphi - \pi - 2\pi y\nu)\}d\nu. \tag{A9.8}$$

Introduction of $2\cos x = \exp(ix) + \exp(-ix)$ results in

$$\begin{aligned} &I(y) - I(\infty) \\ &= \int_0^\infty A^2 r\{\exp i(\varphi - \pi - 2\pi y\nu) + \exp -i(\varphi - \pi - 2\pi y\nu)\}d\nu. \end{aligned} \tag{A9.9}$$

The difference in phase change for reflection on the sample $(\pi - \varphi)$ may also be written as $(\varphi - \pi)$. The sign is of no importance. We can then factor out $\exp i(\varphi - \pi)$ and extend the frequency range formally to negative frequencies (as done above in Section 2.3).

$$I(y) - I(\infty) = \int_{-\infty}^{\infty} \{A^2 r\exp i(\varphi - \pi)\}\{\exp i(2\pi y\nu)\}d\nu. \tag{A9.10}$$

This is the Fourier transform integral in the frequency domain for the function $\{A^2 r(\nu) \exp i(\varphi - \pi)\}$, and the corresponding Fourier transform integral in coordinate space is

$$\{A^2 r(\nu) \exp i(\varphi - \pi)\} = \int_{-\infty}^{\infty} \{I(y) - I(\infty)\} \exp -i(2\pi y\nu) d\nu. \tag{A9.11}$$

Equation (A9.11) is the Fourier transformation of the data $\{I(y) - I(\infty)\}$. Performing the Fourier transformation, we obtain for the integral in Eq. (A9.11) a function of complex numbers depending on ν, and call this function $P(\nu) - iQ(\nu)$,

$$\int_{-\infty}^{\infty} \{I(y) - I(\infty)\} \exp -i(2\pi y\nu) d\nu = P(\nu) - iQ(\nu). \tag{A9.12}$$

The background interferogram is obtained with a mirror in each arm of the Michelson interferometer, and one has for the transformation

$$\begin{aligned} &\{A^2 r \exp i(-\pi)\} \\ &= \int_{-\infty}^{\infty} \{I_B(y) - I_B(\infty)\} \exp -i(2\pi\nu) d\nu = P_B(\nu) - iQ_B(\nu), \end{aligned} \tag{A9.13}$$

where r is assumed to be set to 1 because of the background. Dividing the transformations of the sample and background interferogram we have

$$r(\nu) \exp \varphi(\nu) = \{P(\nu) - iQ(\nu)\}/\{P_B(\nu) - iQ_B(\nu)\}. \tag{A9.14}$$

For the intensity reflection coefficient r^2 we then get

$$r^2 = \{P(\nu)^2 + Q(\nu)^2\}/\{P_B(\nu)^2 + Q_B(\nu)^2\}. \tag{A9.15}$$

Since the background spectrum is real, one may write

$$P(\nu) - iQ(\nu) = (\text{Realnumber}) \exp i\varphi(\nu) \tag{A9.16}$$

and

$$\tan\varphi = Q(\nu)/P(\nu), \tag{A9.17}$$

where the sign of φ, that is, its relation to 2π, has to be chosen.

The optical constants are obtained by using Fresnel's formulas for normal incidence and complex refractive index

$$r^2 = \{(1-n)^2 + K^2\}/\{(1+n)^2 + K^2\} \tag{A9.18}$$

and

$$\tan\varphi = (2K)/(n^2 + K^2 - 1). \tag{A9.19}$$

Explicit calculation of n and K results in

$$n = (1 - r^2)/(1 + 2r\cos\varphi + r^2) \tag{A9.20}$$

$$K = (-2r\sin\varphi)/(1 + 2r\cos\varphi + r^2), \tag{A9.21}$$

where the sign of φ is important in Eq. (A9.21).

We have obtained from the Fourier transformation the complex numbers $P(\nu)$ and $Q(\nu)$ and could use them to calculate r and φ. From Fresnel's formulas we calculate n and K.

C H A P T E R 

Imaging Using Wave Theory

10.1 INTRODUCTION

In geometrical optics we used the thin lens equation to find the image point of an object point when using a thin lens of focal length f. Using wave theory we assume that Huygens' wavelets emerge from each point of the object and travel to the lens. The lens produces the diffraction pattern of the object in its focal plane, which may be seen as the Fourier transformation of the object pattern. The lens also produces the image of the object by refraction. A second Fourier transformation performed on the diffraction pattern results in a pattern having the shape of the object. Since the light travels forward, we associate it with the image. This is schematically shown in Figure 10.1.

The model we use for the description of image formation by a lens is, that one Fourier transformation is applied to the object pattern to obtain the diffraction pattern of the object and a second Fourier transformation is applied to the diffraction pattern to obtain the image pattern. Since we found that the Fourier transform of a Fourier transform is the original, but know that experimentally the image is not exactly the same as the object, we may ask the question, "Where is the perturbation entering the process?" It has been one of the great discoveries of optics (E. Abbe in Born and Wolf, 1964, xxi) that any perturbation of the diffraction pattern modifies the image. In Figure 10.2 we show how changes in the diffraction pattern, such as blocking off certain parts, change the image. We may also introduce phase shifts at some spots and not at others and obtain an image with much stronger contrast (F. Zernicke in Born and Wolf, 1964, xxi).

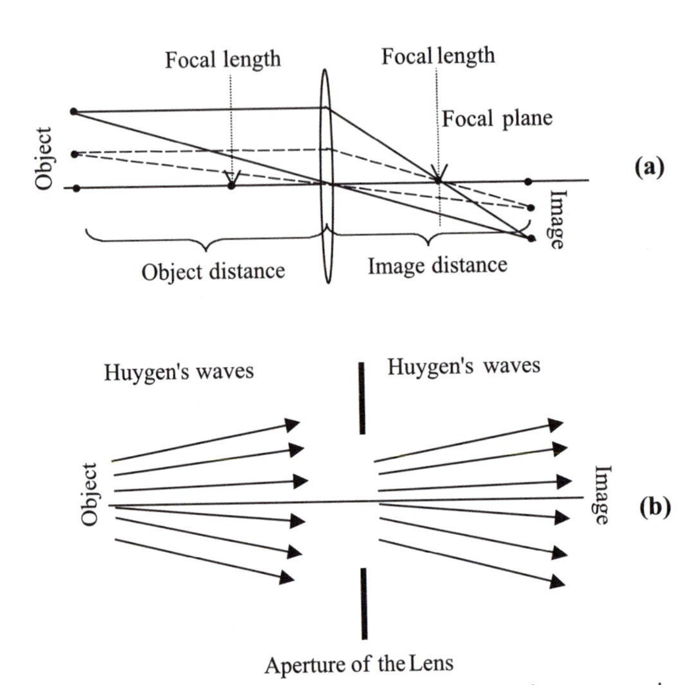

FIGURE 10.1 (a) The geometrical optical imaging process; (b) imaging process using wave theory. The Huygens' wavelets of the object generate the diffraction pattern, and the Huygens' wavelets of the diffraction pattern generate the image.

10.2 SPATIAL WAVES AND BLACKENING CURVES, SPATIAL FREQUENCIES, AND FOURIER TRANSFORMATION

Using scalar diffraction theory, the Kirchhoff–Fresnel integral uses monochromatic light to describe the diffraction pattern of the light emerging from the object. A lens is used in Fraunhofer diffraction to have the diffraction pattern observed in the focal plane of the lens. This same integral may be written as a Fourier transform integral, as done in Chapter 3 for the diffraction on a slit. The coordinates of the object, the slit, are length coordinates in the length domain. The coordinates of the Fourier transformation, the diffraction pattern of the slit, are coordinates in the spatial frequency domain and have 1/length dimensions. We note that we deal with an amplitude diffraction pattern in the frequency domain, which contains phase information, even if we started with a real function in the object plane. After one applies a second Fourier transformation on the diffraction pattern (in the spatial frequency domain), the result is a geometrical image pattern similar to the original object and appearing in the space domain.

In our model description we use the first Fourier transformation from the geometrical space domain into the spatial frequency domain. The object is described by geometrical spatial waves and the Fourier transformation describes the fre-

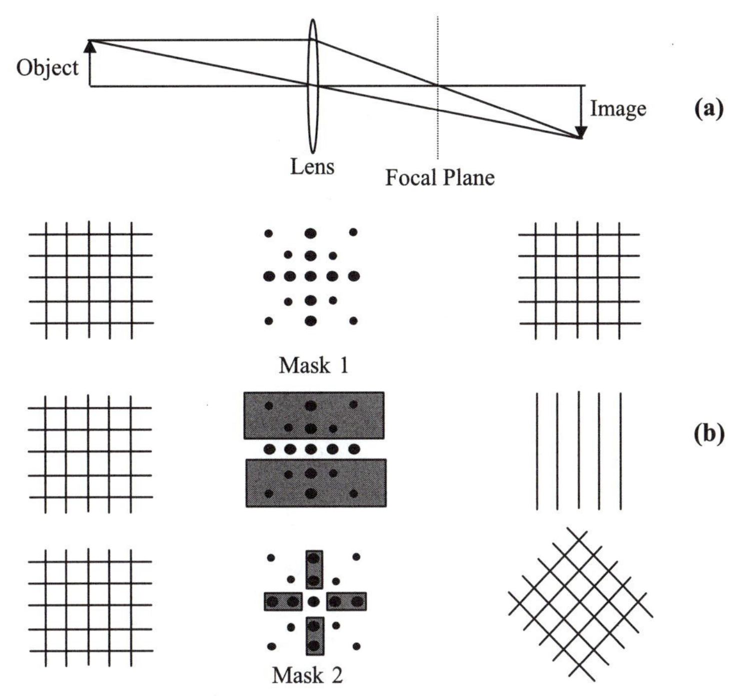

FIGURE 10.2 (a) Schematic of image formation by geometrical optics, the focal plane is indicated; (b) the object is a grid, the diffraction pattern is shown, and the image is again a grid. When masks 1 and 2 are placed at the focal plane of the lens, certain parts of the diffraction pattern are blocked off. The resulting images are shown on the right.

quency spectrum of these spatial waves. The object function $h(y)$ is interpreted as a superposition of spatial waves and the result of the superposition is recorded as the blackening of a photographic plate. The maxima of a spatial wave corresponds to black, the minima corresponds to white, and the gray level indicates zero. By doing so, we attribute phase information to these waves. Superposition of a black maxima $(+A)$ and a white minima $(-A)$ results in gray (0). In the section on holography, we present more details on the presentation of phase information with the black, white, and gray of the photographic plate. In FileFig 10.1 we show this concept in mathematical terms, considering the Fourier series

$$g(x) = \sum_{n=0}^{N} \{[4/((2n+1)\pi)](\sin 2\pi(f_n x))\} \tag{10.1}$$

with frequencies $f_n = (2n+1)/2a$, where a is a certain length constant in the spatial domain.

The spatial waves with wavelengths $\Lambda_0 = 2a$, $\Lambda_1 = 2a/3$, $\Lambda_2 = 2a/5$ correspond to spatial frequencies f_0, f_1, f_2 and are shown in FileFig 10.1. The superposition of these three waves is also shown and one observes that the superposition of a large number of such spatial waves results in a rectangular shape.

FileFig 10.1 (W1FTSERIS)

Fourier series of cosine functions for the composition of a rectangular-shaped object. Different numbers of elements of the sum are plotted separately and as a sum for comparison.

W1FTSERIS

Fourier Series of Spatial Wavelength λ for Interval from −1 to 1 (Shown to 2)

For $N = 0$ the only term is a sine wave from -1 to 1, of wavelength $\lambda = 2$. For $N = 1$ a sine term with 1/3 of λ and smaller amplitude is added. For $N = 2$ a term with 1/5 of λ and smaller amplitude is added, and so on. If N is large, we see a perfect step function. For smaller N (in the 20th), we see *Gibb's phenomenon*, the corners are not round, and there is overshooting. For large N it disappears.

$$x := -1, -/99 \ldots 1.9 \qquad \Lambda := 1 \qquad n := 0, 1 \ldots 200 \qquad N \equiv 100$$

$$g(x) := \sum_{n=0}^{N} \left[\frac{4 \cdot \sin[2 \cdot \pi \cdot x \cdot (f_n)]}{(2 \cdot n + 1) \cdot \pi} \right] \qquad f_n := \frac{2 \cdot n + 1}{2 \cdot \Lambda}.$$

For larger and larger N one can see how more and more waves with shorter and shorter wavelengths are used to build the step function.

$$g0(x) := \frac{4 \cdot \sin\left(\pi \cdot x \cdot \frac{1}{\Lambda}\right)}{\pi} \qquad g1(x) := \frac{4 \cdot \sin\left(\pi \cdot x \cdot \frac{2 \cdot 1+1}{\Lambda}\right)}{(2 \cdot 1 + 1) \cdot \pi}$$

$$g2(x) := \frac{4 \cdot \sin\left(\pi \cdot x \cdot \frac{2 \cdot 2+1}{\Lambda}\right)}{(2 \cdot 1 + 1) \cdot \pi}.$$

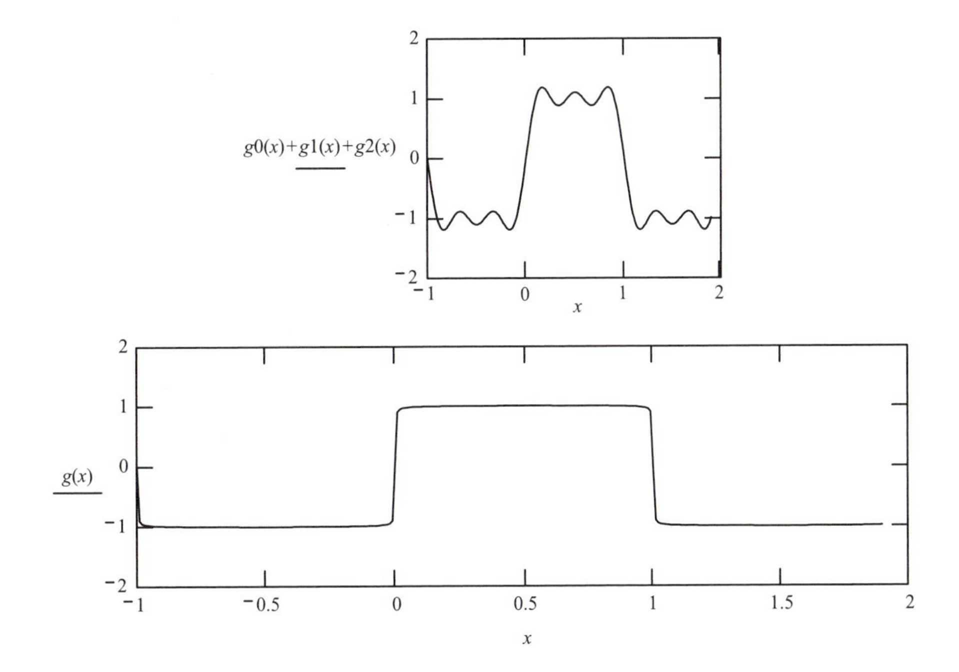

Application 10.1.

1. Use different values for N and NN and compare the resulting rectangular pulse shape.
2. The deviation feature on the edges is called Gibb's phenomenon. Observe the appearance and disappearance of Gibb's phenomena depending on N.
3. Change the sine functions to cosine functions and discuss the result.

In FileFig 10.2 we show schematically the calculation of the spatial frequency spectrum, and the second Fourier transformation for image formation. We also discuss the options for the second Fourier transformation. Could one use the complex Fourier transformation again, or as we should, use the inverse Fourier transformation?

FileFig 10.2 (W2FTCFT)

Object is a composition of step functions. The complex Fourier transformation and the inverse complex Fourier transformation are shown. The complex Fourier transformation may be applied twice, but the image is interchanged left to right.

W2FTCFTS

Example of Real fft and Complex cfft on a Real Object Function

1. The real FT fft
   The Object: $i := 0, 1 \ldots 255$

$A_1 := 33 \quad A_3 := 80 \quad A_4 := 50 \quad A_5 := 20 \quad A_6 := 99 \quad A_7 := 160 \quad A_8 := 200$

$$y_i := \sum_{n=1}^{3} (-\Phi(A_n - i)) + \sum_{n=4}^{8} [\Phi(A_n - i) \cdot (-1)^n].$$

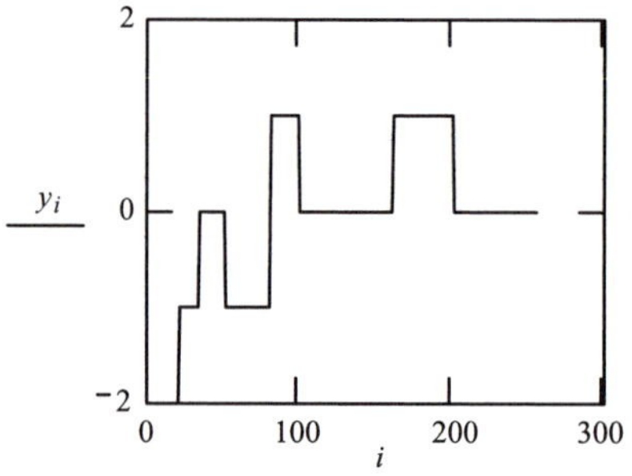

The real Fourier transformation

$$c := fft(y) \quad Nc := \text{last}(c) \quad Nc = 128$$

$$j := 0 \ldots Nc - 1.$$

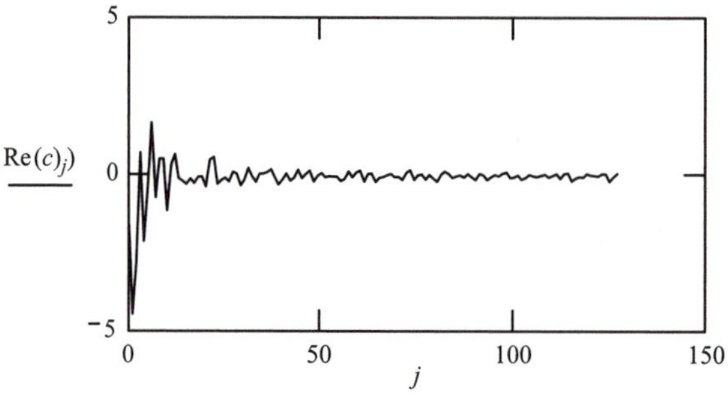

The inverse Fourier transformation

$x := ifft((c))$
$Nx := \text{last}(x)$
$Nx = 255$
$k := 0 \ldots Nx - 1.$

We cannot use $x = fft(c)$; we get the "Error message"; c must be real

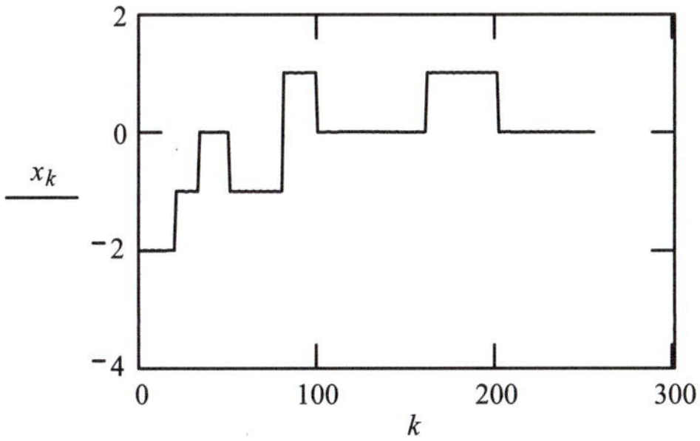

2. The complex Fourier transformation
The Object: $i := 0, 1 \ldots 255$

$$A_1 := 33 \quad A_2 := 80 \quad A_3 := 80 \quad A_4 := 50 \quad A_5 := 20$$
$$A_6 := 99 \quad A_7 := 160 \quad A_8 := 200$$
$$y_i := \sum_{n=1}^{3}(-\Phi(A_n - i)) + \sum_{n=4}^{8}[\Phi(A_n - i) \cdot (-1)^n].$$

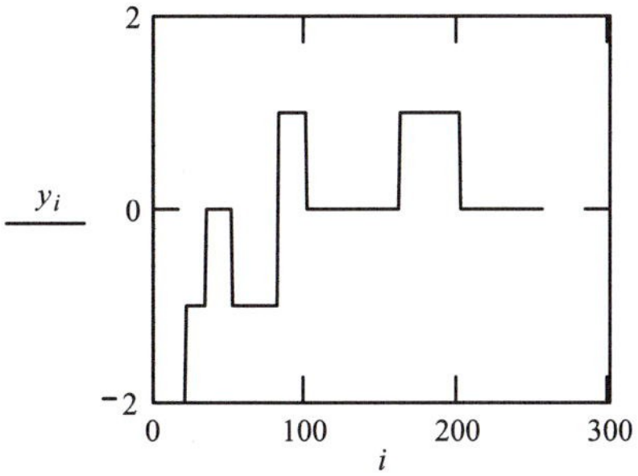

The complex Fourier transformation

$$cc := cfft(y)$$
$$Ncc := \text{last}(cc) \qquad Ncc = 255$$
$$k := 0 \ldots Ncc - 1.$$

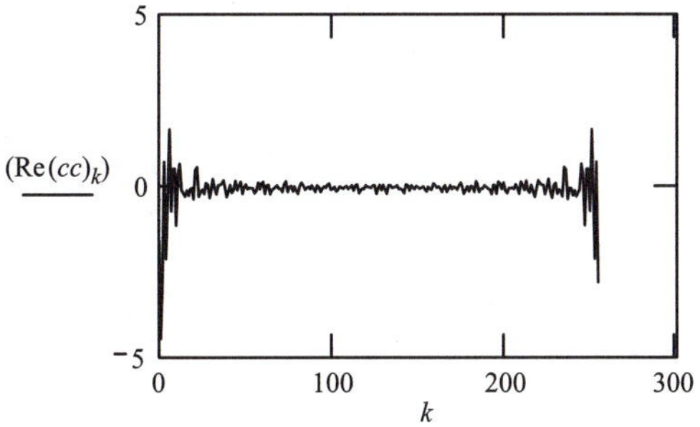

The inverse Fourier transformation

$$xx := icfft(cc) \qquad Nxx := \text{last}(xx)$$
$$Nxx = 255 \qquad k := 0 \ldots Nxx - 1.$$

3. Application of cfft the second time, instead of the inverse icfft results in an image with the left-right interchanged.

$$xxx := cfft(cc)$$
$$Nxxx := \text{last}(xxx) \qquad Nxxx = 255$$
$$f := 0 \ldots Nxxx - 1.$$

10.3 OBJECT, IMAGE, AND THE TWO FOURIER TRANSFORMATIONS

10.3.1 Waves from Object and Aperture Plane and Lens

In geometrical optics, we employed the thin lens equation to find the image point x_i of an object point x_0, when using a thin lens of focal length f. The image of an extended object was obtained by imaging each point of the object to its conjugate point at the image.

The application of wave theory to the image forming process is also done in a point-by-point procedure. The image points are in the y, z plane, and at the distance x a lens is positioned; the finite aperture of the lens is described in the

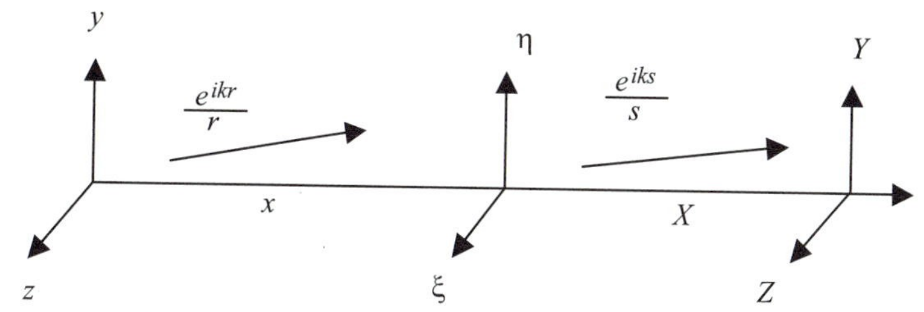

FIGURE 10.3 Coordinate system for object plane, aperture plane, and image plane.

η, ζ plane. At distance X from the lens is the image plane and the image is described in that plane by Y, Z coordinates (see Figure 10.3).[1]

We assume monochromatic light and consider two summation processes and the action of the lens. The action of the lens was discussed in Chapter 3 for the case of Fraunhofer observation. The aperture of the lens is described by $\alpha(\eta, \zeta)$, and the changes of the incident wavefront to the wavefront leaving the lens is given by $\exp(-ik(\eta^2+\zeta^2)/2f)$. The lens bends the wavefront in such a way that it converges to the focal point. The phase factor may be written as $e^{ik\rho}$ where ρ is in the direction of propagation. We obtain $\rho = \eta^2/2f$ and as a result $e^{ik\rho}$ is equal to $\exp(ik\eta^2/2f)$. For simplicity, without losing essential points of importance, we restrict our discussions to a one-dimensional extension of object, lens, and image.

10.3.2 Summation Processes

10.3.2.1 The First Summation Process

The magnitude of the object points is described by $h(y, z)$ and from each point a spherical wave emerges, described by $h(y, z)e^{ikr}/r$. All the waves are summed up for each point η, ζ in the plane of the lens. In the following we restrict the considerations to one dimension.

We call $h(y)$ the magnitude of each object point and calculate r in e^{ikr}/r in terms of x, y, and η (see Figure 10.4)

$$r^2 = (\eta - y)^2 + x^2 = \eta^2 - 2\eta y + y^2 + x^2 = R^2 + \eta^2 - 2\eta y \qquad (10.2)$$
$$= R^2(1 + (\eta^2 - 2\eta y)/R^2),$$

where we call $y^2 + x^2 = R^2$ and we consider R^2 as a constant because $y \ll x$. Developing the square root of $[R^2(1 + (\eta^2 - 2\eta y)/R^2)]$ yields

$$r \approx R\{1 + (\eta^2 - 2\eta y)/2R^2\} \approx R - \eta y/R + \eta^2/2R. \qquad (10.3)$$

[1]See *Physical Optics Notebook* by G. P. Parrent and B. J. Thompson, Society of Photo-Optical Engineers, Redondo Beach, California, 1969.

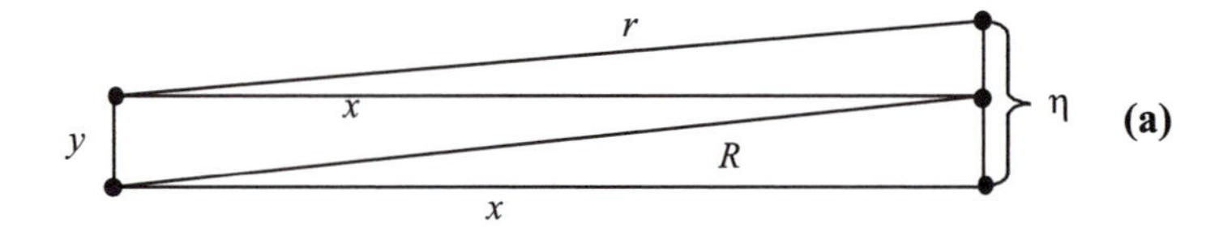

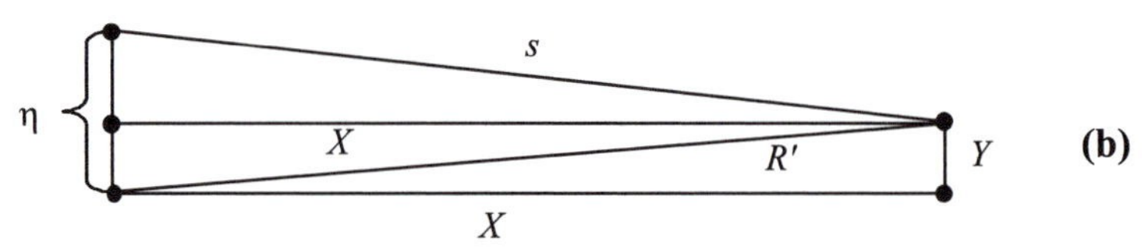

FIGURE 10.4 Coordinates for small angle approximation.

For the first summation process we integrate over all y and have

$$\int h(y)\exp\{ik(R - \eta y/R + \eta^2/2R)\}dy. \tag{10.4}$$

10.3.2.2 Action Lens

The aperture is presented by $\alpha(\eta)$ and from the chapter on diffraction we know that the wavefront is changed by a phase factor of $\exp(-ik\eta^2/2f)$.

10.3.2.3 Second Summation Process

From each point in the plane η, ζ emerges a spherical wave $\alpha(\eta, \zeta)e^{iks}/s$. These amplitudes, including the phase factor introduced by the lens, are summed up for all final image points X, Y.

In a similar way as discussed in Section 10.3.2.1 we may write approximately (Figure 10.4) $s \approx R' - \eta Y/R' + \eta^2/2R'$, with $R'^2 = X^2 + Y^2$ considered as a constant. The amplitude at the image point is called $g(Y)$ for both summation processes

$$g(Y) = \int \left[\int h(y)(1/R)\{\exp ik(R - \eta y/R + \eta^2/2R)\}dy\right]\alpha(\eta)$$
$$\{\exp(-ik\eta^2/2f)(1/R')\}$$
$$\{\exp ik(R' - \eta Y/R' + \eta^2/2R')\}d\eta, \tag{10.5}$$

where one has approximately $1/R$ for $1/r$ and $1/R'$ for $1/s$. All the constants are taken before the integral and one gets

$$g(Y) = (1/RR")\exp ik(R + R')\int\left[\int h(y)\exp\{-ik\eta y/R\}dy\right]$$
$$\alpha(\eta)\{\exp ik(\eta^2/2R - \eta^2/2f + \eta^2/2R')\}\exp ik(-\eta Y/R')d\eta. \tag{10.6}$$

Since we are treating an imaging process, we introduce $R \approx x_0$ and $R' \approx x_i$ into $\exp\{ik \neq \eta^2/2R - \eta^2/2f + \eta^2/R')\}$, and make the factor 1 by using the thin lens equation

$$(ik\eta^2/2)(1/x_0 - 1/f + 1/x_i) = 0. \qquad (10.7)$$

Calling the factor before the integral C, we have further to consider

$$g(Y) = C\int\left[\int h(y)\exp\{-ik\eta y/x_0\}dy\right]\alpha(\eta)\exp\{-ik\eta Y/x_i\}d\eta. \qquad (10.8)$$

10.3.3 The Pair of Fourier Transformations

10.3.3.1 Aperture Function $\alpha(\eta)$ of the Lens Is a Constant

When $\alpha(\eta)$ is independent of η we can include it in the constant before the integral and have

$$g(Y) = C\int\left[\int h(y)\exp\{-ik\eta y/x_0\}dy\right]\exp\{-ik\eta Y/x_i\}d\eta. \qquad (10.9)$$

The integration over y produces a function of η and we call the integral in brackets $\omega(\eta)$

$$\omega(\eta) = \int h(y)\exp\{-ik\eta y/x_0\}dy \qquad (10.10)$$

and write

$$g(Y) = C\int \omega(\eta)\exp\{-ik\eta Y/x_i\}d\eta. \qquad (10.11)$$

The result is: The image amplitude function $g(Y)$ is the Fourier transform of the Fourier transform of the object amplitude function $h(y)$.

Since the Fourier transform of the Fourier transform is the original function, except for a scaling factor, for this idealized case the result is that the function representing the image is the same as the function representing the object, except for a scaling factor and phase factor.

10.3.3.2 Aperture Function $\alpha(\eta)$ of the Lens Is Not a Constant

When $\alpha(\eta)$ is not a constant, we may write, using $\omega(\eta)$,

$$g(Y) = C\int \omega(\eta)\alpha(\eta)\exp\{-ik\eta Y/x_i\}d\eta. \qquad (10.12)$$

We have as the result: The Fourier transform of the modified Fourier transform of the object function is the image function. The function $\alpha(\eta)$ is in most cases an aperture function such as a step function or in two dimensions a function representing a round hole. In that case only the magnitude in the η direction (in two dimensions, the η, ζ plane) is altered. However, $\alpha(\eta)$ may also contain a phase factor representing phase differences between certain points (η) in the plane (η, ζ).

10.4 IMAGE FORMATION USING INCOHERENT LIGHT

10.4.1 Spread Function

We have assumed in the preceding chapters that monochromatic light was used for image formation. We now want to use nonmonochromatic light, (e.g., light from the sun) for image formation. To do this we will assume that there is no fixed phase relation between the light emitted from a pair of object points. Since we want to use only one medium wavelength, the light we assume using is called quasimonochromatic light (see Chapter 4). The case where one uses monochromatic light is discussed in Section 10.5. As in geometrical optics, we consider the object made of points and apply a point-by-point process to obtain the image. The detector used for recording the image will be sensitive to the intensity at each image point, which we assume to be proportional to the square of the amplitude.

When considering one point, we have to replace $h(y)$ (see Section 10.4.2) by a delta function and have from Eq. (10.8),

$$s(Y) = C \int [\int \delta(y) \exp\{-ik\eta y/x_0\} dy] \alpha(\eta) \exp\{-ik\eta Y/x_i\} d\eta. \qquad (10.13)$$

Since

$$\int \delta(y) \exp\{-ik\eta y/x_0\} dy = 1 \qquad (10.14)$$

we get

$$s(Y) = C \int \alpha(\eta) \exp\{-ik\eta Y/x_i\} d\eta. \qquad (10.15)$$

If $\alpha(\eta)$ is a constant and $s(Y)$ is a delta function equal to the object function $h(y)$ we have a point-to-point imaging process. In general, if $\alpha(\eta)$ is not a constant we go back to the first summation process and find for the two-dimensional process that $\alpha(\eta, \zeta)$ describes the amplitude distribution of the light from the point source in the plane of the lens. Therefore $s(Y)$ describes the deviation from a point we would have obtained as the image point.

For the image one needs the intensity and has to square Eq. (10.15) and obtain

$$S(Y) = |C \int \alpha(\eta) \exp\{-ik\eta Y/x_i\} d\eta|^2. \qquad (10.16)$$

We call $S(Y)$ the *spread function*. This is the spread of the object point in the image plane as produced by the lens.

10.4.2 The Convolution Integral

The integral in Eq. eq10.15 may be rewritten using the convolution theorem of Fourier transformations. This theorem tells us that we may write the Fourier transform of a product of two functions, here $\omega(\eta)\alpha(\eta)$, as the convolution integral of the Fourier transformation of $\omega(\eta)$ and of $\alpha(\eta)$.

We consider the integral in Eq. (10.12),

$$g(Y) = C \int \omega(\eta)\alpha(\eta) \exp\{-ik\eta Y/x_i\} d\eta. \tag{10.17}$$

From Eq. (10.10) we have

$$\omega(\eta) = \int h(y) \exp\{-ik\eta y/x_0\} dy. \tag{10.18}$$

The Fourier transform is the original object function

$$h(Y) = C \int \omega(\eta) \exp\{-ik\eta Y/x_i\} d\eta. \tag{10.19}$$

From Eq. (10.15) we have

$$s(Y) = C \int \alpha(\eta) \exp\{-ik\eta Y/x_i\} d\eta \tag{10.20}$$

and may replace $g(Y)$ in Eq. (10.17) by the convolution integral

$$g(Y) = C \int h(Y) s(\tau - Y) d\tau. \tag{10.21}$$

The effect of the aperture function $\alpha(\eta)$ is that the final image function is obtained by convolving the original object function $h(Y)$ with the Fourier transform of the aperture function $\alpha(\eta)$, which is $s(Y)$ (using image coordinates).

The result is that only the coordinates of the image plane are involved.

10.4.3 Impulse Response and the Intensity Pattern

If the object is described by $I_{ob}(Y)$, a point of the object on the object screen can be represented by the product $I_{ob}(Y)$ times a δ function

$$I_{ob}(Y)\delta(Y - Y'), \tag{10.22}$$

which will have the value $I_{ob}(Y')$ for $Y = Y'$.

For a very large aperture, nothing would happen to the δ function and we would have on the observation screen a point. The image point is similar to the object point but we may have to introduce a scaling factor. For a small aperture, the point would spread to a certain pattern and we would have to replace $\delta(Y - Y')$ by the spread function $S(Y - Y')$. The spread function describes the change of the object as it appears on the observation screen. The distortion is produced by the lens.

$$I_{ob}(Y)\delta(Y - Y') \longrightarrow I_{ob}(Y)S(Y - Y'). \tag{10.23}$$

If the object is only one point we call this the *impulse response*. For incoherent light, the spread function is the image of one point on the object screen. If a second point is added on the object screen, it has to be treated similarly and independently and the result has to be added to the image plane. For n points we have for the impulse response

$$I_{im}(Y) = \sum I_{ob}(Y_n)S(Y - Y_n) \tag{10.24}$$

and for a continuous distribution of points we obtain

$$I_{im}(Y) = \int I_{ob}(Y')S(Y - Y')dY'. \tag{10.25}$$

With Eq. (10.25) one is able to calculate the image from an object when the spread function is known. This is demonstrated in the following for a few simple examples.

10.4.4 Examples of Convolution with Spread Function

10.4.4.1 One Bar as an Object and Application of a Cylindrical Lens

We assume that the object is made of just one bar, represented by $h(y)$ as a rectangular pulse of height 1 and width from $y = b1$ to $y = b2$. The lens is a cylindrical lens represented as a pulse by the function $\alpha(\eta)$ of height 1 and width from $\eta = -a$ to $\eta = +a$ (see FileFig 10.3). We call the focal length of the lens f and its diameter 2a. One calls $f/2a$ the $f\#$ of the lens.

The spread, which is produced by the lens, is presented by the spread function $S(Y)$ (Eq. (10.16)) where we have set $X = f$, with f being the focal length of the lens

$$S(Y) = |\int \{(e^{-ikY\eta/f})\}d\eta|^2. \tag{10.26}$$

For a cylindrical lens we have to integrate over $-a$ to a and get

$$S(Y) = 4a^2\{(\sin kaY/f)/(kaY/f)\}^2. \tag{10.27}$$

The image is then obtained from the convolution integral

$$I_{im}(Y) = 4a^2 \int_{Y'=b1}^{Y'=b2} I_{ob}(Y')S(Y - Y')dY', \tag{10.28}$$

where we used $Y' = YY$ in FileFig 10.3. There we show the object and image, with $\gamma = 0.0005$ mm, diameter of the lens $2a$, and choice of $f\# = 10$. The width of the object bar has been assumed to be from $b1 = -0.002$ to $b2 = 0.002$.

FileFig 10.3 (W3IMONES)

The object is one bar. The image is obtained by convolution of the object function with the spread function. The spread function is the Fourier transformation of the pulse, representing the cylindrical lens.

W3IMONEBS is only on the CD.

Application 10.3.

1. Change the values of $b1$, $b2$, and a so that $b2 - b1$ is larger than a.
2. Change the values of $b1$, $b2$, and a so that $b2 - b1$ is smaller than a.

10.4.4.2 Two Bars

Two bars are considered in FileFig 10.4, each of width 0.003, and center-to-center distance of $bb = 0.013$. The lens has width $2a$, and we use for the wavelength $\lambda = 0.0005$ mm and for the $f\# = 10$. The image of each bar is calculated separately with the integral of Eq. (10.28). Because of the linearity of the system the final image is the superposition of the result of imaging of each bar separately.

FileFig 10.4 (W4IMTWOB)

The object is made of two bars. The image is obtained by convolution of each object function with the spread function. The spread function is the Fourier transformation of the pulse, representing the cylindrical lens.

W4IMTWOB is only on the CD.

Application 10.4.

1. Change the distance between the two object bars and study the resolution. If the distance is too small, the image is just one peak.
2. Change the calculation and apply the spread function to the total image over a large range of integration. The image is now more similar to the object.

10.4.4.3 One Round Object and a Circular Lens

In far field diffraction of the round aperture, one finds that the two-dimensional problem is reduced to a one-dimensional circular symmetric problem, solved by using the Bessel function. We may similarly proceed for imaging. The object is a round aperture of diameter $2b = 0.004$ (FileFig 10.5). The round lens has the diameter 2a. For the calculation of the integral, corresponding to the integral of Eq. (10.28), we extend the definition of the radius as it appears in the

Bessel function to negative values. The spread function is calculated using the nonnormalized Bessel function

$$s^2(R) = 4a^2\{J_1(2\pi aR/\lambda f)/(2\pi aR/\lambda f)\}^2. \tag{10.29}$$

One has to simply replace the $\sin y/y$ function by the $J_1(y')/y'$ function. We use for R again Y and for R' the integration variable Y'. For the integral we then have

$$I_{im}(Y) = 4a^2 \int_{Y'=b1}^{Y'=b2} I_{ob}(Y')\left[\frac{J1(2\pi a(Y-Y')/\lambda f)}{(2\pi a(Y-Y')/\lambda f)}\right]^2 dY'. \tag{10.30}$$

The calculation is shown in FileFig 10.5.

FileFig 10.5 (W5IMONEROS)

One round object and a circular lens. The image is obtained by convolution of the object function with the spread function. The spread function is the Fourier transformation of the pulse, representing the lens.

W5IMONEROS is only on the CD.

Application 10.5.

1. Change the values of $b2 - b1$ and a so that $b2 - b1$ is larger than a.
2. Change the values of $b2 - b1$ and a so that $b2 - b1$ is smaller than a.

10.4.4.4 Two Round Objects and a Circular Lens

Two round objects of diameter $2b = 0.004$ and center-to-center distance $bb = 0.014$, are treated in FileFig 10.6.

Since the system is linear, we have to calculate two integrals of the type considered in Section 10.4.4.3 at the given distance.

$$I_{im}(Y) = 4a^2 \int_{Y'b1}^{Y'=b2} I_{ob}(Y')S(Y-Y')dY' + 4a^2 \int_{Y'=b3}^{Y'=b4} I_{ob}(Y')S(Y-Y')dY'. \tag{10.31}$$

The calculation is shown in FileFig 10.6.

FileFig 10.6 (W6IMTWOROS)

Two round objects and a circular lens. The image is obtained by convolution of each object function with the spread function. The spread function is the Fourier transformation of the pulse, representing the lens.

W6IMTWOROS

Imaging: Two Round Apertures and Round Lens (R' is X)

$$Y := -.1, -099 \ldots 6 \qquad b1 \equiv -.002 \qquad b2 \equiv .002 \qquad \lambda := .0005$$

$$\text{Tol} := .1 \qquad f/10 = f/2a$$

$$f := 500 \qquad a := 25 \qquad b3 \equiv .012 \qquad b4 \equiv .016 \qquad k := \frac{2 \cdot \pi}{\lambda}.$$

Object

$$Io1(Y) := (\Phi(b2 - Y) - \Phi(b1 - Y)) \quad Io2(Y) := (\Phi(b4 - Y) - \Phi(b3 - Y))$$

$$lo(Y) := Io1(Y) + Io2(Y).$$

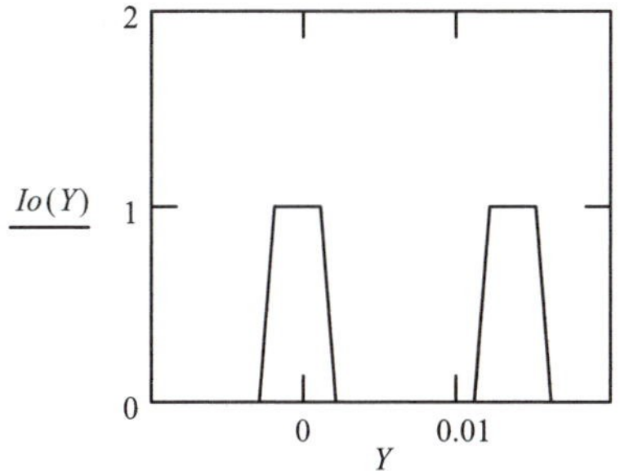

Image

$$\lim(Y) := \int_{b1}^{b2} 4 \cdot a^2 \cdot \left[\frac{J1\left[\frac{k \cdot a(Y - YY)}{f}\right]}{k \cdot a \frac{(Y - YY)}{f}} \right]^2 dYY$$

$$+ \int_{b3}^{b4} 4 \cdot a^2 \cdot \left[\frac{J1\left[\frac{k \cdot a \cdot (Y - YY)}{f}\right]}{k \cdot a \cdot \frac{(Y - YY)}{f}} \right] dYY.$$

Unnormalized

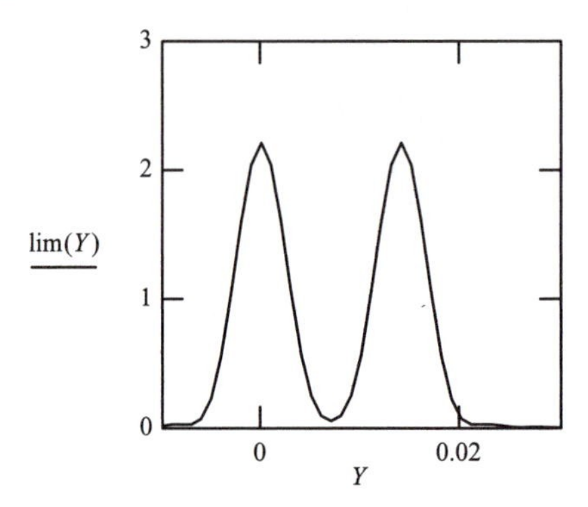

Application 10.6.

1. Change the distance between the two object bars and study the resolution. If the distance is too small, the image is just one peak.
2. Change the calculation and apply the spread function to the total image over a large range of integration. The image is now closer to the object.

10.4.5 Transfer Function

10.4.5.1 Transfer Function and Fourier Transformation

The spread function is used in the coordinate space of the image to calculate how the object appears on the observation screen (image coordinates). If the lens did not alter the object we would observe a perfect image of the object. The image was obtained by convolution of the object function $I_{ob}(Y')$ with the spread function $S(Y - Y')$, both using image space coordinates

$$I_{im}(Y) = \int I_{ob}(Y')S(Y - Y')dY'. \tag{10.32}$$

From the convolution theorem we know that the integral in Eq. (10.32) expresses the Fourier transform of the product of the Fourier transform of $I_{ob}(Y)$ and the Fourier transform of $S(Y)$. If we call these Fourier transforms

$$\omega(\mu) = \int I_{ob}(Y')e^{-i2\pi\mu Y'}dY' \tag{10.33}$$

and

$$\tau(\mu) = \int S(Y')e^{-i2\pi\mu Y'}dY', \tag{10.34}$$

one may write Eq. (10.32) as

$$I_{im}(Y) = C\int \omega(\mu)\tau(\mu)e^{i2\pi\mu Y}d\mu, \tag{10.35}$$

where C contains all constants.

The Fourier transform of $S(Y')$, which is $\tau(\mu)$, is called the (unnormalized) transfer function. Equation (10.35) tells us that the Fourier transform of $I_{im}(Y)$, which we call $\phi_{im}(\mu)$, is equal to the product of the Fourier transforms $\omega(\mu)\tau(\mu)$, which is

$$\phi_{im} = \omega(\mu)\tau(\mu). \tag{10.36}$$

For the coordinates we have $1/Y = \mu$, where μ may be interpreted as the spatial frequency in the frequency domain. The transfer function operates in the spatial frequency domain.

In FileFig 10.7 we calculate the Fourier transform of an object, which is a rectangular pulse, and the Fourier transform of a spread function for a cylindrical lens (transfer function). Then we multiply the Fourier transform of the object with the transfer function and perform the Fourier transformation of the product. The resulting image should look like a rectangular pulse.

FileFig 10.7 (W7PUTRAS)

Demonstration of the use of a transfer function. Object is a pulse and calculation of its Fourier transformation. Spread function and calculation of its Fourier transformation. Product of both Fourier transformations and their Fourier transformation (inverse). The resulting image of these operations looks like the object.

W7PUTRAS is only on the CD.

10.4.5.2 Examples

Fourier Transform of $(\sin x/x)^2$ *as Transfer Function*

We take as an object a grid presented as a series of square pulses (FileFig 10.8) and calculate and perform the Fourier transformation $\omega_{ob}(\mu)$. The spread function for a cylindrical lens is (Eq. (10.27)),

$$S(X) = 4a^2\{\sin(\pi X/\lambda f\#)/(\pi X/\lambda f\#)\}^2, \tag{10.37}$$

where we use $f\# = f/2a$ for X (the space coordinate) $i/255$ and i is the running number of the Fourier transformation from 1 to 256 (0-255). We have to use here $i/255$ in the frequency domain since we used i in the space domain. The Fourier transform of $S(X)$ is $\tau(\mu)$. The product of $\phi(\mu) = \omega_{ob}(\mu)\tau(\mu)$ is shown in FileFig 10.8, and the Fourier transform (inverse) of $\phi(\mu)$ is the final image. In FileFig 10.8 we have used the complex Fourier transform of 256 points for the object and its mirror image. We should remember that the longest wavelength of the spatial waves is 128, and the corresponding smallest frequency 1/128. We also use the scale 1/256 in the transfer function and have the smallest frequency at 1/128. The wavelength is 0.0005 mm and the $f\# = f/2a$ which is 10. Changing the $f\#$ in FileFig 10.8 will change the triangle corresponding to τ_i and will admit more or fewer spatial frequencies for image formation.

FileFig 10.8 (W8TRASIS)

Demonstration of the use of a transfer function. Object is a grid and calculation of it's Fourier transformation. Spread function $(\sin y/y)^2$ and calculation of its Fourier transformation. Product of both Fourier transformations and their Fourier transformation (inverse). The resulting image of these operations looks more or less like the object.

W8TRASIS is only on the CD.

Application 10.8. Change the $f\#$ and observe through the changing triangle that more or fewer frequencies are used for image formation.

Fourier Transformation of the Bessel Function as Transfer Function

We may use as the transfer function $(J1(y)/y)^2$ corresponding to a circular lens instead of $(\sin y/y)^2$, which corresponds to a cylindrical lens. Keeping the argument of the transfer function the same but taking R instead of X we use

$$J1(\pi R/\lambda fn)/(\pi R/\lambda fn), \qquad (10.38)$$

where fn is the f-number. The variable R is replaced by $i/255$ and i is the running number of the Fourier transformation, from 1 to 256 (0–255). We have to use here $i/255$ in the frequency domain, since we used i in the space domain. In FileFig 10.9 the object is a grid presented as a series of square pulses in the space domain, but the spread function is now $(J1(y)/y)^2$, and its Fourier transformation is the transfer function $\tau(\mu)$. The product of $\omega_{ob}(\mu)\tau(\mu)$ is calculated and shown in FileFig 10.9. The Fourier transform (inverse) of $\phi(\mu) = \omega_(ob)(\mu)\tau(\mu)$ is the final image.

Changing the $f\#$ in FileFig 10.9 will change τ_k as shown and will admit more or fewer spatial frequencies for image formation.

FileFig 10.9 (W9TRAJ1S)

Demonstration of the use of a transfer function. Object is a grid and calculation of its Fourier transformation. Spread function $(J1(y)/y)^2$ and calculation of its Fourier transformation. Product of both Fouriers transformations and their

Fourier transformation (inverse). The resulting image of these operations looks more or less like the object.

W9TRAJ1S is only on the CD.

Application 10.9. Change the $f\#$ and observe through the changing transfer function that more or fewer frequencies are used for image formation.

10.4.6 Resolution

The Rayleigh criterion of resolution states that two image peaks are considered resolved if the maximum of the diffraction pattern of one is at the position of the first minimum of the other. This is demonstrated in FileFig 10.10. We consider two round objects and a round lens, and apply the Rayleigh criterion. The object is made of two "rectangles" of width $b1$ - $b2$ and $b3$ - $b4$ (FileFig 10.11). The aperture of the lens of diameter $2a$ forms a diffraction pattern of each point on the image screen.

The argument of the Bessel function $2\pi a R/\lambda f$ is again written as

$$\pi R/\lambda(f/2a) = \pi R/\lambda(f\#), \tag{10.39}$$

where $f\# = f/2a$, and we use Y for R in FileFig 10.11. The first zero of the Bessel function $J_1(\pi R'/\lambda f\#)$ is at $\pi R'/\lambda f\# = 3.83$. Therefore if the maximum of one peak is at the minimum of the other, the distance $2b'$ is $(3.83/\pi)\lambda f\# = 1.22\lambda f\#$. In our example in FileFig 10.11 we have used the diameter 0.0005 mm for the round objects, which is the same as the wavelength. The $f\#$ is equal to 10 and the product $\lambda f\#$ has a value 0.005. The center-to-center distance of the two objects is 0.0061. In FileFig 10.11 we show the image of the two objects at the minimum distance corresponding to the resolution according to the Rayleigh criterion.

FileFig 10.10 (W10BES3DS)

Demonstration of Resolution. Two Bessel functions in 3-D presentation of radius $a = 0.05$, center-to-center distance of $d = 24.5$, wavelength $\lambda = 0.0005$, and distance to screen $X = 4000$.

W10BES3DS

Rayleigh Distance and 3-D Graph for Two Round Apertures at Distance D

Radius of apertures is a, coordinate on the observation screen R, wavelength λ, and distance from aperture to screen is X.

1. Determination of Rayleigh distance for Two Round

$a \equiv .05 \quad X \equiv 4000 \qquad R := 0, 1 \ldots 50$

$$g1(R) := \left[J1 \frac{\left(2 \cdot \pi \cdot a \cdot \frac{R}{X \cdot \lambda}\right)}{\left(2 \cdot \pi \cdot a \cdot \frac{R}{X \cdot \lambda}\right)} \right]^2 \qquad gg1(R) := \left[J1 \frac{\left(2 \cdot \pi \cdot a \cdot \frac{R-d}{X \cdot \lambda}\right)}{\left(2 \cdot \pi \cdot a \cdot \frac{R-d}{X \cdot \lambda}\right)} \right]^2$$

2. 3-D Graph of pattern of two round apertures at distance d

$i := 0 \ldots N \qquad j := 0 \ldots N$

$x_i := (-40) + 2.000 \cdot i \quad yj := -40 + 20001 \cdot j \quad \lambda \equiv .0005$

$RR(x, y) := \sqrt{(x)^2 + (y)^2} \qquad N \equiv 60 \qquad X := 4000$

$$g2(x, y) := \left[J1 \frac{\left(2 \cdot \pi \cdot a \cdot \frac{RR(x,y)}{X \cdot \lambda}\right)}{\left(2 \cdot \pi \cdot a \cdot \frac{RR(x,y)}{X \cdot \lambda}\right)} \right]^2 \qquad gg2(x, y) := \left[J1 \frac{\left(2 \cdot \pi \cdot a \cdot \frac{RR(x,y-d)}{X \cdot \lambda}\right)}{\left(2 \cdot \pi \cdot a \cdot \frac{RR(x,y-d)}{X \cdot \lambda}\right)} \right]^2$$

$$M_{i,j} := g2(x_i, y_j) + gg2(x_i, y_j) \qquad d \equiv 24.5.$$

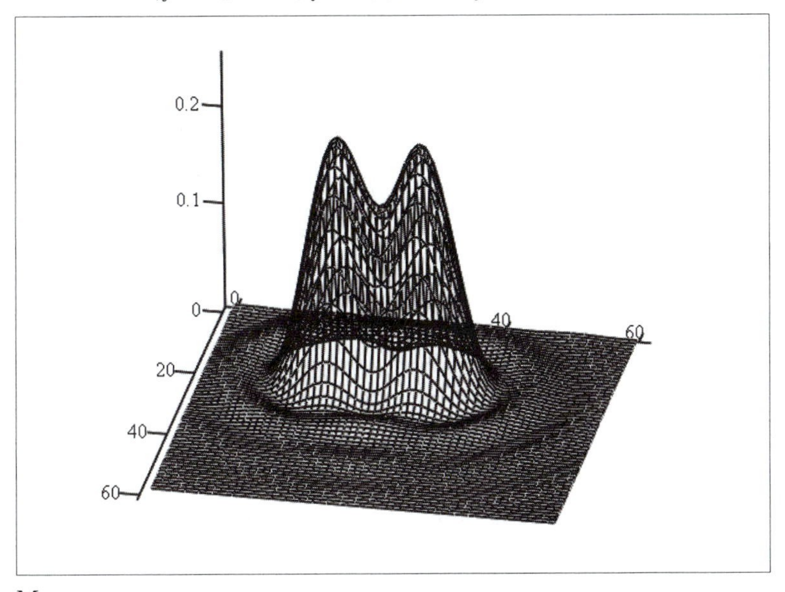

M

FileFig 10.11 (W11TWOROJ1S)

Resolution. Two round objects at diameter 0.0005 and center-to-center distance of 0.0061. The convolution with the spread function $(J1(y)/y)^2$ results in a resolved image.

W11TWOROJ1S

Imaging: Two Round Apertures as Objects

At Rayleigh distance, round lens, and Y used for R.
Object

$$Y := -.01, -.0099 \ldots 02 \qquad \lambda := .0005 \qquad \text{for choice of f\#} = f/2a$$

$$Tol := .1 \qquad k := \frac{2 \cdot \pi}{\lambda} \qquad f := 500\; a := 25$$

$$Io1(Y) := (\Phi(b2 - Y) - \Phi(b1 - Y)) \qquad Io2(Y) := (\Phi(b4 - Y) - \Phi(b3 - Y))$$

$$Io(Y) := Io1(Y)_I o2(Y).$$

Image

$$\lim(Y) := \int_{b1}^{b2} 4 \cdot a^2 \cdot \left[\frac{J1\left[\frac{k \cdot a \cdot (Y-YY)}{f}\right]}{k \cdot a \cdot \frac{(Y-YY)}{f}} \right]^2 dYY$$

$$+ \int_{b3}^{b4} 4 \cdot a^2 \cdot \left[\frac{J1\left[\frac{k \cdot a \cdot (Y-YY)}{f}\right]}{k \cdot a \cdot \frac{(Y-YY)}{f}} \right]^2 dY.$$

Limits of integration

$b1 \equiv -.00025 \quad b2 \equiv +.00025 \quad b3 \equiv +.00585 \quad b4 \equiv +.00635.$

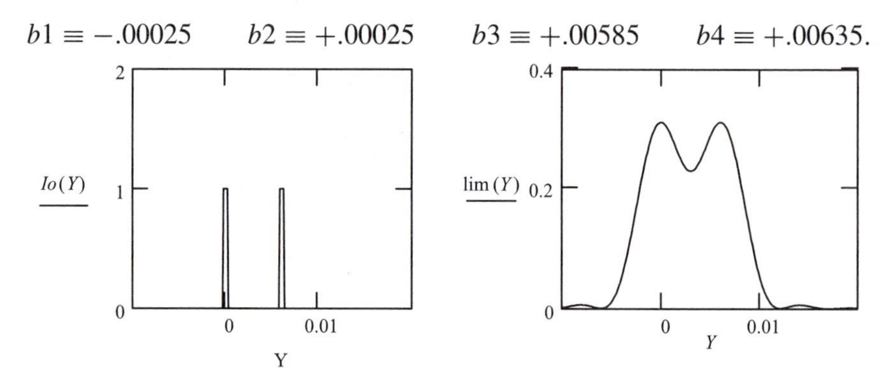

Application 10.11.

1. Read the distance of the peaks of the image from the graph.
2. Change the distance between the objects so that the two are barely not resolved.
3. Change the distance between the objects so that the two are barely separated.
4. Increase the size of the objects until they are no longer resolved.

10.5 IMAGE FORMATION WITH COHERENT LIGHT

10.5.1 Spread Function

We assumed in Section 10.4 that incoherent light is used for image formation. We now discuss the image formation process using coherent light and show that the spread function must be changed. The general process is the same as in Section 10.4. Coherent radiation is now assumed to emerge from an object. Using Huygens' principle in the first process, the wavelets emerging from the object are summed and travel to the lens. In the second process, the wavelets emerging from the lens are summed. Taking the action of the lens into account we found from Eq. (10.8) for the final image,

$$g(Y) = C \int \left[\int h(y) \exp\{-ik\eta y/x_0\} dy \right] \alpha(\eta) \exp\{-ik\eta Y/x_i\} d\eta, \quad (10.40)$$

where the constant C before the integral takes care of all constant factors. Similarly to the way we proceeded for the incoherent case, we use a δ function for a point in the object plane. Substituting for $h(y)$ a delta function into Eq. (10.40) one obtains for the first Fourier transformation

$$\int \delta(y) \exp\{-ik\eta y/x_0\} dy = 1 \quad (10.41)$$

and then has for the remaining part of the integral of Eq. (10.8),

$$s(Y) = C \int \alpha(\eta) \exp\{-ik\eta Y/x_i\} d\eta. \quad (10.42)$$

Equation (10.42) is the spread function $s(Y)$ for the case of coherent light. For the image formation we again use a convolution integral, similar to the incoherent case (see Eq. (10.21)) where $h(Y)$ describes the object. But unlike Eq. (10.21), we use the spread function $s(Y)$, containing phase information. The convolution integral

$$\int h(Y')s(Y - Y')dY' \quad (10.43)$$

has to be squared for the description of the image

$$I_{im}(Y) = |C \int h(Y')s(Y - Y')dY'|^2. \quad (10.44)$$

In contrast to the incoherent case, we have to use $s(Y)$ for the spread function instead of $S(Y)$, which is $s(Y)^2$. The spread function is first convolved with all points of the object and then we have to square the integral to get the image.

10.5.2 Resolution

As an example we discuss the problem of resolution. We consider two round objects of diameter 0.0005 mm, as we did for the incoherent case, and a wavelength of 0.0005 mm and a $f\# = 10$. The distance to be determined should be such that the final image looks similar to the image of the two objects as we found for the incoherent case, when applying the Rayleigh criterion. The calculation is presented in FileFig 10.12.

The image is calculated from the convolution integrals, where we call $h(Y)$ here $iob(Y)$ and omit the constants before the integral,

$$I_{im}(Y) = \left(\int_{Y'=b1}^{Y'=b2} iob(Y')s(Y - Y')dX' + \int_{Y'=b3}^{Y'=b4} iob(Y')s(Y - Y')dY' \right)^2. \quad (10.45)$$

We use for $s(Y)$ the Bessel function similar to Eq. (10.38). The intensity of the image is obtained by squaring the result of the integration (note that $s(Y - Y')$ contains phase information).

The distance of the two peaks in the final image of the two objects for the coherent case is at a center-to-center distance of 0.0082. This is larger than for the incoherent case. The final image is shown for the same distance as used for the incoherent case, and one observes that the two peaks are not resolved. Since the distance used in the coherent case is larger than the one for the incoherent case, for a comparable set of parameters, we have the result: Better resolution is obtained by using incoherent light.

FileFig 10.12 (W12TWOROCOHS)

Two round objects of diameter 0.0005 and center-to-center distance of 0.0061, as used for the incoherent case. The convolution with the spread function, $(J1(y)/y)$, results in an image of one peak, not resolved. For larger center-to-center distance, 0.0082, the peaks are resolved.

W12TWOROCOHS

Imaging with Coherent Light

Two round apertures at Rayleigh distance, round lens, and Y used for R'.

$$Y := -.01, -.0099 \ldots 02 \qquad \lambda := .0005 \qquad f/10 = f/2a$$

$$Tol := .1 \qquad k := \frac{2 \cdot \pi}{\lambda} \qquad f := 500\ a := 25.$$

Object Amplitudes

$$iob(1) := (\Phi(b2 - Y) - \Phi(b1 - Y)) \qquad iob2(Y) := (\Phi(b4 - Y) - \Phi(b3 - Y))$$

$$iob(Y) := iob1(Y) + iob2(Y).$$

Image

$$\lim(Y) := \left[\int_{b1}^{b2} 4 \cdot a^2 \cdot \left[\frac{J1\left[\frac{k \cdot a \cdot (Y - YY)}{f}\right]}{k \cdot a \frac{(Y - YY)}{f}}\right] dYY + \int_{b3}^{b4} 4 \cdot a^2 \cdot \left[\frac{J1\left[\frac{k \cdot a \cdot (Y - YY)}{f}\right]}{k \cdot a \cdot \frac{(Y - YY)}{f}}\right] dYY\right]^2.$$

Integration limits

$$b1 \equiv -.00025 \qquad b2 \equiv +.00025 \qquad b3 \equiv +.00585 \qquad b4 \equiv +.00635.$$

$$\frac{b4 - b3}{2} + b3 = 6.1 \cdot 10^{-3}.$$

Resolution is obtained for $b3 = .00795$, $b4 = .00845$.

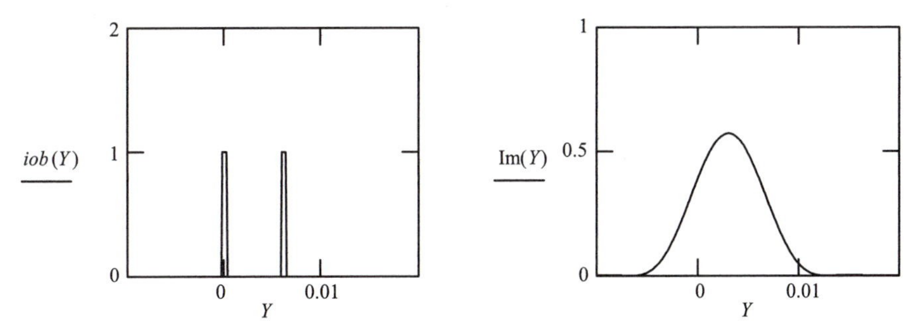

Application 10.12. Introduce the center-to-center distance of 0.0082 between the two objects, $b3 = 0.00795$ and $b4 = 0.00845$ and show that the two objects appear resolved.

10.5.3 Transfer Function

We proceed much as we did in the incoherent case and call

$$\omega(\mu) = \int iob(Y')e^{-i2\pi\mu Y'}dY' \tag{10.46}$$

and

$$\tilde{\tau}(\mu) = \int s(Y')e^{-i2\pi\mu Y'}dY', \tag{10.47}$$

where $\tilde{\tau}(\mu)$ is the transfer function in the coherent case. The difference in Eq. (10.47) with respect to Eq. (10.34) is that $s(Y')$ is an amplitude function that carries phase information. The image is obtained by a Fourier transformation, similar to Eq. (10.35) for the incoherent case. To obtain the description of the intensity of the image, the result has to be squared. We have for the coherent case

$$I_{im}(Y) = |C \int \omega(\mu)\tilde{\tau}(\mu)e^{i2\pi\mu Y}d\mu|^2. \tag{10.48}$$

We call $\phi_{im}(\mu)$ the Fourier transformation of $I_{im}(Y)$ and get

$$I_{im}(Y) = \{\text{FT of the product of the FT of } I_{ob} \text{ and FT of} s\}^2. \tag{10.49}$$

In the frequency domain we have, similar to Eq. (10.36),

$$\phi_{im}(\mu) = \{\omega(\mu)\tilde{\tau}(\mu)\}. \tag{10.50}$$

Comparing (10.50) with (10.36) one sees that a symmetric blocking function that does not carry phase information has the same effect of eliminating certain spatial frequencies for the incoherent and coherent cases.

As examples we consider the imaging of a periodic structure using the spread functions ($\sin y/y$) and ($J1(y)/y$). For the coherent case we use in FileFig 10.13 the spread function ($\sin y/y$) in comparison to the incoherent case of FileFig 10.8 where we used $(\sin y/y)^2$. In FileFig 10.14 the spread function ($J1(y)/y$) is used in comparison to FileFig 10.9 where we used $(J1(y)/y)^2$. For the coherent case the spread functions are not squared. The final image has to be the square of the last Fourier transformation (see Eq. (10.48)). In comparison one sees that the transfer function (i.e., τ) eliminates in the incoherent case higher spatial frequencies in a linear way and in the coherent case in a steplike fashion.

The transfer function τ in FileFig 10.13 is a pulse function and may be interpreted as a blocking function, eliminating parts of the diffraction pattern of the object. Similar action was taken in Figure 10.2 on the diffraction pattern in order to change the image.

FileFig 10.13 (W13TRANCOHSIS)

Coherent case. Transfer function for ($\sin x/x$). Object is a grid and calculation of its Fourier transformation. Spread function ($\sin x/x$) and calculation of its Fourier transformation. Product of both Fourier transformations and their Fourier transformation (inverse). The image as a result of these operations looks more or less like the object.

W13TRANCOHSIS is only on the CD.

Application 10.13.

1. Change the $f\#$; that is, change the width of the spread function and observe that more or fewer frequencies are used for image formation.
2. Compare the incoherent and coherent cases for the same $f\#$. Choose one larger and one smaller $f\#$.

FileFig 10.14 (W14TRANJ1S)

Coherent case. Transfer function for (Bess/arg). Calculation of the transfer function for the coherent case. Object is a grid and calculation of its Fourier transformation. Spread function (Bess/arg) and calculation of its Fourier transformation. Product of both Fourier transformations and their Fourier transformation (inverse). The image as a result of these operations looks more or less like the object.

W14TRANJ1S is only on the CD.

Application 10.14.

1. Change the $f\#$; that is change the width of the spread function and observe that more or fewer frequencies are used for image formation.
2. Compare the incoherent and coherent cases for the same $f\#$. Choose one larger and one smaller $f\#$.

10.6 HOLOGRAPHY

10.6.1 Introduction

When discussing the imaging process for coherent light, the first Fourier transformation produced the diffraction pattern of the object. The result of the first Fourier transformation contained phase information, regardless of whether the object was a real or complex function. The second Fourier transformation needed this phase information to produce the image of the object. To fix the phase information on a photographic plate we follow closely the discussion in Goodmann, 1988, p.198.

10.6.2 Recording of the Interferogram

In holography one uses a photographic film to record the amplitude and phase information necessary for the reconstruction of the image of an object. This is done by interference of the coherent light, scattered from the object with a coherent reference beam (Figure 10.5).

The light scattered by the object is described by the complex amplitude function $\mathbf{a} = a_0 e^{-i\phi}$, where a_0 is a function of y, z and ϕ a function of x, y, z. The reference wave is described by the complex amplitude function $\mathbf{A} = A_0 e^{-i\psi}$,

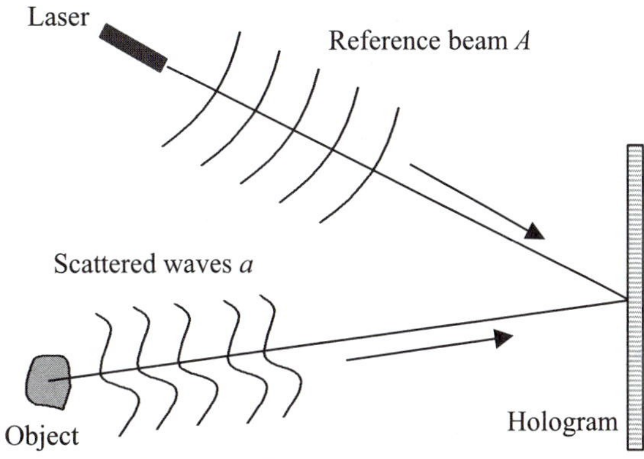

FIGURE 10.5 Production of a hologram. The interference pattern, produced by the waves of the scattered light a of the object and the light of a reference beam A, is recorded on a photographic film, called the hologram.

where A_0 is a constant and ψ contains the coordinates describing the direction of incidence and propagation with respect to the photographic film.

The intensity pattern of the interference of **A** and **a** is

$$
\begin{aligned}
|\mathbf{A}+\mathbf{a}|^2 &= A_0^2 + a_0^2 + A_0 a_0 e^{-i\phi} e^{i\Psi} + A_0 a_0 e^{i\phi} e^{-i\psi} \\
&= A_0^2 + a_0^2 + 2A_0 a_0 \cos(\phi - \psi). \qquad (10.51)
\end{aligned}
$$

The interference pattern is recorded on the photographic film and after development the phase information is contained in the profile of the density of blackening. The relation between the intensity of the light incident on a spot of the film and the resulting blackening is logarithmic. A detailed discussion of the resulting transmission t of the film, which is called a hologram, is presented in Goodmans (1988). Under certain circumstances one can describe the transmission curve in a linear way and have

$$t_{\text{film}} = cA_0^2 + \beta' a_0^2 + \beta' A_0 a_0 e^{-i\phi} e^{i\psi} + \beta' A_0 a_0 e^{i\phi} e^{i\psi} \qquad (10.52)$$

where c and β' are constants. The third and fourth terms are each complex. However, together they are real and therefore t_{film} remains real.

10.6.3 Recovery of Image with Same Plane Wave Used for Recording

10.6.3.1 Virtual Image

We illuminate the hologram with a plane wave equal to the one used for the recording of the hologram. We use in Eq. (10.53) $A_0 e^{-i\psi}$ and in Eq. (10.54) the conjugate $A_0 e^{i\Psi}$ and have

$$A_0 e^{-i\psi} t_{\text{film}} = A_0 e^{-i\psi}(cA_0^2 + \beta' a_0^2 + \beta' A_0 a_0 e^{-i\phi} e^{i\psi} + \beta' A_0 a_0 e^{+i\phi} e^{-i\psi}) \qquad (10.53)$$

$$A_0 e^{i\psi} t_{\text{film}} = A_0 e^{i\psi}(cA_0^2 + \beta' a_0^2 + \beta' A_0 a_0 e^{-i\phi} e^{i\psi} + \beta' A_0 a_0 e^{+i\phi} e^{-i\psi}). \qquad (10.54)$$

The first term in Eqs. (10.53) and (10.54) is a constant term; the second may be neglected if we assume that a_0 is small compared to A. The third term in (10.53) is

$$A_0 e^{-i\psi} \beta' A_0 a_0 e^{-i\phi} e^{i\psi} = (\beta' A_0^2) a_0 e^{-i\phi}. \qquad (10.55)$$

This is the important term for the virtual image. It is the doublet of the wavefront of the original and diverges. In Figure 10.6a. we show the recovery of the image using the beam A. As we know from geometrical optics, the diverging light is traced back to a virtual image of the object.

10.6.3.2 Real Image

Similarly in Eq. (10.54) the fourth term is

$$A_0 e^{i\psi} \beta' A_0 a_0 e^{i\phi} e^{-i\psi} = (\beta' A_0^2) a_0 e^{i\phi} \qquad (10.56)$$

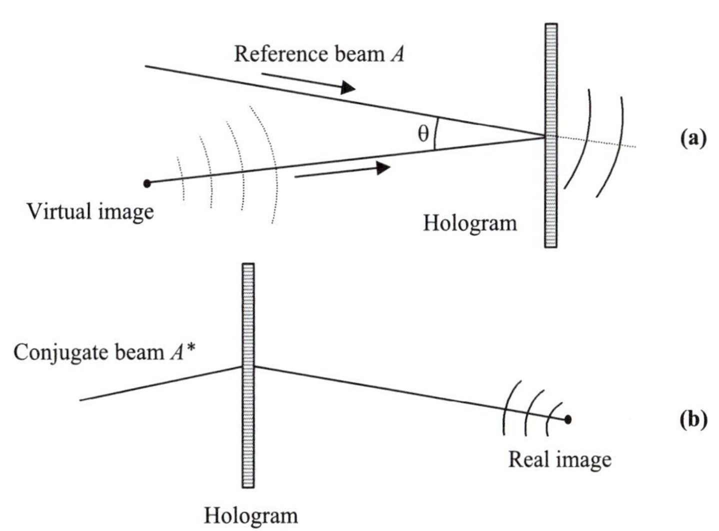

FIGURE 10.6 Recovery of the image: (a) reference beam A illuminates the hologram; wavefront of image diverges. It is traced back to the virtual image (reversed operation of production of hologram); (b) reference beam A^* illuminates the hologram; wavefront of image converges to the real image.

and is the duplication of the conjugate of the wavefront of the original. In Figure 10.6b we show the illumination by the beam A^* and the convergence to the real image, which is in the opposite direction to the virtual image we discussed in Section 10.6.3.1.

10.6.4 Recovery Using a Different Plane Wave

If we produce the hologram with a plane wave of amplitude A and illuminate the hologram with a plane wave of amplitude B, all in the same (horizontal) direction, we get

$$B_0 t_{\text{film}} = cB_0A_0^2 + dB_0a_0^2 + \beta' B_0A_0a_0e^{-i\phi} + \beta' B_0A_0a_0e^{i\phi}. \tag{10.57}$$

The real and virtual image now both appear in the horizontal direction (see Figure 10.7).

10.6.5 Production of Real and Virtual Image Under an Angle

To see the virtual image separately from the real image, one has to use a reference wave under an angle with respect to the normal of the object, around which the scattered light emerges. To do this, we use for A the wave $A_0e^{-i2\pi \sin\ \theta(x/\lambda)}$ and have for the transparency of the film

$$\begin{aligned} t_{\text{film}} = cA_0^2 + \beta' a_0^2 + \beta' A_0a_0e^{-i\phi}e^{i2\pi \sin\theta(x/\lambda)} \\ + \beta' A_0a_0e^{i\phi}e^{-i2\pi \sin\theta(x/\lambda)}. \end{aligned} \tag{10.58}$$

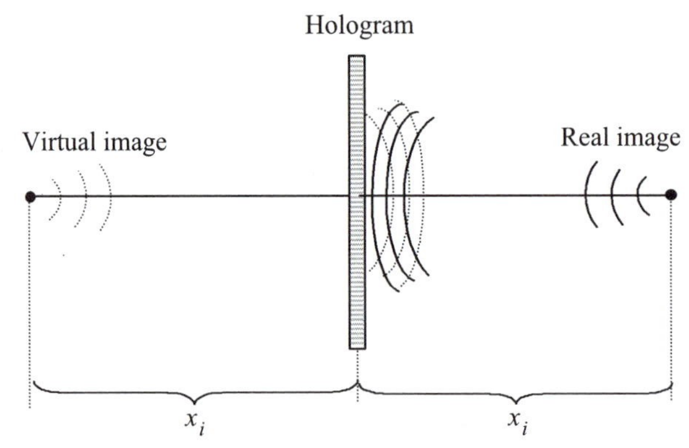

FIGURE 10.7 Real and virtual image appear in one direction, the direction of view.

For recovery we illuminate with a plane wave of amplitude B. Going through the same discussion as presented above, we get for the third (Eq. (10.59)) and fourth terms (Eq. (10.59)a)

$$(\beta' A_0^2)a_0 e^{-i\phi} e^{+i2\pi \sin\theta(x/\lambda)} \tag{10.59a}$$

and

$$(\beta' A_0^2)a_0 e^{i\phi} e^{-i2\pi \sin\theta(x/\lambda)}. \tag{10.59b}$$

The virtual image Eq. (10.59a) is now (after tracing backwards) in the direction of θ. The real image is in the direction of $-\theta$ and that is different from the direction we seek for the virtual image. This is shown in Figure 10.8, where virtual and real images are separated.

10.6.6 Size of Hologram

In the discussion of imaging we found that for a special case using coherent light the image can be calculated from the Fourier transform of the Fourier transform of the object. The hologram may be compared, in a simplified way,

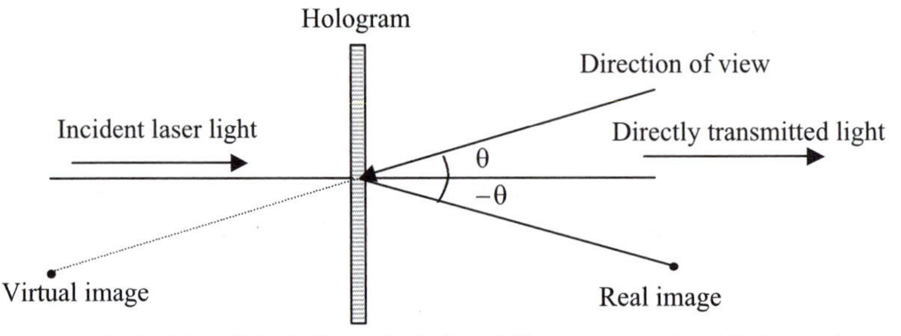

FIGURE 10.8 The incident light is from the left and directly transmitted light continues to the right. The direction of view is at the angle θ, and the virtual image is traced back in that direction. The real image appears on the right at the $angle - \theta$.

with the diffraction pattern of the object. It contains the spatial frequency spectrum including phase information. We may then substitute for the hologram the Fourier transform of the object and obtain the image is by a second Fourier transformation.

Using this simple model we can easily demonstrate that the object can be reconstructed using different sizes of the hologram. Smaller physical sizes will give us more deteriorated images of the object. By cutting a large hologram into small ones, we lose spatial frequency information. For simplicity we consider an object with the first Fourier transformation in such a way that the smallest frequencies are in the center and the largest frequencies at increased distances from the center. To demonstrate the effect of the removal of frequencies we consider the example of a grid as an object. Removing certain sections of the frequency pattern, which *is* accomplished with the first Fourier transformation, will change the image. In FileFig 10.15 we use a blocking function for a low frequency portion, in FileFig 10.16 a blocking function for an intermediate section, and in FileFig 10.17 a blocking function of a grid. In FileFig 10.18 we simulate cutting down the hologram by a blocking function, which symmetrically cuts down the lowest frequencies.

FileFig 10.15 (W15HOGRBLHIS)

The object is a grid. The transfer function removes certain higher frequencies of the first Fourier transformation. The extent of the blocking function depends on a. The modified image is compared with the original object.

W15HOGRBLHIS is only on the CD.

Application 10.15. Observe the changes of the final image by modification of the blocking function, that is, changing a.

FileFig 10.16 (W16HOGRBLLOS)

The object is a grid. The transfer function is a blocking function passing only one portion of the frequencies of the first Fourier transformation. The extent of the blocking function depends on n and a. The modified image is compared to the original object.

W16HOGRBLLOS is only on the CD.

Application 10.16. Observe the changes of the final image by modification of the blocking function, that is, changing n and a.

FileFig 10.17 (W17HOGRPERS)

The object is a grid. The transfer function is a grid-type blocking function blocking periodic parts of the first Fourier transformation. The width of the peaks and the extent of the blocking function depend on q and a. The modified image is compared to the original object.

W17HOGRPERS

Object Is a Periodic Structure

The FT of the object is multiplied by a blocking function. A blocking function has been chosen blocking certain frequencies such that there are twice as many peaks in the image. The Ft (inverse) of (Ft of object) · (blocking function) is the "new" image. The "new" image is compared to the original, that is, the FT of (FT of object). The blocking function removes certain high frequencies of the FT.

Object

$$i := 1, 2 \ldots 127 \qquad b := 2 \qquad q := 7$$

$$y_i := \left[\sum_{n=0}^{q} (\Phi(i - (4 \cdot (2 \cdot n + 1) + 2) \cdot b) - \Phi(i - (4 \cdot (2 \cdot n + 1) + 4) \cdot b)) \right].$$

FT of object

$$\omega := cfft(y) \quad N := \text{last}(\omega)$$

$$N = 127.$$

Blocking function

$$\tau_i := \left[\sum_{n=0}^{q} (\Phi(i - (4 \cdot n + 2) \cdot a) - \Phi(i - (4 \cdot n + 4) \cdot a))\right]$$

$$q \equiv 5 \qquad a \equiv 5.$$

2

τ_i

0

0 50 100 150

i

Product: FT (inverse) of object and blocking function

$$\phi_i := \omega_i \cdot \tau_i$$

$$yy := icfft(\phi)$$

$$N2 := \text{last}(\phi)$$

$$k := 0 \ldots N2.$$

2

1

yy_i

0

−1

0 50 100 150

i

For comparison: original object

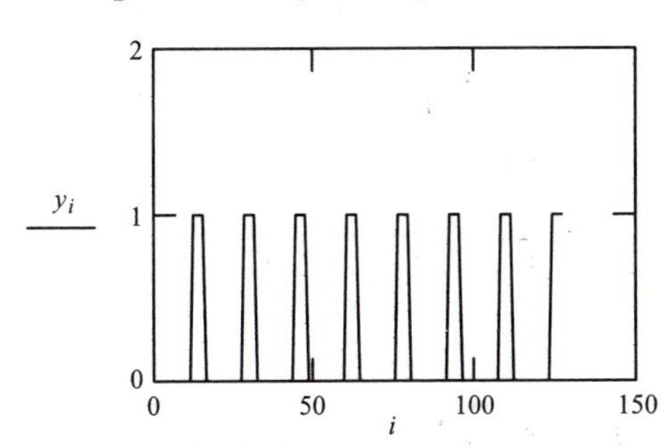

Application 10.17. Observe the changes of the final image by modification of the blocking function, that is, changing q and a.

FileFig 10.18 (W18HOSTEPS)

The object is a step function composition. The transfer function passes the lowest frequencies on both parts of the diffraction pattern, that is, the first Fourier transformation. The number of passing lowest frequencies is monitored by a and b. The hologram is larger when more frequencies are passing. The modified image is compared to the original object.

W18HOSTEPS

Object y

The object y has a complicated shape. Its Ft is the hologram c. It may be produced in the focal plane of a lens, using parallel light. The illumination of the hologram with parallel light will reproduce the object, that is, the FT (inverse) of the FT, called here cc. We study the reproduced object when the information in the hologram is only partly used; that is, we multiply cc with a filder f. We show separately f and the FT of the product of f and cc. The width of the filter f may be changed by using various values for a and b, corresponding to changing the size of the hologram.

The object $i := 0, 1 \ldots 255$.

$$A_1 := 33 \qquad A_2 := 80 \qquad A_3 := 80 \qquad A_4 := 50$$

$$A_5 := 20 \qquad A_6 := 99 \qquad A_7 := 160 \qquad A_8 := 200$$

$$y_i := \left[\sum_{n=1}^{3}(-\Phi(A_n - i))\right] + \left[\sum_{n=4}^{8}[\Phi(A_n - i)\cdot(-1)^n]\right].$$

The hologram

$$c := cfft(y)N := \text{last}(c) \quad N = 255$$

$$k := 0 \ldots 255 \quad j \equiv 0 \ldots 255$$

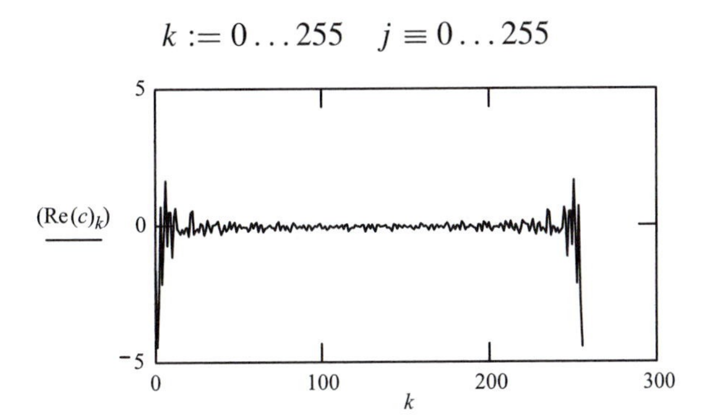

The FT on the FT (hologram)

$$cc_k := c_k yy := icfft(cc)$$

$$N := \text{last}(cc) \quad N = 255 \quad j := 0 \ldots 255$$

Re(yy)_j

j

The filter

$$f_j := \Phi(a - j) + \Phi(j - 255 + b)$$

$$a \equiv 60$$

$$b \equiv 60.$$

f_j

j

The product: hologram x filter

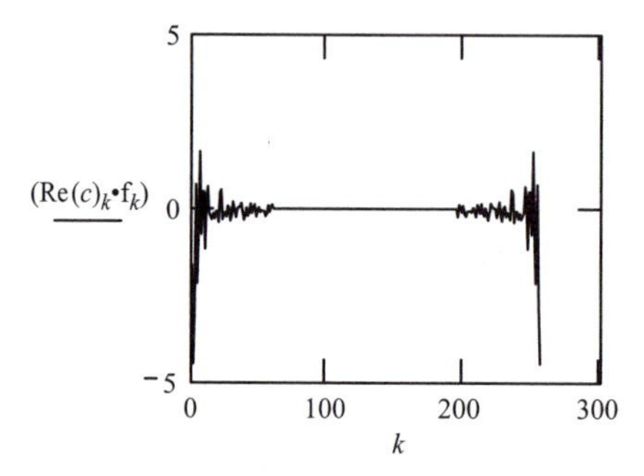

The FT (inverse) of the changed hologram (hologram x filter), similar to the object

$$ccc_k := c_k \cdot f_k \qquad x := icfft(ccc) \qquad N := \text{last}(ccc) \qquad N = 255$$

$$k := 0 \ldots 255.$$

2

0

$(\text{Re}(x)_k)$

−2

−4

0 100 200 300

k

For comparison: the object

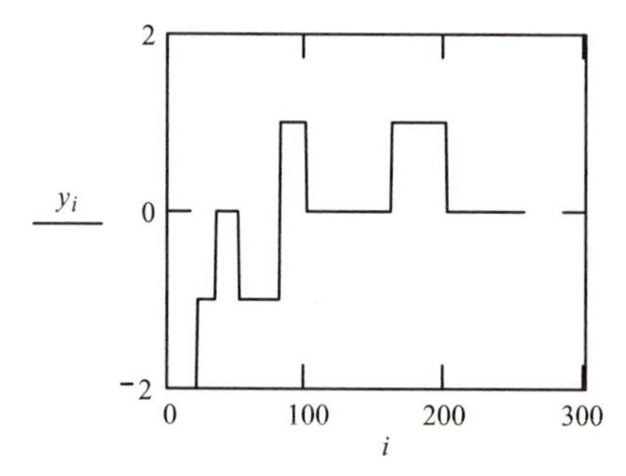

Application 10.18. Observe the changes of the size of the hologram when making the range of passing low frequencies larger or smaller. The blocking function and the final image may be modified by changing a and b.

CHAPTER 11

Aberration

11.1 INTRODUCTION

In Chapter 1 on Geometrical Optics we discussed geometrical image formation by using paraxial theory. The essential assumption of paraxial theory is that the angles between an emerging ray from the object and the axis of the system are small. In general small means that one could replace $\sin\alpha$ by the angle α (in radians). When this assumption can not be made, one obtains a distorted image. There are elaborate computational programs available for lens design, including systematic corrections for the various types of aberrations. To give an introduction to the most commonly known monochromatic aberrations, we discuss spherical aberration of a single refracting surface and a thin lens, and coma, and astigmatism of a spherical surface and a thin lens. At the end we also discuss chromatic aberration.

11.2 SPHERICAL ABERRATION OF A SINGLE REFRACTING SURFACE

In Figure 11.1, two rays are shown from the object point P_1 to the spherical surface. After refraction one ray is connected to the image point P_2, the other to P_2'. When paraxial theory can not be used, the ray with the larger angle α_2 has the image point P_2' closer to the refracting surface. The difference between the points P_2 and P_2' is called the *longitudinal spherical aberration*. For its derivation, we look at Figure 11.2 and derive the relations for image formation at one spherical surface

$$(s_1 + r)/(\sin(180 - \theta_1) = \zeta_1/\sin\beta \tag{11.1}$$

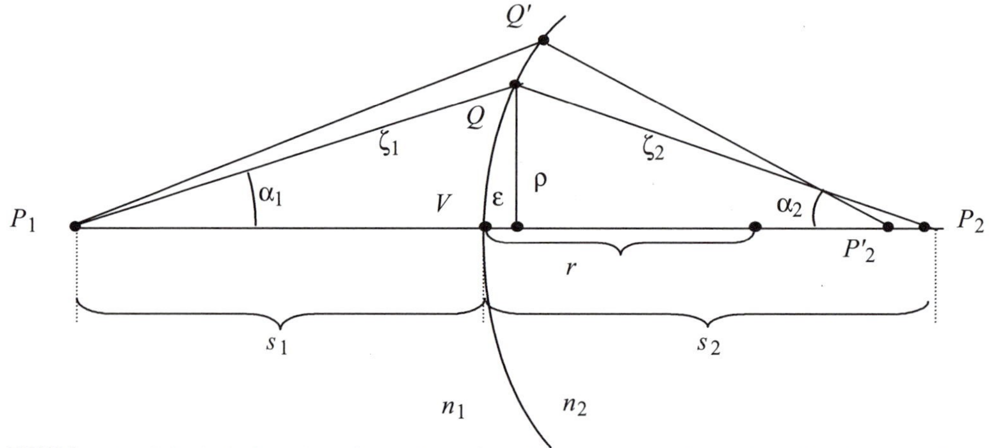

FIGURE 11.1 Spherical aberration of a single surface. The image point P_2 is formed by the paraxial ray P_1QP_2. The marginal ray $P_1Q'P_2'$ forms the image at P_2', a position closer to the lens.

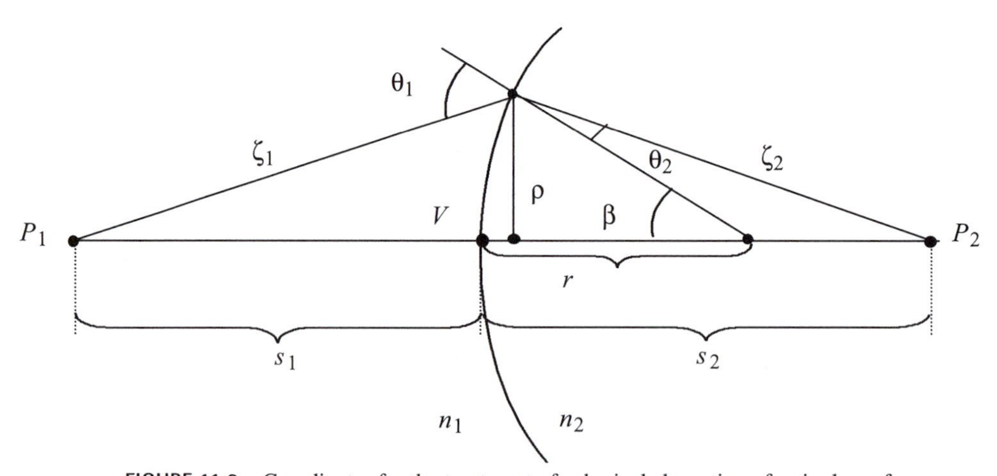

FIGURE 11.2 Coordinates for the treatment of spherical aberration of a single surface.

and

$$(s_2 - r)/\sin\theta_2 = \zeta_2/\sin(180 - \beta). \tag{11.2}$$

Equations (11.1) and (11.2) may be combined to form

$$(s_1 + r)/(s_2 - r) = n\zeta_1/\zeta_2, \tag{11.3}$$

where n is the refractive index of the refracting medium. To get expressions for ζ_1 and ζ_2, we look at the triangle P_1QP_2 (Figure 11.2) and have

$$\zeta_1^2 = r^2 + (s_1 + r)^2 - 2r(s_1 + r)\cos\beta. \tag{11.4}$$

Expanding $\cos\beta = 1 - \beta^2/2$ and setting $\beta = \rho/r$ one gets

$$\zeta_1^2 = r^2 + (s_1 + r)^2 - 2r(s_1 + r)(1 - [\rho^2/2r^2]), \tag{11.5}$$

which results in

$$\zeta_1^2 = s_1^2 + (s_1 + r)[\rho^2/r]. \tag{11.6}$$

Similarly one obtains

$$\zeta_2^2 = s_s^2 - (s_2 - r)[\rho^2/r]. \tag{11.7}$$

Expanding the root one has

$$\zeta_1 = s_1 + (1/s_1 + 1/r)[\rho^2/2] \tag{11.8}$$

$$\zeta_2 = s_2 + (1/s_2 - 1/r)[\rho^2/2]. \tag{11.9}$$

Introduction of Eqs. (11.8) and (11.9) into (11.3) results in

$$(s_1 + r)/(s_2 - r) = n\{s_1 + (1/s_1 + 1/r)[\rho^2/2]\}/\{s_2 + (1/s_2 - 1/r)[\rho^2/2]\}. \tag{11.10}$$

This equation may be rewritten as

$$1/s_1 + n/s_2 + (1-n)/r = (1/r + 1/s_1)(1/r - 1/s_2)(n/s_1 + 1/s_2)[\rho^2/2]. \tag{11.11}$$

Introduction of $s_1 = -x_0$ and $s_2 = x_i$ results in

$$-1/x_0 + n/x_i + (1-n)/r = (1/r - 1/x_0)(1/r - 1/x_i)(1/x_i - n/x_0)[\rho^2/2]. \tag{11.12}$$

In the limit of $\rho \to 0$ we must get back to the imaging equation of paraxial theory,

$$-1/x_0 + n/x_{i1} + (1-n)/r = 0, \tag{11.13}$$

where we have written x_{i1} for the image distance for the paraxial case. The coefficient of correction on the right side of Eq. (11.12) depends on x_0 and x_i. To have it depending only on x_0, we use Eq. (11.13), eliminating x_i and get

$$n/x_{i1sal} = 1/x_0 + (n-1)/r + ((n-1)/n^2)(1/r - 1/x_0)^2(1/r - (n+1)/x_0))[\rho^2/2], \tag{11.14}$$

where we have written x_{i1sal} to indicate the image distance for the longitudinal spherical aberration of a surface of radius of curvature ρ.

The longitudinal spherical aberration (LSA) is defined as $x_{i1} - x_{i1sal}$, which is the difference of the image positions calculated for the paraxial and the spherical aberration cases. In FileFig 11.1, we study the LSA for an object distance, which corresponds to a real image, and two different refractive indices. In FileFig 11.2, we study the dependence of LSA on object distances, which corresponds to real images, and on the refractive index and the radius of curvature. For real images spherical aberration may not be eliminated.

FileFig 11.1 (A1SPHASS)

Calculation of LSA $= x_{i1} - x_{i1sal}$ for a single spherical surface and negative object distance for two different refractive indices.

A1SPHASS is only on the CD.

Application 11.1.

1. Consider positive and negative object distances and choose one refractive index. Show that one can get positive and negative values for the LSA.
2. Decide how small you want to make the LSA and determine the corresponding value of n.

FileFig 11.2 (A2SPASSS)

Calculations for a single spherical surface to demonstrate the dependence of LSA $= x_{i1} - x_{i1sal}$ on the object position, the refractive index, and the radius of curvature. Note that for the choice of parameters used, there is no value of the refractive index for which the LSA is zero.

A2SPASSS is only on the CD.

11.3 LONGITUDINAL AND LATERAL SPHERICAL ABERRATION OF A THIN LENS

In Figure 11.3 spherical aberrations are shown for a positive and a negative thin lens. To calculate the LSA $= x_{i1} - x_{sph}$ for a thin lens we use twice the result obtained for a single spherical surface, as discussed in Section 11.2. There we calculated for a spherical surface with refractive index n, the position x_{i1sal} (see Eq. (11.14)). The light entered from the medium of index 1 and traveled into the medium of index n. We obtained

$$\begin{aligned} &-1/x_0 + n/x_i \qquad (11.15)\\ &= (n-1)/r_1 + ((n-1)/n2)(1/r_1 - 1/x_0)^2(1/r_1 - (n+1)/x_0)[\rho^2/2]. \end{aligned}$$

This result may be used to get a similar expression for light incident on the second surface traveling from the medium with index n into the medium of index 1. We substitute x_0 with $\rightarrow x_{ii}$ and x_i with $\rightarrow x_{00}$ and obtain

$$\begin{aligned} &-n/x_{00} + 1/x_{ii} \qquad (11.16)\\ &= -(n-1)/r_2 - ((n-1)/n2)(1/r_2 - 1/x_{ii})^2(1/r_2 - (n+1)/x_{ii})[\rho^2/2], \end{aligned}$$

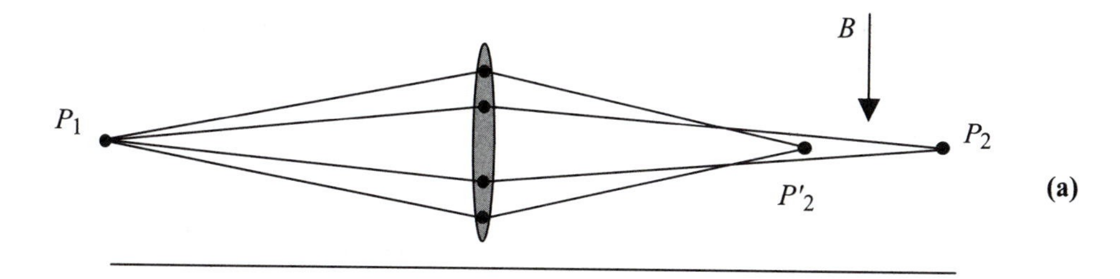

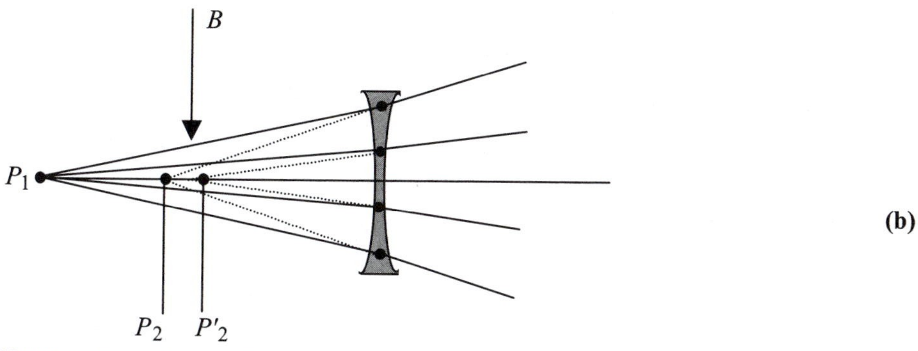

FIGURE 11.3 (a) Spherical aberration for a positive lens, real image points; (b) spherical aberration for a negative lens, virtual image points.

where x_i and x_{00} are the same points when considering the case of a thin lens. Addition of Eqs. (11.15) and (11.16) will eliminate the terms n/x_i and $-n/x_{00}$, and results in

$$\begin{aligned} &-1/x_0 + 1/x_{ii} \qquad (11.17) \\ &= (n-1)/r_1 + ((n-1)/n^2)(1/r_1 - 1/x_0)^2(1/r_1 - (n+1)/x_0)[\rho^2/2] \\ &\quad - (n-1)/r_2 - ((n-1)/n^2)(1/r_2 - 1/x_{ii})^2(1/r_2 - (n+1)/x_{ii})[\rho^2/2]. \end{aligned}$$

To calculate x_{ii} one uses the paraxial equation of the thin lens

$$1/x_{ii} = 1/x_0 + (n-1)(1/r_1 - 1/r_2), \qquad (11.18)$$

where

$$1/f = (n-1)(1/r_1 - 1/r_2) \qquad (11.19)$$

and finally we have

$$\begin{aligned} &-1/x_0 + 1/x_{iisph} \qquad (11.20) \\ &= (n-1)(1/r_1 - 1/r_2) + ((n-1)/n^2)[(1/r_1 - 1/x_0)^2(1/r_1 - (n+1)/x_0) \\ &\quad - (1/x_0 + (n-1)/r_1 - n/r_2)^2(n^2/r_2 - (n+1)/x_0 - (n^2-1)/r_1)][\rho^2/2]. \end{aligned}$$

The longitudinal spherical aberration of a thin lens is defined as LSA $= x_{ii} - x_{iisph}$. We call the right side of Eq. (11.20) the reciprocal focal length for the spherical aberration case, $1/ff(x_0)$, where $ff(x_0)$ is the focal length of the case of spherical aberration. The result is

$$-1/x_0 + 1/x_{iisph} = 1/ff(x_0) \tag{11.21}$$

with

$$ff(x_0) = 1/\{1/f + \{(n-1)/n^2\}[\rho^2/2]\{a(x_0) - b(x_0)c(x_0)\}\} \tag{11.22}$$

and the abbreviations

$$\begin{aligned} a(x_0) &= [(1/r_1 - 1/x_0)^2(1/r_1 - (n+1)/x_0)], \\ b(x_0) &= [1/x_0 + (n-1)/r_1 - n/r_2]^2 \\ c(x_0) &= [n^2/r_2 - (n+1)/x_0 - (n^2-1)/r_1]. \end{aligned} \tag{11.23}$$

In Figure 11.4 we define the lateral spherical aberration as LAT = LSA times ρ/x_{iisph}. In FileFig 11.3 we study the question of elimination of spherical aberration and calculate numerical values of LSA and LAT for a choice of parameters of n, r_1, r_2, ρ, and object distance x_0. The elimination of spherical aberration is further discussed in the next section using the π–σ equation.

FileFig 11.3 (A3SPHTINS)

Calculations of the spherical aberration of a thin lens. Longitudinal spherical aberration and lateral spherical aberration.

A3SPHTINS is only on the CD.

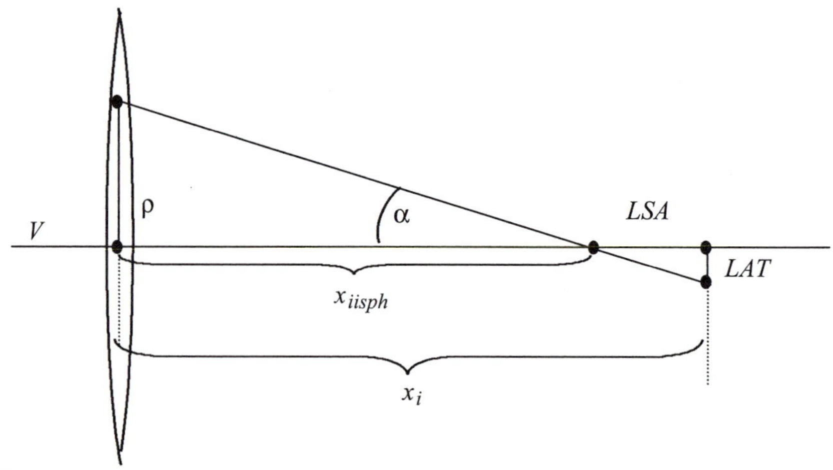

FIGURE 11.4 Longitudinal spherical aberration (LSA). The lateral spherical aberration (LAT) is calculated using tan α.

Application 11.3.

1. Use different values of n and observe that spherical aberration may not be eliminated for real objects and images.
2. Assume x_0 values for virtual images and find positive and negative values for the LSA. Therefore, for this case, spherical aberration may be eliminated.

11.4 THE π-σ EQUATION AND SPHERICAL ABERRATION

We now study whether spherical aberration can be removed when using a special choice of parameters. We introduce the parameters π (position factor) and σ (shape factor),

$$\pi = (x_{ii} + x_0)/(x_{ii} - x_0) \quad \text{and} \quad \sigma = (r_2 + r_1)/(r_2 - r_1). \tag{11.24}$$

Using Eqs. (11.18) and (11.19) we may write

$$1/x_0 = -(\pi + 1)/2f \quad \text{and} \quad 1/x_i = (1 - \pi)/2f \tag{11.25}$$

$$1/r_1 = (\sigma + 1)/\{2f(n-1)\} \quad \text{and} \quad 1/r_2 = (\sigma - 1)/\{2f(n-1)\}. \tag{11.26}$$

Introducing the expressions of Eqs. (11.24) to (11.26) into Eq. (1.17) we get

$$\begin{aligned} &-1/x_0 + 1/x_{iisph} \\ &= (n-1)(1/r_1 - 1/r_2) + [\rho^2/f^3]\{A\sigma^2 + B\sigma\pi + C\pi^2 + D\}, \end{aligned} \tag{11.27}$$

where we abbreviated

$$A = (n+2)/\{8n(n-1)^2\} \qquad B = (n+1)/\{2n(n-1)\} \tag{11.28}$$

$$C = (3n+2)/(8n) \qquad D = n^2/\{8(n-1)^2\}. \tag{11.29}$$

We may look at Eq. (11.27) as the thin lens equation plus a correction term. To study whether spherical aberration can be removed we look at the correction term

$$Y = [\rho^2/f^3]\{A\sigma^2 + B\sigma\pi + C\pi^2 + D\}. \tag{11.30}$$

When Y is equal to zero, spherical aberration is eliminated.

In FileFig 11.4 a graph is shown for $f = 10$, $n = 1.5$, $\rho = 4$, and $x_0 = 4$. There are Y values smaller than zero and, using these parameters, spherical aberration may be eliminated. In FileFig 11.4 one may study an example for positive and negative values of x_0 and ρ between 1 and 4. When Y shows negative values, spherical aberration may be eliminated.

FileFig 11.4 (A4SPHLSIPIS)

The π–σ equation and spherical aberration for the thin lens. The graph shown is for $\rho = 1$ and $x_0 = -11$, and one only has positive values of $Y(\sigma)$. This means spherical aberration is not eliminated.

A4SPHLSIPIS

Spherical Aberration and the $\pi - \sigma$ Equation

We assume $n = 1.5$ and compare the cases of real and virtual images.

1. Image for $f = 10$, and xo to the left of focal point, LSA may not be eliminated

$$f := \frac{1}{(n-1)\cdot\left(\frac{1}{r_1} - \frac{1}{r_2}\right)} \qquad r_1 \equiv 10 \quad r_2 \equiv -10 \qquad n \equiv 1.5$$

$$ro \equiv 4 \qquad xo \equiv 4$$

$$f = 10 \qquad xi := \frac{1}{\left(\frac{1}{f} + \frac{1}{xo}\right)} \qquad xi = 2.857.$$

2. Definitions

$$\sigma = (r_2 + r_1)/(r_2 - r_1) \qquad \sigma := -10, -9.9 \ldots 10$$

$$\pi := \frac{xi + xo}{(xi - xo)} \qquad \pi = -6.$$

3. π–σ Equation

$$A(n) := \frac{n+2}{8\cdot n\cdot(n-1)^2} \qquad B(n) := \frac{n+1}{2\cdot n\cdot(n-1)}$$

$$C(n) := \frac{3\cdot n + 2}{8\cdot n} \qquad D(n) := \frac{n^2}{8\cdot(n-1)^2}$$

$$Y(\sigma) := \frac{(ro)^2}{f^3}\cdot\left(A(n)\cdot\sigma^2 + B(n)\sigma\pi + C(n)\pi^2 + D(n)\right).$$

4. Minimum value of $Y(\sigma)$ The value of $Y(\sigma)$ at the minimum is obtained by differentiation and equal to 0. The result is

$$\sigma \min := -B(n) \cdot \frac{\pi}{2 \cdot A(n)}$$

$$\sigma \min = 4.286.$$

Calculation of the corresponding value of $Y(\sigma \min)$

$$Y(\sigma \min) = -0.013.$$

For our choice of parameters, $Y(\sigma \min)$ is positive and LSA may not be eliminated.

Application 11.4.

1. Study the π–σ equation and give two examples for elimination of spherical aberration, for a positive and a negative value of x_0.
2. Consider a set of lenses all having $f = 10$cm, $n = 1.5$, and radii of curvature $r1$ and $r2$ such that the shape factor σ is between -2 and 2. Plot $fs - f$ depending on σ, where fs is the corrected focal length for spherical aberration. Make sketches of the radii of curvature for values of $\sigma = -2, -1, 0, 1,$ 2 and compare with Jenkins and White (1976, p. 145).

11.5 COMA

So far we have discussed spherical aberration produced by the size of the lens for on-axis points. When the object point is slightly off axis, the resulting aberration is called coma. A new axis appears from the object point through the center of the lens to the center of the image (Figure 11.5). The zones of the lens, indicated by points in Figure 11.5, produce circles instead of image points. Only the center zone produces a point image on the new axis. Larger zones produce circles with larger radii depending on the distance from the new axis. The rings, corresponding to the zones, are arranged like the tail of a comet.

We assume for this discussion of coma that spherical aberration has been eliminated and follow Jenkins and White (1976, p.163). We assume that parallel light is incident on the lens, and the sagittal coma C_S is

$$C_S = [(\rho^2/f^2)\tan\beta][W\sigma + G\pi], \tag{11.31}$$

where ρ is the radius of the largest zone considered, π and σ are defined in the same way as in Eq. (11.24) and β is the angle between the axis of the system and the new axis (Figure 11.6a). For W and G one has

$$W = 3(n+1)/\{4n(n-1)\} \quad \text{and} \quad G = 3(2n+1)/4n. \tag{11.32}$$

The tangential coma C_T is shown in Figure 11.6b and is calculated to be $3C_S$. The condition for elimination of coma is obtained from Eq. (11.31), when $[W\sigma + G\pi]$

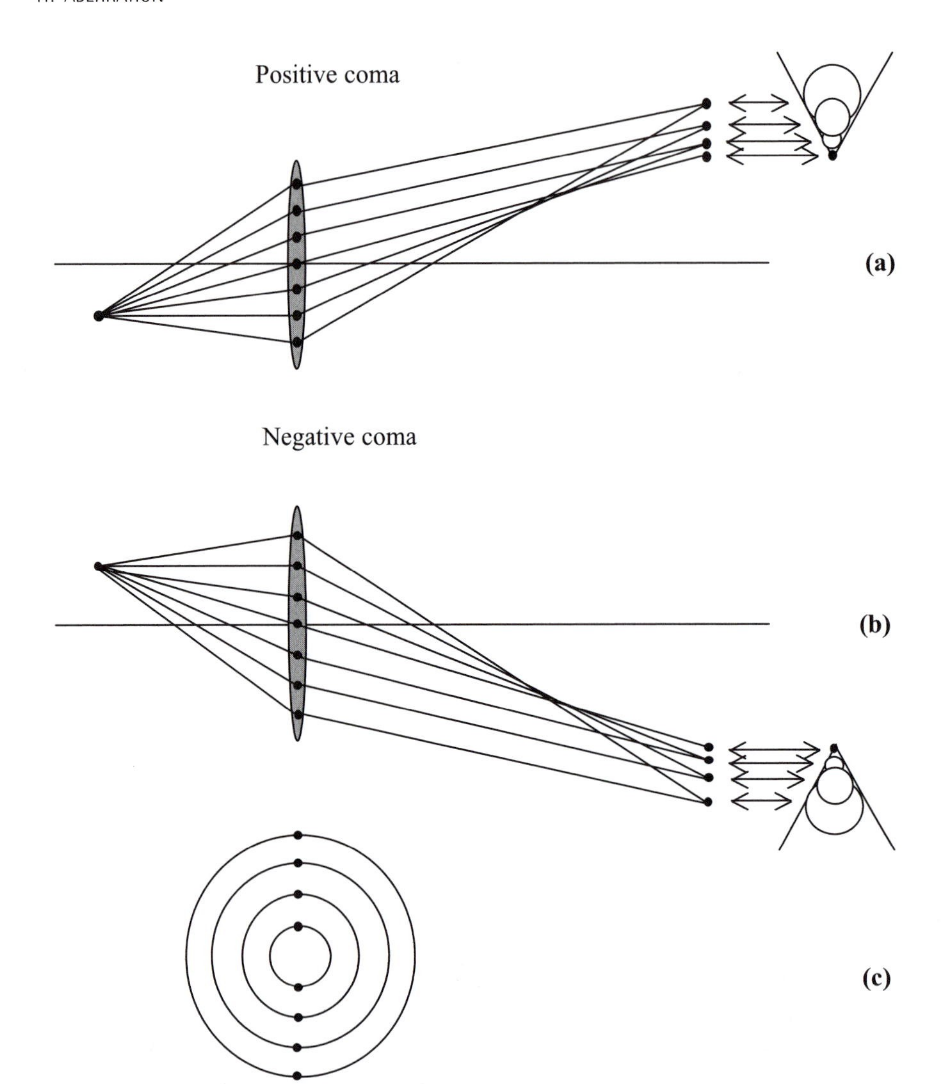

FIGURE 11.5 Circles of coma of increasing radius corresponding to increasing diameters of the zones of the lens: (a) negative coma; (b) positive coma; (c) zones.

is equal to zero or

$$\sigma = -\{(2n+1)(n-1)\pi\}/(n+1). \tag{11.33}$$

In FileFig 11.5, a graph is shown of $CT(\rho)$, calculated depending on the radii of the zones ρ. There are positive and negative values for $CT(\rho)$ depending on the choice of the refractive index and one may choose parameters such that coma is eliminated.

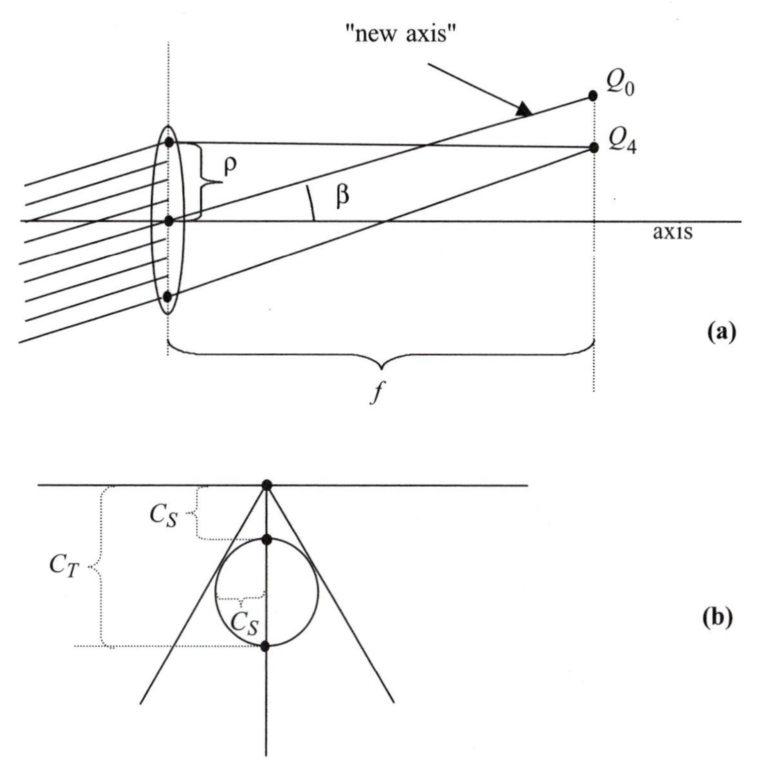

FIGURE 11.6 Coordinates for calculation of coma assuming incident parallel light: (a) Relation of zones of the lens to centers of circles of images. Q_0 is on the new axis. The angle between the system axis and the new axis is β. The point Q_4 is the center of the circle of the largest zone of the lens; (b) tangential coma C_T and sagittal coma C_S.

FileFig 11.5 (A5COMAS)

Calculation of coma of the thin lens. It is assumed that spherical aberration is eliminated. Tangential coma is calculated depending on the zone radius.

A5COMAS is only on the CD.

Application 11.5.

1. Choose values of the refractive index n to obtain positive and negative coma.
2. Find an example of a set of parameters for "no coma."

11.6 APLANATIC LENS

One may design a special lens, called an aplanatic lens, which has no coma and no spherical aberration. We consider the two equations

$$\sigma = -(2n + 1) \quad \text{and} \quad \pi = (n + 1)/(n - 1). \tag{11.34}$$

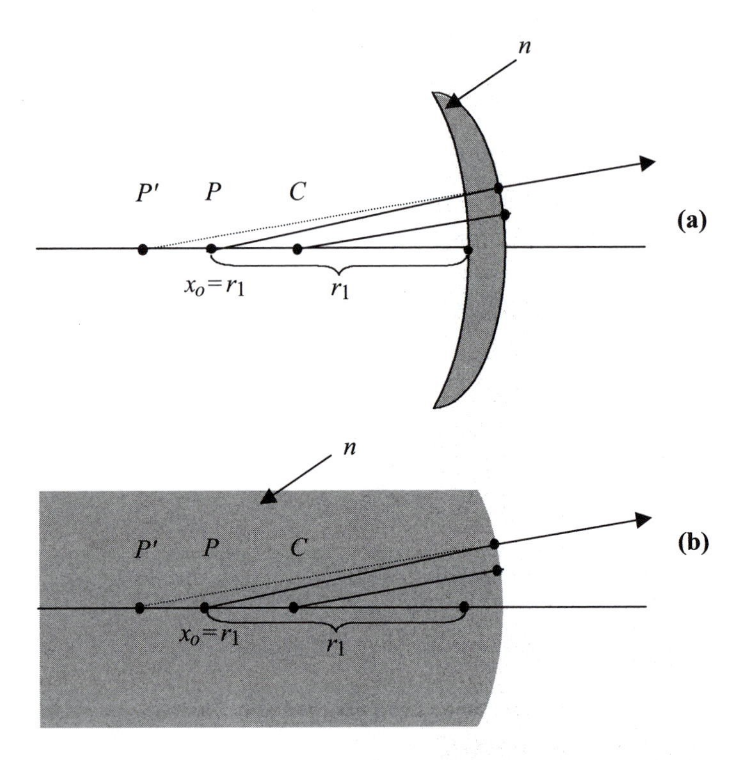

FIGURE 11.7 Aplanatic lens. Light from the point $x_0 = r_1$ is not refracted on the first surface of the lens, but on the second. This is shown in (a) for a lens and in (b) for a sphere.

According to Eq. (11.33) there will be no coma and no spherical aberration because $Y = 0$ (Eq. (11.30)). An example is shown in Figure 11.7. Introduction of σ and π from Eq. (11.34) into the four expressions of Eqs. (11.24) and (11.25) results in

$$r_1 = \{r_2(n+1)\}/n \qquad f = \{r_2(n+1)\}/(1-n) \tag{11.35}$$

$$x_0 = \{r_2(n+1)\}/n \qquad x_i = r_2(n+1). \tag{11.36}$$

The focal length of the aplanatic lens depends only on n and r_2:

$$1/f = (1-n)/\{r_2(n+1)\}. \tag{11.37}$$

To use it as an aplanatic lens, in addition to the thin lens equation, one has to satisfy the relation between x_i and x_0:

$$x_i = n x_0. \tag{11.38}$$

We see from Figure 11.7a that the emerging rays from x_0 are not refracted at the surface of radius r_1, but only when arriving at the surface of radius r_2. Therefore, we may consider x_0 at P as the object point in a medium of refractive index n and spherical surface of radius r_2. This is called an aplanatic sphere, shown in Figure 11.7b. The virtual image point P' is obtained by tracing back the ray refracted at the second surface because the image is virtual.

In FileFig 11.6 we show that coma is eliminated for the aplanatic lens.

FileFig 11.6 (A6COMPLANS)

Calculation of coma for the aplanatic lens. The result is that coma is zero.

A6COMPLANS is only on the CD.

Application 11.6.

1. Make a graph of $ss(nn)$ and find back the value for σ at $nn = 1.5$.
2. Get Y from FileFig 9.4 (A4SPHLSIPIS) and using the values from FF6 show that $Y = 0$.
3. Consider a farsighted eye and assume that the focal length is $f = 2.2$ cm and the distance from eye to retina $d = 2$ cm. Design an applanatic lens for correction, that is, bringing the focal point of the eye to the retina.

11.7 ASTIGMATISM

We have discussed in Sections 11.3 and 11.5 spherical aberration and coma. For spherical aberration the object points were assumed to be on axis, and for coma slightly off axis. When the object point is farther away from the axis, the image points are no longer in one perpendicular plane with respect to the new axis, as we found for coma. The points appear one behind the other on a new axis from the object point through the center of the lens. There are two planes defined with respect to the new axis, called *sagittal* (horizontal) and *meridional* (vertical) *planes* (see Figures 11.8 and 11.9). Each produces its own image point and between these two points is the circle of least confusion.

The difference of the sagittal and meridional image points on the new axis is called the *astigmatic difference*, ASD $= x_{iH} - x_{iV}$. Again we follow Jenkins and White (1976, p. 167).

11.7.1 Astigmatism of a Single Spherical Surface

We first calculate x_{iH} and x_{iV} for a single refracting surface, using the corresponding imaging equations. For the horizontal image points one has

$$-1/x_0 + n/x_{iH} = (n \cos\phi' - \cos\phi)/r \tag{11.39}$$

and for the vertical image point

$$-(\cos\phi)^2/x_0 + n(\cos\phi')/x_{iV} = (n \cos\phi' - cos\phi)/r, \tag{11.40}$$

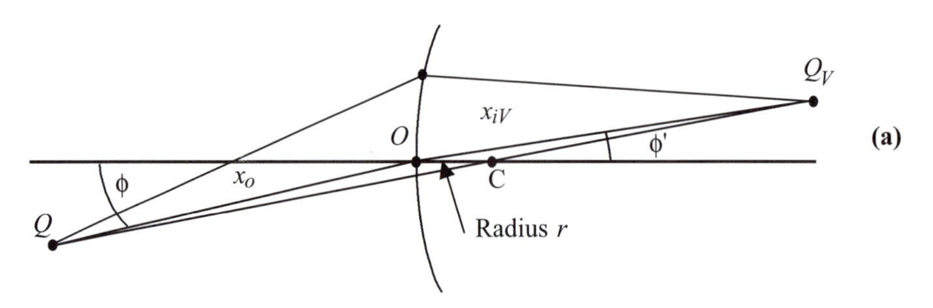

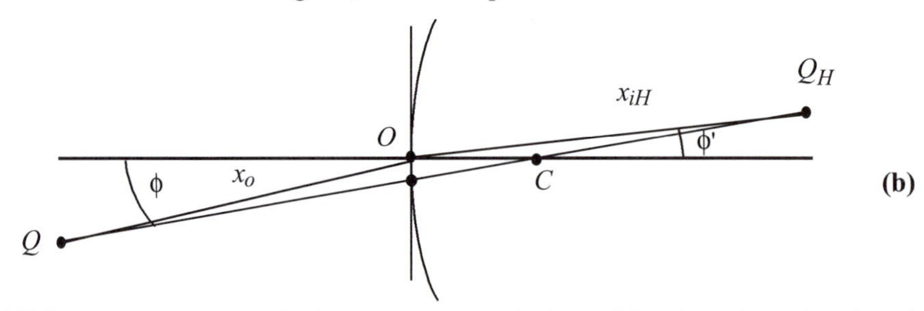

FIGURE 11.8 Astigmatism of a single surface. Rays in the meridional (vertical) plane from the primary focus Q_V, and rays in the sagittal (horizontal) plane form the secondary focus Q_H. The circle of least confusion is between them.

where r is the radius of curvature of the spherical surface. The angles ϕ and ϕ' are shown in Figures 11.8a and b for the horizontal and vertical plane, respectively.

In FileFig 11.7 the astigmatic difference ASD is calculated depending on the angle ϕ. The angle ϕ' may be eliminated using the law of refraction.

11.7.2 Astigmatism of a Thin Lens

For the calculation of the ASD for a thin lens, one has for the sagittal and meridional case each a thin lens equation with a corrected focal length. This is similar to the above discussions of other aberrations. For the sagittal plane one gets

$$-1/x_0 + 1/x_{iH} = (\cos\phi)[(\cos\phi'/\cos\phi) - 1](1/r_1 - 1/r_2) \tag{11.41}$$

and for the meriodinal plane

$$-1/x_0 + 1/x_{iV} = \{1/\cos\phi\}[(\cos\phi'/\cos\phi) - 1](1/r_1 - 1/r_2) \tag{11.42}$$

In FileFig 11.8, we consider a lens with radii of curvature of $r_1 = 10$, $r_2 = -12$, and $n = 1.3$. The first graph shows the calculation of the astigmatic difference ASD depending on the angle ϕ. The second graph shows for the same lens the dependence on the refractive index n for $\phi = 10$ degrees.

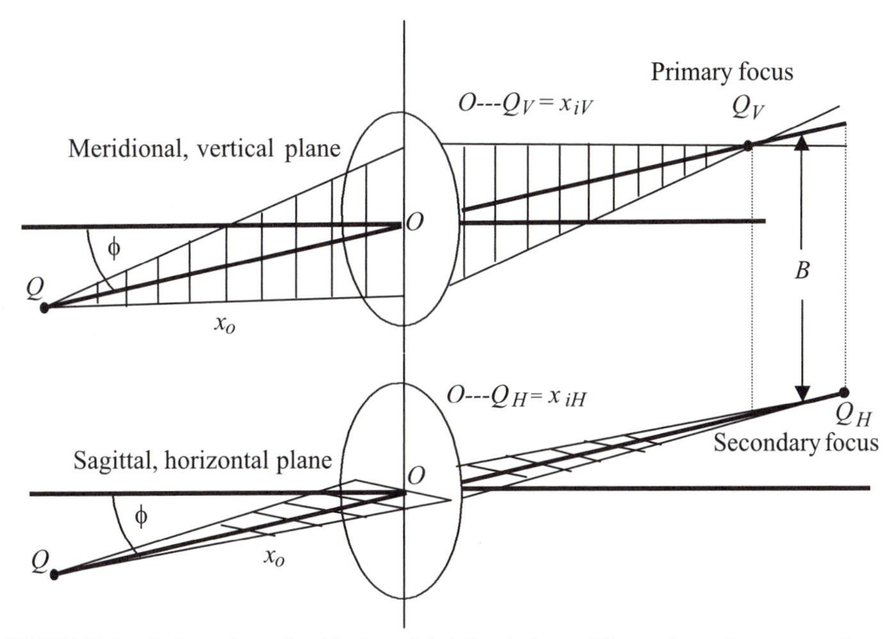

FIGURE 11.9 Astigmatism of a thin lens, Meridional plane: All rays from Q meet at primary focus Q_V. Sagittal plane: All rays from Q meet at point secondary focus Q_H. The circle of least confusion is indicated by B.

FileFig 11.7 (A7ASTSINS)

Astigmatism of a single surface. Calculation of astigmatic difference ASD depending on angle ϕ for a single refracting surface with radius of curvature r.

A7ASTSINS is only on the CD.

Application 11.7.

1. Modify the file for dependence on n for fixed angle ϕ.
2. Study the ASD for small and large values of r.

FileFig 11.8 (A8ASTISMS)

Astigmatism of a thin lens. Calculation of astigmatic difference ASD for a thin lens of radii of curvature r_1 and r_2, depending on angle ϕ or on n.

A8ASTISMS is only on the CD.

Application 11.8.

1. Compare the ASD for lenses with small and large focal lengths.
2. Compare the ASD for a lense with a corresponding lens having one plane surface.

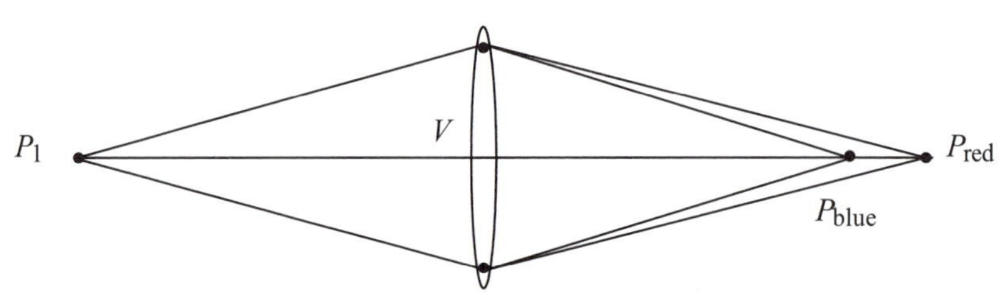

FIGURE 11.10 Chromatic aberration: all rays from P_1 with shorter wavelength (blue light) converge to point P_{blue}; all rays with longer wavelength (red light) converge to P_{red}.

11.8 CHROMATIC ABERRATION AND THE ACHROMATIC DOUBLET

The aberrations discussed so far, spherical aberration, coma, and astigmatism, are called monochromatic aberrations. One assumes that monochromatic light is incident on the lens and therefore the refractive index of the law of refraction is a constant. If the incident light is not monochromatic and the constant of the law of refraction is different for different wavelengths, one speaks of *chromatic aberrations*.

In Figure 11.10 *positive chromatic aberration* is shown. This means the image point of the shorter wavelength light (blue) is closer to the lens than the image point of the longer wavelength light. The reverse case is called *negative chromatic aberration*. Chromatic aberration can be compensated by using two lenses. One has positive and the other negative chromatic aberration. Such a double lens is called an *achromatic doublet*.

An achromatic doublet is made of two lenses in contact. In Chapter 1 we found that the focal length of a compound lens is

$$\begin{aligned} 1/f &= 1/f_1 + 1/f_2 \\ &= (n_1 - 1)(1/r1 - 1/r_2) + (n2 - 1)(1/r_1' - 1/r_2'), \end{aligned} \tag{11.43}$$

where r_1 and r_2 are the radii of curvature of the lens with refractive index n_1, and r_1' and r_2' are the radii of curvature of the lens with n_2. Using the abbreviations

$$k_1 = (1/r_1 - 1/r_2) \quad \text{and} \quad k_2 = (1/r_1' - 1/r_2') \tag{11.44}$$

we may write for $1/f$,

$$1/f = (n_1 - 1)k_1 + (n_2 - 1)k_2. \tag{11.45}$$

The condition to have the focal length independent of the refractive index in the wavelength range λ_b to λ_r is

$$d/d\lambda(1/f) = d/d\lambda[(n_1 - 1)k_1 + (n_2 - 1)k_2] = 0, \tag{11.46}$$

which translates into a condition for the wavelength dependence of the refractive index

$$(dn_1/d\lambda)k_1 + (dn_2/d\lambda)k_2 = 0. \quad (11.47)$$

We simplify Eq. (11.47) by writing $dn_1 = n_{1b} - n_{1r}$ and $dn_2 = n_{2b} = n_{2r}$, where n_b stands for the short wavelength limit of the light (blue) and n_r for the long wavelength limit of the light (red). One then gets

$$k_1/k_2 = -(n_{2b} - n_{2r})/(n_{1b} - n_{1r}). \quad (11.48)$$

For k_1 and k_2 we may write, using Eqs. (11.43) and (11.44),

$$k_1 = 1/\{f_1(n_1 - 1)\} \quad \text{and} \quad k_2 = 1/\{f_2(n_2 - 1)\}. \quad (11.49)$$

Introducing the average refractive indices $n_1 = n_{1D}$ and $n_2 = n_{2D}$, where n_{1D} and n_{2D} are the values of n_1 and n_2 in the middle between n_b and n_r, we define

$$V_1 = (n_{1b} - n_{1r})/(n_{1D} - 1) \quad \text{and} \quad V_2 = (n_{2b} - n_{2r})/(n_{2D} - 1). \quad (11.50)$$

As a result we have

$$f_2/f_1 = -V_2/V_1. \quad (11.51)$$

An example is given in FileFig 11.9 for the calculation of the final focal length f, when the refractive indices and the radii of curvature for calculation of the focal length f_1 are given.

FileFig 11.9 (A9ACHROMS)

Calculation of chromatic aberration. Calculation of elimination of chromatic aberration. The focal length of an achromatic doublet with $t = 0$ is calculated.

A9ACHROMS is only on the CD.

Application 11.9.

1. Do the calculation for a chosen value of f, which means determine the corresponding r_1 and r_2 for f_1.
2. A doublet of two lenses should have flat surfaces in the middle. The doublet should have the final focal length $f = 50$ cm. Use V_1 and V_2 of FF9 and calculate $r1$ and $r4$.
3. Do the calculation for a chosen wavelength range in the visible and for $f =$ 15cm. Use two different materials. The corresponding refractive indices may be obtained from handbooks or Jenkins and White (1976, p. 177, find f_1).

11.9 CHROMATIC ABERRATION AND THE ACHROMATIC DOUBLET WITH SEPARATED LENSES

Chromatic aberration may also be eliminated when using two lenses at distance t. From Chapter 1, Section 1.3, we have for the focal length of such a system

$$1/f = 1/f_1 + 1/f_2 - t(1/f_1 f_2). \tag{11.52}$$

We want to determine t for given focal lengths f_1 and f_2 and choose refractive indices with their wavelength dependence.

For one lens one has

$$1/f_i = (n_i - 1)(1/r_1 - 1/r_2). \tag{11.53}$$

Here i is 1 or 2. Differentiation with respect to n_i yields

$$\Delta(1/f_i) = \Delta n_i(1/r_1 - 1/r_2) = \Delta n_i/\{f_i(n_i - 1)\}. \tag{11.54}$$

From $V_1 = (n_{1b} - n_{1r})/(n_{1D} - 1)$ of Eq. (11.50), which may now be written for i equal to 1 or 2

$$V_i = \Delta n_i/(n_i - 1), \tag{11.55}$$

and we have

$$\Delta(1/f_i) = V_i/f_i, \tag{11.56}$$

one obtains

$$\begin{aligned}\Delta(1/f) &= V_1/f_1 + V_2/f_2 - t\{(1/f_1)\Delta(1/f_2) + (1/f_2)\Delta(1/f_1)\} \\ &= V_1/f_1 + V_2/f_2 - t\{V_1/f_1 f_2 + V_2/f_2 f_1\}. \end{aligned} \tag{11.57}$$

The condition to have no chromatic aberration is obtained from $\Delta(1/f) = 0$ and one gets

$$t = (V_1 f_2 + V_2 f_1)/(V_1 + V_2). \tag{11.58}$$

For two lenses of equal material this reduces to

$$t = (f_1 + f_2)/2. \tag{11.59}$$

An example is given in FileFig 11.10 for the calculation of the distance of the two lenses. To calculate V_i we have used the same values as in FileFig 11.9. The two lenses are assumed to have different radii of curvature and the distance t for no chromatic aberration is calculated.

FileFig 11.10 (A10ACHRTWOS)

Calculation of the distance of separation for two lenses with different radii of curvature in an achromatic doublet.

A10ACHRTWOS is only on the CD.

Application 11.10.

1. Choose other values of n_1 and n_2 and give the distance of the two lenses for no chromatic aberration.
2. Assume two lenses of different focal length and different materials and find the separation for a chosen focal length of the achromatic system.

A P P E N D I X

About Graphs and Matrices in Mathcad

CHANGING NUMBERS IN A FILE AND PLOTTING A GRAPH

$x := 1, 2 \ldots 10 \qquad a := 3 \qquad b := 4 \qquad f(x) := a \cdot x + b$

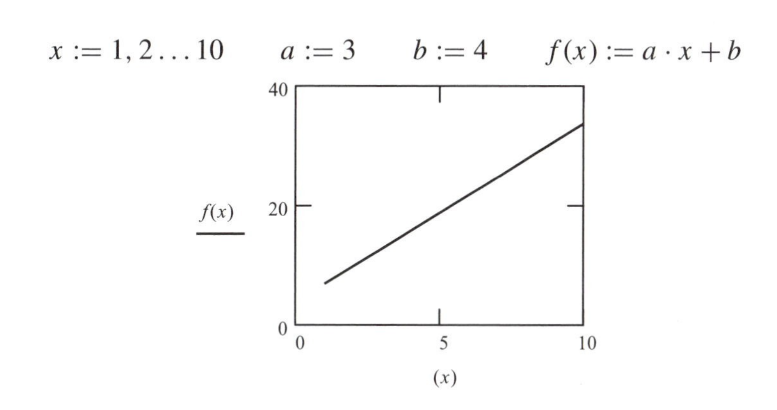

2D GRAPH

$\lambda := 0.5 \quad A := 1 \quad T \equiv 1 \quad \delta 1 \equiv 1 \quad t1 \equiv 0.1$

Specification of the number of x and $t1$ values

$N := 15 \qquad i := 0 \ldots N \qquad j := 0 \ldots N$

Specification of the range

$x_i := -.4 + .025 \cdot i \qquad t1_j := -.4+, 025 \cdot j$

In the specification of the function only x and $t1$ are used

$$uc(x,t1) := \left[2\cdot A\cdot\cos\left[2\cdot\pi\left(\frac{\delta 1}{2\cdot\lambda}\right)\right]\cdot\left[\cos\left[2\cdot\pi\cdot\left(\frac{x}{\lambda}-\frac{t1}{T}\right)-2\cdot\pi\left(\frac{\delta 1}{2\cdot\lambda}\right)\right]\right]\right]^2.$$

In the plotting function one needs the i and j notation

$$M_{i,j} := uc(x_i, t1_j).$$

Call on "Surface plot" and type at the place holder just M and push "F9."

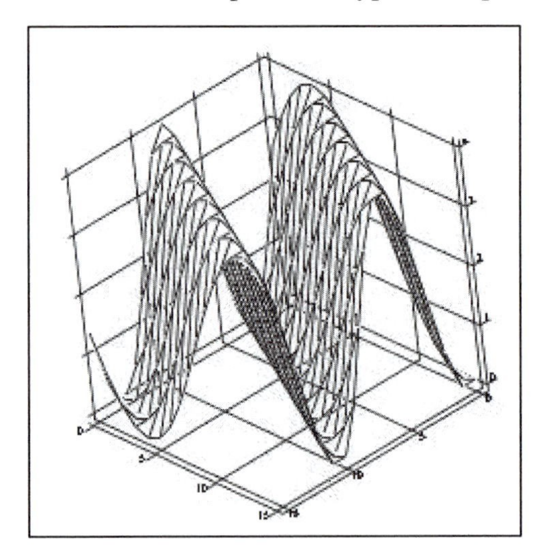

M

Go with the mouse on the graph and change the angle of the "point of view." Click twice on the graph and get "3D-Plot Format" for "graph options." Switch to contour plot.

MATRICES

Go to "Insert" and "Matrix" and select 2 by 2

$$\begin{pmatrix} \blacksquare & \blacksquare \\ \blacksquare & \blacksquare \end{pmatrix}$$

Type $M = \blacksquare$
Indicate the matrix and insert

$$M := \blacksquare \quad \text{to get} \quad M := \begin{pmatrix} \blacksquare & \blacksquare \\ \blacksquare & \blacksquare \end{pmatrix}$$

The manipulation of matrices can easily be seen from files containing matrices. Here we give an example of a matrix composed of functions and how to access the matrix elements after a multiplication has been done.

Fill in functions of x directly and call M no $M(x)$ $\quad x := 0, .1 \ldots 5$

$$M(x) := \begin{pmatrix} \cos(x) & -\sin(x) \\ +\sin(x) & \cos(x) \end{pmatrix}$$

One can access the matrix elements separately. Note that in Mathcad one starts with 0. For the 0, 1 and 1, 1 elements one has

$M(x)_{0,1} =$	$M(x)_{1,1} =$
0	1
-0.1	0.995
-0.199	0.98
-0.296	0.955
-0.389	0.921
-0.479	0.878

Consider the matrix product $M1(x) = M(x)^3$. After multiplication one can again access the matrix elements

$M1(x) := M(x)^3$ one gets for the 0, 1 element

$M1(x)_{0,1} =$
0
-0.296
-0.565
-0.783
-0.932
-0.997

APPENDIX

Formulas

CONSTANTS

$$100\mu\text{m} \Rightarrow 3000 \text{ GHz}$$
$$100\mu\text{m} \Rightarrow 100 \text{ cm}^{-1}$$
$$10\mu\text{m} \Rightarrow 1000 \text{ cm}^{-1}$$
$$1 \text{ meV} = 10^3 \text{ eV} = 1.6 \times 10^{-16}\text{joule} \Rightarrow 8.07 \text{ cm}^{-1}$$
$$100 \text{ nm} = 1000 \text{ Å}$$
$$10000 \text{ Å} = 1\mu$$
$$1 \text{ Å} = 10^{-8} \text{ cm} = 10^{-10} \text{ m}$$

FORMULAS

$$\sqrt{-1} = i \qquad i^2 = -1$$
$$z = a + ib = r(\cos\phi + i\sin\phi) = re^{i\phi}$$
$$\cos x = \frac{e^{ix+e^{ix}}}{2} \qquad \sin x = \frac{e^{ix} - e^{ix}}{2i}$$
$$e^x = 1 + x + \frac{x^2}{2!} + \frac{x^3}{3!} + \cdots, \qquad \sin x = x - \frac{x^3}{3!} + \cdots, \qquad \cos x = 1 - \frac{x^2}{2!} + \cdots$$
$$s_n = a + aq + aq^2 + \cdots\cdots + aq^{n-1} = a\frac{q^n - 1}{q - 1}$$
$$if|q| < 1, N \to \infty \quad s_\infty = \frac{a}{1-q}$$

$$x^2 + ax + b = 0 \qquad x_{1,2} = -\frac{a}{2} \pm \sqrt{\left(\frac{a}{2}\right)^2 - b}$$

$$(1 \pm x)^n \cong 1 \pm nx \qquad |x| << 1$$

$$\begin{pmatrix} a_1 & b_1 \\ a_2 & b_2 \end{pmatrix} \times \begin{pmatrix} c_1 & d_1 \\ c_2 & d_2 \end{pmatrix} = \begin{pmatrix} a_1c_1 + b_1c_2 & a_1d_1 + b_1d_2 \\ a_2c_1 + b_2c_2 & a_2d_1 + b_2d_2 \end{pmatrix}$$

$$\begin{vmatrix} a_1b_1c_1 \\ a_2b_2c_2 \\ a_3b_3c_3 \end{vmatrix} = a_1 \begin{vmatrix} b_2c_2 \\ b_3c_3 \end{vmatrix} - a_2 \begin{vmatrix} b_1c_1 \\ b_3c_3 \end{vmatrix} + a_3 \begin{vmatrix} b_1c_1 \\ b_2c_2 \end{vmatrix}$$

$$\begin{vmatrix} b_1c_1 \\ b_2c_2 \end{vmatrix} = b_1c_2 - b_2c_1$$

TRIGONOMETRIC FORMULAS

	0	30°	45°	60°	90°	180°	270°	360°
sin	0	$\frac{1}{2}$	$\frac{1}{2}\sqrt{2}$	$\frac{1}{2}\sqrt{3}$	1	0	−1	0
cos	1	$\frac{1}{2}\sqrt{3}$	$\frac{1}{2}\sqrt{2}$	$\frac{1}{2}$	0	−1	0	1
tan	0	$\frac{1}{3}\sqrt{3}$	1	$\sqrt{3}$	∞	0	∞	0
cot	∞	$\sqrt{3}$	1	$\frac{1}{3}\sqrt{3}$	0	∞	0	∞

$$\sin^2\alpha + \cos^2\alpha = 1 \qquad \frac{\sin\alpha}{\cos\alpha} = \tan\alpha \qquad \frac{\cos\alpha}{\sin\alpha} = \cot\alpha$$

$$\tan\alpha = \frac{1}{\cos\alpha} \qquad \frac{1}{\cos^2\alpha} = 1 + \tan^2\alpha \qquad \sin\alpha = \frac{\tan\alpha}{\sqrt{1+\tan^2\alpha}}$$

$$\cos\alpha = \frac{1}{\sqrt{1+\tan^2\alpha}}$$

$$\sin(90^\circ \pm \alpha) = +\cos\alpha \qquad \sin(180^\circ \pm \alpha) = \mp\sin\alpha$$

$$\cos(90^\circ \pm \alpha) = \mp\sin\alpha \qquad \cos(180^\circ \pm \alpha) = -\cos\alpha$$

$$\tan(90^\circ \pm \alpha) = \mp\cot\alpha \qquad \tan(180^\circ \pm \alpha) = \pm\tan\alpha$$

$$\cot(90^\circ \pm \alpha) = \mp\tan\alpha \qquad \cot(180^\circ \pm \alpha) = \pm\cot\alpha$$

$$\sin(-\alpha) = -\sin\alpha \qquad \sin(\alpha \pm \beta) = \sin\alpha\cos\beta \pm \cos\alpha\sin\beta$$

$$\cos(-\alpha) = +\cos\alpha \qquad \cos(\alpha \pm \beta) = \cos\alpha\cos\beta \mp \sin\alpha\sin\beta$$

$$\tan(-\alpha) = -\tan\alpha \qquad \tan(+\alpha \pm \beta) = \frac{\tan\alpha \pm \tan\beta}{1 \mp \tan\alpha \cdot \tan\beta}$$

$$\cot(-\alpha) = -\cot\alpha \qquad \cot(\alpha \pm \beta) = \frac{\cot\alpha \cdot \beta \mp 1}{\cot\beta \pm \cot\alpha}$$

$$\sin 2\alpha = 2\sin\alpha\cos\alpha$$

$$\cos 2\alpha = \cos^2\alpha - \sin^2\alpha = 1 - 2\sin^2\alpha = 2\cos^2\alpha - 1$$

$$\sin 2\alpha = \frac{2\tan\alpha}{1+\tan^2\alpha} \qquad \cos 2\alpha = \frac{1-\tan^2\alpha}{1+\tan^2\alpha}$$

$$\tan^2\alpha = \frac{2\tan\alpha}{1-\tan^2\alpha} = \frac{2}{\cot\alpha - \tan\alpha}, \qquad \cot 2\alpha = \frac{\cot^2\alpha - 1}{2\cot\alpha} = \frac{1}{2}(\cot\alpha - \tan\alpha)$$

$$1+\cos\alpha = 2\cos^2\frac{\alpha}{2}, \qquad 1-\cos\alpha = 2\sin^2\frac{\alpha}{2}$$

$$\tan\alpha = \sqrt{\frac{1-\cos 2\alpha}{1+\cos 2\alpha}} = \frac{\sin 2}{1+\cos 2\alpha} = \frac{1-\cos 2\alpha}{\sin 2\alpha} = \frac{2\tan\frac{\alpha}{2}}{1-\tan^2\frac{\alpha}{2}}$$

$$\sin\alpha + \sin\beta = 2\sin\frac{\alpha+\beta}{2}\cdot\cos\frac{\alpha-\beta}{2} \qquad \frac{\sin\alpha+\sin\beta}{\cos\alpha+\cos\beta} = \tan\frac{\alpha+\beta}{2}$$

$$\sin\alpha - \sin\beta = 2\cos\frac{\alpha+\beta}{2}\cdot\sin\frac{\alpha-\beta}{2}$$

$$\cos\alpha + \cos\beta = 2\cos\frac{\alpha+\beta}{2}\cdot\cos\frac{\alpha-\beta}{2} \qquad \frac{\sin\alpha-\sin\beta}{\cos\alpha+\cos\beta} = \tan\frac{a-\beta}{2}$$

$$\cos\alpha - \cos\beta = -2\sin\frac{\alpha+\beta}{2}\cdot\sin\frac{\alpha-\beta}{2}$$

$$\frac{\tan\alpha+\tan\beta}{\cot\alpha+\cot\beta} = \tan\alpha\cdot\tan\beta$$

$$\tan\alpha+\tan\beta = \frac{\sin(\alpha\pm\beta)}{\cos\alpha\cos\beta} \qquad \frac{1+\tan\alpha}{1-\tan\alpha} = (\tan 45^\circ + \alpha)$$

$$\cot\alpha\pm\cot\beta = \frac{\pm\sin(\alpha\pm\beta)}{\sin\alpha\sin\beta} \qquad \frac{\cot\alpha+1}{\cot\alpha-1} = \cot(45^\circ-\alpha)$$

$$\cos\alpha+\sin\alpha = \sqrt{2}\sin(45^\circ+\alpha) = \sqrt{2}\cos(45^\circ-\alpha)$$

$$\cos\alpha-\sin\alpha = \sqrt{2}\cos(45^\circ+\alpha = \sqrt{2}\sin(45^\circ-\alpha)$$

$$\cot\alpha+\tan\alpha = \frac{2}{\sin 2\alpha} \qquad \cot\alpha-\tan\alpha = 2\cot 2\alpha$$

DIFFERENTIATION

$$(u \cdot v)' = uv' + u'v$$

$$\left(\frac{u}{v}\right)' = \frac{u'v - v'u}{v^2}$$

$$(\sin x)' = \cos x \qquad (e^x)' = e^x$$

$$(\cos x)' = -\sin x \qquad (\ln x)' = \frac{1}{x}$$

$$(\tan x)' = \frac{1}{\cos^2 x} \qquad (\arcsin x)' = \frac{1}{\sqrt{1-x^2}}$$

$$(\cot x)' = \frac{-1}{\sin^2 x} \qquad (\arccos x)' = -\frac{1}{\sqrt{1-x^2}}$$

INTEGRATION

$$\int u\,dv = uv - \int v\,du$$

$$\int dx = x \qquad \int x^n dx = \frac{x^{n+1}}{n+1} (n \neq -1) \qquad \int \frac{dx}{x} = \ln x$$

$$\int \sin x\,dx = -\cos x \qquad \int \cos x\,dx = \sin x$$

$$\int \cot x\,dx = \ln \sin x \qquad \int \frac{dx}{\sin^2 x} = -\cot x$$

$$\int \frac{dx}{\cos^2 x} = \tan x$$

$$\int \frac{dx}{1-x^2} = \ln \sqrt{\frac{1+x}{1-x}}$$

$$\int \frac{dx}{\sqrt{1-x^2}} = \arcsin x \qquad \int \frac{dx}{1+x^2} = +\arctan x$$

$$\int e^x dx = e^x \qquad \int a^x dx = \frac{a^x}{\ln a} \qquad \int \frac{dx}{x \pm a} = \ln(x \pm a)$$

References

M. Born and E. Wolf, *Principles of Optics*, 2nd ed. Pergamon Press, New York, 1964.

M. Cagnet, M. Francon, and J. C. Thrierr, *Atlas of Optical Phenomena*, Springer-Verlag, Heidelberg, 1962.

P. Feynman, R. B. Leighton, and M. Sands, *The Feynman Lecture on Physics*, 6th ed. Addison-Wesley Publishing Company, Reading, Massachusetts, 1977.

J. W. Goodman, *Introduction to Fourier Optics*, McGraw-Hill, Inc., New York, 1988.

J. D. Jackson, *Classical Electrodynamics*, 2nd ed. John Wiley & Sons, New York, 1975.

F. A. Jenkins and H. E. White, *Fundamentals of Optics*, 4th ed. McGraw-Hill Inc., New York, 1976.

C. Kittel, *Introduction to Solid State Physics*, 3rd ed. John Wiley & Sons, Inc., New York, 1967.

H. Kogelnik and T. Li, *Appl. Opt.* Vol 5, p. 1550 (1966).

G. P. Parrent and B. J. Thompson, *Physical Optics Notebook*, Society of Photo-Optical Engineers, Redondo Beach, California, 1969.

F. K. Kneubühl and M. W. Sigrist, *Laser*, B. G. Teubner, Stuttgart, 1988.

E. D. Palik, *Handbook of Optical Constants of Solids II*, Academic Press, Inc., New York, 1991.

R. W. Pohl, *Einführung in die Optik*, Springer-Verlag, Heidelberg, 1948.

Index